Lecture Notes in Computer Science 4036

Commenced Publication in 1973
Founding and Former Series Editors:
Gerhard Goos, Juris Hartmanis, and Jan van Leeuwen

Oscar H. Ibarra   Zhe Dang (Eds.)

Developments in Language Theory

10th International Conference, DLT 2006
Santa Barbara, CA, USA, June 26-29, 2006
Proceedings

 Springer

Volume Editors

Oscar H. Ibarra
University of California, Department of Computer Science
Santa Barbara, CA 93106, USA
E-mail: ibarra@cs.ucsb.edu

Zhe Dang
Washington State University, School of Electrical Engineering and Computer Science
Pullman, WA 99163, USA
E-mail: zdang@eecs.wsu.edu

Library of Congress Control Number: 2006927282

CR Subject Classification (1998): F.4.3, F.4.2, F.4, F.3, F.1, G.2

LNCS Sublibrary: SL 1 - Theoretical Computer Science and General Issues

ISSN   0302-9743
ISBN-10   3-540-35428-X Springer Berlin Heidelberg New York
ISBN-13   978-3-540-35428-4 Springer Berlin Heidelberg New York

Springer is a part of Springer Science+Business Media

springer.com

© Springer-Verlag Berlin Heidelberg 2006
Printed in Germany

Typesetting: Camera-ready by author, data conversion by Scientific Publishing Services, Chennai, India
Printed on acid-free paper   SPIN: 11779148   06/3142   5 4 3 2 1 0

Preface

The 10th International Conference on Developments in Language Theory (DLT 2006) was held at the University of California, Santa Barbara, USA on June 26–29, 2006. This was the first DLT conference to take place in North America. Past meetings were held in Turku (1993), Magdeburg (1995), Thessaloniki (1997), Aachen (1999), Vienna (2001), Kyoto (2002), Szeged (2003), Auckland (2004), and Palermo (2005). The conference series is under the auspices of the European Association for Theoretical Computer Science.

The scope of the conference includes topics in the following areas: grammars, acceptors and transducers for strings, trees, graphs, arrays; efficient text algorithms; algebraic theories for automata and languages; combinatorial and algebraic properties of words and languages; variable-length codes; symbolic dynamics; decision problems; relations to complexity theory and logic; picture description and analysis; polyominoes and bidimensional patterns; cryptography; concurrency; bio-inspired computing; quantum computing.

This volume of *Lecture Notes in Computer Science* contains the papers that were presented at DLT 2006, including the abstracts or full papers of four invited lectures presented by Rajeev Alur, Yuri Gurevich, Gheorghe Paun, and Grzegorz Rozenberg.

The 36 contributed papers were selected from 63 submissions. Each submitted paper was reviewed by three Program Committee members, with the assistance of external referees. The authors of the papers came from the following countries and regions: Canada, Czech Republic, Finland, France, Germany, Greece, Hong Kong, Hungary, India, Italy, Japan, Korea, The Netherlands, Poland, Romania, Russia, South Africa, Spain, Sweden, the UK, and the USA.

A great many contributed to the success of DLT 2006. We extend our sincere thanks to the authors who submitted papers and to all those who gave presentations. We express our appreciation to the members of the Program Committee and their colleagues who assisted in the review process. To our invited speakers, we thank you for sharing your insights and expertise. We would like to acknowledge the work of the Organizing Committee, who thoughtfully and energetically planned the event over the course of months. To members of the Steering Committee, we are grateful for your counsel. Finally, we wish to recognize the kind support of the conference sponsors: Ask.com, Citrix, Google, and UCSB's Department of Computer Science, College of Engineering, and Graduate Division.

June 2006   Oscar H. Ibarra

Organization

Program Committee

Jean Berstel (Marne-la-Vallée, France)
Cristian Calude (Auckland, New Zealand)
Erzsebet Csuhaj-Varju (Budapest, Hungary)
Zhe Dang (Pullman, USA)
Volker Diekert (Stuttgart, Germany)
Omer Egecioglu (Santa Barbara, USA)
Zoltan Esik (Szeged, Hungary & Tarragona, Spain)
Juraj Hromkovic (Zurich, Switzerland)
Oscar H. Ibarra, Chair (Santa Barbara, USA)
Masami Ito (Kyoto, Japan)
Natasha Jonoska (Tampa, USA)
Juhani Karhumaki (Turku, Finland)
Lila Kari (London, Canada)
Werner Kuich (Vienna, Austria)
Giancarlo Mauri (Milan, Italy)
Jean-Eric Pin (Paris, France)
Bala Ravikumar (Sonoma, USA)
Antonio Restivo (Palermo, Italy)
Kai Salomaa (Kingston, Canada)
Jeffrey Shallit (Waterloo, Canada)
Wolfgang Thomas (Aachen, Germany)
Hsu-Chun Yen (Taipei, Taiwan)
Sheng Yu (London, Canada)

Organizing Committee

Omer Egecioglu, Co-chair (Santa Barbara, USA)
Bee Jay Estalilla (Santa Barbara, USA)
Cagdas Gerede (Santa Barbara, USA)
Oscar H. Ibarra, Co-chair (Santa Barbara, USA)
Jianwen Su (Santa Barbara, USA)
Sara Woodworth (Santa Barbara, USA)

Proceedings Committee

Zhe Dang (Pullman, USA)
Oscar H. Ibarra (Santa Barbara, USA)

Publicity Committee

Kai Salomaa (Kingston, Canada)
Sheng Yu (London, Canada)

Steering Committee

Jean Berstel (Marne-la-Vallé, France)
Cristian Calude (Auckland, New Zealand)
Volker Diekert (Stuttgart, Germany)
Juraj Hromkovic (Zurich, Switzerland)
Oscar H. Ibarra (Santa Barbara, USA)
Masami Ito (Kyoto, Japan)
Werner Kuich (Vienna, Austria)
Gheorghe Paun (Bucharest, Romania & Seville, Spain)
Antonio Restivo (Palermo, Italy)
Grzegorz Rozenberg, Chair (Leiden, The Netherlands)
Arto Salomaa (Turku, Finland)
Sheng Yu (London, Canada)

External Referees

Marcella Anselmo, Marie-Pierre Béal, Franziska Biegler, Hans-Joachim Boeck-
enhauer, Bernd Borchert, Henning Bordihn, Giusi Castiglione, Mark Daley,
Jürgen Dassow, Jörg Desel, Mike Domaratzki, Pal Domosi, Claudio Ferretti,
Wan Fokkink, Zoltan Fulop, Stefan Göller, Valentin Goranko, Vesa Halava, Tero
Harju, Markus Holzer, Geza Horvath, Lucian Ilie, Jarkko Kari, Joachim Kupke,
Martin Kutrib, A. Lepistö, Christof Löding, Markus Lohrey, Sylvain Lombardy,
Robert Lorenz, Marina Madonia, Stuart W. Margolis, Tobias Moemke, Anca
Muscholl, Loránd Muzamel, Benedek Nagy, Zoltan L. Nemeth, Alexander
Okhotin, Friedrich Otto, Holger Petersen, Narad Rampersad, Grigore Rosu,
Nicolae Santean, Stefan Schwoon, Marinella Sciortino, Sebastian Seibert, Olivier
Serre, Petr Sosik, Howard Straubing, Simayk Taati, Siegbert Tiga, Sandor
Vagvolgyi, György Vaszil, Bow-Yaw Wang, Pascal Weil, Gaoyan Xie, Claudio
Zandron

Sponsors

Computer Science Department, University of California, Santa Barbara
College of Engineering, University of California, Santa Barbara
Graduate Division, University of California, Santa Barbara
Ask.com
Citrix
Google

Table of Contents

Invited Lectures

Papers

Adding Nesting Structure to Words

Rajeev Alur[1] and P. Madhusudan[2]

[1] University of Pennsylvania, USA
[2] University of Illinois at Urbana-Champaign, USA

1   Introduction

We propose *nested words* to capture models where there is *both* a natural linear sequencing of positions and a hierarchically nested matching of positions. Such dual structure exists for executions of structured programs where there is a natural well-nested correspondence among entries to and exits from program components such as functions and procedures, and for XML documents where each open-tag is matched with a closing tag in a well-nested manner.

We define and study finite-state automata as acceptors of nested words. A nested-word automaton is similar to a classical finite-state word automaton, and reads the input from left to right according to the linear sequence. However, at a position with two predecessors, one due to linear sequencing and one due to a hierarchical nesting edge, the next state depends on states of the run at both these predecessors. The resulting class of *regular* languages of nested words has all the appealing theoretical properties that the class of classical regular word languages enjoys: deterministic nested word automata are as expressive as their nondeterministic counterparts; the class is closed under operations such as union, intersection, complementation, concatenation, and Kleene-∗; decision problems such as membership, emptiness, language inclusion, and language equivalence are all decidable; definability in monadic second order logic of nested words corresponds exactly to finite-state recognizability; and finiteness of the congruence induced by a language of nested words is a necessary and sufficient condition for regularity.

The motivating application area for our results has been software verification. Given a sequential program P with stack-based control flow, the execution of P is modeled as a nested word with nesting edges from calls to returns. Specification of the program is given as a nested word automaton A, and verification corresponds to checking whether every nested word generated by P is accepted by A. Nested-word automata can express a variety of requirements such as stack-inspection properties, pre-post conditions, and interprocedural data-flow properties. If we were to model program executions as words, all of these properties are non-regular, and hence inexpressible in classical specification languages based on temporal logics, automata, and fixpoint calculi (recall that context-free languages cannot be used as specification languages due to nonclosure under intersection and undecidability of key decision problems such as language inclusion). In finite-state software model checking, the data variables in the program are abstracted into a set of boolean variables, and in that case, the set of nested words generated by the abstracted program is regular. This implies that algorithmic software verification is possible for all regular specifications of nested

O.H. Ibarra and Z. Dang (Eds.): DLT 2006, LNCS 4036, pp. 1–13, 2006.

words. We believe that the nested-word view will provide a unifying basis for the next generation of specification logics for program analysis, software model checking, and runtime monitoring. As explained in Section 3, another potential area of application is XML document processing.

1.1 Related Work

The finite automata on nested words that we study here have been motivated by our recent work on *visibly pushdown automata* [6]. A visibly pushdown automaton is one in which the input alphabet Σ is partitioned into three parts, $\langle \Sigma_c, \Sigma_i, \Sigma_r \rangle$ such that the automaton pushes exactly one symbol when reading symbols from Σ_c, pops one symbol from the stack when reading a symbol in Σ_r, and does not touch the stack when reading letters of Σ_i. The input word hence has an implicit nesting structure defined by matching occurrences of symbols in Σ_c with symbols in Σ_r. In nested words, this nesting is given explicitly, and this lets us define an automaton without a stack[1]. We believe that nested words is a more appealing and simpler formulation of the insights in the theory of visibly pushdown languages. However, in terms of technical results, this paper only reformulates the corresponding results for visibly pushdown languages in [6].

Visibly pushdown languages are obviously related to Dyck languages, which is the class of languages with well-bracketed structure. The class of *parenthesis* languages studied by McNaughton comes closest to our notion of visibly pushdown languages [16]. A parenthesis language is one generated by a context free grammar where every production introduces a pair of parentheses that delimit the scope of the production. Viewing the nesting relation as that defined by the parentheses, parenthesis languages are a subclass of visibly pushdown languages. In [16, 11], it was shown that parenthesis languages are closed under union, intersection and difference, and that the equivalence problem for them is decidable. However, parenthesis languages are a strict subclass of visibly pushdown languages, and are not closed under Kleene-*.

The class of visibly pushdown languages, was considered in papers related to parsing *input-driven languages* [22, 9]. Input-driven languages are precisely visibly pushdown languages (the stack operations are *driven* by the input). However, the papers considered only the membership problem for these languages (namely showing that membership is easier for these languages than for general context-free languages) and did not systematically study the *class* of languages defined by such automata.

2 Nested Words

2.1 Definition

A *nested relation* ν of width k, for $k \geq 0$, is a binary relation over $\{1, 2 \ldots k\}$ such that (1) if $\nu(i, j)$ then $i < j$; (2) if $\nu(i, j)$ and $\nu(i, j')$ then $j = j'$, and if

---

[1] It is worth noting that most of the algorithms for inter-procedural program analysis and context-free reachability compute summary edges between control locations to capture the computation of the called procedure (see, for example [18]).

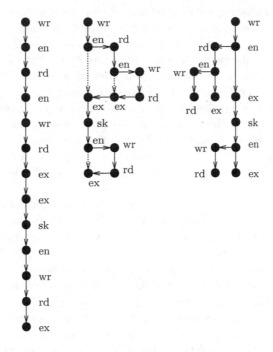

Fig. 1. Execution as a word, as a nested word, and as a tree

$\nu(i,j)$ and $\nu(i',j)$ then $i = i'$; (3) if $\nu(i,j)$ and $\nu(i',j')$ and $i < i'$ then either $j < i'$ or $j' < j$.

Let ν be a nested relation. When $\nu(i,j)$ holds, the position j is called a *return-successor* of the position i, and the position i is called a *call-predecessor* of the position j. Our definition requires that a position has at most one return-successor and at most one call-predecessor, and a position cannot have both a return-successor and a call-predecessor. A position is called a *call* position if it has a return successor, a *return* position if it has a call-predecessor, and an *internal* position otherwise.

A *nested word* nw over an alphabet Σ is a pair $(a_1 \ldots a_k, \nu)$, for $k \geq 0$, such that a_i, for each $1 \leq i \leq k$, is a symbol in Σ, and ν is a nested relation of width k. Let us denote the set of nested words over Σ as $NW(\Sigma)$. A language of nested words over Σ is a subset of $NW(\Sigma)$.

2.2   Example: Program Executions as Nested Words

Execution of a program is typically modeled as a word over an alphabet Σ. The choice of Σ depends on the desired level of detail. As an example, suppose we are interested in tracking read/write accesses to a program variable x. The variable x may get redefined, for example, due to a declaration of a local variable within a called procedure, and we need to track the scope of these definitions. For simplicity, let's assume every change in context redefines the variable. Then, we can choose the following set of symbols: rd to denote a read access to x, wr to denote a write access to x, en to denote beginning of a new scope (such as a call to a function or a procedure), ex to denote the ending of the current scope, and sk to denote all other actions of the program. Note that in any structured programming language, in a given execution, there is a natural nested matching of the symbols en and ex. Figure 1 shows a possible execution as a word as well as a nested word. The nesting edges are shown as dotted edges. A vertical path can be interpreted as a local path through a procedure. There is a natural connection between nested words and binary trees, and is also depicted in Figure 1. In this view, at a call node, the left subtree encodes the computation within the called procedure, while a path along the right children gives the local computation within a procedure.

In modeling the execution as a word, the matching between calls and returns is only implicit, and a pushdown automaton is needed to reconstruct the matching. The tree view makes the hierarchical structure explicit: every matching exit is a right-child of the corresponding entry node. However, this view loses linearity: the left and right subtrees of an entry node are disconnected, and (top-down) tree automata need nondeterminism to relate the properties of the subtrees[2]. Our hypothesis is that the nested-word view is the most suitable view for program verification. In this view, a program will be a *generator* of nested words, and will be modeled as a language of nested words. For acceptors, linearity is used to obtain a left-to-right deterministic acceptor, while nesting is exploited to keep the acceptor finite state.

2.3   Operations on Nested Words

Analogs of a variety of operations on words and word languages can be defined for nested words and corresponding languages. We describe a few of the interesting ones here.

Given two nested words $nw_1 = (w_1, \nu_1)$ and $nw_2 = (w_2, \nu_2)$, of lengths k_1 and k_2, respectively, the *concatenation* of nw_1 and nw_2 is the nested word $nw_1.nw_2 = (w_1.w_2, \nu)$ of length $k_1 + k_2$, where ν is the nested relation $\nu_1 \cup \{(k_1 + i, k_1 + j) \mid (i, j) \in \nu_2\}$. The concatenation extends to languages of nested words. The Kleene-$*$ operation is defined as usual: if L is a language of nested words over Σ, then L^* is the set of nested words $nw_1.nw_2 \ldots nw_i$, where $i \in \mathbb{N}$, and each $w_j \in L$.

Given a nested word $nw = (a_1 \ldots a_k, \nu)$ of length k, its *reverse* is $nw^r = (a_k \ldots a_1, \nu^r)$ where $\nu^r = \{(i, j) \mid (k + 1 - j, k + 1 - i) \in \nu\}$.

Finally, we define a notion of *insertion* for nested words. A *context* is a pair (nw, i) where nw is a nested word of length k, and $0 \leq i \leq k$. Given a context (nw, i), for $nw = (a_1 \ldots a_k, \nu)$, and a nested word nw', with $nw' = (w', \nu')$, $(nw, i) \oplus nw'$ is the nested word obtained by inserting the

---

[2] It is worth mentioning that in program verification, trees are used for a different purpose: an execution tree encodes all possible executions of a program, and branching corresponds to the choice within the program. It is possible to define *nested trees* in which each path encodes a structured execution as a nested word [3].

nested word nw' at position i in nw. More precisely, $(nw, i) \oplus nw'$ is the nested word $(a_1 \ldots a_i.w'.a_{i+1} \ldots a_k, \nu'')$, where $\nu'' = \{(\pi_1(j), \pi_1(j')) \mid (j, j') \in \nu\} \cup \{(\pi_2(j), \pi_2(j')) \mid (j, j') \in \nu'\}$ where $\pi_1(j)$ is j, if $j \leq i$, and $|w'| + j$ otherwise, and $\pi_2(j) = i + j$.

Note that our definition of nested word requires one-to-one matching between call and return positions. It is possible to generalize this definition and allow a nested relation to contain pairs of the form $(i, \bot)$ and $(\bot, j)$ corresponding to unmatched call and return positions, respectively. Concatenation of two nested words would match the last unmatched call in the first word with the first unmatched return in the second one. Natural notions of prefix and suffix exist in this generalized definition. The results of this paper can be adapted to this general definition also.

3 Regular Languages of Nested Words

3.1 Automata over Nested Words

A *nested word automaton* (NWA) A over an alphabet Σ is a structure (Q, Q_{in}, δ, Q_f) consisting of

- a finite set Q of states,
- a set of initial states $Q_{in} \subseteq Q$,
- a set of final states $Q_f \subseteq Q$,
- a set of transitions $\langle \delta_c, \delta_r, \delta_i \rangle$ where
  - $\delta_c \subseteq Q \times \Sigma \times Q$ is a transition relation for call positions, and
  - $\delta_i \subseteq Q \times \Sigma \times Q$ is a transition relation for internal positions, and
  - $\delta_r \subseteq Q \times Q \times \Sigma \times Q$ is a transition relation for return positions.

The automaton A starts in an initial state, and reads the word from left to right. At a call or an internal position, the next state is determined by the current state and the input symbol at the current position, while at a return position, the next state can additionally depend on the state of the run just before the matching call-predecessor. Formally, a *run* ρ of the automaton A over a nested word $nw = (a_1 \ldots a_k, \nu)$ is a sequence $q_0, \ldots, q_k$ over Q such that $q_0 \in Q_{in}$, and for each $1 \leq i \leq k$,

- if i is a call position of ν, then $(q_{i-1}, a_i, q_i) \in \delta_c$;
- if i is a internal position, then $(q_{i-1}, a_i, q_i) \in \delta_i$;
- if i is a return position with call-predecessor is j, then $(q_{i-1}, q_{j-1}, a_i, q_i) \in \delta_r$.

The automaton A accepts the nested word nw if it has a run $q_0, \ldots, q_k$ over nw such that $q_k \in Q_f$. The language $L(A)$ of a nested-word automaton A is the set of nested words it accepts.

A language L of nested words over Σ is *regular* if there exists a nested-word automaton A over Σ such that $L = L(A)$.

Observe that if L is a regular language of words over Σ, then $\{(w, \nu) \mid w \in L\}$ is a regular language of nested words. Conversely, if L is a regular language of nested words, then $\{w \mid (w, \nu) \in L \text{ for some } \nu\}$ is a context-free language of words, but need not be regular.

The fact that a nested automaton at a return position can look at the state at the corresponding call position is, of course, crucial to expressiveness as it allows a run to implicitly encode a stack. In automata theory, such a definition that allows combining of states is quite common. For example, bottom-up tree automata allow such a join. Various notions of automata on partial-orders and graphs are also defined this way [20]. In fact, one can define a more general notion of automata on nested words by giving *tiling systems* that tile the positions using a finite number of tiles with local constraints that restrict the tiles that can occur at a position, given the tiles in its neighborhood. The notion of neighborhood for a node in a nested word would be its linear successor and predecessor, and its return-predecessor or call-successor. It turns out that automata defined in this fashion also define regular nested languages.

3.2   Determinization

A nested-word automaton $A = (Q, Q_{in}, (\delta_c, \delta_i, \delta_r), Q_f)$ is said to be *deterministic* if $|Q_{in}| = 1$, and for every $a \in \Sigma$ and $q, q' \in Q$, $|\{q'' \mid (q, a, q'') \in \delta_c\}| = 1$ and $|\{q'' \mid (q, a, q'') \in \delta_i\}| = 1$ and $|\{q'' \mid (q, q', a, q'') \in \delta_r\}| = 1$. Thus, a deterministic nested-word automaton has a single initial state, and the transition relations δ_c and δ_i are functions from $Q \times \Sigma$ to Q, and the transition relation δ_r is a function from $Q \times Q \times \Sigma$ to Q. Given a nested word nw, a deterministic nested-word automaton has exactly one run over nw.

Adapting the classical subset construction for determinizing finite automata over words turns out to be slightly tricky, but possible:

Theorem 1. *Given a nested-word automaton A over Σ, there exists a deterministic nested-word automaton A' over Σ such that $L(A) = L(A')$. Furthermore, if A has n states, then A' has at most 2^{n^2} states.*

Proof. The deterministic automaton will keep track of *summaries* of state-transitions, rather than just the states reached. More precisely, after reading the first i positions of a nested word $nw = (w, \nu)$, if j is the last call position at or before i (if there is none, choose $j = 1$), then the automaton will be in a state $S \subseteq Q \times Q$ where S is the set of pairs of states (q, q') such that A has a run from q to q' on reading the nested word starting at position j to i. It hence starts in the initial state $\{(q, q) \mid q \in Q\}$. At an internal position labeled a, the automaton replaces each pair (q, q') in the current state by pairs of the form (q, q'') such that $(q', a, q'') \in \delta_i$. At a call position labeled a, the summary gets reinitialized: the new state contains pairs of the form (q, q'), where $(q, a, q') \in \delta_c$. Consider a return position labeled a, and suppose S denotes the current state and S' denotes the state just before the call-predecessor. Then (q, q') belongs to the new state, provided there exist states q_1, q_2 such that $(q, q_1) \in S'$ and $(q_1, q_2) \in S$ and $(q_2, q_1, a, q') \in \delta_r$. A state is final if it contains a pair of the form (q, q') with $q \in Q_{in}$ and $q' \in Q_f$.   □

Since the call and internal transition relations are separate, our definition allows the automaton to check whether the current position is a call or an internal position. It is easy to verify that this distinction is not necessary for nondeterministic automata. However, for deterministic automata, removing this distinction will reduce expressiveness. On the other hand, as the above proof shows, a deterministic NWA can accept all regular languages of nested words, even if we restrict the call transition function to depend only on the current symbol.

3.3   Closure Properties

The class of regular nested languages enjoy many closure properties, similar to the class of regular languages over words.

Theorem 2. *The class of regular languages of nested words is (effectively) closed under union, intersection, complementation, concatenation, Kleene-∗, and reverse.*

3.4   Application: Software Model Checking and Program Analysis

In the context of software verification, a popular paradigm to verification is through *data abstraction*, where the data in a program is abstracted using a finite set of boolean variables that stand for predicates on the data-space [7, 10]. The resulting models hence have finite-state but stack-based control flow (see Boolean programs [8] and recursive state machines [1] as concrete instances of pushdown models of programs). Given a program P modeled as a pushdown automaton, we can view P as a generator of nested words in the following manner. We choose an observation alphabet Σ, and associate an element of Σ with every transition of P. At every step of the execution of P, if the transition of P is a *push* transition, then the corresponding position is a call position; if the transition of P does not update the stack, then the corresponding position is an internal position; and if P executes a *pop* transition, then the corresponding position is a return, with a nesting edge from the position where the corresponding element was pushed. We assume that P pushes or pops at most one element, and halts when the stack is empty. Then, the nesting edges satisfy the desired constraints. Let $L(P)$ be the set of nested words generated by a pushdown model P. Then, $L(P)$ is a regular language of nested words.

The requirements of a program can also be described as regular languages of nested words. For instance, consider the example of Section 2. Suppose we want to specify that within each scope (that is, between every pair of matching entry and exit), along the local path (that is, after deleting every enclosed matching subword from an entry to an exit), every write access is followed by a read access. Viewed as a property of words, this is not a regular language, and thus, not expressible in the specification languages supported by existing software model checkers such as SLAM [8] and BLAST [10]. However, over nested words, there is a natural two-state deterministic nested-word automaton. The initial state is q_0, and has no pending obligations, and is the only final state. The state q_1 denotes that along the local path of the current scope, a write-access has been encountered, with no following read access. The transitions are: for $j = 0, 1$, $\delta_i(q_j, rd) = q_0$; $\delta_i(q_j, wr) = q_1$; $\delta_i(q_j, sk) = q_j$; $\delta_c(q_j, en) = q_0$; and $\delta_r(q_0, q_j, ex) = q_j$. The automaton reinitializes the state to q_0 upon entry, while processing internal read/write symbols, it updates the state as in a finite-state word automaton, and at a return, if the current state is q_0 (meaning the called context satisfies the desired requirement), it restores the state of the calling context. (Formally, we need one more state q_3 in order to make the automaton complete; when in state q_1 and reading a return, the automaton will go to state q_3, and all transitions from q_3 will go to q_3.)

Further, we can build *specification* logics for programs that exploit the nested structure. An example of such a temporal logic is CARET [4], which extends linear temporal logic by *local* modalities such as $\langle a \rangle \varphi$, which holds at a call if the return-successor of the call satisfies φ. CARET can state many interesting properties of programs, including stack-inspection properties, pre-post conditions of programs, local flows in programs, etc. Analogous to the theorem that a linear temporal formula can be compiled into an automaton that accepts its models [21], any CARET formula can be compiled into a nested word automaton that accepts its models. Decidability of inclusion then yields a decidable model-checking problem for program models against CARET [6, 4].

3.5 Application: XML Document Processing

Turning to XML, XML documents (which resemble HTML documents in structure) are hierarchically structured data with open- and close-tag constructs used to define the hierarchy. An XML document is naturally a nested word, where each open-tag is matched with its corresponding closing tag. Document type definitions (DTDs) and their specialized counterparts (SDTDs) are used to define classes of documents, using a grammar. The grammar however is special in that the non-terminals always stand for tags. Consequently, type definitions can be encoded using nested word automata. Though trees and automata on unranked trees are traditionally used in the study of XML (see [17, 14] for recent surveys), nested word automata lend more naturally to describing the document especially when the document needs to be processed as a word being read from left to right (as in the case of processing streaming XML documents). The closure and determinization theorems for nested word automata have immediate consequences in checking type-inclusion and in checking streaming XML documents against SDTDs. Further, minimization theorems for nested word automata can be exploited to construct minimal machines to process XML documents [13].

4   Alternative Characterizations

We now show alternate characterizations of the class of regular nested word languages.

4.1   Monadic Second Order Logic of Nested Words

Let us fix a countable set of first-order variables FV and a countable set of monadic second-order (set) variables SV. We denote by x, y, x', etc., elements in FV and by X, Y, X', etc., elements of SV.

The *monadic second-order logic of nested words* is given by the syntax:

$$\varphi := Q_a(x) \mid x = y \mid x \leq y \mid \nu(x, y) \mid \varphi \vee \varphi \mid \neg\varphi \mid \exists x.\varphi \mid \exists X.\varphi,$$

where $a \in \Sigma$, $x, y \in FV$, and $X \in SV$.

The semantics is defined over nested words in a natural way. The first-order variables are interpreted over positions of the nested word, while set variables are interpreted over sets of positions. $Q_a(x)$ holds if the letter at the position interpreted for x is a, $x \leq y$ holds if the position interpreted for x is before the position interpreted for y, and $\nu(x, y)$ holds if the positions x and y are ν-related. For example,

$$\forall x.\forall y. \ (Q_a(x) \wedge \nu(x, y)) \Rightarrow Q_b(y)$$

holds in a nested word iff for every call labeled a, the corresponding return-successor is labeled b.

For a sentence φ (a formula with no free variables), the language it defines is the set of all nested words that satisfy φ. The corresponding result for visibly pushdown languages [6] can be used to show that:

Theorem 3. *A language L of nested words over Σ is regular iff there is an MSO sentence φ over Σ that defines L.*

4.2   Finite Congruence

Let L be a language of nested words. Define the following relation on nested words. For two nested words nw_1 and nw_2, $nw_1 \sim_L nw_2$ if for every context (nw, i), $(nw, i) \oplus nw_1 \in L$ iff $(nw, i) \oplus nw_2 \in L$. Note that $\sim_L$ is an equivalence relation and a congruence (i.e. if $nw_1 \sim_L nw_2$ and $nw_1' \sim_L nw_2'$, then $nw_1.nw_1' \sim_L nw_2.nw_2'$). We can now show that the finiteness of this congruence characterizes regularity for nested-word languages using the corresponding result for visibly pushdown languages [5].

Theorem 4. *For a set L of nested words, L is regular iff $\sim_L$ has finitely many congruence classes.*

Proof. Let L be a regular language of nested words. Let A be a NWA that accepts L, and let its set of states be Q. Now, define the following relation on nested words: $nw \approx_A nw'$ if for every $q, q' \in Q$, A has a run from q to q' on nw if and only if A has a run from q to q' on nw'. It is easy to verify that $\approx_A$ is an equivalence relation and, in fact, a congruence. Clearly there are no more than $|Q|^2$ congruence classes defined by it. Also, it is easy to see that whenever $nw \approx_A nw'$, it is the case that $nw \sim_L nw'$. It follows that $\sim_L$ is of finite index.

For the converse, assume L is such that $\sim_L$ is of finite index, and let us denote by $[nw]$ the equivalence class of $\sim_L$ that a nested word nw belongs to. Then let $A = (Q, Q_{in}, \delta, Q_f)$, where:

- $Q = \{[nw] \mid nw \ is \ a \ nested \ word\}$,
- $Q_{in} = \{[nw_0]\}$, where nw_0 is the empty nested word,
- $Q_f = \{[nw] \mid nw \in L\}$, and
- $\delta = \langle \delta_c, \delta_i, \delta_r \rangle$ where
  - $\delta_c = \{([nw], a, [(a, \emptyset)]) \mid nw \in NW(\Sigma), a \in \Sigma\}$
  - $\delta_i = \{([nw], a, [nw.a]) \mid nw \in NW(\Sigma), a \in \Sigma\}$
  - $\delta_r([(w_1, \nu_1)], [(w_2, \nu_2)], a) = [w_2.w_1.a, \nu]$, where, if $(w_2, \nu_2).(w_1, \nu_1).a = (w_1.w_2, \nu'$, then $\nu = \nu' \cup \{(|w_2| + 1, |w_2| + |w_1| + 1)\}$.

It can then be proved that A is well-defined and accepts L. □

4.3   Visibly Pushdown Word Languages

A nested word over Σ can be encoded as a word over a finite *structured* alphabet in the following manner. Let $\Sigma' = \{c, int, r\} \times \Sigma$. Let the set of *well-matched* words over Σ' (denoted $WM(\Sigma)$) be the words generated by the grammar

$$W := \epsilon \mid (int, a)W \mid (c, a)W(r, a') \mid W.W,$$

for $a, a' \in \Sigma$. Given a nested word $(a_1 \ldots a_k, \nu)$ over Σ, we will encode it as the well-matched word u over Σ' by setting $u = (x_1, a_1) \ldots (x_k, a_k)$ where $x_i = c$ if i is a call, $x_i = r$ if i is a return, and $x_i = int$ otherwise. Let us call this mapping $nw2w : NW(\Sigma) \to WM(\Sigma)$. It is also clear that for every well-matched word over Σ', there is a unique nested word over Σ that corresponds to it. Consequently, we can treat languages of nested words over Σ as languages of words over the structured alphabet Σ'.

A finite automaton on nested words over Σ can be simulated by a *pushdown* automaton on the corresponding word over Σ'. The pushdown automaton would simply push the current state at each call position, and at return positions it would pop the state to retrieve the state at the corresponding call. Note that this pushdown automaton is restricted in that it pushes exactly one symbol when reading symbols of the form (c, a), pops the stack when reading symbols of the form (r, a), and does not touch the stack when reading symbols (int, a). This kind of pushdown automaton is called a *visibly pushdown automaton* [6]. The automaton accepts if it reaches a final state and the stack is empty.

Proposition 1. *A language L of nested words over Σ is regular iff $nw2w(L)$ is accepted by a visibly pushdown automaton over the structured alphabet Σ'.*

4.4   Regular Tree Languages

Given a nested word over Σ, we can associate it with a Σ-labeled (ranked) binary tree that represents the nested word, where each position in the word corresponds to a node of the tree. Further, the tree will encode the return position corresponding to a call right next to the call. See Figure 1 for an example of a tree-encoding of a nested word. Formally, we define the map from nested words to trees using the function $nw2t$ that maps nested words to sets of trees (we allow more than one tree to correspond to a nested word since we do not differentiate the left-child from a right-child when a node has only one child):

- For the empty nested word $nw = (\epsilon, \nu)$, $nw2t(nw)$ is the empty tree.
- For a nested word $nw = (a_1.w_1, \nu)$ such that the first position is internal, let nw_1 be the nested word corresponding to w_1. Then $nw2t(nw)$ is any tree whose root is labeled a_1, the root has one child, and the subtree at this child is in $nw2t(nw_1)$.
- For a nested word $nw = (a_1.w_1.a_2.w_2, \nu)$ such that $(1, |w_1| + 2) \in \nu$, let $nw_1 = (w_1, \nu_1)$ be the nested word corresponding to the w_1 portion, and $nw_2 = (w_2, \nu_2)$ be the nested word corresponding to the w_2 portion. Then $nw2t(nw)$ is any tree whose root is labeled a_1, the subtree rooted at its left-child is in $nw2t(nw_1)$, its right-child u is labeled a_2, and u has one child and the the subtree rooted at this child is in $nw2t(nw_2)$.

We can now show that the class of regular nested word languages precisely corresponds to regular languages of trees:

Theorem 5. *A language T of trees is a regular tree language iff the set of nested words $\{nw2t^{-1}(t) \mid t \in T\}$ is a regular nested word language.*

Note that the closure of nested languages under various operations as stated in Theorem 2 can be proved using this connection to regular tree languages. However, the determinization result (Theorem 1) does not follow from the theory of tree automata.

5   Decision Problems

The emptiness problem (given A, is $L(A) = \emptyset$?) and the membership problem (given A and nw, is $nw \in L(A)$?) for nested word automata are solvable since we can reduce it to the emptiness and membership problems for pushdown automata (using Proposition 1).

  If the automaton A is *fixed*, then we can solve the membership problem in simultaneously linear time and linear space, as we can determinize A and simply simulate the word on A. In fact, this would be a *streaming* algorithm that uses at most space $O(d)$ where d is the *depth* of nesting of the input word. A streaming algorithm is one where the input must be read left-to-right, and can be read only once. Note that this result comes useful in type-checking streaming XML documents, as the depth of documents is often not large [19, 13]. When A is fixed, the result in [22] exploits the visibly pushdown structure to solve the membership problem in logarithmic space, and [9] shows that membership can be checked using boolean circuits of logarithmic depth. These results lead to:

Theorem 6. *The emptiness problem for nested word automata is decidable in time $O(|A|^3)$.*

  The membership problem for nested word automata, given A and w, can be solved in time $O(|A|^3.|w|)$. When A is fixed, it is solvable (1) in time $O(|w|)$ and space $O(d)$ (where d is the depth of the nesting in w) in a streaming setting; (2) in space $O(\log |w|)$ and time $O(|w|^2.\log |w|)$; and (3) by (uniform) Boolean circuits of depth $O(\log |w|)$.

The inclusion problem (and hence the equivalence problem) for nested word automata is decidable. Given A_1 and A_2, we can check $L(A_1) \subseteq A_2$ by checking if $L(A_1) \cap \overline{L(A_2)}$ is empty, since regular nested languages are effectively closed under complement and intersection. It follows from the results in [6] that:

Theorem 7. *The inclusion and equivalence problems for nested word automata are* EXPTIME-*complete.*

6   Conclusions

Nested words allow capturing linear and hierarchical structure simultaneously, and automata over nested words lead to a robust class of languages with appealing theoretical properties. This theory offers a way of extending the expressiveness of specification languages supported in model checking and program analysis tools: instead of modeling a boolean program as a context-free language of words and checking regular properties, one can model both the program and the specification as regular languages of nested words.

The theory of regular languages of nested words is a reformulation of the theory of visibly pushdown languages by moving the nesting structure from labeling to the underlying shape. Besides the results reported here, many results already exist for visibly pushdown automata: visibly pushdown languages over infinite words have been studied in [6]; games on pushdown graphs against visibly pushdown winning conditions are decidable [15]; congruence based characterizations and minimization theorems for visibly pushdown automata exist [5]; and active learning, conformance testing, and black-box checking for visibly pushdown automata are studied in [12]. The nested structure on words can be extended to trees, and automata on nested trees are studied in [3, 2]. Finally, a version of the μ-calculus on nested structures has been defined in [3], and is shown to be more powerful than the standard μ-calculus, while at the same time remaining robust and tractable [3, 2].

Acknowledgments. We thank Swarat Chaudhuri, Kousha Etessami, Viraj Kumar, Leonid Libkin, Christof Löding, Mahesh Viswanathan, and Mihalis Yannakakis for fruitful discussions related to this paper. This research was partially supported by ARO URI award DAAD19-01-1-0473 and NSF award CCR-0306382.

References

1. R. Alur, M. Benedikt, K. Etessami, P. Godefroid, T. Reps, and M. Yannakakis. Analysis of recursive state machines. *ACM Transactions on Programming Languages and Systems*, 27(4):786–818, 2005.
2. R. Alur, S. Chaudhuri, and P. Madhusudan. Automata on nested trees. Under submission, 2006.

3. R. Alur, S. Chaudhuri, and P. Madhusudan. A fixpoint calculus for local and global program flows. In *ACM POPL*, pages 153–165, 2006.

4. R. Alur, K. Etessami, and P. Madhusudan. A temporal logic of nested calls and returns. In *TACAS*, LNCS 2988, pages 467–481, 2004.

5. R. Alur, V. Kumar, P. Madhusudan, and M. Viswanathan. Congruences for visibly pushdown languages. In *ICALP*, LNCS 3580, pages 1102–1114, 2005.

6. R. Alur and P. Madhusudan. Visibly pushdown languages. In *ACM STOC*, pages 202–211, 2004.

7. T. Ball, R. Majumdar, T.D. Millstein, and S.K. Rajamani. Automatic predicate abstraction of C programs. In *AACM PLDI*, pages 203–213, 2001.

8. T. Ball and S. Rajamani. Bebop: A symbolic model checker for boolean programs. In *SPIN Workshop*, LNCS 1885, pages 113–130, 2000.

9. P. Dymond. Input-driven languages are in *log n* depth. *Inf. Process. Lett.*, 26(5):247–250, 1988.

10. T.A. Henzinger, R. Jhala, R. Majumdar, G.C. Necula, G. Sutre, and W. Weimer. Temporal-safety proofs for systems code. In *CAV*, LNCS 2404, 526–538, 2002.

11. D.E. Knuth. A characterization of parenthesis languages. *Information and Control*, 11(3):269–289, 1967.

12. V. Kumar, P. Madhusudan, and M. Viswanathan. Minimization, learning, and conformance testing of boolean programs. Under submission, 2006.

13. V. Kumar, P. Madhusudan, and M. Viswanathan. Visibly pushdown languages for XML. Technical Report UIUCDCS-R-2006-2704, UIUC, 2006.

14. L. Libkin. Logics for unranked trees: An overview. In *ICALP*, LNCS 3580, pages 35–50, 2005.

15. C. Löding, P. Madhusudan, and O. Serre. Visibly pushdown games. In *FSTTCS*, LNCS 3328, pages 408–420, 2004.

16. R. McNaughton. Parenthesis grammars. *Journal of the ACM*, 14(3):490–500, 1967.

17. F. Neven. Automata, logic, and XML. In *CSL*, pages 2–26, 2002.

18. T. Reps, S. Horwitz, and S. Sagiv. Precise interprocedural dataflow analysis via graph reachability. In *ACM POPL*, pages 49–61, 1995.

19. L. Segoufin and V. Vianu. Validating streaming XML documents. In *ACM PODS*, pages 53–64, 2002.

20. W. Thomas. On logics, tilings, and automata. In *ICALP*, LNCS 510, pages 441–454, 1991.

21. M.Y. Vardi and P. Wolper. Reasoning about infinite computations. *Information and Computation*, 115(1):1–37, 1994.

22. B. von Braunmühl and R. Verbeek. Input-driven languages are recognized in *log n* space. In *FCT*, LNCS 158, pages 40–51, 1983.

Can Abstract State Machines Be Useful in Language Theory?

(Extended Abstract)

Yuri Gurevich[1] and Charles Wallace[2]

[1] Microsoft Research, Redmond WA 98052, USA
gurevich@microsoft.com
[2] Michigan Technological University, Houghton MI 49931, USA
wallace@mtu.edu

1 Introduction

Abstract state machines (originally called evolving algebras) constitute a modern computation model [8]. ASMs describe algorithms without compromising the abstraction level. ASMs and ASM based tools have been used in academia and industry to give precise semantics for computing artifacts and to specify software and hardware [1, 2, 6]. In connection to the conference on Developments in Language Theory, we consider how and whether ASMs could be useful in language theory.

The list of topics on the conference site starts with "grammars, acceptors and transducers for strings, trees, graphs, arrays". The conventional computation models cannot deal directly with graphs or other abstract structures. For example, you cannot put an abstract graph on the tape of a Turing machine. Instead, the conventional models deal with presentation of abstract structures. Accordingly, when people speak about graphs they often mean ordered graphs, that is, graphs with a linear order on vertices. This seems to be the case with the current research on grammars, acceptors and transducers for graphs [13]. If there were indeed interest in grammars, acceptors and transducers for (unordered) graphs or other abstract structures, ASMs would be indispensible. The current models for computations with abstract structures and the related complexity theory build upon ASMs [4, 5].

Another possible application of ASMs is to write language algorithms on their natural abstraction level, devoid of unnecessary details. We give one illustrative example below: an ASM program of the well-known algorithm for minimizing a deterministic finite-state automaton [12]. More examples will be presented during the conference talk. Before turning to the example, we point out some properties of ASMs that from our outsiders' point of view appear to have relevance to language theory: their universality, their facility for abstraction, and their ability to capture concurrency and non-determinism.

2 The ASM Computation Model: How Is It Different?

The original definition of ASMs in [8] is still valid. But there has been a substantial advance in the meantime. See in particular the ASM based specification

O.H. Ibarra and Z. Dang (Eds.): DLT 2006, LNCS 4036, pp. 14–19, 2006.
© Springer-Verlag Berlin Heidelberg 2006

language AsmL [2]. We do not define ASMs here; instead, we discuss some of the distinctive qualities of ASMs that may be of interest to this audience.

2.1 A Richer Notion of Universality

Turing's original model of computation was intended to capture the notion of computable function from strings to strings. Turing convincingly argued that every string-to-string computable function is computable by some Turing machine [14]. His thesis is now widely accepted.

The string-to-string restriction does not seem severe because usually inputs and outputs can be satisfactorily coded by strings. But the restriction is not innocuous. First, some well-known (already in Turing's time and even earlier) algorithms work with inputs that do not admit string encoding. Think for example of the Gaussian elimination algorithm, which deals with genuine reals, or of the ruler-and-compass constructions of classical geometry, which deal with continuous objects. Second, some inputs can be coded by strings but not in a satisfactory way. Consider graphs for example. Graphs can be represented by adjacency matrices, and matrices are perfectly string codable. But there is no known canonical and feasible adjacency-matrix presentation of graphs. It is essentially the well-known problem of database theory: how to deal with databases in an implementation independent way?

One can argue that, taken literally, the Gaussian elimination algorithm is too abstract, that in any actual computation one deals with finite approximations of reals which are perfectly represented by strings. But in many cases, it is desirable to deal with abstract algorithms. This brings us to an essential drawback of Turing's computation model. While it is perfect for the intended purpose, its abstraction level is essentially that of single bits.

There is another drawback of Turing's model. A Turing machine simulation of a given algorithm is guaranteed to preserve only the input/output behavior. There may be more to an algorithm than the function it computes.

The ASM thesis asserts that, for every algorithm A, there is an ASM B that is behaviorally equivalent to A. In particular, B step-for-step simulates A. Substantial parts of the thesis have been proved from first principles [9, 3, 11].

2.2 Abstraction

We have mentioned already that the single, low abstraction level of the Turing machine inhibits its ability to faithfully simulate algorithms, and that an appropriate ASM simulator operates at the abstraction level of the original algorithm. Each ASM is endowed with a fixed vocabulary of function names. A state of the ASM is a (first-order) structure of that vocabulary: a collection of elements, along with interpretations of the function names as functions over the elements. The author of an ASM program has flexibility in choosing the level of abstraction. For example, atoms, sets of atoms, sets of sets of atoms, *etc.* can be treated as elements of a structure of the vocabulary with the containment relation ∈. Similarly, other complex data — maps, sequences, trees, graphs, sets of maps, *etc.* — can be treated as elements without any encoding. This makes

ASMs appropriate for various applications — *e.g.*, specifications of software — dealing with high-level abstractions. It remains to be seen whether there are areas of language theory that can take advantage of it.

rule *MinDFA*:

```
1              if not initialized then
2                  Q' := {F, Q − F}, F' := ∅
3                  initialized := true
4              else
5                  choose X ∈ Q', σ ∈ Σ where splits(X, σ)
6                      do-forall Y ∈ Q' where reaches(X, σ, Y)
7                          add {q : δ(q, σ) ∈ Y} to Q'
8                      remove X from Q'
9                  ifnone
10                     choose X ∈ Q' where q₀ ∈ X
11                         q'₀ := X
12                     do-forall X ∈ Q'
13                         choose q ∈ X
14                             do-forall σ ∈ Σ
15                                 choose Y ∈ Q' where δ(q, σ) ∈ Y
16                                     δ'(X, σ) := Y
17                             if q ∈ F then add X to F'
18                     halt
```

Fig. 1. ASM program: DFA minimization

2.3   Concurrency and Non-determinism

We distinguish between sequential-time ASMs and distributed ASMs. A sequential time ASM computes in a step-after-step manner. As in the case of Turing machines, the program describes a single step. Already in the case of Turing machines, a single step may involve several operations executed in parallel: change the control state, change the content of the observed cell, move the head on the tape. In the case of an ASM, there may be no *a priori* bound on the amount of work done in parallel during one step. In particular, the **do-forall** rule provides a powerful form of concurrency; see Figure 1. In the ASM world, parallelism is the default. If you have a rule R_1 followed by rule R_2, it is presumed that they are executed in parallel. For example, the three assignments in lines 2 and 3 of Figure 1 are executed in parallel. In the ASM based specification language AsmL [2] mentioned earlier, you pay a syntactic price for requiring that the rule R_1 is executed first, and the rule R_2 is executed second.

A sequential-time ASM can be non-deterministic. This is achieved by means of the **choose** rule that allows you to non-deterministically choose an element from a finite set. You may require that the chosen element satisfies a specified condition; see Figure 1.

A distributed ASM is a dynamic set of agents operating asynchronously over a global state. The global state could reflect a physical common memory or be a pure mathematical abstraction with no physical counterpart [8, 1].

There are various new computational paradigms that exploit the possibility of massive parallelism: quantum computing, DNA computing, membrane computing, evolutionary computing. All of these approaches seem well suited for description in terms of ASMs. In fact, Grädel and Nowack proved that every model of quantum computing in the literature can be viewed as a specialized sequential-time ASM model, some steps of which are hugely parallel [7].

3 Example: Minimization Algorithm

We illustrate the use of ASMs with an example taken from elementary automata theory: the well-known algorithm for minimizing a deterministic finite automaton [12]. This example is instructive in that it uses the power of the **do-forall** and **choose** rules, and it demonstrates the ability to compute with sets.

3.1 Informal Description

We recall a version of the minimization algorithm (without proving its correctness). Let Σ be a finite alphabet. Given a finite automaton $A = \langle Q, \Sigma, q_0, \delta, F \rangle$, the algorithm computes a minimal finite automaton $A' = \langle Q', \Sigma, q'_0, \delta', F' \rangle$. A' is equivalent to A in the sense that $L(A) = L(A')$. And it is minimal in the sense that Q' is as small as possible. Here Q is the set of states of A, q_0 is the initial state of A, $\delta : Q \times \Sigma \to Q$ is the transition function of A, and F is the set of final states of A. And Q', q'_0, δ' and F' play the same roles respectively for A'.

The algorithm computes Q' and then uses Q' to compute q'_0, δ' and F'. Q' is computed by means of successive approximations. Q' is initialized to $\{F, Q - F\}$, and then the splitting process starts. View the current members of Q' as candidates for the membership in the ultimate Q'; every candidate X is a subset of Q. A candidate Y is σ-next for a candidate X if $\delta(q, \sigma) \in Y$ for some $q \in X$, and a candidate X σ-splits if there exist distinct candidates Y, Z that are σ-next for X. If X splits on σ, replace it with new candidates $\{q \in X : \delta(q, \sigma) \in Y\}$ where Y ranges over σ-next candidates for X. The splitting process stops when no candidate splits on any letter. At this point we have the desired Q'. The new initial state q'_0 is the candidate that contains q_0. Now consider a member X of the ultimate Q' and let q be any element of X. For any letter σ, $\delta'(X, \sigma)$ is the unique candidate Y such that $\delta(q, \sigma) \in Y$ for some $q \in X$. And $F' = \{X \in Q' : X \subseteq F\}$. This does not depend on the choice of q as X does not split on any letter σ.

3.2 Minimizing ASM

Figure 1 gives an ASM form of the algorithm. For the reader's convenience we number the lines of the program. Lines 1–3 reflect the initialization process. We use two auxiliary Boolean terms, next and splits, defined as follows.

next(X, σ, Y): $(\exists q \in X)\ \delta(q, \sigma) \in Y$.

splits(X, σ): $(\exists Y, Z \in Q' : Y \neq Z)$ next(X, σ, Y) and next(X, σ, Z).

Lines 5–8 reflect the splitting process. Lines 10–17 reflect the computation of the remaining components of A'. Note the significant degree of parallelism through the use of **do-forall** and **choose**. Furthermore, there is implicit parallelism in every sequence of instructions. For instance, splitting a candidate X consists of two actions: creating the appropriate subsets of X (lines 5–6) and removing X (line 7), which occur simultaneously.

The ASM program in Figure 1 looks like pseudo-code but it has a well-defined semantics. It is a simple exercise to rewrite the program in AsmL [2, 10]. The result will be an executable version of the program.

4   Conclusion

Our intention is to raise awareness of the ASM model as a potential tool for language theory. Whether it is a tool of real value for this area is a question that can only be answered by language theory people. We hope that this little exposition attracts their attention.

References

1. Abstract State Machines. http://www.eecs.umich.edu/groups/gasm/
2. Abstract State Machine Language (AsmL). http://research.microsoft.com/fse/asml/
3. Blass, A., Gurevich, Y.: Algorithms: A Quest for Absolute Definitions. Bulletin of the European Association for Theoretical Computer Science **81** (2003) 195–225
4. Blass, A., Gurevich, Y., Shelah, S.: Choiceless Polynomial Time. Annals of Pure and Applied Logic **100** (1999) 141–187
5. Blass, A., Gurevich, Y., Shelah, S.: On Polynomial Time Computation Over Unordered Structures. Journal of Symbolic Logic **67** (3) (2002) 1093–1125
6. Börger, E.: Abstract State Machines: A Unifying View of Models of Computation and of System Design Frameworks. Annals of Pure and Applied Logic **133** (2005) 149–171
7. Grädel, E., Nowack, A.: Quantum Computing and Abstract State Machines. In Börger, E. (ed.): Abstract State Machines: Advances in Theory and Applications. Lecture Notes in Computer Science, Vol. 2589. Springer-Verlag (2003) 309–323
8. Gurevich, Y.: Evolving Algebras 1993: Lipari Guide. In: Börger, E.: Specification and Validation Methods. Oxford University Press (1995) 9–36
9. Gurevich, Y.: Sequential Abstract State Machines Capture Sequential Algorithms. ACM Transactions on Computational Logic **1** (1) (2000) 77–111
10. Gurevich, Y., Rossman, B., Schulte, W.: Semantic Essence of AsmL. Theoretical Computer Science **343** (3) (2005) 370–412
11. Gurevich, Y: Interactive Algorithms 2005. In: Jedrzejowicz, J., Szepietowski, A.: Mathematical Foundations of Computer Science 2005. Lecture Notes in Computer Science, Vol. 3618. Springer-Verlag (2005) 26–38

12. Nerode, A: Linear Automaton Transformations. Proceedings of the American Mathematical Society **9** (1958) 541–544
13. Rozenberg, G. (ed.): Handbook of Graph Grammars and Computing by Graph Transformation. Volume 1 - Foundations. World Scientific (1997)
14. Turing, A. M: On computable numbers, with an application to the Entscheidungsproblem. Proceedings of London Mathematical Society, series 2, **42** (1936–1937) 230–265. Correction, *ibidem* **43** 544–546

Languages in Membrane Computing: Some Details for Spiking Neural P Systems

Gheorghe Păun

Institute of Mathematics of the Romanian Academy
PO Box 1-764, 014700 Bucharest, Romania, and
Department of Computer Science and AI
University of Sevilla
Avda Reina Mercedes s/n, 41012 Sevilla, Spain
george.paun@imar.ro, gpaun@us.es

Abstract. After a brief introduction to membrane computing, pointing out the more important intersections with formal language theory, we survey a series of recent results related to spiking neural P systems used as devices for handling languages. Several open problems are formulated.

1   Introduction: MC Versus LT

Membrane computing is a branch of natural computing which abstracts computing models from the structure and the functioning of living cells and from the way the cells cooperate in tissues or in higher order structures. This phrase is sort of a slogan of the domain – in the same extent as Figure 1 is used as sort of logo of it. Indeed, after adding that the computing models investigated in this area (called P systems) are parallel and distributed cell-like or tissue-like devices, processing multisets of objects in compartments defined by membranes, we have already a rather exact, although general, description of membrane computing. (It is maybe necessary to also add that most classes of P systems are synchronized models, that they are Turing complete as computing power, and that, if enhanced parallelism is provided, e.g., by means of membrane division, then polynomial solutions of computationally hard problems – **NP**-complete, but also **PSPACE**-complete problems – can be devised.)

Still, there are here several keywords which need clarification and which are central to the scope of our notes – a discussion about the relationships between membrane computing and formal language theory. The starting point should be the syntagma *multisets of objects*.

The membranes of a cell have two main roles: to delimit "protected reactors", where specific reactions take place, and to provide support for certain reactants, especially enzymes/proteins, which control part of the reactions from the compartments and the passage of molecules across membranes. Reactants, molecules, chemicals, from ions to large macromolecules – in short and more general, *objects*. Then, because in bio-chemistry the numbers matter (actually, mainly the abundance), membrane computing considers the *multiset* as the basic data structure. Multisets, i.e., sets with multiplicities associated with their elements. Specifically, in the compartments of a membrane structure as that from Figure 1 one places multisets of abstract objects; the objects are represented by symbols of a given alphabet O, hence the multisets can be represented by strings over O, with the obvious convention that all permutations of a string represent the same multiset. Then, as basic rules for making multisets evolve we can use "chemical reactions", which are nothing else than multiset rewriting rules, of the form $u \rightarrow v$ (the objects specified by multiset u react, are consumed, and as a result of the reaction we get the objects indicated by v). Using the rules in a synchronized way (a global clock is assumed, and in each step each rule which can be applied should be applied – of course, in competition for objects with other rules), we obtain computations. A result is associated with halting computations, as provided by the multiset of objects placed in a specified region of the system – and this result can be the vector describing the multiplicity of objects in that multiset or the total number of objects in the multiset.

We started already to speak in terms of formal languages: the multiplicity of objects in a multiset described by a string is given by the Parikh vector of the string, while sets of multisets are nothing else than commutative languages. The first Workshop on Membrane Computing, organized in August 2000 in Curtea de Argeş, Romania, was actually explicitly devoted to multiset processing – see [2], where a series of details can be found about the multiset bridge between membrane computing and computer science and general, including a proposal to start a more general theory of families of sets of multisets, of the AFL theory type.

There are several other aspects related to the early years of membrane computing (well, five to seven years ago...) which made people consider this area as a branch of language theory: (i) The approach was initially "grammatical-like", with the P systems used to generate sets of numbers (due to the inherent non-determinism in using the rules for multiset processing, the computations branch and in the end of halting computations one gets sets of numbers or of vectors of numbers). (ii) The main tools used in investigating the computing power of P systems were matrix grammars with appearance checking in the binary normal form (which was improved in several stages with new motivations coming from membrane computing, for instance, in what concerns the number of symbols used in appearance checking rules – see, e.g., [8]), grammar systems, L systems, and later register machines; all these are classic devices in language (and automata) theory. (iii) The types of results were also of a known type: comparisons with families of languages in Chomsky and Lindenmayer hierarchies, modulo ordering of symbols (hence stated in terms of Parikh images of known families of languages)[1].

The "second generation" of membrane computing has departed from language theory – as we will see immediately, still keeping a close contact with it. First, besides multiset rewriting rules there were considered symport and antiport rules,

---

[1] By the way: is the inclusion $PsE0L \subseteq PsET0L$ proper? (For a family FL of languages, $PsFL$ denotes the family of Parikh images of languages in FL).

of a direct biological inspiration, which are only moving objects across membranes, never changing them. This is still very much in the spirit of register machines – which became the main tool in investigating such classes of P systems (and this made again necessary clarifications concerning the relationships between various types of register machines, matrix grammars, vector addition systems – see [9]). Then, mainly based on membrane division, which makes possible the creation of an exponential workspace in a linear time, by means of a natural (in both senses: coming from nature, and simple-basic-intuitive) operation, membrane computing became much interested in solving hard problems in a feasible time, thus getting closer to computational complexity. In this moment, this is one of the most active directions of research, with a series of nice results (e.g., characterizations of classic complexity classes) and intriguing open problems (mainly concerning the borderline between efficiency and non-efficiency: which ingredient makes possible a polynomial solution to an **NP**-complete problem, such that without this ingredient only problems from class **P** can be solved in polynomial time by a given type of P systems?)[2].

Still, languages are always present also in these frameworks, of symport/antiport rules and of complexity investigations – let us mention only the automata approach (see an overview in [7]), which will be considered in more details below.

It is worth returning now to the basic ingredients of a P system: the "objects" of bio-chemistry are not only "symbols" (atoms, in the etymological sense), but also "words", molecules with a structure which plays a role in the behavior inside the cell. Thus, we can consider string-objects (actually, this was already done in the initial paper, [18]), with two possibilities: working in the multiset sense, or in the standard language sense. An important detail appears here: what means "standard language sense"? Sets when we discuss operations with languages, multisets with infinite multiplicity when defining the derivations in a grammar: starting from a sentential form, all continuations made possible by the rules of the grammar (which rule to apply and where in the string to apply it) are simultaneously possible. This does not make a difference when defining the language generated by a Chomsky grammar, but it is crucial in situations where the strings interact (like in DNA computing) in order to produce new strings, and where it is important which strings are available in each step.

Both sets and multisets of strings were considered in membrane computing, processes by means of usual rewriting rules, by splicing rules, or by other types of rules inspired from the genome area. The multiplicity matters when considering P systems with string-objects for solving computationally hard problems: using rewriting with replication (when applying a rule $a \rightarrow u_1||u_2||\ldots||u_n$ to x_2ax_2 we get n strings $x_1u_1x_2,\ldots,x_1u_nx_2$ at the same time) we can again solve **NP**-complete problems in polynomial time, like in the case of using membrane division.

---

[2] I am wondering whether similar problems were investigated for Lindenmayer systems: can we make use of the possibility of growing exponentially the strings of, e.g., an ET0L system, in order to solve **NP**-complete problems using ET0L systems?

This close initial relationship of membrane computing with language theory stays nicely in balance with recent developments (we may say that we have now P systems of a "third generation") where tissue-like and neural-like P systems are much investigated. Besides the next two sections, all the remaining sections of this paper will be devoted to presenting such a model, the spiking neural P systems, mainly as they are used for handling languages.

2   A Glimpse to Membrane Computing

For the sake of completeness, we recall here some basic ideas, definitions, and results from membrane computing. Further details can be found in [19] (with applications presented in [6]), while updated information can be found at the web site [23].

Actually, we discuss only cell-like P systems, and we refer the reader to the cited sources for tissue-like P systems and population P systems.

The basic ingredient of the computing device we consider is that of a membrane structure; the related terminology and representation are given in Figure 1. Mathematically, we can represent a membrane structure as a rooted unordered tree, or as a string of matching parentheses.

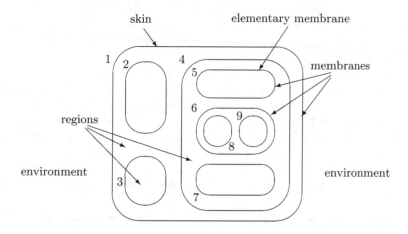

Fig. 1. A membrane structure

By placing multisets of objects and evolution rules in the regions of a membrane structure, we get a P system of the basic type, called *transition P system*. Formally, such a device (of degree $m \geq 1$) is a construct of the form

$$\Pi = (O, C, \mu, w_1, w_2, \ldots, w_m, R_1, R_2, \ldots, R_m, i_o),$$

where:

1. O is the (finite and nonempty) alphabet of *objects*;
2. $C \subset O$ is the set of *catalysts*;

3. μ is a membrane structure, consisting of m membranes, labeled $1, 2, \ldots, m$; we say that the membrane structure, and hence the system, is *of degree m*;

4. $w_1, w_2, \ldots, w_m$ are strings over O representing the *multisets of objects* present in regions $1, 2, \ldots, m$ of the membrane structure;

5. $R_1, R_2, \ldots, R_m$ are finite *sets of evolution rules* associated with regions $1, 2, \ldots, m$ of the membrane structure;

6. i_o is either one of the labels $1, 2, \ldots, m$, and then the respective region is the *output region* of the system, or it is 0, and then the result of a computation is collected in the environment of the system.

The rules are of the form $u \to v$ or $u \to v\delta$, with $u \in O^+$ and $v \in (O \times Tar)^*$, where $Tar = \{here, in, out\}$. The rules can be cooperative (with u arbitrary), non-cooperative (with $u \in O - C$), or catalytic (of the form $ca \to cv$ or $ca \to cv\delta$, with $a \in O - C, c \in C$, and $v \in ((O - C) \times Tar)^*$); note that the catalysts never evolve and never change the region, they only help the other objects to evolve.

When using a rule $u \to v$, the objects of u are consumed and those of v are produced. An object a appearing in v in a pair (a, tar) is placed in the region specified by tar: if $tar = here$, then the object remains in the same region where the rule is used, if $tar = in$, then the object goes in any of the directly inner membranes, and if $tar = out$, then object a exits the membrane and becomes an element of the immediately upper region (this is the environment if the rule was used in the skin region). If δ is present, then the membrane is dissolved, its rules are removed and its objects become elements of the immediately upper membrane; the skin membrane is never dissolved.

As suggested in the previous section, the rules are used in a non-deterministic way, choosing randomly the objects to evolve and the rules by which they evolve, but in such a way that the choice is *maximal*, no further rule can be applied in the same step to the remaining objects. This is an essential point, important in establishing the computing power of these devices (the maximal parallelism, combined with the halting restriction in defining successful computations, is closely related to the appearance checking in regulated grammars, and to the check for zero in register machines – see a more detailed discussion about this point in [11]).

By using the rules in this way, we get transitions among configurations of the system Π. A sequence of transitions starting from the initial configuration of the system forms a computation. With a halting computation we associate a result, for instance, in the form of the number of objects present in the halting configuration in region i_o. We denote by $N(\Pi)$ the set of numbers computed/generated in this way, and by $NOP_m(type\text{-}of\text{-}rules)$ the family of sets $N(\Pi)$ for systems Π with at most m membranes and using rules of the *type-of-rules* specified.

We have given this general notation, because the number of types of rules which can be used in a P system is very large, with motivations coming both from biology and from computer science. The rules can be cooperative, non-cooperative (context-free), or catalytic, they can have promoters or inhibitors, can control the membrane permeability, can dissolve or create membranes, and so on and so forth. In particular, they can correspond to the biological phenomena of passing molecules through membranes in couples, in the same direction (this is called *symport*) or in opposite directions (*antiport*).

Such rules can be formalized as (x, in) and (x, out) (symport) and $(z, out; w, in)$ (antiport), where x, z, and w are multisets of objects. The meaning is that all objects of x are moved across the membrane with which the rule is associated, going inside in the case of (x, in) and outside in the case of (x, out), while the application of the antiport rule $(z, out; w, in)$ moves the objects of z out of the membrane at the same time with moving in the objects of w. The length of x is called the *weight* of the symport rule, and $\max(|z|, |w|)$ is the *weight* of the antiport rule. The formal definition of a symport/antiport P system is the same as above, with the important additional detail that we need an inexhaustible supply of objects in the environment, otherwise we can handle only a finite number of objects, those present initially in the system. The transition among configurations is defined in the same way (using the rules in the non-deterministic maximally parallel manner). The families of numbers generated in this context are denoted by $NOP_m(sym_p, anti_q)$ when considering systems of degree at most m, using symport rules of weight at most p and antiport rules of weight at most q.

We mention here only two universality results related to these basic types of P systems (cat_2 indicates the fact that one uses catalytic rules, with only two catalysts present in the system, and NRE is the family of Turing computable sets of numbers, hence the length sets of recursively enumerable languages; an intriguing current open problem in this area is whether or not systems with one catalyst are universal, with the conjecture that this is not the case – see papers and references about this subject in [23], for instance, from the Fourth Brainstorming Week on Membrane Computing, Sevilla, 2006):

1. $NRE = NOP_1(cat_2)$, [10].
2. $NRE = NOP_3(sym_1, anti_1) = NOP_3(sym_2, anti_0)$, [1].

We do not continue this general presentation of membrane computing, in particular, not touching the important complexity related area of research (neither the applications area, maybe still more important/promising in this moment), because we want to return to languages, first as they already appear in relation with "classic" types of P systems, and then as they appear in the recently introduced spiking neural P systems.

3 Languages in Membrane Computing

As said before, we can work with objects described by strings, and then the languages appear in a direct way – with several possible types of systems: using rewriting, splicing, or other operations with strings; rewriting can be with replication; the rules can be context-free or controlled by promoters or inhibitors; the strings can have multiplicities of not; as successful computations we can consider only halting computations or all computations, in the latter case selecting the strings, e.g., by sending them outside the system or by using a terminal alphabet.

Still, there are more interesting ways to associate strings with computations in a P system which have inside multisets of symbol objects, and the interest comes from the essential difference between the "internal" data structure, multisets, and the "external" data structure, that of the result of computations, strings (with syntax, positional information).

Three main ways to compute strings were investigated for P systems with symbol objects: (i) considering an external output, (ii) using P systems in the accepting mode, and (iii) following the traces of a traveler across membranes.

The first idea was considered already at the beginning of research in membrane computing, [22]: let us stay outside the system and record all objects which exit it, in the order they are expelled from the system; if several symbols exit at the same time, then any permutation of them is allowed in the generated string. For a computation to be successful, it has to halt, otherwise we do not know when a string is completed. The definition is clear for P systems with multiset rewriting rules, because the objects sent into the environment are never retrieved (and not considered yet for symport/antiport systems, because the symbols can come back into the system, but also for this case we can investigate such externally generated strings, allowing or not the symbols already introduced in the string to return into the system).

Dual to this idea is the second one mentioned above, also natural in view of the grammar-automata duality in language and automata theory: using a P system not to generate the strings of a language, but to recognize them. This is possible also for numbers: introduce a number in a region, in the form of the multiplicity of a specified object, start the computation and accept the number if and only if the computation halts. For strings it is necessary to involve again the environment: the symbols of the strings are supposed to be available in the environment, in the order they appear in the string, and then rules for bringing them inside are necessary. Thus, accepting P systems were considered only for symport/antiport systems (and for certain related systems with the computation based on communication only), with several variations in the definition: the symbols should be introduced in the system one by one, in the first steps of the computation (this is very restrictive, because no intermediate computations are possible, e.g., for encoding the positional information of the string in the numbers available inside the system), or only from time to time, at the request of the system; restrictions can be imposed on the form of the used rules; we can consider the string not as provided in the environment, but constructed from the symbols imported from the environment, in the order they are brought in, maybe associating a label to a multiset (then the accepted string is that of labels of imported multisets); a computation can be considered successful if the system is halting or also using states, like in automata theory, but with states defined here in terms of the multisets present in the regions of the system; a system can be deterministic or non-deterministic, etc.

We refer to [7] for further details about the many variants of accepting P systems (sometimes called *P automata*), and to results in this area (mainly universality), and we only point out here a recent result, which establishes a nice link with Chomsky hierarchy, part of a larger research topic which still waits for extended efforts to be fully covered: looking for characterizations (or at least representations via specified operations with languages) of families of languages from classic language theory (Chomsky, Lindenmayer, Marcus hierarchies) in terms of membrane computing. This is a challenging question in general in natural computing: it seems that the most (only?) "natural" classes of languages are the regular and the recursively enumerable ones, which have direct and sometimes easy characterizations (e.g., in terms of splicing systems), while the families of linear, context-free, or context-sensitive languages do not have a direct counterpart in natural computing.

This result concerns a characterization of context-sensitive languages, recently obtained in [13], using a restricted class of accepting symport/antiport P systems: one considers accepting one-membrane systems, with an input alphabet $\Sigma \subseteq O$ containing a distinguished symbol \$ (the end marker), the environment containing all objects from $O - \Sigma$ (and no object from Σ), and rules of the following four types:

1. $(u, out; v, in)$, where $u, v \in (V - \Sigma)^*$ with $|u| \geq |v|$.

2. $(u, out; va, in)$, where $u, v \in (V - \Sigma)^*$ with $|u| \geq |v|$, and $a \in \Sigma$. A rule of this type is called a *read-rule*.

3. $(u, out; v, in)|_a$, where $u, v \in (V - \Sigma)^*$, and $a \in \Sigma$ (a is a promoter). Note that there is no restriction on the relative lengths of u and v.

4. For every $a \in \Sigma$, there is at least one rule of the form $(a, out; v, in)$ in the set R_1, where $v \in (V - \Sigma)^*$. Moreover, this is the only type of rules for which a can appear on the left part of the rule.

Such a system accepts a string $x = a_1 \ldots a_n\$$, where a_i is in $\Sigma - \{\$\}$ for $1 \leq i \leq n$, if its symbols are brought into the system in the order they appear in x by means of read-rules, and, after reading the string completely, the computation eventually halts.

P systems of this type characterize the context-sensitive languages; if some of the restrictions 1–4 are removed or certain changes are made in the definition, then characterizations of regular or of recursively enumerable languages are obtained.

The third idea mentioned above for defining a string associated with a computation in a P system with symbol-objects is specific to membrane computing, it has no counterpart in classic grammar and automata theory: in communicative P systems (those with the computation based on moving objects from a region to another one, like in symport/antiport P systems), we can consider a distinguished object (a "traveler"), and the trajectory of this object through the system provides a string, by recording certain *events* related to the traveler journey, for instance, writing the symbol b_i when the traveler enters membrane i. When the computation halts, the trace-string is completed.

We do not discuss further the trace languages for standard P systems, but we will consider this idea in more details for spiking neural P systems, in Section 6.

4   Spiking Neural P Systems

The question of incorporating ideas from neuro-biology into membrane computing was formulated several times as a research topic, and there are several contributions to this issue, including a chapter in [19]. Still, this research direction
is not at all explored as it deserves to be; the neurons functioning and especially
their cooperation in various constructions, the brain included, is a huge source
of ideas. Recent contributions were added to this topic by introducing so-called
spiking neural P systems (in short, SN P systems), which capture the important
idea of neural biology concerning the way the neurons communicate by means
of "spikes", electrical impulses of identical intensity and shape, but occurring
at time moments which are carrying information in the distance between them.
We refer to [16] for details and further references about the biological processes
related to spiking and about the way they are used in neural computing (one
speaks in the last years about a neural computing of the "third generation"
based on spiking neurons).

   The way the idea is modeled in terms of P systems is rather simple: one
considers only one type of objects, the spike, denoted by a, and neurons linked
by synapses (elementary membranes placed in the nodes of a directed graph),
containing spikes and rules for handling them. These rules are of two forms:

(1) $E/a^c \to a; t$, where E is a regular expression over $\{a\}$, $c \geq 1$, and $t \geq 0$;
(2) $a^s \to \lambda$, for some $s \geq 1$, with the restriction that $a^s \notin L(E)$ for any rule
    $E/a^c \to a; t$ of type (1) from the same neuron.

   The rules of type (1) are *firing* (we also say *spiking*) *rules*, the rules of type
(2) are *forgetting rules*.

   A neuron gets fired when using a rule $E/a^c \to a; t$, and this is possible only if
the neuron contains n spikes such that $a^n \in L(E)$ and $n \geq r$. This means that
the regular expression E "covers" exactly the contents of the neuron. The use
of a rule $E/a^c \to a; t$ in a step q means firing in step q, consuming c spikes, and
spiking in step $q+t$. That is, if $t = 0$, then the spike is produced immediately, in
the same step when the rule is used. If $t = 1$, then the spike will leave the neuron
in the next step, and so on. In the interval between using the rule and releasing
the spike, the neuron is assumed *closed* (in the refractory period), hence it cannot
receive further spikes, and, of course, cannot fire again. This means that if $t \geq 1$
and another neuron emits a spike in any moment $q, q+1, \ldots, q+t-1$, then its
spike will not pass to the neuron which has used the rule $E/a^c \to a; t$ in step
q. In the moment when the spike is emitted, the neuron can receive new spikes.
This means that if $t = 0$, then no restriction is imposed, the neuron can receive
spikes in the same step when using the rule. Similarly, the neuron can receive
spikes in moment t, in the case $t \geq 1$.

   If a neuron σ_i spikes, its spike is replicated in such a way that one spike is
sent to *each* neuron σ_j such that there is a synapse from σ_i to σ_j (we write
$(i, j) \in syn$, with the set syn specified in advance), *and* σ_j is open at that
moment. If a neuron σ_i fires and either it has no outgoing synapse, or all neurons

σ_j such that $(i, j) \in syn$ are closed, then the spike of neuron σ_i is lost; the firing is allowed, it takes place, but it produces no spike.

By using a forgetting rule $a^s \to \lambda$, s spikes are simply removed ("forgotten"). Like in the case of spiking rules, the left hand side of a forgetting rule must "cover" the contents of the neuron, that is, $a^s \to \lambda$ is applied only if the neuron contains exactly s spikes.

An SN P system is said to be *finite* if $L(E)$ is a finite language for each rule $E/a^c \to a; t$.

Note that each neuron uses at most one rule at a time – hence the neurons work in a sequential manner (but the system itself is synchronized: in each time unit, each neuron which can use a rule should do it).

One of the neurons is designated as the output neuron of the system and when it spikes, besides spikes sent to other neurons along synapses, a spike is also sent to the environment. In this way, the system produces a *spike train*, a sequence of time units when we have spikes leaving the system. The number of steps elapsed between two consecutive spikes can be considered as being computed by the system, with many possibilities to define the computed set of numbers: taking all intervals, taking only the interval between the first two spikes, considering alternately the intervals (we take the first interval, we ignore the second one, we take the third interval, and so on), considering all computations or only the halting ones. In this way, the spiking neural P systems behave as number computing devices. It is also possible to consider the spike train itself as the output of a computation, codified as a binary sequence: we write 1 for a time unit when the system sends a spike into the environment and 0 for a time unit when no spike is sent out. In the non-halting case we get then an infinite binary sequence. If also input neurons are considered, then we can work in the accepting case or even with spiking neural P systems as transducers of binary strings/sequences.

We do not give here examples of SN P systems, because in the next sections we will present specific P systems for various language processing tasks.

Many of the above mentioned possibilities of using a spiking neural P system for computing numbers were considered in [14], [20], [21], and part of them were also investigated in some detail. In particular, two main results were proved in [14] for the case of considering only the distance between the first two spikes, and then extended in [20] to many other cases:

1. arbitrary SN P systems are Turing complete,
2. finite SN P systems characterize the semilinear sets of numbers.

The universality result is obtained for systems with a small number of rules in each neuron (at most two), and small numbers of spikes consumed in firing and forgetting rules (at most three), but without any bound on the number of neurons. Such bounds were produced in [17], starting from simulating (universal) register machines as those from [15]: about 50–80 neurons suffice (this depends on the type of rules used and on the definition of universality). Several improvements of the results from [14], [20] in the form of rules (no delay, or no forgetting rules, or particular forms of regular expressions) were given in [12].

The paper [21] considers the SN P systems as string and (infinite) sequence processors, for the case of the binary alphabet: both an input and an output neuron are provided, with spikes coming from the environment to the input neuron; the input sequence of bits is thus processed by the system, in the automaton style (accepting or not the input, depending on whether the computation halts or not), or in the transducer style (an output string is produced, which is thus the translation of the input string). A tool-kit for handling binary strings is provided in [21]: computing Boolean operations, taking prefixes, suffixes, substrings, computing a guided crossover, and so on.

Still, a lot of problems are open about SN P systems as string/sequence processors, for instance, in what concerns the case of infinite sequences, starting with the necessary comparison of the family of sequences computed in this framework with infinite sequences recognized by finite automata, by Turing machines, or appearing in other areas (for instance, generated by L systems).

5 SN P Systems as Language Generators

A very natural way of using an SN P system is as a string generator, taking the binary encoding of a spike trace associated with a halting computation as the string generated by that computation. Instead of a formal definition, let us examine an example from [3], the system Π_1 from Figure 2. This also gives us the opportunity to illustrate the way to graphically represent an SN P system: as a directed graph, with the neurons as nodes and the synapses indicated by arrows; an arrow also exits from the output neuron, pointing to the environment; in each neuron we specify the rules and the spikes present in the initial configuration.

We have $L(\Pi_1) = \{0^{n+4}1^{n+4} \mid n \geq 0\}$. The reader can check that in $n \geq 0$ steps when neuron σ_1 uses the rule $a^2/a \to a; 0$ the output neuron accumulates $2n + 6$ spikes. When neuron σ_1 uses the rule $a^2/a \to a; 1$, one more spike will arrive in neuron σ_9 (in step $n + 4$). In this way, the number of spikes present in neuron σ_9 becomes odd, and the rule $a(a^2)^+/a^2 \to a; 0$ can be repeatedly used until only one spike remains; this last spike is used by the rule $a \to a; 0$, thus $n + 4$ occurrences of 1 are produced.

Let us denote by $LSNP_m(rule_k, cons_p, forg_q)$ the family of languages $L(\Pi)$, generated by SN P systems Π with at most m neurons, each neuron having at most k rules, each of the spiking rules consuming at most p spikes, and each forgetting rule removing at most q spikes. When using only finite systems, we write $LFSNP_m(rule_k, cons_p, forg_q)$ for the corresponding family. As usual, a parameter m, k, p, q is replaced with $*$ if it is not bounded.

Here are some of the results obtained in [3] (B denotes the binary alphabet, $B = \{0, 1\}$, and FIN, REG, REC, RE are the families of finite, regular, recursive, and recursively enumerable languages).

Theorem 1. (i) *There are finite languages (for instance, $\{0^k, 10^j\}$, for any $k \geq 1$, $j \geq 0$) which cannot be generated by any SN P system, but for any $L \in FIN$, $L \subseteq B^+$, we have $L\{1\} \in LFSNP_1(rule_*, cons_*, forg_0)$, and if $L =$*

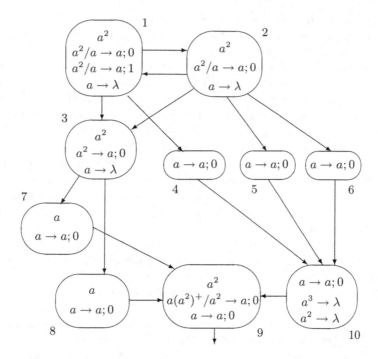

Fig. 2. An SN P system generating a non-regular language

$\{x_1, x_2, \ldots, x_n\}$, *then we also have* $\{0^{i+3}x_i \mid 1 \leq i \leq n\} \in LFSNP_*(rule_*,$ $cons_1, forg_0)$.

(ii) *The family of languages generated by finite SN P systems is strictly included in the family of regular languages over the binary alphabet, but for any regular language* $L \subseteq V^*$ *there is a finite SN P system* Π *and a morphism* $h : V^* \longrightarrow B^*$ *such that* $L = h^{-1}(L(\Pi))$.

(iii) $LSNP_*(rule_*, cons_*, forg_*) \subset REC$, *but for every alphabet* $V = \{a_1, a_2, \ldots, a_k\}$ *there are a morphism* $h_1 : (V \cup \{b, c\})^* \longrightarrow B^*$ *and a projection* $h_2 : (V \cup \{b, c\})^* \longrightarrow V^*$ *such that for each language* $L \subseteq V^*$, $L \in RE$, *there is an SN P system* Π *such that* $L = h_2(h_1^{-1}(L(\Pi)))$.

These results show that the language generating power of SN P systems is rather ex-centric; on the one hand, finite languages (like $\{0, 1\}$) cannot be generated, on the other hand, we can represent any RE language as the direct morphic image of an inverse morphic image of a language generated in this way. This ex-centricity is due mainly to the restricted way of generating strings, with one symbol added in each computation step. A natural idea to avoid this restriction is to use *extended* spiking rules, as already considered in [17], i.e., rules of the form $E/a^c \rightarrow a^p$, with the meaning that c spikes are consumed and p are produced, with $p \geq 0$. In this way, a common generalization is obtained for both spiking and forgetting rules – also a delay can be considered, but, because it was not necessary in the proofs of the results below, this feature was not introduced for extended rules.

Such rules are used as in restricted SN P systems, and a language can now be generated by associating the symbol b_i with a step when the output neuron sends out i spikes, with an important decision to take in the case $i = 0$: we can either consider b_0 as a separate symbol, or we can assume that emitting 0 spikes means inserting λ in the generated string. Thus, we both obtain strings over arbitrary alphabets, not only over the binary one, and, in the case where we ignore the steps when no spike is emitted, a considerable freedom is obtained in the way the computation proceeds. This latter variant (with λ associated with steps when no spike exits the system) is considered below.

We denote by $LSN^eP_m(rule_k, cons_p, prod_q)$ the family of languages $L(\Pi)$, generated by SN P systems Π using extended rules, with at most m neurons, each neuron having at most k rules, each rule consuming at most p spikes and producing at most q spikes. Again, the parameters m, k, p, q are replaced by $*$ if they are not bounded.

The next results were obtained in [5], as counterparts of the results from Theorem 1; as expected, the extended rules are useful, the obtained families of languages are larger, and finite, regular, and recursively enumerable can be directly obtained, without additional symbols and squeezing mechanisms.

Theorem 2. (i) $FIN = LSN^eP_1(rule_*, cons_*, prod_*)$ and this result is sharp, because $LSN^eP_2(rule_2, cons_2, prod_2)$ contains infinite languages.

(ii) $LSN^eP_2(rule_*, cons_*, prod_*) \subseteq REG \subset LSN^eP_3(rule_*, cons_*, prod_*)$; the second inclusion is proper, because $LSN^eP_3(rule_3, cons_4, prod_2)$ contains non-regular languages; actually, the family $LSN^eP_3(rule_3, cons_6, prod_4)$ contains non-semilinear languages.

(iii) $RE = LSN^eP_*(rule_*, cons_*, prod_*)$.

As in the case of symport/antiport P systems used in the accepting mode, it is an open problem to find characterizations (even only representations) of other families of languages in the Chomsky hierarchy.

6   Trace Languages of SN P Systems

The idea of following the traces of a distinguished object in its journeys through the system can be considered also for SN P systems: we distinguish one of the neurons of the system as the *input* one and in the initial configuration of the system we "mark" one spike from this neuron – the intuition is that this spike has a "flag" – and we follow the path of this flag during the computation, *recording the labels of the neurons where the flag is present in the end of each step.* (When presenting the spikes of a neuron, one of them is primed.) If a rule consumes the marked spike, then one of the spikes which leave the neuron is marked (because we may have several outgoing synapses and one spike goes along each of them, the "flag" is taken by only one of the spikes which exit the neuron where the rule was used). If a neuron uses a rule, but the marked spike is not consumed (e.g., we have three spikes in a neuron, aaa', and we use the rule $a^3/a \rightarrow a; 0$, then two spikes remain and one of them can be the marked one), then it remains in the neuron.

The trace of the marked spike can be recorded as a string of symbols b_i, with b_i associated with neuron σ_i. The set of all strings obtained in this way along all halting computations of Π is denoted by $T(\Pi)$. By $TSNP_m(rule_k, cons_p, forg_q)$ we denote the family of languages $T(\Pi)$, generated by systems Π with at most m neurons, each neuron having at most k rules, each of the spiking rules consuming at most p spikes, and each forgetting rule removing at most q spikes. A parameter m, k, p, q is replaced with $*$ if it is not bounded.

Before recalling some results from [4], let us consider an example, the system Π_2 whose initial configuration is given in Figure 3. We have $T(\Pi_3) = (b_7 b_6)^+ b_7 \cup (b_7 b_6)^+ b_7 b_6$.

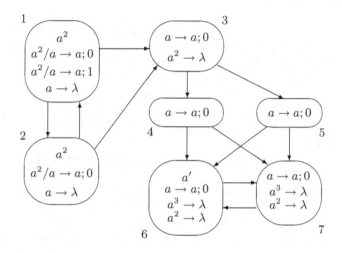

Fig. 3. The initial configuration of system Π_2

The neurons σ_6, σ_7 exchange the marked spike among them as long as they do not get "flooded" by neurons σ_4, σ_5, and this happens when a spike comes from neuron σ_3. In turn, this neuron spikes only in the step when neuron σ_1 uses the rule $a^2/a \to a; 1$ (if neuron σ_1 uses the rule $a^2/a \to a; 0$, then the cooperation of neurons σ_1, σ_2 continues, both of them returning to the initial state by exchanging spikes). When neurons σ_6, σ_7 have inside two or three spikes, they forget them and the computation stops. Depending on the step when neuron σ_1 uses the rule $a^2/a \to a; 1$ (an odd or even step), the string ends with b_7 or with $b_7 b_6$.

The trace families are again ex-centric, like in the case of languages generated by SN P systems in the restricted case (the extended rules were not considered yet also for SN P systems used as trace generators):

Theorem 3. (i) *There are singleton languages which are not in* $TSNP_*(rule_*, cons_*, forg_*)$.

(ii) *The family of trace languages generated by SN P systems by means of computations with a bounded number of spikes present in their neurons is strictly included in the family of regular languages, but for each regular language $L \subseteq V^*$ there is an SN P system Π such that each neuron from any computation of Π contains a bounded number of spikes, and $L = h(\partial_c^r(T(\Pi))$ for some coding h, and symbol c not in V. In turn $TSNP_{12}(rule_2, cons_2, forg_1) - REG \neq \emptyset$.*

(iii) *Every unary language $L \in RE$ can be written in the form $L = h(L') = (b_1^* \backslash L') \cap b_2^*$, where $L' \in TSNP_*(rule_2, cons_3, forg_3)$, and h is a projection.*

It is not know whether the last result above can be proved for languages over arbitrary alphabets – maybe extended rules are useful/necessary in this respect.

7   Final Remarks

The aim of this paper was only to point out some of the several (direct) bridges between membrane computing and formal language theory, especially considering those situations where P systems are used as language processing (generating, accepting, translating) devices, with some further details given for the recently introduced spiking neural P systems. Many topics remain to be investigated in this area: improving existing results from the point of view of the complexity of the used systems, extending these results to other classes of P systems or to other families of languages, examining in a closer extent the case of infinite sequences, looking for sub-universal systems and the properties of the associated language families, and so on and so forth.

Returning to the initial question, whether or not membrane computing is part of formal language theory, it should be now obvious that the answer is a qualified yes (this is the same with a qualified no...): the intersection/interaction of the two research areas is considerable, in the benefit of both of them. New motivation and applications for old notions, tools, and results of formal language theory arise, as well as new information (e.g., characterizations) about old families of languages, while membrane computing takes a considerable advantage from using the well developed tools and techniques of formal language (and automata) theory in investigating the "computing cell" (in the form modeled by P systems of various types).

References

1. A. Alhazov, R. Freund, Y. Rogozhin: Computational power of symport/antiport: history, advances and open problems. *Membrane Computing. 6th Intern. Workshop, WMC2005, Vienna, Austria, July 2005. Revised Selected and Invited Papers*, LNCS 3850, Springer, Berlin, 2006, 1–30.
2. C.S. Calude, Gh. Păun, G. Rozenberg, A. Salomaa, eds.: *Multiset Processing. Mathematical, Computer Science, and Molecular Computing Points of View.* LNCS 2235, Springer, Berlin, 2001.
3. H. Chen, R. Freund, M. Ionescu, Gh. Păun, M.J. Pérez-Jiménez: On string languages generated by spiking neural P systems. *Proc. Fourth Brainstorming Week on Membrane Computing*, Sevilla, 2006.

4. H. Chen, M. Ionescu, A. Păun, Gh. Păun, B. Popa: On trace languages generated by spiking neural P systems. *Proc. Fourth Brainstorming Week on Membrane Computing*, Sevilla, 2006.
5. H. Chen, Gh. Păun, M.J. Pérez-Jiménez: Spiking neural P systems with extended rules. *Proc. Fourth Brainstorming Week on Membrane Computing*, Sevilla, 2006.
6. G. Ciobanu, Gh. Păun, M.J. Pérez-Jiménez, eds.: *Applications of Membrane Computing*. Springer, Berlin, 2006.
7. E. Csuhaj-Varju: P automata. *Membrane Computing. International Workshop WMC5, Milano, Italy, 2004*, LNCS 3365, Springer, 2005, 19–35.
8. H. Fernau, K. Reinhardt, R. Freund, M. Oswald: Refining the nonterminal complexity of graph-controlled, programmed and matrix grammars. *J. Automata, Languages, and Combinatorics*, to appear.
9. R. Freund, O.H. Ibarra, Gh. Păun, H.-C. Yen: Matrix languages, register machines, vector addition systems. *Proc. Third Brainstorming Week on Membrane Computing*, Sevilla, 2005, RGNC Report 01/2005, 155–168.
10. R. Freund, L. Kari, M. Oswald, P. Sosik: Computationally universal P systems without priorities: Two catalysts are sufficient. *Theoretical Computer Science*, 330, 2 (2005), 251–266.
11. P. Frisco: P systems, Petri nets and program machines, *Membrane Computing, 6th Intern. Workshop, WMC2005, Vienna, Austria, July 2005, Revised Selected and Invited Papers*, LNCS 3850, Springer, Berlin, 2006, 209–223.
12. O.H. Ibarra, A. Păun, Gh. Păun, A. Rodríguez-Patón, P. Sosik, S. Woodworth: Normal forms for spiking neural P systems. *Proc. Fourth Brainstorming Week on Membrane Computing*, Sevilla, 2006.
13. O.H. Ibarra, Gh. Păun: Characterizations of context-sensitive languages and other language classes in terms of symport/antiport P systems. *Theoretical Computer Sci.*, 2006 (in press).
14. M. Ionescu, Gh. Păun, T. Yokomori: Spiking neural P systems. *Fundamenta Informaticae*, 71 (2006).
15. I. Korec: Small universal register machines. *Theoretical Computer Science*, 168 (1996), 267–301.
16. W. Maass, C. Bishop, eds.: *Pulsed Neural Networks*. MIT Press, Cambridge, 1999.
17. A. Păun, Gh. Păun: Small universal spiking neural P systems. *Proc. Fourth Brainstorming Week on Membrane Computing*, Sevilla, 2006.
18. Gh. Păun: Computing with membranes. *Journal of Computer and System Sciences*, 61, 1 (2000), 108–143, and *Turku Center for Computer Science-TUCS Report* No 208, 1998 (www.tucs.fi).
19. Gh. Păun: *Membrane Computing – An Introduction*. Springer, Berlin, 2002.
20. Gh. Păun, M.J. Pérez-Jiménez, G. Rozenberg: Spike trains in spiking neural P systems. *Intern. J. Found. Computer Sci.*, to appear (also available at [23]).
21. Gh. Păun, M.J. Pérez-Jiménez, G. Rozenberg: Infinite spike trains in spiking neural P systems. Submitted, 2006.
22. Gh. Păun, G. Rozenberg, A. Salomaa: Membrane computing with an external output. *Fundamenta Informaticae*, 41, 3 (2000), 313–340
23. The P Systems Web Page: http://psystems.disco.unimib.it.

Computational Nature of Biochemical Reactions

A. Ehrenfeucht[1] and G. Rozenberg[1,2]

[1] Department of Computer Science
University of Colorado at Boulder
Boulder, CO 80309, U.S.A.
[2] Leiden Institute of Advanced Computer Science (LIACS)
Leiden University
Niels Bohrweg 1, 2300 RA Leiden
The Netherlands, rozenber@liacs.nl

The functioning of a living cell consists of a (huge) number of individual biochemical reactions. These reactions are regulated, where the two main regulation mechanisms are: facilitation/acceleration and inhibition/retardation. The interaction between individual biochemical reactions takes place through their influence on each other, and this influence is through facilitation or inhibition (or both).

We present a formal model of reaction systems – its goal is to analyze/ understand, on an abstract level, some aspects of the functioning of a living cell, and in particular to analyze the interaction between biomolecular reactions. Therefore in the theory of reaction systems the formalization of an individual reaction relies on the above mentioned two regulation mechanisms, while the interaction between individual reactions does not have to be formalized: it is there "for free".

In this approach reactions are primary while structures are secondary: reactions *create* states (rather than *transform* states as is the case in traditional approaches in computer science). We also assume the "threshold" supply of elements (molecules) – if an element is present, then there is "enough" of it; thus we perform a qualitative rather than quantitative analysis of functioning of reaction systems. Moreover we do not assume permanency of elements but rather their sustainability: if nothing happens to an element (it is not a reactant for any active reaction), then it ceases to exist ("life must be sustained").

Altogether we argue that the axioms/assumptions underlying models of biochemical reactions and their interactions are *very* different from the axioms underlying models of computation in computer science. Thus, although reaction systems formalize (massively) concurrent systems, such as living cells, the basic set up here is very different than, e.g., for Petri Nets.

We present the basic theory of reaction systems, and illustrate it through examples coming both from biology and computer science. We demonstrate how the investigation of (suitably defined) computations in reaction systems allows one to understand their functioning. In particular one can define and investigate in this framework topics such as malfunctioning, or the formation of functional units (modules).

O.H. Ibarra and Z. Dang (Eds.): DLT 2006, LNCS 4036, p. 36, 2006.
© Springer-Verlag Berlin Heidelberg 2006

Polynomials, Fragments of Temporal Logic and the Variety DA over Traces

Manfred Kufleitner

Universität Stuttgart, FMI, Germany
kufleitner@fmi.uni-stuttgart.de

Abstract. We show that some language theoretic and logical characterizations of recognizable word languages whose syntactic monoid is in
the variety **DA** also hold over traces. To this aim we give algebraic characterizations for the language operations of generating the polynomial
closure and generating the unambiguous polynomial closure over traces.

We also show that there exist natural fragments of local temporal logic
that describe this class of languages corresponding to **DA**. All characterizations are known to hold for words.

1 Introduction

Traces were introduced by Mazurkiewicz as a generalization of words to describe the behavior of concurrent processes [4]. Since then traces have become a
rather popular setting to study concurrency. A lot of aspects of traces and trace
languages have been researched, see [1] for an overview.

Over words it has turned out that finite semigroups are a powerful technique
to refine the class of recognizable languages [2]. Two natural operations on classes
of languages are the polynomial closure and the unambiguous polynomial closure. For particular classes of languages, so called language varieties, it has been
shown that there exist algebraic counterparts in terms of the so-called Mal'cev
product [10]. In Section 3 (resp. Section 4) we will show that this correspondence
between the Mal'cev product and the polynomial closure (resp. the unambiguous
polynomial closure) for restricted varieties also holds over traces.

In Section 5 we tighten these results in the particular case of the class **DA**
of finite monoids to get two language theoretic characterizations of the class of
trace languages whose syntactic monoid is in **DA**. In Section 6 we show that
over traces the fragments of local temporal logic TL[XF, YP], TL[XF, YP, M] and
TL[X_a, Y_a] also express exactly these languages. All three characterizations are
known to hold for words [11, 12].

2 Preliminaries

A set S is a *semigroup* if it is equipped with an associative binary operation. The
set S forms a *monoid* if it is a semigroup and if there exists a neutral element,
i.e. an element denoted by 1 and satisfying $1a = a = a1$ for all $a \in S$. An element

O.H. Ibarra and Z. Dang (Eds.): DLT 2006, LNCS 4036, pp. 37–48, 2006.
© Springer-Verlag Berlin Heidelberg 2006

e of a semigroup is called *idempotent* if $e^2 = e$. A mapping $\eta : S \to T$ between two semigroups S and T is a *semigroup homomorphism* if $\eta(ab) = \eta(a)\eta(b)$ for all $a, b \in S$. If furthermore S and T are monoids and $\eta(1) = 1$, then η is *monoid homomorphism*. A relation $\tau \subseteq S \times T$ is a *relational semigroup morphism* between semigroups S and T if $\tau(a) \neq \emptyset$ and $\tau(ab) \subseteq \tau(a)\tau(b)$ for all $a, b \in S$ where $\tau(a) = \{ c \in T \mid (a, c) \in \tau \}$. In the context of monoids we additionally require $1 \in \tau(1)$ and then τ is called a *relational monoid morphism*. If there is no confusion or if the statement holds in either case we omit the terms *relational*, *semigroup* and *monoid* and only use the words *morphism* and *homomorphism*. As for functional homomorphisms, we also use the notation $\tau : S \to T$ for morphisms. For two (homo)morphisms $\eta : S \to T$ and $\nu : S \to R$ we define their *product* $\eta \times \nu : S \to T \times R : a \mapsto (\eta(a), \nu(a))$.

The *graph* of a morphism $\tau : S \to T$ is defined as $\mathrm{graph}(\tau) = \{ (a, c) \mid c \in \tau(a) \}$. It is easy to see that $\mathrm{graph}(\tau)$ is a subsemigroup (resp. submonoid) of $S \times T$. For any relational morphism $\tau : S \to T$ the projections $\pi_1 : \mathrm{graph}(\tau) \to S$ and $\pi_2 : \mathrm{graph}(\tau) \to T$ satisfy $\tau(a) = \pi_2(\pi_1^{-1}(a))$ for all $a \in S$, i.e. $\tau = \pi_2 \circ \pi_1^{-1}$. The condition $\tau(a) \neq \emptyset$ for all $a \in S$ implies that π_1 is onto. In fact, whenever we have two homomorphisms $\alpha : R \to S$ and $\beta : R \to T$ and α is onto, the composition $\beta \circ \alpha^{-1} : S \to T$ forms a relational morphism [6].

An *ordered semigroup* is a semigroup S equipped with a partial order relation $\leq$ such that $a \leq b$ implies $ca \leq cb$ and $ac \leq bc$ for all $a, b, c \in S$. Every semigroup S forms also an ordered semigroup $(S, =)$. For homomorphisms between ordered semigroups $\eta : (S, \leq) \to (T, \preceq)$ we additionally require $a \leq b \Rightarrow \eta(a) \preceq \eta(b)$ for all $a, b \in S$. More details can be found in [7].

We are interested in the interplay between classes of finite monoids and classes of recognizable subsets of infinite monoids. The connection between them is the syntactic congruence. Let L be a subset of a monoid $\mathbb{M}$. Then the *syntactic congruence* $\sim_L \subseteq \mathbb{M} \times \mathbb{M}$ of L is defined by

$$p \sim_L q \Leftrightarrow (\forall u, v \in \mathbb{M} : upv \in L \Leftrightarrow uqv \in L).$$

The natural homomorphism $\mu_L : \mathbb{M} \to \mathbb{M}/\!\!\sim_L : p \mapsto [p]_{\sim_L}$ is called the *syntactic homomorphism* of L. The monoid $M(L) = \mathbb{M}/\!\!\sim_L$ is called the *syntactic monoid* of L. The *syntactic quasiordering* $\leq_L$ of L is defined by $p \leq_L q \Leftrightarrow (\forall u, v \in \mathbb{M} : uqv \in L \Rightarrow upv \in L)$. The relation $\leq_L$ induces a partial order on $M(L)$ such that $(M(L), \leq_L)$ forms an ordered monoid. It is called the *syntactic ordered monoid* of L.

Equations are one possibility to describe classes of finite semigroups. Let Ω be a finite set and let $w, v \in \Omega^+$ (resp. Ω^* for monoids). A semigroup S *satisfies* the equation $w = v$, if for all homomorphisms $\eta : \Omega^+ \to S$ we have $\eta(w) = \eta(v)$. In a finite semigroup the unique idempotent power of an element a is denoted by a^ω. We also allow the ω-operator in equalities and define $\eta(w^\omega) = \eta(w)^\omega$. By $[\![w = v]\!]$ we denote the set of finite semigroups (resp. finite monoids) satisfying $w = v$. Analogously, we can define the class of finite ordered semigroups satisfying an inequality $w \leq v$.

The next possibility we will need in order to define classes of finite semigroups is the Mal'cev product. Let $\mathbf{V}$ and $\mathbf{W}$ be two classes of finite semigroups. A semigroup S is contained in the *Mal'cev product* $\mathbf{W} \,\textcircled{M}\, \mathbf{V}$ of $\mathbf{V}$ by $\mathbf{W}$ if there exists a semigroup $T \in \mathbf{V}$ and a relational morphism $\tau : S \to T$ such that for each idempotent $e \in T$ the set $\tau^{-1}(e)$ forms a semigroup in $\mathbf{W}$.

Let Σ be a finite alphabet and $I \subseteq \Sigma \times \Sigma$ be a symmetric and irreflexive relation. The *trace monoid* generated by (Σ, I) is the quotient $\mathbb{M}(\Sigma, I) = \Sigma^* / \sim_I$ where $\sim_I$ is the congruence generated by $\{\, (ab, ba) \mid (a, b) \in I \,\}$, i.e. for $v, w \in \Sigma^*$ we have $v \sim_I w$ if and only if v can be transformed into w using only commutations of contiguous letters a and b with $(a, b) \in I$. The elements of $\mathbb{M}(\Sigma, I)$ are called *traces*, I is called the *independence relation* and $D = \Sigma^2 \setminus I$ is the *dependence relation*. Let $w \in \Sigma^*$. By $[w]_I$ we denote the trace $[w]_I = \{\, v \in \Sigma^* \mid v \sim_I w \,\}$. The word $w \in \Sigma^*$ is called a *word representative* of a trace $t \in \mathbb{M}(\Sigma, I)$ if $t = [w]_I$. As for words, $|t| \in \mathbb{N}$ is the *length* of the trace $t \in \mathbb{M}(\Sigma, I)$ and $\mathrm{alph}(t) \subseteq \Sigma$ is its *alphabet*, i.e. the set of letters which occur in it. To each trace t we can associate a graph.

Let w be a word representative of a trace t. With t we can associate a graph $(V_t, <_t, \mathrm{label}_t)$ where $V_t = \{\, \nu \mid \nu \text{ is a position of } w \,\}$ is the set of vertices and

$$\mathrm{label}_t : V_t \to \Sigma : \nu \mapsto \text{``letter of } w \text{ at position } \nu\text{''}$$

is a labeling of the vertices. Let

$$\to_t \;=\; \{\, (\nu, \chi) \in V_t^2 \mid \nu \text{ occurs before } \chi \text{ in } w \text{ and } (\mathrm{label}_t(\nu), \mathrm{label}_t(\chi)) \in D \,\}.$$

The set of edges $<_t$ is now defined as the transitive closure of $\to_t$. The relation $<_t$ is a (strict) partial order on V_t. Up to isomorphism, the definition of this graph is independent of the choice of the word representative. The linearizations of this graph are exactly the word representatives of t. Therefore, by abuse of notation we will identify the word representative w, the trace t and its graph $(V_t, <_t, \mathrm{label}_t)$.

3   Polynomial Closure

In the following, we fix the trace monoid $\mathbb{M}(\Sigma, I)$ over a non-empty finite alphabet Σ. A class of finite monoids $\mathbf{V}$ is called *variety* if it is closed under finite products, if it is closed under taking submonoids and if it is closed under homomorphic images [6]. We will also use this notion if $\mathbf{V}$ is a class of finite ordered monoids [7]. By **Com** we denote the class of finite commutative monoids $[\![\, xy = yx \,]\!]$ and by $\mathbf{J_1}$ we denote the class of idempotent and commutative monoids $[\![\, x^2 = x \,]\!] \cap \mathbf{Com}$.

Lemma 1. *Let $\mathbf{V}$ be a variety of monoids with $\mathbf{J_1} \subseteq \mathbf{V}$ and let $M_0, \ldots, M_n \in \mathbf{V}$. For all $i \in \{\, 0, \ldots, n \,\}$ let $\mu_i : \mathbb{M}(\Sigma, I) \to M_i$ be homomorphisms. Then there exists a monoid $N \in \mathbf{V}$ and a homomorphism $\eta : \mathbb{M}(\Sigma, I) \to N$ such that for all $x, y \in \mathbb{M}(\Sigma, I)$ satisfying $\eta(x) = \eta(y)$ the following conditions hold:*

(a) For all homomorphisms μ_i, $0 \leq i \leq n$, we have $\mu_i(x) = \mu_i(y)$.
(b) $\mathrm{alph}(x) = \mathrm{alph}(y)$.

(c) Let x' and y' be connected components of x and of y respectively such that $\mathrm{alph}(x') = \mathrm{alph}(y')$. Then we have $\eta(x') = \eta(y')$.

(d) If $\eta(x)$ is idempotent then $\eta(x')$ is idempotent for every connected component x' of x.

Proof. The power set 2^Σ of Σ using union $\cup$ as operation forms a commutative and idempotent monoid. We set

$$N = 2^\Sigma \times \prod_{\Gamma \subseteq \Sigma} M_0 \times \cdots \times M_n.$$

Since $2^\Sigma \in \mathbf{J_1} \subseteq \mathbf{V}$ and since $\mathbf{V}$ is a variety, we have $N \in \mathbf{V}$. Next we define

$$\eta = \mathrm{alph} \times \prod_{\Gamma \subseteq \Sigma} ((\mu_0 \times \cdots \times \mu_n) \circ \pi_\Gamma) : \mathrm{M}(\Sigma, I) \to N,$$

where π_Γ is the natural projection $\mathrm{M}(\Sigma, I) \to \mathrm{M}(\Gamma, \Gamma^2 \cap I) : x \mapsto \pi_\Gamma(x)$. Note that $\mathrm{M}(\Gamma, \Gamma^2 \cap I) \subseteq \mathrm{M}(\Sigma, I)$. Condition (a) is verified in the components of N and ν corresponding to $\Gamma = \Sigma$ and condition (b) is fulfilled by reason of the first component. Let x' and y' be connected components of x and of y with $\mathrm{alph}(x') = \mathrm{alph}(y')$. Let $\Gamma \subseteq \Sigma$ and let $i \in \{0, \ldots, n\}$. To conclude (c) we have to show $\mu_i(\pi_\Gamma(x')) = \mu_i(\pi_\Gamma(y'))$. Since $x' = \pi_{\mathrm{alph}(x')}(x)$ we have $\pi_\Gamma(x') = \pi_\Gamma(\pi_{\mathrm{alph}(x')}(x)) = \pi_{\Gamma'}(x)$ with $\Gamma' = \Gamma \cap \mathrm{alph}(x')$. A similar argument for y' and y and $\eta(x) = \eta(y)$ implies $\mu_i(\pi_\Gamma(x')) = \mu_i(\pi_{\Gamma'}(x)) = \mu_i(\pi_{\Gamma'}(y)) = \mu_i(\pi_\Gamma(y'))$. Now let $\eta(x)$ be idempotent. This means that every component of $\eta(x)$ is idempotent and since every component of $\eta(x')$ is also a component of $\eta(x)$, we have that $\eta(x')$ is also idempotent. $\qquad\square$

We say that a trace t_1 is a *factor* of a trace t_2 if there exist traces s_1 and s_2 such that $t_2 = s_1 t_1 s_2$.

Lemma 2. *Let $a \in \Sigma$, let $t_0, t_1 \in \mathrm{M}(\Sigma, I)$ and let $x \in \mathrm{M}(\Sigma, I)$ be connected. If $x^{|\Sigma|+m}$ is a factor of $t_0 a t_1$ for $m \in \mathbb{N}$ then there exist $m_0, m_1 \in \mathbb{N}$ such that $m_0 + m_1 = m$ and x^{m_i} is a factor of t_i for $i = 0$ and $i = 1$.*

Proof. The proof is similar to the proof that x^* is recognizable if $x \in \mathrm{M}(\Sigma, I)$ is connected [5, Proposition 6.3.11].

Since x is connected, between any two letters of $\mathrm{alph}(x)$ we have an undirected path in the dependence graph (Σ, D) of length at most $|\Sigma|$ such that all vertices on this path are in $\mathrm{alph}(x)$. Directed paths following the same labels also exist in $x^{|\Sigma|}$ between all vertices of the first x and all vertices of the last x in this product. There could be some x's that have vertices in t_0 as well as vertices in $a t_1$. The above argument shows that starting with the first x with this property we could lose at most $|\Sigma| - 1$ many of the x's of $x^{|\Sigma|+m}$ as factors of t_0 or $a t_1$. The letter a could be a factor of one x. It follows that there remain m many x's as factors of either t_0 or t_1. $\qquad\square$

For a set $\mathcal{V}$ of trace languages over $M(\Sigma, I)$ we define the *polynomials* $\mathrm{Pol}\mathcal{V}$ over $\mathcal{V}$ as the set of trace languages that are finite unions of languages of the form

$$L_0 a_1 L_1 \cdots a_n L_n,$$

where $n \in \mathbb{N}$ and $L_i \in \mathcal{V}$ for all $0 \leq i \leq n$. We say that a set $\mathcal{V}$ of trace languages over $M(\Sigma, I)$ *corresponds* to a class of monoids **V**, if $\mathcal{V} = \{ L \subseteq M(\Sigma, I) \mid M(L) \in \mathbf{V} \}$, resp. $\mathcal{V} = \{ L \subseteq M(\Sigma, I) \mid (M(L), \leq_L) \in \mathbf{V} \}$ for ordered monoids **V**.

Theorem 1. *Let* **V** *be a variety of monoids such that* $\mathbf{J_1} \subseteq \mathbf{V}$ *and let* $\mathcal{V}$ *be the set of trace languages corresponding to* **V**. *Then the syntactic ordered monoid of every language* $L \in \mathrm{Pol}\mathcal{V}$ *is in* $[\![x^\omega y x^\omega \leq x^\omega]\!] \, \textcircled{M} \, \mathbf{V}$.

Proof. We modify the proof for words in [10]. Let $L = L_0 a_1 L_1 \cdots a_n L_n$, where $L_i \in \mathcal{V}$ for all $0 \leq i \leq n$. Let $\eta : M(\Sigma, I) \to N$ be as in Lemma 1 with $M_i = M(L_i)$ for $i \in \{ 0, \ldots, n \}$. Let $(M(L), \leq_L)$ be the syntactic ordered monoid of L and $\mu : M(\Sigma, I) \to M(L)$ its syntactic homomorphism. We obtain the relational morphism $\tau = \eta \circ \mu^{-1} : M(L) \to N$.

Let $e \in N$ be idempotent, let $x, y, u, v \in M(\Sigma, I)$ such that $\eta(x) = e = \eta(y)$ and let $m \geq n|\Sigma| + 1$. The trace x can be decomposed into connected components $x = x_1 \cdots x_\ell$ such that $\mathrm{alph}(x_i) \times \mathrm{alph}(x_j) \subseteq I$ for all $1 \leq i \neq j \leq \ell$. Lemma 1 (b) implies $\mathrm{alph}(x) = \mathrm{alph}(y)$. Hence, the trace y can also be decomposed into connected components $y = y_1 \cdots y_\ell$ such that $\mathrm{alph}(y_j) = \mathrm{alph}(x_j)$ for all $1 \leq j \leq \ell$. Suppose $ux^m v \in L$. By applying Lemma 2 up to n times we can conclude that for every $j \in \{ 1, \ldots, \ell \}$ there exists $i \in \{ 0, \ldots, n \}$ and a factorization $ux^m v = z_0 z_1 x_j z_2 z_3$ such that

$$z_0 \in L_0 a_1 L_1 \cdots L_{i-1} a_i$$
$$z_1 x_j z_2 \in L_i$$
$$z_3 \in a_{i+1} L_{i+1} \cdots a_n L_n.$$

By Lemma 1 we have that $\mu_i(x_j) = \mu_i(y_j)$ is idempotent and therefore we have $z_1 x_j x_j^{k_1} y_j x_j^{k_2} z_2 \in L_i$ for all $k_1, k_2 \in \mathbb{N}$. By applying this pumping argument to all connected components of x, by a suitable choice of the exponents we can conclude $ux^m y x^m v \in L$.

Thus for all $m \geq n|\Sigma| + 1$ we have $\mu(x^m y x^m) \leq_L \mu(x^m)$ and therefore $\tau^{-1}(e) \in [\![x^\omega y x^\omega \leq x^\omega]\!]$. This shows $(M(L), \leq_L) \in [\![x^\omega y x^\omega \leq x^\omega]\!] \, \textcircled{M} \, \mathbf{V}$. This Mal'cev product forms a variety of ordered semigroups. Language classes corresponding to varieties are closed under finite unions. Hence, we can conclude that $\mathrm{Pol}\mathcal{V}$ is a subset of the trace languages corresponding to $[\![x^\omega y x^\omega \leq x^\omega]\!] \, \textcircled{M} \, \mathbf{V}$. $\square$

By $\pi_I : \Sigma \to M(\Sigma, I) : w \mapsto [w]_I$ we denote the canonical projection from Σ^* to $M(\Sigma, I)$.

Lemma 3. *Let* $\eta : \Sigma^* \to M$ *be a homomorphism from* Σ^* *to a commutative monoid* M. *Then there exists a unique homomorphism* $\nu : M(\Sigma, I) \to M$ *such that* $\nu \circ \pi_I = \eta$.

The proof of the following lemma can be found in [2] as a special case of Proposition 1.1, page 186.

Lemma 4. *Let $L \subseteq \mathrm{M}(\Sigma, I)$ be a trace language and let $\eta : \Sigma^* \to M$ be the syntactic homomorphism of $\pi_I^{-1}(L)$. Then M is isomorphic to the syntactic monoid of L.*

Let $\mathbf{V}$ be a variety of finite monoids. We say that $\mathcal{V}$ is the corresponding *-variety* if $\mathcal{V} = \{ K \subseteq \Sigma^* \mid M(K) \in \mathbf{V} \}$. As for sets of trace languages, we define Pol$\mathcal{V}$ as the (word) languages that are finite unions of languages of the form $K_0 a_1 K_1 \cdots a_n K_n$, where $n \in \mathbb{N}$ and $K_i \in \mathcal{V}$ for all $0 \le i \le n$. For a partial converse of Theorem 1 we will use the following theorem from [10].

Theorem 2 (Pin/Weil, 1997). *Let $\mathbf{V}$ be a variety of finite monoids, let $\mathcal{V}$ be the corresponding *-variety and let $K \subseteq \Sigma^*$. If the syntactic ordered monoid of K is in $[\![x^\omega y x^\omega \le x^\omega]\!] \textcircled{M} \mathbf{V}$, then $K \in$ Pol$\mathcal{V}$.*

For commutative varieties we can state this theorem for traces.

Theorem 3. *Let $\mathbf{V} \subseteq \mathbf{Com}$ be a variety of finite commutative monoids, let $\mathcal{V}$ be the set of trace languages corresponding to $\mathbf{V}$ and let $L \subseteq \mathrm{M}(\Sigma, I)$ be a trace language. If the syntactic ordered monoid of L is in $[\![x^\omega y x^\omega \le x^\omega]\!] \textcircled{M} \mathbf{V}$, then $L \in$ Pol$\mathcal{V}$.*

Proof. Let $K = \pi_I^{-1}(L)$. By Theorem 2 we can conclude that

$$K = \bigcup_{1 \le i \le m} K_{i,0} \, a_{i,1} \, K_{i,1} \cdots a_{i,n_i} \, K_{i,n_i}$$

for $m, n_1, \ldots, n_m \in \mathbb{N}$, $a_{i,j} \in \Sigma$ and $K_{i,j} \subseteq \Sigma^*$ such that $M(K_{i,j}) \in \mathbf{V}$. By Lemma 3 we have $\pi_I^{-1} \pi_I(K_{i,j}) = K_{i,j}$ and by Lemma 4 we can conclude that the syntactic monoid of $L_{i,j} = \pi_I(K_{i,j})$ is in $\mathbf{V}$. Hence

$$L = \pi_I(K) = \pi_I \left(\bigcup_{1 \le i \le m} K_{i,0} \, a_{i,1} \, K_{i,1} \cdots a_{i,n_i} \, K_{i,n_i} \right)$$

$$= \bigcup_{1 \le i \le m} \pi_I(K_{i,0}) \, a_{i,1} \, \pi_I(K_{i,1}) \cdots a_{i,n_i} \, \pi_I(K_{i,n_i})$$

$$= \bigcup_{1 \le i \le m} L_{i,0} \, a_{i,1} \, L_{i,1} \cdots a_{i,n_i} \, L_{i,n_i} \in \text{Pol}\mathcal{V}.$$

$\square$

Corollary 1. *Let $\mathbf{J_1} \subseteq \mathbf{V} \subseteq \mathbf{Com}$ be a variety of finite monoids and let $\mathcal{V}$ be the corresponding variety of trace languages. Then Pol$\mathcal{V}$ corresponds to $[\![x^\omega y x^\omega \le x^\omega]\!] \textcircled{M} \mathbf{V}$.*

Let $\mathcal{V}$ be a set of trace languages. By coPol$\mathcal{V}$ we denote the set of trace languages L whose complement $\overline{L}$ is in Pol$\mathcal{V}$. Since the syntactic ordered monoid of the complement $\overline{L}$ of a trace language L is $(M(\overline{L}), \le_{\overline{L}}) = (M(L), \le_L^{-1})$ we obtain the following corollary.

Corollary 2. *Let* $\mathbf{J_1} \subseteq \mathbf{V} \subseteq \mathbf{Com}$ *be a variety of finite monoids and let* $\mathcal{V}$ *be the corresponding trace languages. Then* $\mathrm{coPol}\mathcal{V}$ *corresponds to* $[\![\, x^\omega y x^\omega \geq x^\omega \,]\!] \,\textcircled{M}\, \mathbf{V}$.

A class $\mathcal{V}$ of trace languages is a *language variety* if it is closed under boolean operations, under inverse homomorphism and under quotients. A *left quotient* of L is $a^{-1}L = \{\, t \mid at \in L \,\}$ for $a \in \Sigma$. *Right quotients* are symmetric. It is well known that a set of languages that corresponds to a variety of monoids forms a language variety [2, page 186f]. Since

$$[\![\, x^\omega y x^\omega \leq x^\omega \,]\!] \,\textcircled{M}\, \mathbf{V} \cap [\![\, x^\omega y x^\omega \geq x^\omega \,]\!] \,\textcircled{M}\, \mathbf{V} \;=\; [\![\, x^\omega y x^\omega = x^\omega \,]\!] \,\textcircled{M}\, \mathbf{V}$$

and since $[\![\, x^\omega y x^\omega = x^\omega \,]\!] \,\textcircled{M}\, \mathbf{V}$ is a variety if $\mathbf{V}$ is a variety [9, 10], we obtain the following corollary.

Corollary 3. *Let* $\mathbf{J_1} \subseteq \mathbf{V} \subseteq \mathbf{Com}$ *be a variety of finite monoids and let* $\mathcal{V}$ *be the corresponding variety of trace languages. Then* $\mathrm{Pol}\mathcal{V} \cap \mathrm{coPol}\mathcal{V}$ *is a (trace) language variety that corresponds to the variety* $[\![\, x^\omega y x^\omega = x^\omega \,]\!] \,\textcircled{M}\, \mathbf{V}$.

4   Unambiguous Polynomial Closure

For a position ν of $t \in \mathbb{M}(\Sigma, I)$ we define the following factors:

$$\mathrm{pre}(\nu) = \{\, \chi \in t \mid \chi <_t \nu \,\} \qquad\qquad \text{is the past of } \nu,$$
$$\mathrm{par}(\nu) = \{\, \chi \in t \mid \nu \not<_t \chi, \chi \not<_t \nu, \nu \neq \chi \,\} \quad \text{is the parallel part of } \nu,$$
$$\mathrm{suf}(\nu) = \{\, \chi \in t \mid \nu <_t \chi \,\} \qquad\qquad \text{is the future of } \nu.$$

We now have the following two factorizations $t = \mathrm{pre}(\nu)\,\mathrm{label}(\nu)\,\mathrm{par}(\nu)\,\mathrm{suf}(\nu)$ and $t = \mathrm{pre}(\nu)\,\mathrm{par}(\nu)\,\mathrm{label}(\nu)\,\mathrm{suf}(\nu)$. We say that a product $L = L_1 a L_2$ of trace languages $L_1, L_2 \subseteq \mathbb{M}(\Sigma, I)$, $a \in \Sigma$ is *left unambiguous* if for all $t \in L$ there exists a unique position ν in t such that

- $\mathrm{label}(\nu) = a$ and
- $\mathrm{pre}(\nu) \in L_1$ and $\mathrm{par}(\nu)\,\mathrm{suf}(\nu) \in L_2$.

Right unambiguous products are defined symmetrically, i.e. the parallel part $\mathrm{par}(\nu)$ is related to L_1. A product $L_1 a L_2$ is *unambiguous* if it is left unambiguous *or* right unambiguous. Let $\mathcal{V}$ be a set of trace languages. Then we define $\mathrm{UPol}\mathcal{V}$ as the closure of $\mathcal{V}$ under boolean operations and unambiguous products. Note that the unambiguous product for traces is not associative.

Theorem 4. *Let* $\mathbf{V}$ *be a variety of monoids such that* $\mathbf{J_1} \subseteq \mathbf{V}$ *and let* $\mathcal{V}$ *be the set of trace languages corresponding to* $\mathbf{V}$. *Then the syntactic monoid of every trace language in* $\mathrm{UPol}\mathcal{V}$ *is in* $[\![\, x^\omega y x^\omega = x^\omega \,]\!] \,\textcircled{M}\, \mathbf{V}$.

Proof. By **LI** we denote the semigroup variety $[\![\, x^\omega y x^\omega = x^\omega \,]\!]$. Let $L = L_1 a L_2$ be a left unambiguous product of L_1 and L_2 and let their syntactic monoids be $M_1, M_2 \in \mathbf{LI} \,\textcircled{M}\, \mathbf{V}$. Let $\eta : \mathbb{M}(\Sigma, I) \to N$ be as in Lemma 1 and let $M(L)$ be the syntactic monoid of L and $\mu : \mathrm{M}(\Sigma, I) \to M(L)$ its syntactic homomorphism. We obtain the relational morphism $\tau = \eta \circ \mu^{-1} : M(L) \to N$. Since $\mathbf{LI} \, \textcircled{m} \, \mathbf{V}$ forms a variety [6] and since $\mathbf{LI} \, \textcircled{m} \, (\mathbf{LI} \, \textcircled{m} \, \mathbf{V}) = \mathbf{LI} \, \textcircled{m} \, \mathbf{V}$, see [8], it is sufficient to show that for all idempotents $e \in N$ we have $\tau^{-1}(e) \in \mathbf{LI}$. The theorem then follows by left-right symmetry and from the fact that classes of languages corresponding to varieties of monoids are closed under boolean operations [2].

Let $e^2 = e \in N$, let $x, y, u, v \in \mathrm{M}(\Sigma, I)$ such that $\eta(x) = e = \eta(y)$ and let $m \geq |\Sigma| + 1$ such that $\mu(x)^m$ is idempotent. We will show that $ux^m v \in L$ if and only if $ux^m y x^m v \in L$. The direction from left to right is the same as in Theorem 1. Suppose $ux^m y x^m v \in L$. Then there exists a left unambiguous factorization $ux^m y x^m v = z_1 a z_2$ with $z_1 \in L_1$ and $z_2 \in L_2$. Let $x = x_1 \cdots x_\ell$ and $y = y_1 \cdots y_\ell$ be factorizations into connected components such that $\mathrm{alph}(x_j) = \mathrm{alph}(y_j)$ for all $1 \leq j \leq \ell$. Suppose the connected component x_1 of x from the left x^m block matches with a factor of z_1 and the same connected component x_1 of x from the right x^m block matches with a factor of z_2. Since $\eta(x_1) = \eta(y_1)$ is idempotent, we can arbitrarily pump x_1 and y_1 at these two positions without changing membership to L. The possibility of pumping at both position leads to two different factorizations of $ux^m y x^m y_1 x_1^m v \in L$. This contradicts the choice of L_1 and L_2 such that L is left unambiguous. The same argument holds for all connected components of x.

Together with Lemma 2 it follows that for every index $j \in \{1, \ldots, \ell\}$ of a connected component there exists $i \in \{1, 2\}$ such that the last occurrence of x_j in the left x^m block and the first occurrence of x_j in the right x^m block in $ux^m y x^m v$ are factors of z_i. Hence, the component y_j of y lies between these two occurrences of x_j, i.e. $x_j y_j x_j$ is a factor of z_i. Since $\eta(x_j y_j x_j) = \eta(x_j x_j)$, we can remove all connected components of y without changing membership to L. Therefore, $ux^m x^m v \in L$ and by idempotency of $\mu(x)^m$ we can conclude $ux^m v \in L$. $\qquad\square$

In the next section we will present a converse of the previous theorem for the variety $\mathbf{V} = \mathbf{J_1}$.

5 The Variety DA

The variety $\mathbf{DA}$ is defined as $\mathbf{DA} = [\![(xy)^\omega x (xy)^\omega = (xy)^\omega]\!]$. It is known that $\mathbf{DA} = [\![x^\omega y x^\omega = x^\omega]\!] \, \textcircled{m} \, \mathbf{J_1}$, see [6].

A factorization $t = t_- a t_+$ is a *left factorization* if $a \notin \mathrm{alph}(t_-)$ and if $t_- = sb$ implies $(a, b) \in D$, i.e. in this factorization a is the first occurrence of the letter a in t and no minimal element of t_- is independent of a. Symmetrically, we say that a factorization $t = t_- a t_+$ is a *right factorization* if $a \notin \mathrm{alph}(t_+)$ and if $t_+ = bs$ implies $(a, b) \in D$.

Definition 1. *We define the relation $\equiv_{A,k} \subseteq \mathrm{M}(\Sigma, I)^2$ for $A \subseteq \Sigma$ and $k \in \mathbb{N}$:*

- $t \equiv_{A,0} s$ *if* $\mathrm{alph}(t) \subsetneq A \supsetneq \mathrm{alph}(s)$ *or* $\mathrm{alph}(t) \subseteq A \supseteq \mathrm{alph}(s)$.
- $t \equiv_{A,k} s$ *for $k > 0$ if* $\mathrm{alph}(t) \subsetneq A \supsetneq \mathrm{alph}(s)$ *or the following three conditions hold:*

- $\mathrm{alph}(t) = \mathrm{alph}(s) \subseteq A$.
- *For all $a \in \mathrm{alph}(t)$ and all left factorizations $t = t_- a t_+$ and $s = s_- a s_+$ the conditions $t_- \equiv_{A \backslash \{a\}, k-1} s_-$ and $t_+ \equiv_{A, k-1} s_+$ hold.*
- *For all $a \in \mathrm{alph}(t)$ and all right factorizations $t = t_- a t_+$ and $s = s_- a s_+$ the conditions $t_- \equiv_{A, k-1} s_-$ and $t_+ \equiv_{A \backslash \{a\}, k-1} s_+$ hold.*

It is clear that for all $A \subseteq \Sigma$ and all $k \in \mathbb{N}$ the relation $\equiv_{A,k}$ is an equivalence relation of finite index. The analog of the following lemma in the case of words was shown in [12].

Lemma 5. *Let $\gamma \subseteq \mathrm{M}(\Sigma, I)^2$ be a congruence of finite index such that the monoid $\mathrm{M}(\Sigma, I)/\gamma$ is in **DA**. Then there exists $k \in \mathbb{N}$ such that $\equiv_{\Sigma, k} \subseteq \gamma$.*

It follows that every trace language L with $M(L) \in \mathbf{DA}$ is the disjoint union of $\equiv_{A,k}$-classes. We define the set of trace languages $\mathcal{A} = \{ A^* \mid A \subseteq \Sigma \}$. Clearly, $\mathcal{A}$ is a subset of the trace languages corresponding to $\mathbf{J_1}$.

Lemma 6. *Let $A \subseteq \Sigma$ and $k \in \mathbb{N}$. Every equivalence class of $\equiv_{A,k}$ is in $\mathrm{UPol}\mathcal{A}$.*

For every set of letters $A \subseteq \Sigma$ we have

$$\{ t \in \mathrm{M}(\Sigma, I) \mid \mathrm{alph}(t) = A \} \quad = \bigcup_{\{a_1, \ldots, a_n\} = A} A^* a_1 A^* \cdots a_n A^*,$$

where $n = |A|$. Therefore, the language variety corresponding to $\mathbf{J_1}$ is contained in $\mathrm{Pol}\mathcal{A}$. Together with Corollary 3 and Theorem 4 we can conclude:

Corollary 4. *The language variety $\mathrm{UPol}\mathcal{A} = \mathrm{Pol}\mathcal{A} \cap \mathrm{coPol}\mathcal{A}$ corresponds to the variety **DA**.*

6   Temporal Logic

In this section we introduce two characterizations of **DA** with temporal logics. In this paper, a *temporal formula* is a term of the form

$$\varphi ::= a \mid \neg\varphi \mid (\varphi_1 \vee \varphi_2) \mid (\varphi_1 \wedge \varphi_2) \mid \mathsf{XF}\varphi \mid \mathsf{YP}\varphi \mid \mathsf{M}\varphi \mid \mathsf{X}_a \varphi \mid \mathsf{Y}_a \varphi$$

where $a \in \Sigma$. The operators XF, YP, M, X_a and Y_a are called *temporal operators*. The letter X comes from the word neXt, Y stands for Yesterday, F for Future, P for Past and M for soMetime. By $\mathrm{TL}[\mathsf{XF}, \mathsf{YP}]$ we denote the set of temporal formulae where XF and YP are the only temporal operators and in the fragment $\mathrm{TL}[\mathsf{XF}, \mathsf{YP}, \mathsf{M}]$ we additionally allow the M operator. By $\mathrm{TL}[\mathsf{X}_a, \mathsf{Y}_a]$ we denote the set of temporal formulae where all temporal operators are of the form X_a or Y_a for $a \in \Sigma$. Next we define, when a trace $t = (V, <, \mathrm{label})$ at position $\nu \in V$ models a temporal formula:

$$t, \nu \models a \qquad \Leftrightarrow \; \mathrm{label}(\nu) = a, \quad \text{for } a \in \Sigma.$$

$$t, \nu \models \neg\varphi \qquad \Leftrightarrow \; \text{not } t, \nu \models \varphi.$$

$$t, \nu \models \varphi_1 \vee \varphi_2 \; \Leftrightarrow \; t, \nu \models \varphi_1 \text{ or } t, \nu \models \varphi_2.$$

$$t, \nu \models \varphi_1 \wedge \varphi_2 \; \Leftrightarrow \; t, \nu \models \varphi_1 \text{ and } t, \nu \models \varphi_2.$$

$$t, \nu \models \mathsf{XF}\varphi \qquad \Leftrightarrow \; \exists \chi \in V: \; \nu < \chi \text{ and } t, \chi \models \varphi.$$

$$t, \nu \models \mathsf{YP}\varphi \qquad \Leftrightarrow \; \exists \chi \in V: \; \chi < \nu \text{ and } t, \chi \models \varphi.$$

$$t, \nu \models \mathsf{M}\varphi \qquad \Leftrightarrow \; \exists \chi \in V: \; t, \chi \models \varphi.$$

$$t, \nu \models \mathsf{X}_a\varphi \qquad \Leftrightarrow \; \exists \chi \in V: \; \nu < \chi \text{ and } t, \chi \models a \wedge \varphi \text{ and}$$
$$\big(\forall \xi \in V: \; \nu < \xi < \chi \Rightarrow \mathrm{label}(\xi) \neq a\big).$$

$$t, \nu \models \mathsf{Y}_a\varphi \qquad \Leftrightarrow \; \exists \chi \in V: \; \chi < \nu \text{ and } t, \chi \models a \wedge \varphi \text{ and}$$
$$\big(\forall \xi \in V: \; \chi < \xi < \nu \Rightarrow \mathrm{label}(\xi) \neq a\big).$$

The usage of "alphabetic filters" (as in X_a and Y_a) has been introduced in [3] for local temporal logic over traces.

An *outer* temporal formula is a boolean combination of formulae of the form $\mathsf{XF}\varphi$, $\mathsf{YP}\varphi$, $\mathsf{M}\varphi$, $\mathsf{X}_a\varphi$ or $\mathsf{Y}_a\varphi$ where φ is an arbitrary temporal formula. Next we define when a trace $t = (V, <, \mathrm{label})$ models an outer temporal formula. Boolean combinations are defined straightforwardly.

$$t \models \mathsf{XF}\varphi \quad \Leftrightarrow \; \exists \nu \in V: \; t, \nu \models \varphi.$$

$$t \models \mathsf{YP}\varphi \quad \Leftrightarrow \; \exists \nu \in V: \; t, \nu \models \varphi.$$

$$t \models \mathsf{M}\varphi \quad \Leftrightarrow \; \exists \nu \in V: \; t, \nu \models \varphi.$$

$$t \models \mathsf{X}_a\varphi \quad \Leftrightarrow \; \exists \nu \in V: \; t, \nu \models a \wedge \varphi \text{ and}$$
$$\big(\forall \xi \in V: \; \xi < \nu \Rightarrow \mathrm{label}(\xi) \neq a\big).$$

$$t \models \mathsf{Y}_a\varphi \quad \Leftrightarrow \; \exists \nu \in V: \; t, \nu \models a \wedge \varphi \text{ and}$$
$$\big(\forall \xi \in V: \; \nu < \xi \Rightarrow \mathrm{label}(\xi) \neq a\big).$$

Note that $\mathsf{XF}\varphi$, $\mathsf{YP}\varphi$ and $\mathsf{M}\varphi$ as outer formulae are equivalent. The idea is that when evaluating XF and X_a we start at a position in front of the trace and when evaluating YP and Y_a we start at a position behind the trace. The trace language generated by an outer temporal formula φ is

$$L(\varphi) = \{ t \in \mathbb{M}(\Sigma, I) \mid t \models \varphi \}.$$

We say that a trace language $L \subseteq \mathbb{M}(\Sigma, I)$ is expressible in $\mathrm{TL}[\mathsf{XF}, \mathsf{YP}]$ (resp. in $\mathrm{TL}[\mathsf{XF}, \mathsf{YP}, \mathsf{M}]$ or $\mathrm{TL}[\mathsf{X}_a, \mathsf{Y}_a]$) if there exists an outer temporal formula $\varphi \in \mathrm{TL}[\mathsf{XF}, \mathsf{YP}]$ (resp. $\mathrm{TL}[\mathsf{XF}, \mathsf{YP}, \mathsf{M}]$ or $\varphi \in \mathrm{TL}[\mathsf{X}_a, \mathsf{Y}_a]$) such that $L = L(\varphi)$.

Lemma 7. *Let φ be an outer temporal formula in $\mathrm{TL}[\mathsf{XF}, \mathsf{YP}, \mathsf{M}]$. Then the syntactic monoid of $L(\varphi)$ is in* **DA**.

Lemma 8. *Let φ be an outer temporal formula in $\mathrm{TL}[\mathsf{X}_a, \mathsf{Y}_a]$. Then the syntactic monoid of $L(\varphi)$ is in* **DA**.

We will show that all trace languages L with $M(L) \in \mathbf{DA}$ can be expressed by a formula in $\mathrm{TL}[\mathsf{XF}, \mathsf{YP}]$. By Lemma 5 it suffices to show that all equivalence classes of $\equiv_{A,k}$ are expressible in $\mathrm{TL}[\mathsf{XF}, \mathsf{YP}]$.

Lemma 9. *Let $A \subseteq \Sigma$ and $k \in \mathbb{N}$. Every equivalence class of $\equiv_{A,k}$ is expressible in $\mathrm{TL}[\mathsf{XF}, \mathsf{YP}]$.*

Lemma 10. *Let $A \subseteq \Sigma$ and $k \in \mathbb{N}$. Every equivalence class of $\equiv_{A,k}$ is expressible in $\mathrm{TL}[\mathsf{X}_a, \mathsf{Y}_a]$.*

In the next theorem we summarize the characterizations of trace languages whose syntactic monoid is in the variety **DA**.

Theorem 5. *Let $L \subseteq \mathbb{M}(\Sigma, I)$. Then the following are equivalent:*

(i) $M(L) \in \mathbf{DA}$.
(ii) $L \in \mathrm{UPol}\mathcal{A}$.
(iii) $L \in \mathrm{Pol}\mathcal{A}$ *and* $\overline{L} \in \mathrm{Pol}\mathcal{A}$.
(iv) L *is expressible in* $\mathrm{TL}[\mathsf{XF}, \mathsf{YP}]$.
(v) L *is expressible in* $\mathrm{TL}[\mathsf{XF}, \mathsf{YP}, \mathsf{M}]$.
(vi) L *is expressible in* $\mathrm{TL}[\mathsf{X}_a, \mathsf{Y}_a]$.

Proof. The equivalence of *(i)*, *(ii)* and *(iii)* is Corollary 4. The direction "*(i)* $\Rightarrow$ *(iv)*" follows from Lemma 5 and Lemma 9. Since $\mathrm{TL}[\mathsf{XF}, \mathsf{YP}] \subseteq \mathrm{TL}[\mathsf{XF}, \mathsf{YP}, \mathsf{M}]$ we have "*(iv)* $\Rightarrow$ *(v)*". The implication "*(v)* $\Rightarrow$ *(i)*" is Lemma 7. The direction "*(i)* $\Rightarrow$ *(vi)*" follows from Lemma 5 and Lemma 10 and the implication "*(vi)* $\Rightarrow$ *(i)*" is Lemma 8. $\square$

7   Conclusion

We have given an algebraic characterization of $\mathrm{Pol}\mathcal{V}$ and $\mathrm{Pol}\mathcal{V} \cap \mathrm{coPol}\mathcal{V}$ in the case that $\mathcal{V}$ corresponds to a variety of commutative monoids that contains the monoid $(2^\Sigma, \cup, \emptyset)$ over subsets the alphabet Σ. We have also shown that all trace languages in $\mathrm{UPol}\mathcal{V}$ satisfy a particular algebraic property if $\mathcal{V}$ corresponds to a variety that contains the monoid $(2^\Sigma, \cup, \emptyset)$. That this property is sufficient for $\mathrm{UPol}\mathcal{V}$ has been shown in the case that $\mathcal{V}$ corresponds to $\mathbf{J_1}$. This leads to two language-theoretic characterizations of the variety **DA**: $\mathrm{Pol}\mathcal{A} \cap \mathrm{coPol}\mathcal{A}$ and $\mathrm{UPol}\mathcal{A}$ where $\mathcal{A} = \{\, A^* \mid A \subseteq \Sigma \,\}$. Then we have given two logical characterizations of **DA**: the fragments $\mathrm{TL}[\mathsf{XF}, \mathsf{YP}]$ and $\mathrm{TL}[\mathsf{X}_a, \mathsf{Y}_a]$ and we have shown that additionally allowing the operator M does not change the expressive power of the first fragment.

Two interesting open problems are whether it is possible to proof that the algebraic characterization of $\mathrm{Pol}\mathcal{V}$ also holds for larger classes of trace languages $\mathcal{V}$ and whether it is possible to give a language-theoretic characterization of $\mathrm{Pol}\mathcal{V} \cap \mathrm{coPol}\mathcal{V}$ in terms of disjoint unions of unambiguous polynomials as for words.

Acknowledgement. The author would like to thank Martin Müller and Pascal Tesson for many helpful discussions and the anonymous referees for valuable comments.

References

1. Volker Diekert and Grzegorz Rozenberg, editors. *The Book of Traces*. World Scientific, Singapore, 1995.
2. Samuel Eilenberg. *Automata, Languages, and Machines*, volume B. Academic Press, New York and London, 1976.
3. Paul Gastin and Madhavan Mukund. An elementary expressively complete temporal logic for Mazurkiewicz traces. In Peter Widmayer et al., editors, *Proc. 29th International Colloquium on Automata, Languages and Programming (ICALP'2002), Málaga (Spain), 2002*, number 2380 in Lecture Notes in Computer Science, pages 938–949. Springer-Verlag, 2002.
4. Antoni Mazurkiewicz. Concurrent program schemes and their interpretations. DAIMI Rep. PB 78, Aarhus University, Aarhus, 1977.
5. Edward Ochmański. Recognizable trace languages. In V. Diekert and G. Rozenberg, editors, *The Book of Traces*, chapter 6, pages 167–204. World Scientific, Singapore, 1995.
6. Jean-Éric Pin. *Varieties of Formal Languages*. North Oxford Academic, London, 1986.
7. Jean-Éric Pin. A variety theorem without complementation. In *Russian Mathematics (Izvestija vuzov.Matematika)*, volume 39, pages 80–90, 1995.
8. Jean-Éric Pin, Howard Straubing, and Denis Thérien. Locally trivial categories and unambiguous concatenation. *Journal of Pure and Applied Algebra*, 52:297–311, 1988.
9. Jean-Éric Pin and Pascal Weil. Profinite semigroups, Mal'cev products and identities. *Journal of Algebra*, 182:604–626, 1996.
10. Jean-Éric Pin and Pascal Weil. Polynominal closure and unambiguous product. *Theory Comput. Syst*, 30(4):383–422, 1997.
11. Pascal Tesson. Personal communication.
12. Pascal Tesson and Denis Thérien. Diamonds are Forever: The Variety **DA**. In Gracinda Maria dos Gomes Moreira da Cunha, Pedro Ventura Alves da Silva, and Jean-Eric Pin, editors, *Semigroups, Algorithms, Automata and Languages, Coimbra (Portugal) 2001*, pages 475–500. World Scientific, 2002.

Weighted Automata and Weighted Logics on Infinite Words

Manfred Droste[1,*] and George Rahonis[2]

[1] Institut für Informatik, Universität Leipzig
D-04109 Leipzig, Germany
droste@informatik.uni-leipzig.de
[2] Department of Mathematics, Aristotle University of Thessaloniki
54124 Thessaloniki, Greece
grahonis@math.auth.gr

Abstract. We introduce weighted automata over infinite words with Muller acceptance condition and we show that their behaviors coincide with the semantics of weighted restricted MSO-sentences. Furthermore, we establish an equivalence property of weighted Muller and weighted Büchi automata over certain semirings.

Keywords: Weighted logics, Weighted Muller automata, Infinitary formal power series.

1 Introduction

One of the cornerstones of automata theory is Büchi's theorem [6] on the co-incidence of the class of regular languages of infinite words with the family of languages definable by monadic second order logic (MSO logic for short). This led to the development of several models of automata acting on infinite words, like Büchi, Muller, Rabin and Streett, cf. [29, 33, 34] for surveys; it also led to practical applications in model checking and for non-terminating processes, cf. [1, 25, 26]. On the other hand, Schützenberger [32] introduced finite automata with weights which can model quantitative aspects of transitions like use of resources, reliability or capacity. Schützenberger characterized the behavior of such automata as rational formal power series. For the theory of weighted automata, see [3, 21, 24, 31] for surveys. Recently, weighted automata were applied in digital image compression [7, 17, 18, 19] as well as in speech-to-text processing [27, 28].

It is the goal of this paper to extend Büchi's theorem mentioned above into the context of weighted automata, thereby obtaining a quantitative version. Furthermore, we obtain an equivalence result to a model investigated recently by Ésik and Kuich [16]. The last few years weighted automata over infinite words have attracted the interest of several researchers. This effort is not a simple generalization of the finitary case since convergence problems arise depending

---

* Part of this work was done during a research stay of this author at Aristotle University of Thessaloniki in October 2005. He would like to thank his colleagues for their hospitality.

O.H. Ibarra and Z. Dang (Eds.): DLT 2006, LNCS 4036, pp. 49–58, 2006.
© Springer-Verlag Berlin Heidelberg 2006

on the underlying semiring. This issue is dealt with either by considering special classes of automata [8, 11] or by restricting the underlying semirings so that convergence problems can be solved [12, 16, 20, 30].

Very recently, Droste and Gastin [10] extended the result of Büchi and Elgot [5, 14] to weighted automata over finite words. They introduced an MSO logic with weights and described the semantics of the formulas obtained as formal power series. The main result of their paper states that the recognizable formal power series over commutative semirings coincide with the series definable by certain weighted MSO-sentences.

In this paper, we will introduce weighted Muller automata acting on infinite words, and we will extend the weighted MSO logic of [10] to infinite words. We describe the behavior of weighted Muller automata as formal power series on infinite words. Our first main result states the coincidence of these ω-Muller-recognizable series with the semantics of a restricted weighted MSO logic and also with the semantics of a restricted existential MSO logic. Furthermore, we prove an equivalence to the important model of weighted Büchi automata investigated in Ésik and Kuich [16]. They have characterized the behaviors of weighted Büchi automata precisely as the ω-rational formal power series; for further work on this model, see [22, 23]. Combining these results, we thus obtain a robust notion of weighted automata, logics and rational series on infinite words. As in [16], we assume our semiring of weights to permit infinite sum and product operations. Such "complete" semirings have been investigated in detail in the literature, cf. [4, 13, 21]. However, we derive from this a version of our result for semirings which are *not* complete; this includes all Boolean algebras and also max-min semirings used for capacity models. In particular, when considering the Boolean semiring, we obtain Büchi's result as a very special consequence.

Next we briefly describe the structure of our paper. In Section 2, we introduce the notions of totally commutative complete semirings and weighted Muller automata and we state their basic properties. In Section 3 we recall weighted MSO logic from [10], but we interpret the semantics of weighted MSO-formulas as formal power series over infinite words. The main result of the paper in Section 4 states that a formal power series is Muller recognizable iff it is definable in our restricted weighted MSO logic iff it is definable in existential restricted weighted MSO logic. Its proof requires, in particular, a construction of specific weighted Muller automata for the universal quantifier. Then in Section 5, we relate our weighted Muller automata to the weighted Büchi automata of Ésik and Kuich [16], and we show that these two models are equivalent. Finally, in Section 6, we deal with bi-aperiodic semirings which were introduced by Droste and Gastin in [9, 10]. We show that our main result remains true if the underlying semiring is just commutative and weakly bi-aperiodic. Büchi's classical theorem follows as a very special case.

2   Semirings and Weighted Muller Automata

In this section, we introduce totally commutative complete semirings, infinitary formal power series and weighted Muller automata. The reader is referred to [3, 13, 21, 24, 31] for semirings, and to [29, 33, 34] for classical Muller automata.

Let $(K, +, \cdot, 0, 1)$ be a *complete semiring*[13, 21, 16], i.e, a semiring that permits infinite sums extending the associativity, the commutativity and the distributivity laws of the finite sum operation. Then K is called *totally complete*[15], if it is endowed with countably infinite product operations satisfying for all sequences $(a_i \mid i \geq 1)$ of elements of K the following conditions:

$$a_1 \cdot \prod_{i \geq 1} a_{i+1} = \prod_{i \geq 1} a_i, \qquad \prod_{i \geq 1} a_i = \prod_{i \geq 1} (a_{n_{i-1}+1} \cdots a_{n_i}),$$

$$\prod_{i \geq 0} 1 = 1, \qquad \prod_{j \geq 1} \sum_{i_j \in I_j} a_{i_j} = \sum_{(i_1, i_2, \ldots) \in I_1 \times I_2 \times \ldots} \prod_{j \geq 1} a_{i_j}$$

where $0 = n_0 \leq n_1 \leq n_2 \leq \ldots$ and $I_1, I_2, \ldots$ are arbitrary index sets.

Furthermore, we will call a totally complete semiring *totally commutative complete* if it is commutative and satisfies the statement:

$$\prod_{j \in J} \left(\prod_{i_j \in I_j} a_{i_j} \right) = \prod_{i \geq 0} a_i$$

where $\bigcup_{j \in J} I_j = \mathbb{N}$ and $I_j \cap I_{j'} = \varnothing$ for $j \neq j'$.

Concrete examples of totally commutative complete semirings are the semiring $(\mathbb{N} \cup \{\infty\}, +, \cdot, 0, 1)$ of *extended natural numbers* [16], the *fuzzy semiring* $F = ([0, 1], \sup, \inf, 0, 1)$ [30, 23], and each *completely distributive lattice* (cf. [2]) with the operations *supremum* and *infimum*. Further examples will be given in Section 6.

Let A be a finite alphabet. As usual we denote by A^ω the set of all infinite words over A. An infinite word $w = a_0 a_1 \ldots \in A^\omega$ is written as $w = w(0)w(1)\ldots$ with $w(i) = a_i$, $i \geq 0$. We shall denote the set of natural numbers $\mathbb{N}$ also by ω.

Given a finite alphabet A and a semiring K, an *infinitary formal power series* or *series* for short, is a mapping $S : A^\omega \to K$. The class of all power series over A and K is denoted by $K \langle\langle A^\omega \rangle\rangle$. The *sum* $S + T$, the *scalar products* kS and Sk, the *Hadamard product* $S \odot T$ for $S, T \in K \langle\langle A^\omega \rangle\rangle$ and $k \in K$, as well as the *characteristic series* $1_L : A^\omega \to K$ of $L \subseteq A^\omega$, are defined in $K \langle\langle A^\omega \rangle\rangle$ pointwise as in the finitary case.

Consider two alphabets A, B and an homomorphism $h : A^* \to B^*$. Then h can be extended to a mapping $h : A^\omega \to B^\omega$ in the obvious way. For any power series $T \in K \langle\langle B^\omega \rangle\rangle$ the series $h^{-1}(T) \in K \langle\langle A^\omega \rangle\rangle$ is defined by $(h^{-1}(T), u) = (T, h(u))$ for $u \in A^\omega$. Furthermore, if h is non-deleting, i.e., $h(a) \neq \varepsilon$ for each $a \in A$, and K is complete, then for any $S \in K \langle\langle A^\omega \rangle\rangle$ the series $h(S) \in K \langle\langle B^\omega \rangle\rangle$ is specified by $(h(S), w) = \sum_{u \in h^{-1}(w)} (S, u)$ for $w \in B^\omega$. The homomorphism h is *strict alphabetic*, if $h(a) \in B$ for each $a \in A$.

For the rest of this section, let A be a finite alphabet and K be a totally complete semiring. We shall simply denote the operation $\cdot$ by concatenation.

Definition 1. *A weighted Muller automaton (WMA for short) over A and K is a quadruple $\mathcal{A} = (Q, in, wt, \mathcal{F})$, where Q is the finite state set, $in : Q \to K$ is* *the* initial distribution, $wt : Q \times A \times Q \to K$ *is a* mapping assigning weights to the transitions *of the automaton, and* $\mathcal{F} \subseteq 2^Q$ *is the* family of final state sets.

Let $w = a_0 a_1 \ldots \in A^\omega$. A *path of* $\mathcal{A}$ *over* w is an infinite sequence of transitions $P_w := (t_i)_{i \geq 0}$, so that $t_i = (q_i, a_i, q_{i+1})$ for all $i \geq 0$. The *weight of* P_w is defined by $weight(P_w) := in(q_0) \cdot \prod_{i \geq 0} wt(t_i)$. The path P_w is called *successful* if the set of states that appear infinitely often along P_w constitute a final state set. The *behavior* of $\mathcal{A}$ is the formal power series $\|\mathcal{A}\| : A^\omega \to K$ whose coefficients are determined by $(\|\mathcal{A}\|, w) = \sum_{P_w} weight(P_w)$ for $w \in A^\omega$, where the sum is taken over all successful paths P_w of $\mathcal{A}$ over w. A series $S : A^\omega \to K$ is said to be *Muller recognizable* if there is a WMA $\mathcal{A}$ so that $S = \|\mathcal{A}\|$. We shall denote the family of all such series over A and K by $K^{M-rec} \langle\langle A^\omega \rangle\rangle$.

The next result states closure properties of the family $K^{M-rec} \langle\langle A^\omega \rangle\rangle$.

Theorem 1. *The class* $K^{M-rec} \langle\langle A^\omega \rangle\rangle$ *is closed under:*

- *sum and scalar products; furthermore, if K is totally commutative complete, then $K^{M-rec} \langle\langle A^\omega \rangle\rangle$ is also closed under Hadamard products,*
- *non-deleting homomorphisms,*
- *inverse image of strict alphabetic homomorphisms.*

Proposition 1. *The characteristic series* $1_L : A^\omega \to K$ *of any ω-recognizable language* $L \subseteq A^\omega$ *is Muller recogizable.*

We will call a power series $S : A^\omega \to K$ a *Muller recognizable step function* if $S = \sum_{1 \leq j \leq n} k_j 1_{L_j}$ where $k_j \in K$ and $L_j \subseteq A^\omega$ ($1 \leq j \leq n$ and $n \in \mathbb{N}$) are ω-recognizable languages. Then by Theorem 1 and Proposition 1, S is Muller recognizable.

3   Weighted Monadic Second Order Logic

Weighted MSO logic was introduced by Droste and Gastin in [10] in order to obtain a logical characterization of recognizable formal power series over finite words.

Let A be a finite alphabet and $\mathcal{V}$ a finite set of first and second order variables. An infinite word $w \in A^\omega$ is represented by the relational structure $(\omega, \leq, (R_a)_{a \in A})$ where $R_a = \{i \mid w(i) = a\}$ for $a \in A$. A $(w, \mathcal{V})$-*assignment* σ is a mapping associating first order variables from $\mathcal{V}$ to elements of ω, and second order variables from $\mathcal{V}$ to subsets of ω. If x is a first order variable and $i \in \omega$, then $\sigma[x \to i]$ denotes the $(w, \mathcal{V} \cup \{x\})$-assignment which associates i to x and acts as σ on $\mathcal{V} \setminus \{x\}$. For a second order variable X and $I \subseteq \omega$, the notation $\sigma[X \to I]$ has a similar meaning. In order to encode pairs (w, σ) for all $w \in A^\omega$ and any $(w, \mathcal{V})$-assignment σ, we use an extended alphabet $A_\mathcal{V} = A \times \{0, 1\}^\mathcal{V}$ (cf. [33, 34, 10])

It is not difficult to see that the set $N_\mathcal{V} = \{(w, \sigma) \in A_\mathcal{V}^\omega \mid \sigma$ is a valid $(w, \mathcal{V})$-assignment$\}$ is ω-recognizable. Let now φ be an MSO-formula [33, 34]. We shall write A_φ for $A_{Free(\varphi)}$ and $N_\varphi = N_{Free(\varphi)}$. The fundamental Büchi theorem [6] states that for $Free(\varphi) \subseteq \mathcal{V}$ the language $\mathcal{L}_\mathcal{V}(\varphi) = \{(w, \sigma) \in N_\mathcal{V} \mid (w, \sigma) \models \varphi\}$ defined by φ over $A_\mathcal{V}$ is ω-recognizable. We simply write $\mathcal{L}(\varphi) = \mathcal{L}_{Free(\varphi)}(\varphi)$. Conversely, each ω-recognizable language $L \subseteq A^\omega$ is definable by an MSO-sentence φ, i.e., $L = \mathcal{L}(\varphi)$.

Now we turn to weighted logics.

Definition 2. *The syntax of formulas of the weighted MSO logic is given by*

$$\varphi := k \mid P_a(x) \mid \neg P_a(x) \mid x \leq y \mid \neg(x \leq y) \mid x \in X \mid \neg(x \in X)$$
$$\mid \varphi \vee \psi \mid \varphi \wedge \psi \mid \exists x \cdot \varphi \mid \exists X \cdot \varphi \mid \forall x \cdot \varphi \mid \forall X \cdot \varphi$$

where $k \in K$, $a \in A$. We shall denote by $MSO(K, A)$ the set of all such weighted MSO-formulas φ.

Next we represent the semantics of the formulas in $MSO(K, A)$ as formal power series over the extended alphabet $A_\mathcal{V}$ and the semiring K. We assume K to be a totally commutative complete semiring.

Definition 3. *Let $\varphi \in MSO(K, A)$ and $\mathcal{V}$ be a finite set of variables with $Free(\varphi) \subseteq \mathcal{V}$. The semantics of φ is a formal power series $\|\varphi\|_\mathcal{V} \in K \langle\langle A_\mathcal{V}^\omega \rangle\rangle$. Consider an element $(w, \sigma) \in A_\mathcal{V}^\omega$. If σ is not a valid assignment, then we put $\|\varphi\|_\mathcal{V}(w, \sigma) = 0$. Otherwise, we inductively define $\|\varphi\|_\mathcal{V}(w, \sigma) \in K$ as follows:*

- $\|k\|_\mathcal{V}(w, \sigma) = k$
- $\|P_a(x)\|_\mathcal{V}(w, \sigma) = \begin{cases} 1 & \text{if } w(\sigma(x)) = a \\ 0 & \text{otherwise} \end{cases}$
- $\|x \leq y\|_\mathcal{V}(w, \sigma) = \begin{cases} 1 & \text{if } \sigma(x) \leq \sigma(y) \\ 0 & \text{otherwise} \end{cases}$
- $\|x \in X\|_\mathcal{V}(w, \sigma) = \begin{cases} 1 & \text{if } \sigma(x) \in \sigma(X) \\ 0 & \text{otherwise} \end{cases}$
- $\|\neg\varphi\|_\mathcal{V}(w, \sigma) = \begin{cases} 1 & \text{if } \|\varphi\|_\mathcal{V}(w, \sigma) = 0 \\ 0 & \text{if } \|\varphi\|_\mathcal{V}(w, \sigma) = 1 \end{cases}$, *provided that φ is of the form $P_a(x)$, $(x \leq y)$ or $(x \in X)$*
- $\|\varphi \vee \psi\|_\mathcal{V}(w, \sigma) = \|\varphi\|_\mathcal{V}(w, \sigma) + \|\psi\|_\mathcal{V}(w, \sigma)$
- $\|\varphi \wedge \psi\|_\mathcal{V}(w, \sigma) = \|\varphi\|_\mathcal{V}(w, \sigma) \cdot \|\psi\|_\mathcal{V}(w, \sigma)$
- $\|\exists x \cdot \varphi\|_\mathcal{V}(w, \sigma) = \sum_{i \in \omega} \|\varphi\|_{\mathcal{V} \cup \{x\}}(w, \sigma[x \to i])$
- $\|\exists X \cdot \varphi\|_\mathcal{V}(w, \sigma) = \sum_{I \subseteq \omega} \|\varphi\|_{\mathcal{V} \cup \{X\}}(w, \sigma[X \to I])$
- $\|\forall x \cdot \varphi\|_\mathcal{V}(w, \sigma) = \prod_{i \in \omega} \|\varphi\|_{\mathcal{V} \cup \{x\}}(w, \sigma[x \to i])$ *and*
- $\|\forall X \cdot \varphi\|_\mathcal{V}(w, \sigma) = \prod_{I \subseteq \omega} \|\varphi\|_{\mathcal{V} \cup \{X\}}(w, \sigma[X \to I])$.

The reader may notice that the product in universal second order quantification is uncountable. But this is not a problem since later we exclude it from our constructions. Also as in [10], we have restricted negation to atomic formulas.

The reason is that if K is not a Boolean algebra, then it is difficult to define the semantics of the negation of an arbitrary formula. Our restriction is not essential in comparison to classical MSO logics, since any MSO-formula φ is equivalent (both logically and in the sense of defining the same ω-language) to one in which negation is applied only to atomic formulas. We simply write $\|\varphi\|$ for $\|\varphi\|_{Free(\varphi)}$. If φ has no free variables, i.e., if it is a sentence, then $\|\varphi\| \in K \langle\langle A^\omega \rangle\rangle$. Next, we present several examples of possible interpretations for weighted formulas, for details see [10].

(i) Consider the semiring $K = (\mathbb{N} \cup \{\infty\}, +, \cdot, 0, 1)$ and assume that φ does not contain constants $k \in K$. Then we may interpret $\|\varphi\| (w, \sigma)$ as the number of proofs we have that (w, σ) satisfies formula φ.

(ii) The formula $\exists x \cdot P_a(x)$ counts *how often* (depending on the semiring) the letter a occurs in the word.

(iii) For any formula φ over the fuzzy semiring F, we have that $\|\varphi\| (w, \sigma) \neq 0$ iff (w, σ) satisfies φ.

(iv) Let K be an arbitrary Boolean algebra $(B, \vee, \wedge, ^-, 0, 1)$. In this case, infinite sums correspond to suprema and infinite products to infima. For any formula φ, we can define the semantics of $\neg\varphi$, by $\|\neg\varphi\| (w, \sigma) := \overline{\|\varphi\| (w, \sigma)}$. Especially, for $K = \mathbf{B}$ the 2-valued Boolean algebra our semantics coincides with the usual semantics of classical MSO-formulas, identifying characteristic series with their supports.

The reader may observe that the above definition is valid for each formula $\varphi \in MSO(K, A)$ and each finite set $\mathcal{V}$ of variables containing $Free(\varphi)$. According to the next proposition the semantics $\|\varphi\|_{\mathcal{V}}$ depends only on $Free(\varphi)$.

Proposition 2. *Let $\varphi \in MSO(K, A)$ and $\mathcal{V}$ be a finite set of variables such that $Free(\varphi) \subseteq \mathcal{V}$. Then*

$$\|\varphi\|_{\mathcal{V}} (w, \sigma) = \|\varphi\| (w, \sigma|_{Free(\varphi)})$$

for each $(w, \sigma) \in A_{\mathcal{V}}^\omega$, where σ is a valid $(w, \mathcal{V})$-assignment. Furthermore, $\|\varphi\|$ is Muller recognizable iff $\|\varphi\|_{\mathcal{V}}$ is Muller recognizable.

Let now $Z \subseteq MSO(K, A)$. A series $S : A^\omega \to K$ is called Z-*definable* if there is a sentence $\varphi \in Z$ so that $S = \|\varphi\|$.

It has been proved in [10] that universal quantifiers do not preserve in general the recognizability property of power series over finite words. Thus the authors worked on a restricted framework of weighted MSO logics, which we also adopt here. More precisely, a formula $\varphi \in MSO(K, A)$ will be called *restricted* (cf. [10]) if it contains no universal quantification of the form $\forall X \cdot \psi$, and whenever φ contains a universal first order quantification $\forall x \cdot \psi$, then $\|\psi\|$ is a Muller recognizable step function. The subclass of all restricted formulas of $MSO(K, A)$ will be denoted by $RMSO(K, A)$. Moreover, a formula $\varphi \in RMSO(K, A)$ is *restricted existential* if it is of the form $\exists X_1, \ldots, X_n \cdot \psi$ with $\psi \in RMSO(K, A)$ and ψ contains no set quantification. All such restricted existential formulas will compose the class $REMSO(K, A)$. We let $K^{rmso} \langle\langle A^\omega \rangle\rangle$ (resp. $K^{remso} \langle\langle A^\omega \rangle\rangle$) comprise all series from $K \langle\langle A^\omega \rangle\rangle$ which are definable by some sentence in $RMSO(K, A)$ (resp. in $REMSO(K, A)$).

4   The Main Result

In this section we establish our main result:

Theorem 2. *Let A be an alphabet and K any totally commutative complete semiring. Then*

$$K^{M-rec}\langle\langle A^\omega \rangle\rangle = K^{rmso}\langle\langle A^\omega \rangle\rangle = K^{remso}\langle\langle A^\omega \rangle\rangle.$$

Proof. Let us present a sketch of the proof. First, using Theorem 1 and Proposition 1, we show by induction on the structure of $RMSO$-formulas that K^{rmso} $\langle\langle A^\omega \rangle\rangle \subseteq K^{M-rec}\langle\langle A^\omega \rangle\rangle$. The most difficult case arises with first order universal quantification. Let $\mathcal{W} = Free(\varphi)$ and $\mathcal{V} = Free(\forall x \centerdot \varphi) = \mathcal{W} \setminus \{x\}$. Let also $\|\varphi\| = \sum\limits_{1 \leq j \leq n} k_j 1_{L_j}$ with ω-recognizable languages $L_j \subseteq A_{\mathcal{W}}^\omega$ ($1 \leq j \leq n$). We claim that $\|\forall x \centerdot \varphi\|$ is Muller recognizable. Without any loss, we can assume that the family $(L_j)_{1 \leq j \leq n}$ is a partition of $A_{\mathcal{W}}^\omega$. We distinguish two cases.

Case 1: $x \in \mathcal{W}$.
We consider the alphabet $\widetilde{A} = A \times \{1, \dots, n\}$, and the language $\widetilde{L} \subseteq \widetilde{A}_{\mathcal{V}}^\omega$ to be the collection of all words $(w, v, \sigma) \in \widetilde{A}_{\mathcal{V}}^\omega$, so that for all $i \in \omega$ and $j \in \{1, \dots, n\}$, then $v(i) = j$ implies $(w, \sigma[x \to i]) \in L_j$. The languages L_j are ω-recognizable by deterministic Muller automata, and from these we construct a deterministic Muller automaton $\widetilde{A}$ recognizing $\widetilde{L}$. In the sequel, we convert $\widetilde{A}$ to a WMA $\mathcal{A}$ over $\widetilde{A}_{\mathcal{V}}$, and we show that $\|\forall x \centerdot \varphi\| = h(\|\mathcal{A}\|)$, where h is the projection of $\widetilde{A}_{\mathcal{V}}^\omega$ to $A_{\mathcal{V}}^\omega$. Thus by Theorem 1 the series $\|\forall x \centerdot \varphi\|$ is Muller recognizable.

Case 2: If $x \notin \mathcal{W}$, then we consider the formula $\varphi' = \varphi \wedge (x \leq x)$, and the result comes by Case 1.

Now we state the converse inclusion. Given a WMA $\mathcal{A}$ over A and K we effectively construct a $RMSO(K, A)$-formula ψ representing the paths of $\mathcal{A}$. Next we equip ψ with weights, so that the semantics $\|\varphi\|$ of the obtained formula φ takes as values the weights of the corresponding paths of $\mathcal{A}$. Finally, from φ we obtain a formula ξ in $REMSO(K, A)$ whose semantics equals the behavior of the automaton $\mathcal{A}$.

5   Weighted Büchi Automata

In this section, we consider weighted automata over infinite words with Büchi acceptance condition which were introduced by Ésik and Kuich [16]. We show their equivalence to our model with Muller acceptance condition.

Let A be any alphabet and K be a totally complete semiring.

Definition 4. *([16]) A weighted Büchi automaton (WBA for short) over A and K is a quadruple $\mathcal{A} = (Q, in, wt, F)$, where Q is the finite state set, $in : Q \to K$ is the initial distribution, $wt : Q \times A \times Q \to K$ is a mapping assigning weights to the transitions of $\mathcal{A}$, and F is the final state set.*

Given an infinite word $w = a_0 a_1 \dots \in A^\omega$, a *path* P_w of $\mathcal{A}$ over w and its *weight* are defined as for weighted Muller automata. The path P_w is called *successful* if at least one final state appears infinitely often. The *behavior of* $\mathcal{A}$ is the infinitary power series $\|\mathcal{A}\| : A^\omega \to K$, with coefficients specified for $w \in A^\omega$ by $(\|\mathcal{A}\|, w) = \sum_{P_w} weight(P_w)$, where the sum is taken over all successful paths P_w of $\mathcal{A}$ over w. A series $S : A^\omega \to K$ is called ω-*recognizable* if there is a WBA $\mathcal{A}$ such that $S = \|\mathcal{A}\|$. The class of all ω-recognizable series over A and K is denoted by $K^{\omega-rec} \langle\langle A^\omega \rangle\rangle$.

Theorem 3. *Let A be an alphabet and K any totally complete semiring. Then*

$$K^{\omega-rec} \langle\langle A^\omega \rangle\rangle = K^{M-rec} \langle\langle A^\omega \rangle\rangle.$$

6   Bi-aperiodic Semirings

In this section we state our main result for weakly bi-aperiodic [9, 10] and commutative semirings. A semiring $(K, +, \cdot, 0, 1)$ is called *bi-aperiodic* if there exists an integer $m \geq 0$ such that for all $a \in K$ $(m+1)a = ma$ and $a^{m+1} = a^m$. All distributive lattices with 0 and 1 are bi-aperiodic semirings with supremum and infimum as operations. Furthermore, all Boolean algebras, and the reals with max-min or min-max, constitute bi-aperiodic semirings. In this case, for any $a \in K, m \geq 0$ as above, and any infinite index set I, we can define the infinite sum and product of a's by letting $\sum_{i \in I} a = ma$ and $\prod_{i \in I} a = a^m$. Now assume that our semiring K is finite, bi-aperiodic and commutative. Then we can also define the infinite sum and product of any family of elements of K, by splitting them suitably and then taking the corresponding finite sums and products. We obtain:

Proposition 3. *Each finite bi-aperiodic commutative semiring $(K, +, \cdot, 0, 1)$ is totally commutative complete.*

Next, a semiring $(K, +, \cdot, 0, 1)$ is called *weakly bi-aperiodic* iff for each element $a \in K$ there exists $m \geq 0$ such that $(m+1)a = ma$ and $a^{m+1} = a^m$. Trivially, each bi-aperiodic semiring is weakly bi-aperiodic and each finite weakly bi-aperiodic semiring is bi-aperiodic. We refer the reader to [10] for examples of weakly bi-aperiodic semirings. Furthermore,

Example 1. Let $0 < c < 1$ and $K = \{0\} \cup [c, 1]$. We define in K the truncated multiplication $\cdot_c$ in the following way: for $x \neq 0$ and $y \neq 0$, $x \cdot_c y := x \cdot y$ if $x \cdot y \geq c$ and $x \cdot_c y := c$ if $x \cdot y \leq c$. The semiring $(K, \max, \cdot_c, 0, 1)$ is called the *truncated probabilistic semiring*. Obviously, it is weakly bi-aperiodic but not aperiodic.

Next we state that

Theorem 4. *Let A be an alphabet and K any weakly bi-aperiodic commutative semiring. Then*

$$K^{M-rec} \langle\langle A^\omega \rangle\rangle = K^{rmso} \langle\langle A^\omega \rangle\rangle = K^{remso} \langle\langle A^\omega \rangle\rangle.$$

Corollary 1 (Büchi's Theorem). *An infinitary language is ω-recognizable iff it is definable by a MSO-sentence.*

7   Conclusion

We introduced weighted Muller automata over totally complete semirings. We verified that the family of their behaviors coincides with the class of infinitary formal power series obtained as semantics of weighted restricted MSO-sentences, provided that the underlying semiring is totally commutative complete and also with the family of behaviors of weighted Büchi automata investigated by Ésik and Kuich [16]. We do not know if this family coincides with the class of series specified by all weighted MSO-sentences. Also, the question arises whether Theorem 2, in particular the construction of a WMA $\mathcal{A}$ for a given MSO-formula φ can be made effective. The problem is the universal quantifier: Given a WMA for φ as described in the proof of Theorem 2, how do we obtain the values k_j and WMA for the languages L_j? In the case of finite words and given a field K, Droste and Gastin [10] could use results from the literature on formal power series to obtain a construction. Therefore, also in our situation we should consider specific semirings.

References

1. A. Arnold, Finite transition systems, *International Series in Computer Science*, Prentice Hall, 1994.
2. R. Balbes, P. Dwinger, Distributive Lattices, University of Missouri Press, 1974.
3. J. Berstel, C. Reutenauer, *Rational Series and Their Languages*. EATCS Monographs in Theoretical Computer Science, vol. 12, Springer-Verlag, 1988.
4. S. L. Bloom, Z. Ésik, Iteration Theories, *EATCS Monographs on Theoretical Computer Science*, Springer, 1993.
5. J. R. Büchi, Weak second-order arithmetic and finite automata, *Z. Math. Logik Grundlager Math.* 6(1960) 66-92.
6. J. R. Büchi, On a decision method in restricted second order arithmetic, in: *Proc. 1960 Int. Congr. for Logic, Methodology and Philosophy of Science*, (1962), pp.1-11.
7. K. Culik II, J. Kari, Image compression using weighted finite automata, *Computer and Graphics*, 17(1993) 305-313.
8. K. Culik, J. Karhumäki, Finite automata computing real functions, *SIAM J. Comput.* 23(4)(1994) 789-814.
9. M. Droste, P. Gastin, On aperiodic and star-free formal power series in partially commuting variables. In: *Proceedings of FPASAC 00*, Springer 2000, pp.158-169., and full version: Research Report LSV-05, ENS de Cachan, France, 2005.
10. M. Droste, P. Gastin, Weighted automata and weighted logics, in: *32nd ICALP, LNCS* 3580(2005) 513-525. Full version in: http://www.informatik.uni-leipzig.de/theo/pers/droste/publications.html.
11. M. Droste, D. Kuske, Skew and infinitary formal power series, in: *30th ICALP, LNCS* 2719(2003) 426-438.
12. M. Droste, U. Püschmann, Weighted Büchi Automata, (submitted).
13. S. Eilenberg, *Automata, Languages and Machines, vol. A*, Academic Press 1974.
14. C. Elgot, Decision problems of finite automata design and related arithmetics, *Trans. Amer. Math. Soc.* 98(1961) 21-52.
15. Z. Ésik, W. Kuich, On iteration semiring-semimodule pairs. To appear.

16. Z. Ésik, W. Kuich, A semiring-semimodule generalization of ω-regular languages I and II. Special issue on "Weighted automata" (M. Droste, H. Vogler, eds.) *J. of Automata Languages and Combinatorics*, to appear.

17. U. Hafner, Low Bit-Rate Image and Video Coding with Weighted Finite Automata, *PhD thesis*, Universität Würzburg, Germany, 1999.

18. Z. Jiang, B. Litow and O. de Vel, Similarity enrichment in image compression through weighted finite automata, in: *COCOON 00, LNCS* 1858(2000) 447-456.

19. F. Katritzke, Refinements of data compression using weighted finite automata, *PhD thesis*, Universität Siegen, Germany, 2001.

20. K. Krithivasan, K. Sharda, Fuzzy ω-automata, *Inf. Sci.* 138(2001) 257-281.

21. W. Kuich, Semirings and formal power series: Their relevance to formal languages and automata theory. In: *Handbook of Formal Languages* (G. Rozenberg, A. Salomaa, eds.), vol. 1, Springer, 1997, pp. 609–677.

22. W. Kuich, On skew formal power series, in: *Proceedings of the Conference on Algebraic Informatics* (S. Bozapalidis, A. Kalampakas, G. Rahonis, eds.), Thessaloniki 2005, pp. 7-30.

23. W. Kuich, G. Rahonis, Fuzzy regular languages over finite and infinite words, *Fuzzy Sets and Systems*, to appear.

24. W. Kuich, A. Salomaa, Semirings, Automata, Languages, *EATCS Monographs in Theoretical Computer Science, vol. 5*, Springer-Verlag, 1986.

25. R. P. Kurshan, Computer-Aided Verification of Coordinating Processes, *Princeton Series in Computer Science*, Princeton University Press, 1994.

26. K. McMillan, Symbolic Model Checking, *Kluwer Academic Publishers*, 1993.

27. M. Mohri, Finite-state transducers in language and speech processing, *Computational Linguistics* 23(1997) 269-311.

28. M. Mohri, F. Pereira and M. Riley, The design principles of a weighted finite-state transducer library, *Theoret. Comput. Sci.* 231(2000) 17-32.

29. D. Perrin, J. E. Pin, Infinite Words, Elsevier 2004.

30. G. Rahonis, Infinite fuzzy computations, *Fuzzy Sets and Systems*, 153(2005) 275-288.

31. A. Salomaa, M. Soittola, *Automata-Theoretic Aspects of Formal Power Series*. Texts and Monographs in Computer Science, Springr-Verlag, 1978.

32. M. Schützenberger, On the definition of a family of automata, *Inf. Control* 4(1961) 245-270.

33. W. Thomas, Automata on infinite objects, in: *Handbook of Theoretical Computer Science, vol. B* (J. v. Leeuwen, ed.), Elsevier Science Publishers, Amsterdam 1990, pp. 135-191.

34. W. Thomas, Languages, automata and logic, in: *Handbook of Formal Languages* vol. 3 (G. Rozenberg, A. Salomaa, eds.), Springer, 1997, pp. 389-485.

Simulation Relations for Alternating Parity Automata and Parity Games*

Carsten Fritz and Thomas Wilke

Christian-Albrechts-Universität zu Kiel
{fritz, wilke}@ti.informatik.uni-kiel.de

Abstract. We adapt the notion of delayed simulation to alternating parity automata and parity games. On the positive side, we show that (i) the corresponding simulation relation can be computed in polynomial time and (ii) delayed simulation implies language inclusion. On the negative side, we point out that quotienting with respect to delayed simulation does not preserve the language recognized, which means that delayed simulation cannot be used for state-space reduction via merging of simulation equivalent states. As a remedy, we introduce finer, so-called biased notions of delayed simulation where we show quotienting does preserve the language recognized. We propose a heuristic for reducing the size of alternating parity automata and parity games and, as an evidence for its usefulness, demonstrate that it is successful when applied to the Jurdziński family of parity games.

1 Introduction

The motivation for studying simulation relations for automata is, in general, two-fold: First, simulation relations are an appropriate means for comparing the structure of automata. They formalize the idea that one automaton is capable of mimicking the behavior of another automaton. In other words, they are useful for identifying structural similarities of automata. This is also true in the context of transition systems and processes, in fact, simulation relations were introduced in a wider context [1]. Second, simulation relations have proved to be very useful for efficiently reducing the size of (finite-state) automata, the basic idea being to merge states which simulate each other. This method is also known as quotienting, and it is a well-established method for reducing the number of states of a given Büchi automaton, especially in the context of generating small Büchi automata from formulas in linear temporal logic, see [2, 3, 4, 5, 6, 7, 8, 9].

The objective of the work presented in this paper is to extend the notion of simulation to alternating parity automata (also known as alternating Rabin chain automata), see [10, 11], and to identify how simulation can be used for state-space reduction. From the simpler scenario with Büchi automata it is known, see [12, 13], that it is reasonable to distinguish different types of simulations: direct, delayed, and fair simulation. Direct simulation is less interesting from the

---

* Project funded by the Deutsche Forschungsgemeinschaft under no. 223228.

O.H. Ibarra and Z. Dang (Eds.): DLT 2006, LNCS 4036, pp. 59–70, 2006.

point of view of state-space reduction, because it yields the finest of the three relations and thus pays off the least. Fair simulation, on the other hand, is too coarse, for quotienting with respect to it may change the recognized language [13, 14]. That is why we focus on delayed simulation.

There are two major technical problems to overcome for delayed simulation. The first problem is that a priori it is not at all clear how the different priorities of the states of a parity automaton should be taken into account in a definition of simulation where delays are allowed. We try to give a definition as general as possible in the sense that the resulting simulation relation is as coarse as possible. Our approach is game-theoretic, just as in [13], and allows us to prove that our notion has the basic properties of a simulation relation. When it comes to quotienting, our definition, however, turns out to be too general: quotienting does not preserve the language recognized, which is then the second technical problem to overcome. We describe two ways of strengthening our notion of delayed simulation, so-called biased simulations, for which we can then show that quotienting still works.

To achieve our goal of developing an efficient heuristic for reducing the state spaces of alternating parity automata, we combine quotienting with respect to the biased relations with basic simplification methods and simplification methods involving our general delayed simulation relation. The heuristic we propose turns out to be successful when applied to parity games (a parity game is simply an alternating parity automaton over a unary alphabet): It reduces parity games which have been shown to be difficult instances for a certain game solving algorithm quite fast to games with just two positions, see [15].

The paper is structured as follows. In Sect. 2, we briefly describe our notation and terminology. In Sect. 3, we then introduce our general definition of delayed simulation for alternating parity automata, describe its main properties, and explain why quotienting does not work. In Sect. 4, we explain how the biased variants of the general delayed simulation are obtained and how they can be used for quotienting. Before we conclude, we describe our heuristic for state-space reduction in Sect. 5.

The proofs of most of the results are very involved; the reader is referred to [16] for details. For background on games and accepting/winning conditions, see [17].

2   Basic Notation and Terminology

An *infinite game* is a tuple

$$\mathcal{G} = (P, P_0, P_1, p_I, Z, W) \tag{1}$$

where P is a set of *positions*, $P_0 \subseteq P$ are the *positions of Player 0*, $P_1 \subseteq P$ are the *positions of Player 1* such that $P = P_0 \cup P_1$ and $P_0 \cap P_1 = \emptyset$, $p_I \in P$ is an *initial position*, $Z \subseteq P \times P$ is the set of *moves*, and $W \subseteq P^\omega$ is the *winning condition (for Player 0)*. For convenience, we require that the set of moves is *complete*, that is, for every $p \in P$ there is some $p' \in P$ such that $(p, p') \in Z$ is a move.

A *play* of such a game is an infinite path $\pi = p_0 p_1 p_2 \ldots$ through the *game graph* (P, Z) starting in p_I. It is *winning* for Player 0 if $\pi \in W$.

A *parity game* is an infinite game as in (1) where W is specified indirectly by a *priority function* $\Omega \colon P \to \omega$ which is required to have a finite image. The winning condition associated with Ω, denoted $W(\Omega)$, is the set which contains a sequence $p_0 p_1 p_2 \ldots$ if $\min\{m \mid \exists^\infty i (\Omega(p_i) = m)\}$ (which is well-defined because Ω is required to have a finite image) is even. That is, Player 0 wins if the minimum priority occurring infinitely often is even.

An *alternating parity automaton (APA)* is a tuple

$$\mathcal{Q} = (Q, \Sigma, q_I, \Delta, E, U, \Omega) \tag{2}$$

where Q is a finite set of *states*, Σ is an *alphabet*, $q_I \in Q$ is the *initial state*, $\Delta \subseteq Q \times \Sigma \times Q$ is the *transition relation*, $E \subseteq Q$ is the set of *existential states*, $U \subseteq Q$ is the set of *universal states* such that $E \cup U = Q$ and $E \cap U = \emptyset$, and $\Omega \colon Q \to \omega$ is the *priority function*. Without loss of generality, we require that Δ is complete, that is, for every $q \in Q$, $a \in \Sigma$ there must exist a state $q' \in Q$ such that $(q, a, q') \in \Delta$.

Acceptance of an APA is best explained using games. Given an APA $\mathcal{Q}$ as in (2) and an ω-word $w_0 w_1 w_2 \cdots \in \Sigma^\omega$, the *word game*

$$G(\mathcal{Q}, w) = (P, P_0, P_1, p_I, Z, \Omega') \tag{3}$$

is the parity game where $P = Q \times \omega$, $P_0 = E \times \omega$, $P_1 = U \times \omega$, $p_I = (q_I, 0)$, $Z = \{((q, i), w_i, (q', i+1)) \mid i \in \omega \wedge (q, w_i, q') \in \Delta\}$ and $\Omega'((q, i)) = \Omega(q)$.

In this game Player 0 and Player 1 are called *Automaton* and *Pathfinder*, respectively, following the terminology from [18]. The word w is *accepted* by $\mathcal{Q}$ if Automaton has a winning strategy in $G(\mathcal{Q}, w)$. The set of all words accepted by $\mathcal{Q}$ is denoted $L(\mathcal{Q})$.

We will use the following ordering on natural numbers (also used in, e.g., [19]), which reflects the parity winning condition. The *reward order* $\preceq$ is the total order on ω defined by $m \preceq n$ if and only if m is even and n is odd, or m and n are even and $m \leq n$, or m and n are odd and $n \leq m$. That is, $0 \prec 2 \prec 4 \prec \ldots \prec 5 \prec 3 \prec 1$. When $n \prec m$, we will say n *is better than* m, while terms like *minimum* and *smaller than* will always be used w.r.t. the standard order $\leq$.

3   Delayed Simulation for the Parity Condition

On a very abstract level, delayed simulation can be explained as follows. A state s simulates a state q *directly*, if everything that can be done starting from q can be mimicked step-by-step starting from s. Here, mimicking means that the simulating step must be as good as (with respect to acceptance) the simulated step. For *delayed* simulation, direct simulation is relaxed: A step need not be simulated directly, but a finite delay is allowed.

Delays lead to "pending obligations", every one of which has to be fulfilled eventually. Our definition of delayed simulation will therefore have mechanisms for keeping track of pending obligations and for checking that pending obligations have been fulfilled.

3.1   Formal Definition of Delayed Simulation

Suppose $\mathcal{Q}$ and $\mathcal{S}$ are APA's of the form $\mathcal{Q} = (Q, \Sigma, q_I, \Delta, E^{\mathcal{Q}}, U^{\mathcal{Q}}, \Omega^{\mathcal{Q}})$ and $\mathcal{S} = (S, \Sigma, s_I, \Delta^{\mathcal{S}}, E^{\mathcal{S}}, U^{\mathcal{S}}, \Omega^{\mathcal{S}})$. We want to define what it means for $\mathcal{Q}$ to be simulated by $\mathcal{S}$ in a delayed fashion. We do this by first describing an infinite game, named *simulation game* and denoted $\mathcal{G}^{de}(\mathcal{Q}, \mathcal{S})$, where Player 0 and Player 1 are called *Duplicator* and *Spoiler*, respectively. We then say that $\mathcal{S}$ simulates $\mathcal{Q}$ if Duplicator wins the game.

The idea of the game, similar to other game-based definitions of simulation, see, e.g., [13], is as follows. Throughout the game, there is one pebble on a state of $\mathcal{Q}$ and another pebble on a state of $\mathcal{S}$. The game proceeds in rounds in such a way that in every round first Spoiler chooses a letter and then Spoiler and Duplicator move the pebbles along transitions labeled with the chosen letter according to rules which depend on the modes of the states the pebbles are on. The outcome of a play are two infinite sequences of states in $\mathcal{Q}$ and $\mathcal{S}$, respectively. Duplicator wins the play if each obligation built up during the course of the game is eventually fulfilled. For an easier formal description, information about the pending obligations will be built into the game graph.

Formally, $\mathcal{G}^{de}(\mathcal{Q}, \mathcal{S}) = (P, P_{Du}, P_{Sp}, p_I, Z, W)$ with components defined as follows. The **positions** contain information on where the pebbles are on, whether Spoiler has already chosen a letter and if so, which letter, what the pending obligations are, and who is going to move next and where (if that cannot be deduced from the other components). We use the elements of $K = \Omega^q(Q) \cup \Omega^s(S) \cup \{\surd\}$ to describe the pending obligations in terms of a priority to be met by Duplicator, where the check mark stands for "all obligations fulfilled".

The set of all positions is the union of P^0, P^1, and P^2 defined by

$$P^0 = Q \times S \times K , \qquad\qquad P^1 = Q \times S \times K \times \Sigma ,$$
$$P^2 = Q \times S \times K \times \Sigma \times \{Sp, Du\} \times \{\mathcal{Q}, \mathcal{S}\} ,$$

where positions in P^i describe configurations of the game at the beginning of a round, after Spoiler has chosen a letter, and after one of the players has moved and one player is still to move, respectively; a position in P^2 explicitly specifies who is to move and which pebble.

The **initial position** is determined by: If $\Omega^{\mathcal{Q}}(q_I) \prec \Omega^{\mathcal{S}}(s_I)$, then the initial position is given by $p_I = (q_I, s_I, \min\{\Omega^{\mathcal{Q}}(q_I), \Omega^{\mathcal{S}}(s_I)\})$, else it is $p_I = (q_I, s_I, \surd)$.

Depending on the modes of the states the pebbles are on at the beginning of a round, the players **move** according to the following table:

q	s	1st player plays on		2nd player plays on	
$E^{\mathcal{Q}}$	$E^{\mathcal{S}}$	Sp	$\mathcal{Q}$	Du	$\mathcal{S}$
$E^{\mathcal{Q}}$	$U^{\mathcal{S}}$	Sp	$\mathcal{Q}$	Sp	$\mathcal{S}$
$U^{\mathcal{Q}}$	$E^{\mathcal{S}}$	Du	$\mathcal{S}$	Du	$\mathcal{Q}$
$U^{\mathcal{Q}}$	$U^{\mathcal{S}}$	Sp	$\mathcal{S}$	Du	$\mathcal{Q}$

For instance, if both, q and s, are existential, Spoiler moves first on $\mathcal{Q}$ and then Duplicator moves on $\mathcal{S}$.

The table explains why P_{Du} is defined by

$$P_{Du} = (U^{\mathcal{Q}} \times E^{\mathcal{S}} \times K \times \Sigma) \cup (Q \times S \times K \times \Sigma \times \{Du\} \times \{\mathcal{Q}, \mathcal{S}\})$$

and that P_{Sp} is defined to be the set of the remaining positions.

The important part of a move is how the pending obligations are updated. To describe this, we define a function $\gamma \colon \omega \times \omega \times (\omega \cup \{\sqrt{}\}) \to \omega \cup \{\sqrt{}\}$ as follows. First, we set $\gamma(i, j, \sqrt{}) = \sqrt{}$ if $j \preceq i$, and $\gamma(i, j, \sqrt{}) = \min\{i, j\}$ otherwise. When the third argument is a natural number k, we set

$$\gamma(i, j, k) = \begin{cases} \sqrt{} \, , & \text{if} \begin{cases} j \preceq i, \, i \text{ odd}, \, i \le k, \text{ or} \\ j \preceq i, \, j \text{ even}, \, j \le k, \end{cases} \\ \min\{i, j, k\} \, , & \text{otherwise.} \end{cases} \tag{4}$$

For instance, the first clause says that if Duplicator is supposed to meet k, q has priority i, s has priority j, then it is enough when $j \preceq i$ and $i \le k$ for an odd i.

That is, we store an obligation $\min\{i, j\}$ if the priority i of $\mathcal{Q}$ is better than the priority j of $\mathcal{S}$. One possibility to meet this obligation in a future round is that the priority i of $\mathcal{Q}$ in that round is odd and less than or equal to the obligation (especially, i is at most as good as the obligation) while at the same time, j is at least as good as i. In that case, the stored obligation no longer is a witness for a, so to say superior acceptance behavior of $\mathcal{Q}$ as compared to $\mathcal{S}$, which means it can be erased. Symmetrically, the obligation can be met by a small even value of j in $\mathcal{S}$.

The set Z of all moves is the union of the sets Z^0, Z^1, and Z^2 defined below, where $2\mathrm{pl}(q, s)$ and $2\mathrm{au}(q, s)$ are determined by the above table (last two columns):

$$Z^0 = \{((q, s, k), (q, s, k, a)) \mid (q, s, k) \in P^0 \wedge a \in \Sigma\} \, ,$$

$$Z^1 = \{((q, s, k, a), (q', s, k, a, 2\mathrm{pl}(q, s), 2\mathrm{au}(q, s))) \mid q \in E^{\mathcal{Q}} \wedge (q, a, q') \in \Delta^{\mathcal{Q}}\}$$
$$\cup \{((q, s, k, a), (q, s', k, a, 2\mathrm{pl}(q, s), 2\mathrm{au}(q, s))) \mid q \in U^{\mathcal{Q}} \wedge (s, a, s') \in \Delta^{\mathcal{S}}\},$$

$$Z^2 = \{((q, s, k, a, x, \mathcal{Q}), (q', s, \gamma(\Omega(q'), \Omega(s), k))) \mid x \in \{Sp, Du\} \wedge (q, a, q') \in \Delta^{\mathcal{Q}}\}$$
$$\cup \{((q, s, k, a, x, \mathcal{S}), (q, s', \gamma(\Omega(q), \Omega(s'), k))) \mid x \in \{Sp, Du\} \wedge (s, a, s') \in \Delta^{\mathcal{S}}\}.$$

A play $\pi = p_0 p_1 p_2 \ldots$ of the delayed simulation game is a win for Duplicator iff there are infinitely many i such that the check mark occurs in the third component of p_i (every obligation is eventually satisfied), that is, the winning condition can be expressed as a Büchi condition and, of course, as a parity condition.

If Duplicator has a winning strategy in $\mathcal{G}^{de}(\mathcal{Q}, \mathcal{S})$, we write $\mathcal{Q} \le_{de} \mathcal{S}$ and say "$\mathcal{S}$ de-simulates $\mathcal{Q}$". By abuse of notation, when q and q' are states of the same automaton $\mathcal{Q}$, we write $q \le_{de} q'$ and say q' de-simulates q if $\mathcal{Q}[q] \le_{de} \mathcal{Q}[q']$ where $\mathcal{Q}[q]$ and $\mathcal{Q}[q']$ are obtained from $\mathcal{Q}$ by making q and q', respectively, the initial state.

3.2   Basic Properties of $\leq_{de}$

Delayed simulation for APA's has a number of useful and important properties, which we now describe. We start with some definitions.

We say that a relation $\leq$ between automata *implies language containment* if $\mathcal{Q} \leq \mathcal{S}$ implies $L(\mathcal{Q}) \subseteq L(\mathcal{S})$. This is one of the basic properties one expects of a simulation relation.

The *dual* of an APA $\mathcal{Q}$, denoted $\tilde{\mathcal{Q}}$, is obtained from $\mathcal{Q}$ by exchanging the roles of existential and universal states and replacing Ω by $\Omega + 1$. (Observe that $\tilde{\tilde{\mathcal{Q}}}$ is the same as $\mathcal{Q}$ modulo reducing all priorities by 2. Also, $L(\tilde{\mathcal{Q}}) = \Sigma^{\omega} \setminus L(\mathcal{Q})$.)

Theorem 1 (properties of delayed simulation)
1. *On APA's, the relation $\leq_{de}$*
   (a) is a preorder (reflexive and transitive) and
   (b) implies language containment.
2. *The relation $\leq_{de}$ can be computed in time $O(n^3 l^2 m)$ and space $O(mnl)$ on an APA with n states, m transitions, and l priorities.*
3. *For APA's $\mathcal{Q}$ and $\mathcal{S}$, we have $\mathcal{Q} \leq_{de} \mathcal{S}$ iff $\tilde{\mathcal{S}} \leq_{de} \tilde{\mathcal{Q}}$.*

The proof of the transitivity of $\leq_{de}$ is quite technical; it involves a notion of strategy composition, in analogy to what is explained in [20]. Since the delayed simulation game has a Büchi winning condition, it can be solved using the approach of [12], which gives the desired bound on the running time and space.

3.3   Quotienting is a Problem for $\leq_{de}$

Recall that a major motivation for studying simulation relations is state-space reduction, the basic idea being that states that simulate each other are merged and thus incurring a reduction in the number of states. This process is usually referred to as *quotienting*.

More precisely, let $\leq$ be a preorder on the state space of an automaton $\mathcal{Q}$ and $\equiv$ the corresponding equivalence relation defined by $q \equiv q'$ iff $q \leq q'$ and $q' \leq q$. Then the states of a quotient of $\mathcal{Q}$ with respect to $\leq$ are the equivalence classes of $\equiv$. But this does not fully determine a quotient. In addition, one has to specify: how the equivalence classes are connected by transitions, for any automaton; how the the set of equivalence classes is partitioned into universal and existential classes, for any alternating automaton; how priorities are assigned to equivalence classes, for any parity automaton. The overall objective is to define a quotient in such a way that it is simulation equivalent to the given automaton—we call this a *simulation preserving quotient*—and recognizes the same language. (Formally, two APA's $\mathcal{Q}$ and $\mathcal{S}$ are simulation equivalent if $\mathcal{Q} \leq \mathcal{S}$ and $\mathcal{S} \leq \mathcal{Q}$.)

Several quotients have been discussed in the literature for various types of automata and simulation relations, for instance, *naive quotients*, where every transition in the given automaton induces a transition in the quotient (representative-wise), *minimax quotients*, and *semi-elective quotients*, see, for instance, [20].

Unfortunately, it turns out to be quite difficult to find a working definition of a simulation preserving quotient with respect to delayed simulation. In fact, it is

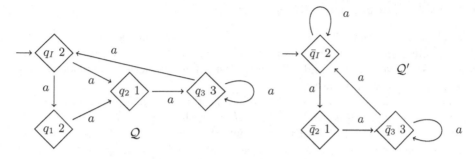

Fig. 1. An APA $\mathcal{Q}$ and a naive delayed simulation quotient $\mathcal{Q}'$

not at all clear how such a quotient should be defined, as we will argue in what follows. Note that in all the examples below not even the language is preserved.

For a start, consider the APA $\mathcal{Q}$ in Figure 1. As usual, existential states are shown as diamonds (there are no universal states). The labels of the states give the name and the priority of a state. It is easy to check that $q_I \equiv_{de} q_1 <_{de} q_2 <_{de} q_3$. The automaton $\mathcal{Q}'$ in Figure 1 is a naive quotient of $\mathcal{Q}$ where the equivalence class of a state q with respect to $\equiv$ is denoted $\bar{q}$.

Now observe that $L(\mathcal{Q})$ is empty, while $L(\mathcal{Q}') = \{a^\omega\}$. Further, note that $|\Sigma| = 1$ (the automaton is merely a parity game), all states have the same mode, only states in the same strongly connected component (SCC) are merged, and there are only three different priorities, which shows the situation is not complicated at all.

For $\mathcal{Q}$, one gets a simulation-equivalent quotient if the transition $(\bar{q}_I, a, \bar{q}_I)$ is removed from $\mathcal{S}$. This might suggest to use a minimax quotient as advocated in [20] for alternating Büchi automata and direct simulation. But note that, in general, minimax quotients do not even work for alternating Büchi automata and delayed simulation. In addition, the semi-elective quotient introduced in [20] for alternating Büchi automata and delayed simulation does not work for alternating parity automata, because in semi-elective quotients, transitions originating from existential states must not be removed, which means the semi-elective quotient of $\mathcal{Q}$ does not preserve the language.

The studies on alternating Büchi automata might suggest an approach in which existential states retain only their maximal transitions provided their priority is even. This would be correct for $\mathcal{Q}$, but we have other examples that exclude this approach as well, see [16].

4 Biased Delayed Simulations

We have just seen that our delayed simulation relation makes it difficult to merge equivalent states. As a remedy, we present two simulation relations, denoted $\leq^e_{de}$ and $\leq^o_{de}$, and corresponding quotienting constructions that are simulation preserving.

4.1   Definition and Basic Properties

The relations $\leq_{de}^e$ and $\leq_{de}^o$ are finer than $\leq_{de}$. They are defined just as $\leq_{de}$ with the only difference that the function γ which accumulates the pending obligations is replaced by variants, γ^e and γ^o, respectively. These functions coincide with γ except for the following cases:

- if $j \preceq i$, i odd, $i \leq k$, and (j odd or $k < j$), then $\gamma(i,j,k)^e = k$, and
- if $j \preceq i$, j even, $j \leq k$, and (i even or $k < i$), then $\gamma(i,j,k)^o = k$.

That is, in the case of $\leq_{de}^e$ (e reminiscent of "even") , once the value of the priority memory is not $\sqrt{}$, it will change back to the value $\sqrt{}$ only if this is triggered by a small even priority in the simulating automaton, while small odd priorities in the simulated automaton are ignored.

We call $\leq_{de}^e$ and $\leq_{de}^o$ the *even-biased delayed simulation relation* and *odd-biased delayed simulation relation*, respectively.

These two new relations have all basic properties of a simulation relation:

Theorem 2 (properties of biased delayed simulations)
1. *On APA's, the simulation relations $\leq_{de}^e$ and $\leq_{de}^o$*
   (a) are preorders (reflexive and transitive),
   (b) imply language containment, and
   (c) are at least as fine as $\leq_{de}$.
2. *The relations $\leq_{de}^e$ and $\leq_{de}^o$ can be computed in time $O(n^3 l^2 m)$ and space $O(mnl)$ on an APA with n states, m transitions, and l priorities.*
3. *For APA's $\mathcal{Q}$ and $\mathcal{S}$, we have $\mathcal{Q} \leq_{de}^e \mathcal{S}$ iff $\tilde{\mathcal{S}} \leq_{de}^o \tilde{\mathcal{Q}}$.*

The proofs are similar to the proofs of the assertions in Theorem 1, but in some places even more complicated because of the asymmetry in the two definitions.

We note that, in general, $\leq_{de}^e$ and $\leq_{de}^o$ are strictly finer than $\leq_{de}$, and even the reflexive-transitive closure of $\leq_{de}^l \cup \leq_{de}^r$ is strictly finer than $\leq_{de}$.

4.2   Quotienting

We next present the quotients which can be used with the biased delayed simulation relations; they are enhanced versions of the semi-elective quotient introduced in [20] for alternating Büchi automata.

Let $\mathcal{Q}$ be an APA as in (2), $\leq$ a preorder on its state space, and $\equiv$ the corresponding equivalence relation. The *min semi-elective quotient* of $\mathcal{Q}$ with respect to $\leq$ is

$$\mathcal{Q}_{min}^{se} = (Q/{\equiv}, \Sigma, \bar{q}_I, \Delta^{\min}, E^{\min}, U^{\min}, \Omega^{\min}) \tag{5}$$

defined by

$$\Delta^{\min} = \{(\bar{q}, a, \bar{q}') \mid (q, a, q') \in \Delta, q \in E\} \cup \{(\bar{q}, a, \bar{q}') \mid \bar{q} \subseteq U \wedge q' \in \min_a(q)\} \ ,$$

$$U^{\min} = \{\bar{q} \mid \bar{q} \subseteq U\} \ , \qquad E^{\min} = \{\bar{q} \mid \bar{q} \cap E \neq \emptyset\} \ ,$$

$$\Omega^{\min} : \bar{q} \mapsto \min\{\Omega(q') \mid q' \in \bar{q}\} \ ,$$

where $\min_a(q) = \{q' \mid (q, a, q') \in \Delta \land \forall q''((q, a, q'') \in \Delta \land q'' \leq q' \to q' \leq q'')\}$, that is, $\min_a(q)$ denotes the set of minimum a-successors of q. The *max semi-elective quotient* is defined in the same way with the only exceptions that min is replaced by max and the roles of U and E are exchanged.

We next apply these quotients to our biased delayed simulations. Given an APA $\mathcal{Q}$, the *even semi-elective quotient* of $\mathcal{Q}$ is the min semi-elective quotient with respect to $\leq_{de}^e$, and the *odd semi-elective quotient* of $\mathcal{Q}$ is the max semi-elective quotient with respect to $\leq_{de}^o$, denoted $\mathcal{Q}_{de}^e$ and $\mathcal{Q}_{de}^o$, respectively.

Theorem 3 (even and odd semi-elective quotients) *For every APA $\mathcal{Q}$, the even and odd semi-elective quotients of $\mathcal{Q}$ are simulation preserving, in particular, $L(\mathcal{Q}) = L(\mathcal{Q}_{de}^e) = L(\mathcal{Q}_{de}^o)$.*

The proof of this theorem is similar to the proof of the correctness of other quotients, but, technically, it is more involved.

We mention that, in general, $\mathcal{Q} \not\equiv_{de}^o \mathcal{Q}_{de}^e$ and $\mathcal{Q} \not\equiv_{de}^e \mathcal{Q}_{de}^o$. Therefore, there is no obvious way to combine the two quotients or to perform one after the other without recomputing one of the biased simulation relations.

5 A Simulation-Based Simplification Algorithm

We conclude by describing a heuristic for reducing the number of states of a given APA. There are three parts to this heuristic: the quotienting procedures from the previous section; simplification methods based on the general delayed simulation relation from Sect. 3; basic simplification methods.

5.1 Basic Simplification Methods

There are two basic simplification methods. To describe them, let $\mathcal{Q}$ be an APA as in (2) and let $q \rightsquigarrow q'$ denote that there is a path from q to q' (with any labeling).

Reachability reduction. Remove all states q where $q_I \not\rightsquigarrow q$.

Normalization. Repeatedly redefine Ω by $\Omega(q) := \Omega(q) - 2$ for a state q such that $\Omega(q) \geq 2$ and there is no state q' with $\Omega(q') = \Omega(q) - 1$ in the same SCC.

Clearly, these methods are correct, that is, they do not change the language of the given APA. Also, they can be implemented quite efficiently.

5.2 Delayed Simplifications

Next, we describe simplification methods based on delayed simulation. Let $\bar{q}$ be the equivalence class of a state q with respect to the equivalence relation $\equiv_{de}$ corresponding to $\leq_{de}$.

Homogenization. Redefine Ω by $\Omega(q) = \min\{\Omega(q') \mid q \equiv_{de} q'\}$.

Stretching. For every equivalence class C of $\equiv_{de}$ choose $\mu(C) \in C$ such that there exists no $q' \in C$ satisfying $\mu(C) \rightsquigarrow q' \not\rightsquigarrow \mu(C)$, that is, a maximum representative with respect to reachability. Replace every transition (q, a, q') by $(q, a, \mu(\bar{q}'))$.

0-1-minimaxing. For every $q \in E$ with $\Omega(q) = 0$, remove every transition (q, a, q') where $q' \notin \max_a(q)$. Symmetrically, for every $q \in U$ with $\Omega(q) = 1$, remove every transition (q, a, q') where $q' \notin \min_a(q)$.

In addition, repeatedly remove a transition (q, a, q') with $q \in E$ if there exists $q'' \neq q'$ such that (q, a, q'') is a transition, $q' \leq_{de} q''$, and $\Omega(q'') = 0$. Symmetrically, repeatedly remove a transition (q, a, q') with $q \in U$ if there exists $q'' \neq q'$ such that (q, a, q'') is a transition, $q' \geq_{de} q''$ and $\Omega(q'') = 1$.

Reachability minimaxing. For every $q \in E$ remove a transition (q, a, q') if there exists a transition (q, a, q'') such that $q' \leq_{de} q''$ and $q'' \not\rightsquigarrow q$. Symmetrically, for every $q \in U$ remove a transition (q, a, q') if there exists a transition (q, a, q'') such that $q' \geq_{de} q''$ and $q'' \not\rightsquigarrow q$.

We prove that all these methods are correct. It is easy to see that they can be implemented quite efficiently, once $\leq_{de}$ has been computed.

5.3   Heuristic for State-Space Reduction

The simplification methods described above can be combined in many different reasonable ways to reduce the number of states of a given APA, but they only allow to remove states and edges. For a good state-space reduction heuristic, we need to combine them with quotienting as described in the previous section. One reasonable way to do this, which has proved to be useful, is the following:

State-Space Reduction Heuristic SSRH for an APA $\mathcal{Q}$

1. Choose a cut-off threshold $t > 0$.
2. Normalize the APA.
3. Compute $\leq_{de}$ and perform the delayed simplifications from Subsection 5.2 in the same order as described. Delete unreachable states whenever possible (reachability reduction).
4. Compute $\leq_{de}^e$ and $\leq_{de}^o$ and pass to the even or the odd semi-elective quotient, whichever results in fewer states. (See remark at the end of Sect. 4.)
5. Let $t := t - 1$. If $t > 0$ and the number of states has been reduced, then go to (2), that is, start all over again, else stop.

Parity games are APA's over a unary alphabet. Thus they are useful for testing how well SSRH performs. In general, one cannot expect that SSRH, even when restricted to parity games, will give perfect results, because the computational complexity of determining the winner in a parity game is still unknown. The best one knows is that it belongs to **UP** $\cap$ **co-UP**.

We have studied how SSRH performs when applied to the Jurdziński family of parity games, $\{H_{m,n}\}_{m,n}$, which have proved to be hard for Jurdziński's algorithm for solving parity games, see [15].

Our algorithm performs well on this family:

Theorem 4 *For every choice of m and n and for $t = 2$, on input $H_{m,n}$, SSRH yields a parity game with two positions in a number of steps polynomial in the input size.*

6   Conclusion

We have shown how the concept of delayed simulation can be adapted to alternating parity automata, that it is hardly useful for state-space reduction via quotienting, but that it can also be modified appropriately in order to arrive at a useful state-space reduction heuristic.

One interesting question that we do not know how to answer is whether there is an intermediate relation $\leq$ such that $(\leq_{de}^e \cup \leq_{de}^o) \subseteq \leq \subseteq \leq_{de}$ and where quotienting is simulation preserving. Also, it would be useful to have a rigorous proof that quotienting with respect to $\leq_{de}$ does not work; one would have to come up with an APA Q such that no APA with fewer states accepts the same language, but where two different states simulate each other.

References

1. Milner, R.: An algebraic definition of simulation between programs. In Cooper, D.C., ed.: Proc. 2nd Internat. Joint Conf. on Artificial Intelligence, London, UK, William Kaufmann (1971) 481–489
2. Etessami, K., Holzmann, G.: Optimizing Büchi automata. In Palamidessi, C., ed.: 11th Int. Conf. on Concurrency Theory (CONCUR 2000), University Park, PA, USA. Vol. 1877 of LNCS, Springer, Berlin (2000) 153–167
3. Etessami, K.: (Temporal massage parlor) available at http://www.bell-labs.com/project/TMP/.
4. Somenzi, F., Bloem, R.: Efficient Büchi automata from LTL formulae. In Emerson, E.A., Sistla, A.P., eds.: Computer Aided Verification, 12th Internat. Conf., CAV 2000, Chicago, IL, USA. Vol. 1855 of LNCS, Springer, Berlin (2000) 248–263
5. Gurumurthy, S., Bloem, R., Somenzi, F.: Fair simulation minimization. In Brinksma, E., Guldstrand Larsen, K., eds.: Computer Aided Verification, 14th Internat. Conf., CAV 2002, Copenhagen, Denmark. Vol. 2404 of LNCS, Springer, Berlin (2002) 610–623
6. Bloem, R.: (Wring: an LTL to Buechi translator) available at http://www.ist.tugraz.at/staff/bloem/wring.html.
7. Fritz, C.: Constructing Büchi automata from linear temporal logic using simulation relations for alternating Büchi automata. In Ibarra, O.H., Dang, Z., eds.: Implementation and Application of Automata, 8th Internat. Conf., CIAA 2003, Santa Barbara, CA, USA. Vol. 2759 of LNCS, Springer, Berlin (2003) 35–48
8. Fritz, C., Teegen, B.: (LTL → NBA (improved version)) available at http://www.ti.informatik.uni-kiel.de/~fritz/ABA-Simulation/ltl.cgi.
9. Fritz, C.: Concepts of automata construction from LTL. In Sutcliffe, G., Voronkov, A., eds.: LPAR. Vol. 3835 of LNCS, Springer (2005) 728–742
10. Mostowski, A.W.: Regular expressions for infinite trees and a standard form of automata. In: Computation Theory. Vol. 208 of LNCS. Springer (1984) 157–168
11. Emerson, E.A., Jutla, C.S.: Tree automata, mu-calculus and determinacy (extended abstract). In: Proc. 32nd Ann. Symp. on Foundations of Computer Science (FoCS '91), San Juan, Puerto Rico, IEEE Computer Society Press (1991) 368–377
12. Etessami, K., Wilke, Th., Schuller, R.A.: Fair simulation relations, parity games, and state space reduction for büchi automata. In Orejas, F., Spirakis, P.G., van Leeuwen, J., eds.: ICALP. Vol. 2076 of LNCS, Springer (2001) 694–707

13. Etessami, K., Wilke, Th., Schuller, R.A.: Fair simulation relations, parity games, and state space reduction for Büchi automata. SIAM J. Comput. **34**(5) (2005) 1159–1175
14. Henzinger, T.A., Rajamani, S.K.: Fair bisimulation. In Graf, S., Schwartzbach, M.I., eds.: TACAS. Vol. 1785 of LNCS, Springer (2000) 299–314
15. Jurdziński, M.: Small progress measures for solving parity games. In Reichel, H., Tison, S., eds.: STACS 2000, 17th Ann. Symp. on Theoretical Aspects of Computer Science, Lille, France. Vol. 1770 of LNCS, Springer, Berlin (2000) 290–301
16. Fritz, C.: Simulation-Based Simplification of omega-Automata. PhD thesis, Technische Fakultät der Christian-Albrechts-Universität zu Kiel (2005) available at `http://e-diss.uni-kiel.de/diss_1644/`.
17. Grädel, E., Thomas, W., Thomas Wilke, eds.: Automata, Logics, and Infinite Games: A Guide to Current Research [outcome of a Dagstuhl seminar, February 2001]. In Grädel, E., Thomas, W., Thomas Wilke, eds.: Automata, Logics, and Infinite Games. Vol. 2500 of LNCS, Springer (2002)
18. Gurevich, Y., Harrington, L.: Trees, automata, and games. In: 14th ACM Symp. on the Theory of Computing, San Francisco, CA, USA, ACM Press (1982) 60–65
19. Vöge, J., Jurdziński, M.: A discrete strategy improvement algorithm for solving parity games. In Emerson, E.A., Sistla, A.P., eds.: CAV. Vol. 1855 of LNCS, Springer (2000) 202–215
20. Fritz, C., Wilke, Th.: Simulation relations for alternating Büchi automata. Theoretical Computer Science **338**(1–3) (2005) 275–314
21. Calude, C., Calude, E., Khoussainov, B.: Finite nondeterministic automata: Simulation and minimality. Theor. Comput. Sci. **242**(1-2) (2000) 219–235

Equivalence of Functions Represented by Simple Context-Free Grammars with Output

Cédric Bastien[1], Jurek Czyzowicz[1], Wojciech Fraczak[1,2], and Wojciech Rytter[3]

[1] Dépt d'informatique, Université du Québec en Outaouais, Gatineau PQ, Canada
[2] IDT Canada Incorporation, Ottawa ON, Canada
[3] Institute of Informatics, Warsaw University, Warsaw, Poland

Abstract. A partial function $F : \Sigma^* \to \Omega^*$ is called a *simple* function if $F(w) \in \Omega^*$ is the output produced in the generation of a word $w \in \Sigma^*$ from a nonterminal of a simple context free grammar G with output alphabet Ω. In this paper we present an efficient algorithm for testing equivalence of simple functions. Such functions correspond also to one-state deterministic pushdown transducers. Our algorithm works in time polynomial with respect to $|G| + v(G)$, where $|G|$ is the size of the textual description of G, and $v(G)$ is the maximum of the shortest lengths of words generated by nonterminals of G.

1 Introduction

The decidability problem of equivalence for functions defined by different classes of deterministic push-down automata and pushdown transducers (dpdt) was studied extensively, see for example [8, 11], leading eventually to a proof of the decidability of the equivalence problem for deterministic pushdown transducers. The main issue was decidability, and little was said about the effective algorithms for the equivalence of pushdown transducers.

In this paper we present an efficient and easy to implement algorithm for deciding the equivalence of simple functions, i.e., functions defined by one-state dpdts. Simple functions were initially defined in [5] as the semantic domain of a network packet classification engine developed at IDT Canada, Inc. Simple functions are applied by IDT Canada to perform packet classification at wire speed. Classification policies are described with the aid of a class of context free grammars and implemented as so called *Concatenation State Machines*, a hardware implementation of single-state dpdts. In order to manage large sets of those classification policies in memory, it is useful to be able to identify if two classification policies are semantically the same. This was our motivation to investigate the problem of the equivalence of simple functions from a practical point of view and to develop an efficient and easy to implement algorithm for this task. The algorithm we propose in this paper is a nontrivial extension of the simple languages equivalence algorithms from [2, 7, 1] to the case of simple functions.

Simple functions can be seen as a proper extension of sequential functions (functions realized by deterministic finite transducers) and simple languages.

O.H. Ibarra and Z. Dang (Eds.): DLT 2006, LNCS 4036, pp. 71–82, 2006.
© Springer-Verlag Berlin Heidelberg 2006

Simple languages were introduced in [9] as languages recognized by a dpda with a single state, also called *simple dpda*, or, equivalently, as languages generated by *simple grammars*. We extend the definition of simple grammars to functions defined by grammars with output.

A **simple function grammar** is formally described by a 4-tuple:

$$G = (\Sigma, \Omega, N, P),$$

where Σ, Ω, N are disjoint sets of *input symbols*, *output symbols*, and *nonterminals*, and $P \subset N \times \Sigma \times (N \cup \Omega)^*$ is a finite set of *production rules* with output. Moreover, we require that for given $X \in N$ and $a \in \Sigma$ there is at most one $\alpha \in (N \cup \Omega)^*$, such that $(X, a, \alpha) \in P$.

Each production can be written as $A \to s\alpha$, where $\alpha \in (N \cup \Omega)^*$. We also write $A \xrightarrow{s} \alpha$. The relation $\xrightarrow{s}$ is extended in the following way.

We write $\alpha_1 \xrightarrow{s} \alpha_2$, iff $\alpha_1 = \beta_1 A \beta_2$, $\beta_1 \in \Omega^*$, $A \in N$, $\alpha_2 = \beta_1 \gamma \beta_2$ and $A \to s\,\gamma$ is a production. Intuitively, relation $\alpha_1 \xrightarrow{s} \alpha_2$ corresponds to a single-step leftmost derivation.

For $w = s_1 s_2 \ldots s_n$ and $\alpha_i \in (N \cup \Omega)^*$ we write $\alpha_0 \xrightarrow{w} \alpha_n$ iff

$$\alpha_0 \xrightarrow{s_1} \alpha_1, \ \alpha_1 \xrightarrow{s_2} \alpha_2, \ \alpha_2 \xrightarrow{s_3} \alpha_3, \ \ldots \alpha_{n-1} \xrightarrow{s_n} \alpha_n.$$

For $w \in \Sigma^*$ we write:

$$\beta = \mathtt{Derived}(\alpha, w) \ \Leftrightarrow \ \alpha \xrightarrow{w} \beta, \text{ where } \beta \in (N \cup \Omega)^*.$$

If there is no derivation $\alpha \xrightarrow{w} u$ for any u then we write $\mathtt{Derived}(\alpha, w) = \bot$. The input-output relation corresponding to a sequence $\alpha \in (N \cup \Omega)^*$ is defined in the following way:

$$F_G(\alpha) \overset{\mathtt{def}}{=} \{ \ (w, u) \in \Sigma^* \times \Omega^* \mid u = \mathtt{Derived}(\alpha, w), \ u \neq \bot \ \}.$$

A relation $F_G(\alpha)$, for any given $\alpha \in (\Omega \cup N)^*$, over input and output strings, which can be defined by a simple function grammar, is called a **simple function**. We use also function terminology, i.e., $F_G(\alpha)(w) = u$ iff $(w, u) \in F_G(\alpha)$.

The domain of a simple function is a simple language, i.e., if Ω is empty then G is just a simple grammar.

We define the **simple function equivalence problem** as follows.

Input
    a simple function grammar G and two nonterminals $A, B \in N$;

Output
    SUCCESS if $F_G(A) = F_G(B)$, and FAILURE otherwise.

Example 1. Let us consider the simple grammar with output $G = (\{0, 1\}, \{a, b\}, \{S_1, S_2, A_1, A_2\}, P)$, where P is given by rules:

$$S_1 \to 0aS_1A_1b, \quad S_1 \to 1, \quad A_1 \to 1, \quad S_2 \to 0aS_2A_2, \quad S_2 \to 1, \quad A_2 \to 1b$$

and consider the equivalence problem $"F_G(S_1) = F_G(S_2)$?". We have SUCCESS since $F_G(S_1) = F_G(S_2)$. For w which are not of the form 0^n1^{n+1}, $F_G(S_1)(w)$ and $F_G(S_2)(w)$ are both undefined. Otherwise, we have:

$$F_G(S_1)(0^n1^{n+1}) = F_G(S_2)(0^n1^{n+1}) = a^nb^n.$$

Let $\alpha \in (N \cup \Omega)^*$. By $||\alpha||$ we denote the *shortest-word complexity* of α defined as the length of a shortest $w \in \Sigma^*$ such that $F_G(\alpha)(w)$ is defined. The *shortest-word complexity* of grammar G is defined as $v(G) \stackrel{\text{def}}{=} \max\{||A|| \mid A \in N \}$. $|G|$ denotes the size of the textual description of G.

Our main result is the constructive proof of the following theorem.

Theorem 1. *Assume A, B are two nonterminals of a simple grammar G with output. Then we can test if $F_G(A) = F_G(B)$ in time polynomial with respect to $|G| + v(G)$.*

2 Free Group over Ω and Properties of Simple Functions

In the course of the algorithm we consider, as intermediate data, output sequences which are to be compensated later. For example we could know that the output for A is the same as for B, except for a prefix u that must be cut off from every output for B. Then, we formally write $uA = B$, or equivalently $A = u^{-1}B$. This motivates the introduction of the free group $\Omega^\otimes$ over the output alphabet Ω. The concept of this group and the operation Derived of taking a syntactic remainder are among our basic tools. First we introduce some basic properties and definitions related to the free group over Ω.

By ε we denote an empty sequence. Simple functions together with concatenation defined by $fg \stackrel{\text{def}}{=} \{(x_1x_2, y_1y_2) \mid (x_1, y_1) \in f, (x_2, y_2) \in g\}$, constitute a monoid with $\{(\varepsilon, \varepsilon)\}$ acting as unit[1] and with $\bot \stackrel{\text{def}}{=} \{\}$ acting as zero. More details about simple functions seen as a monoid can be found in [5]. We will write w and u instead of $\{(w, \varepsilon)\}$ and $\{(\varepsilon, u)\}$, with $w \in \Sigma^*$ and $u \in \Omega^*$, respectively. In particular, the unit function $\{(\varepsilon, \varepsilon)\}$ will be denoted by ε.

Let f, g be simple functions. By $g^{-1}f$ we will denote the unique, if it exists, simple function h such that $f = gh$.

As mentioned above, for technical reasons we extend the image of simple functions to the free group generated by Ω, denoted by $\Omega^\otimes$, so $w^{-1}f$ would be such that $w(w^{-1}f) = f$, for all $w \in \Omega^\otimes$. More precisely, $\Omega^\otimes \stackrel{\text{def}}{=} (\Omega \cup \overline{\Omega})^*_{/\{a\bar{a}=\bar{a}a=\varepsilon|a\in\Omega\}}$, where $\overline{\Omega}$ is a copy of Ω with bijection $^{-} : \Omega \mapsto \overline{\Omega}$ playing the role of the inverse. Therefore, apart from monoid properties, i.e., $x(yz) = (xy)z$, $x\varepsilon = \varepsilon x = x$, we have $a\bar{a} = \bar{a}a = \varepsilon$, for every $a \in \Omega$. For example, $(ab\bar{c})^{-1} = c\bar{b}\bar{a}$, or $bc(abc)^{-1} = \bar{a}$.

Given two strings u, v over $\Omega \cup \overline{\Omega}$, we write $u = v$ to say that they are equivalent in $\Omega^\otimes$. When we want to underline that u and v are identical as strings, we write $u \equiv v$. A usual way of representing an element of $\Omega^\otimes$, i.e., an

---

[1] The unit simple function corresponds to $F_G(\varepsilon)$.

equivalence class over $(\Omega \cup \overline{\Omega})^*$, is to choose the shortest string from the class (such a word does not contain subwords $a\overline{a}$ or $\overline{a}a$, for any $a \in \Omega$). Given a string $u \in (\Omega \cup \overline{\Omega})^*$, by $\texttt{reduce}(u)$ we denote the shortest string over $\Omega \cup \overline{\Omega}$, such that $\texttt{reduce}(u) = u$.

If $\texttt{reduce}(u) \equiv u$ then u is called *reduced*. The reduced form can be easily computed in linear time with respect to $|u|$. We say that $u \in \Omega^{\otimes}$ is not *primitive* if there is an $x \in \Omega^{\otimes}$ and $k > 1$ such that $u = x^k$; otherwise u is *primitive*. For every u there exists a unique primitive $x \in \Omega^{\otimes}$, denoted $\texttt{root}(u)$, and a $k > 0$, denoted $\texttt{power}(u)$, such that $u = x^k$.

Lemma 1. *Let $u \in (\Omega \cup \overline{\Omega})^*$. There is an algorithm for calculating $\texttt{power}(u)$ and $\texttt{root}(u)$ running in $O(n)$, where $n = |u|$.*

Proof. We assume that u is given in the reduced form. Let u be written as $u_1 u_2 u_1^{-1}$, where u_1 is the maximal length prefix of u such that its inverse is a suffix of u. The value of u_1 can be easily computed in linear time. Note that u is a power of a primitive word x iff u_2 is a power of a primitive word y such that $x = u_1 y u_1^{-1}$. By the choice of u_2, the first and the last symbols of y are not inverse of each other, therefore $\texttt{reduce}(y^k) \equiv y^k$ for all $k \geq 1$ and u_2 can be treated as a word in a free monoid generated by the alphabet $(\Omega \cup \overline{\Omega})$. In this context, computing $\texttt{power}(u_2)$ and $\texttt{root}(u_2)$ can be done by finding the occurrences of u_2 in the word $u_2 u_2$ using any linear time pattern matching algorithm (see [4] for details), from which we can deduce the values of $\texttt{power}(u)$ and $\texttt{root}(u)$. □

Lemma 2. *Let $X \subseteq \Omega^*$, $r_1, r_2 \in \Omega^{\otimes}$, such that $|X| \geq 2$, $r_1 X = X r_2$, and r_1, r_2 being primitive. For every $u, v \in \Omega^{\otimes}$, $uX = Xv$ iff $\texttt{power}(u) = \texttt{power}(v)$ and $(\texttt{root}(u), \texttt{root}(v)) \in \{(r_1, r_2), (r_1^{-1}, r_2^{-1})\}$.*

Proof. Firstly, we show that in $\Omega^{\otimes}$, $uw = wv$ iff $u = st$, $v = ts$ and $w = (st)^k s$ for some $s, t \in \Omega^{\otimes}$ and $k \in \mathbb{Z}$. Furthermore, if u and v are primitives, then s and t are unique. Using these facts, it is straightforward to prove the *if* part of the lemma.

Secondly, we observe that if $uX = Xv$ then $\texttt{root}(u)X = X\texttt{root}(v)$ and $\texttt{power}(u) = \texttt{power}(v)$. We elaborate a deterministic method of finding primitive words s and t from two words $w_1, w_2 \in X$ such that $r_1 = st, r_2 = ts$ and $w_i = (st)^{k_i} s$, for $i \in 1, 2$ and $k_i \in \mathbb{Z}$. Hence, there exists exactly one (modulo inverse) pair of primitive words r_1 and r_2 such that $r_1 X = X r_2$. □

Example 2. Note that, if $|X| = 1$ then the lemma is not true. E.g., for $X = \{\varepsilon\}$, $uX = Xu$ for all $u \in \Omega^{\otimes}$.

From this point on, we extend the definition of the output alphabet to $\Omega \cup \overline{\Omega}$ and we will assume $G = (\Sigma, \Omega \cup \overline{\Omega}, N, P)$ is a simple grammar with output.

We distinguish two types of sequences α over $N \cup \Omega \cup \overline{\Omega}$: α is of *output* type if $\alpha \in (\Omega \cup \overline{\Omega})^*$; and α is of *general* type, when α is of form $uA\alpha'$, for some $u \in (\Omega \cup \overline{\Omega})^*$, $A \in N$, and $\alpha' \in (N \cup \Omega \cup \overline{\Omega})^*$. In the case of general type of α, we refer to u, A, and α' by $\texttt{OutPref}(\alpha)$, $\texttt{First}(\alpha)$, and $\texttt{Tail}(\alpha)$, respectively.

Example 3. Let $\alpha = baabaAbaCAb$, where $a, b \in \Omega$, then $\texttt{First}(\alpha) = A$, $\texttt{OutPref}(\alpha) = baaba$, and $\texttt{Tail}(\alpha) = baCAb$.

For every simple function $F_G(A)$, denoted by $\min F_G(A)$ the unique element $(w, u) \in F_G(A)$ such that $w \in \Sigma^*$ is the shortest and lexicographically smallest input word generated by G from A. For every nonterminal $A \in N$ we can precompute $(w_A, u_A) \stackrel{\text{def}}{=} \min F_G(A)$ in time polynomial with respect to $||A||$.

Moreover, we compute the set $\texttt{SingleOut}(G) \subseteq N$ of all non-terminals, each of them producing only one output string, i.e.,

$$\texttt{SingleOut}(G) \stackrel{\text{def}}{=} \{A \in N \mid \forall(x, y), (x', y') \in F_G(A), \ y = y'\}.$$

Lemma 3. *We can calculate $\mathit{SingleOut}(G)$ in time $O(|G| + v(G))$.*

Proof. We assume G to be reduced. Associate with each rule $A \to a\alpha$ an integer value $n_{A \to a\alpha}$, initialized to the number of occurrences of different nonterminals which are in α, and associate also a boolean flag which can take either the value *marked* or *unmarked*, initially set to *unmarked*. Furthermore, assume that we have for each $A \in N$ a reference to all the rules $B \to a\alpha$ such that $\alpha = \alpha_1 A \alpha_2$, for some α_1, α_2. We also associate a word w_A to each A, initially set to *nil*, and a boolean flag specifying whether or not we have found that A produces more than one output. This information can be precomputed in $O(|G|)$.

Following this preprocessing, we can find for each nonterminal whether it generates more than one output word by iterating the following procedure:

> Consider all the *unmarked* rules $A \to a\alpha \in P$ such that $n_{A \to a\alpha} = 0$. If there is no such rule, terminate. Otherwise, set each such rule as *marked*. Then, for each of these rules, we know that $w_B \neq$ *nil* for every nonterminal B in α, and we can easily compute a word w' for A.
> If A already has an output word $w_A \neq w'$ associated with it, we have found that A generates more than one output word and we can mark it as such. We also recursively propagate this information to any nonterminal B such that $B \to a\alpha$, with $\alpha = \alpha_1 A \alpha_2$ (unless B is already marked as such), since B must necessarily generate at least 2 different output words.
> Otherwise, if $w_A =$ *nil*, set $w_A := w'$ and subtract 1 from $n_{B \to a\alpha}$.

This procedure takes time $O(|G| + v(G))$ since we consider an occurrence of a nonterminal in a rule a constant number of times. □

Proposition 1. *Let $G = (\Sigma, \Omega \cup \overline{\Omega}, N, P)$ be a simple function grammar, α, α', β, $\beta' \in (N \cup \Omega \cup \overline{\Omega})^*$ such that $||\alpha|| \leq ||\beta||$, $A \in N$, and $u, v \in (\Omega \cup \overline{\Omega})^*$.*

1. *$F_G(\alpha) = F_G(\beta)$ iff $(\alpha, \beta \in (\Omega \cup \overline{\Omega})^*$ and $\alpha = \beta)$ or $\forall a \in \Sigma \ F_G(\texttt{Derived}(\alpha, a)) = F_G(\texttt{Derived}(\beta, a))$*
2. *$F_G(\alpha \alpha') = F_G(\beta \beta')$ iff $F_G(\alpha \gamma) = F_G(\beta)$ and $F_G(\alpha') = F_G(\gamma \beta')$, where $(w_\alpha, u_\alpha) = \min F_G(\alpha)$ and $\gamma = u_\alpha^{-1} \texttt{Derived}(\beta, w_\alpha)$.*

Proof. The first "if and only if" statement is obvious. The second statement follows from the fact that the monoid of simple functions is cancellative. The following cases are possible:

1. γ is such that $F_G(\alpha\gamma) = F_G(\beta)$, i.e., $F_G(\gamma) = (F_G(\alpha))^{-1}F_G(\beta)$. In this case the "if and only if" statement is straightforward.
2. If $F_G(\alpha\gamma) \neq F_G(\beta)$ then $(F_G(\alpha))^{-1}F_G(\beta)$ is not defined, and thus, assuming $||\alpha|| \leq ||\beta||$, $F_G(\alpha\alpha')$ cannot be equal to $F_G(\beta\beta')$. □

Corollary 1. *Let G be a simple function grammar, $\alpha, \beta \in (N \cup \Omega \cup \overline{\Omega})^*$, $A \in N$, and $u, v \in (\Omega \cup \overline{\Omega})^*$.*

$$F_G(uA\alpha) = F_G(vA\beta) \quad \text{iff} \quad F_G(\alpha) = F_G(\gamma\beta) \text{ and } F_G(uA\gamma) = F_G(vA),$$

where $(w_A, u_A) = \min F_G(A)$ and $\gamma = (uu_A)^{-1}\text{Derived}(vA, w_A) = u_A^{-1}u^{-1}vu_A$. Notice that $A \in \text{SingleOut}(G)$ implies $F_G(uA\gamma) = F_G(vA)$.

3   Equivalence Algorithm

The algorithm EQUIVALENCE which checks for the equality of A and B, consists of constructing a relation $\mathcal{R} \subset (N \cup \Omega \cup \overline{\Omega})^+ \times (N \cup \Omega \cup \overline{\Omega})^+$, which implies $F_G(A) = F_G(B)$. In terminology of [3], $\mathcal{R}$ would be called *self-proving relation*. In our case $\mathcal{R}$ will consist of two relations $\mathcal{D}$, called *decomposition* relation, and $\mathcal{C}$, called *conjugation* relation.

Let $\leq$ be a total order over nonterminals verifying $A \leq B \Rightarrow ||A|| \leq ||B||$.

Decomposition Relation and *Unfolding*. Decomposition relation is a partial mapping $\mathcal{D} : N \to (N \cup \Omega \cup \overline{\Omega})^+$ such that $\mathcal{D}(A) \in (\{X \in N \mid X < A\} \cup \Omega \cup \overline{\Omega})^+$, i.e., $\mathcal{D}(A)$ contains only nonterminals smaller than A. By $\mathcal{D}^*(\beta)$ we denote the *complete unfolding* of β. This means that if $(A, \alpha) \in \mathcal{D}$ then A is replaced in β by α, such an operation is iterated until the resulting string β stabilizes. e.g., if $\mathcal{D} = \{(A, BcB), (B, Cb)\}$ with $A, B, C \in N$ and $a, b, c \in (\Omega \cup \overline{\Omega})$, then $\mathcal{D}^*(aAA) = aCbcCbCbcCb$.

Conjugation Relation. The relation $\mathcal{C}$ contains *conjugation equations* of form $r_1A = Ar_2$, where $A \in N$ and $r_1, r_2 \in \Omega^{\otimes}$. In $\mathcal{C}$ we will keep only reduced nontrivial conjugation equations, i.e., we will assume that the nonterminal A present in the equation generates at least two different elements, $A \notin \text{SingleOut}(G)$, and that r_1 and r_2 are primitive and reduced. By Lemma 2, it is enough to keep in $\mathcal{C}$ only one such conjugation equation per A. Hence, the size of conjugation relation $|\mathcal{C}|$ is bounded by $|N|$.

Description of the Algorithm. The algorithm is presented in Fig. 1. Intuitively, the algorithm constructs $\mathcal{R} = \mathcal{C} \cup \mathcal{D}$, maintaining a list Q of equations on sequences over $N \cup \Omega \cup \overline{\Omega}$, called *targets*. The targets are processed within a while-loop one by one until the set Q becomes empty, which is equivalent to

Algorithm EQUIVALENCE(A, B); $\{A, B \in N;\}$
$\{$*the algorithm returns* SUCCESS *iff* $F_G(A) = F_G(B)\}$
$\quad Q := \{(A, B)\}; \mathcal{C} := \emptyset; \mathcal{D} := \emptyset;$

while Q is not empty **do:**
$\quad (\alpha_1, \alpha_2) := delete(Q);$
$\quad$ **If** $\alpha_1 = \bot$ and $\alpha_2 = \bot$ **then** start the next iteration.
$\quad$ **If** $\alpha_1 = \bot$ or $\alpha_2 = \bot$ **then return** FAILURE.

$\quad \alpha_1 \leftarrow \mathcal{D}^*(\alpha_1), \alpha_2 \leftarrow \mathcal{D}^*(\alpha_2) \quad$ — *Unfolding* α_1 *and* α_2 *by* $\mathcal{D}$.
$\quad$ Simplify (α_1, α_2) by eliminating the common prefix.
$\quad$ **If** $\alpha_1 = \alpha_2 = \varepsilon$ **then** start the next iteration. $\hfill (1)$
$\quad$ **If** α_1 or α_2 is of *output* type **then return** FAILURE. $\hfill (2)$

$\qquad\qquad$ — **Comment**: *At this stage* α_1 *and* α_2 *are of general type, i.e.,*
$\qquad\qquad \alpha_1 = u_1 A \alpha_1'$ *and* $\alpha_2 = u_2 B \alpha_2'$, *and they differ syntactically*
$\qquad\qquad$ *on the first (nonterminal or output) symbol.*

$\quad u_1 \leftarrow \texttt{OutPref}(\alpha_1), A \leftarrow \texttt{First}(\alpha_1), \alpha_1' \leftarrow \texttt{Tail}(\alpha_1)$
$\quad u_2 \leftarrow \texttt{OutPref}(\alpha_2), B \leftarrow \texttt{First}(\alpha_2), \alpha_2' \leftarrow \texttt{Tail}(\alpha_2)$
$\qquad\qquad$ — **Comment**: *Without loss of generality, assume* $A \leq B$.
$\quad (w_A, u_A) \leftarrow \min F_G(A); \ \gamma \leftarrow u_A^{-1} u_1^{-1} \texttt{Derived}(u_2 B, w_A)$
$\quad$ **If** $\gamma = \bot$ **then return** FAILURE. $\hfill (3)$
$\quad$ Add $(\alpha_1', \gamma \alpha_2')$ to Q. $\hfill (4)$

$\quad$ **If** $A = B$ **then:**
$\qquad$ **If** $A \in \texttt{SingleOut}(G)$ **then** start the next iteration.
$\qquad$ **If** $\texttt{power}(u_1^{-1} u_2) \neq \texttt{power}(\gamma)$ **then return** FAILURE. $\hfill (5)$
$\qquad x \leftarrow \texttt{root}(u_1^{-1} u_2), y \leftarrow \texttt{root}(\gamma)$
$\qquad\qquad$ — **Comment**: *Equation* $u_1 A \gamma = u_2 A$ *corresponds to conjuga-*
$\qquad\qquad$ *tion* $u_1^{-1} u_2 A = A \gamma$, *and, by Lemma 2, to* $xA = Ay$.

$\qquad$ **If** $(r_1 A, A r_2)$ is in $\mathcal{C}$ **then**
$\qquad\qquad$ **If** $(x = r_1$ and $y = r_2)$ or $(x = r_1^{-1}$ and $y = r_2^{-1})$ **then**
$\qquad\qquad\quad$ start the next iteration **else return** FAILURE.
$\qquad$ Add (xA, Ay) to $\mathcal{C}$
$\qquad$ **For each** $a \in \Sigma$ **do:**
$\qquad\qquad \beta_1 \leftarrow \texttt{Derived}(xA, a), \beta_2 \leftarrow \texttt{Derived}(Ay, a)$, Add (β_1, β_2) to Q

$\quad$ **else** $\qquad\qquad$ — **Comment**: $A < B$ $\hfill (6)$
$\qquad$ Add $(B, u_2^{-1} u_1 A \gamma)$ to $\mathcal{D}$
$\qquad$ **For each** $a \in \Sigma$ **do:**
$\qquad\qquad \beta_1 \leftarrow \texttt{Derived}(B, a), \beta_2 \leftarrow \texttt{Derived}(u_2^{-1} u_1 A \gamma, a)$, Add (β_1, β_2) to Q
end $\{$of while$\}$

return SUCCESS.

Fig. 1. SIMPLE FUNCTION EQUIVALENCE algorithm

a proof that the initial equation is true, or until a counter-example disproving the equivalence is found and **FAILURE** is reported. Intuitively, the processing of a target (α_1, α_2) from Q is as follows. Firstly, the target is normalized through the unfolding by $\mathcal{D}$ and the removal of the common prefix from $\mathcal{D}^*(\alpha_1)$ and $\mathcal{D}^*(\alpha_2)$.

If the normalized target is trivialy true or false it is immediatly treated as such. Otherwise, the target, which can be written $(u_1 A \alpha_1', u_2 B \alpha_2')$, is split right after the first nonterminals creating new targets $(u_1 A \gamma, u_2 B)$ and $(\alpha_1', \gamma \alpha_2')$, assuming $A \leq B$. The second target is put back into Q for a later processing. The former target $(u_1 A \gamma, u_2 B)$ is considered immediatly, and is processed according to its format: if $A = B$, it is added to C unless it is already present, in which case it is compared to the existing value; if $A \neq B$, it is added to D. If the target is added to either C or D, the target is also derived by all terminal symbols and the resulting targets are added to Q.

3.1 Correctness of the Equivalence Algorithm

In order to demonstrate that the algorithm is correct we will show that:

1. The algorithm always terminates.
2. The validity of the set of equations corresponding to $Q \cup D \cup C$ is an invariant at every iteration of the while loop.
3. The value FAILURE is reported only if the chosen target $(\alpha_1, \alpha_2) \in Q$ is such that $F_G(\alpha_1) \neq F_G(\alpha_2)$.
4. If $F_G(\alpha_1) \neq F_G(\alpha_2)$ for some $(\alpha_1, \alpha_2) \in Q$ then FAILURE is reported.

Let $||Q||$ denote the *shortest-word complexity* of Q, i.e.,

$$||Q|| \overset{\text{def}}{=} \sum \{||\alpha|| + ||\beta|| \mid (\alpha, \beta) \in Q\}.$$

At every iteration which does not add anything to D nor to C, the value $||Q||$ strictly decreases. The algorithm terminates since the number of insertions into D and C is bounded, hence $||Q||$ eventually decreases to 0 or FAILURE is reported.

The invariant of point 2 follows from Proposition 1, Corollary 1, and Lemma 2.

Point 3 can be checked by examining all five FAILURE reports present in the algorithm. First two are obvious. The third occurrence follows from Proposition 1(2). The forth and fifth ones follow from Lemma 2.

The last point, item 4, stating that FAILURE is reported whenever Q contains a pair of sequences which are not equivalent, is argued using the following proposition.

Proposition 2. *Let $\alpha_1, \alpha_2 \in (N \cup \Omega \cup \bar{\Omega})^*$ and $w \in \Sigma^*$ such that $F_G(\alpha_1)(w) \neq F_G(\alpha_2)(w)$. If at some point of the execution of the algorithm (α_1, α_2) appears in Q then the algorithm reports FAILURE.*

Proof. By induction on the length of w.

If $|w| = 0$ then α_1 or α_2 is in $(\Omega \cup \bar{\Omega})^*$. Therefore, if $\alpha_1 \neq \alpha_2$ then FAILURE is reported in (2).

Assume that FAILURE is reported whenever $F_G(\alpha_1)(w) \neq F_G(\alpha_2)(w)$ with $|w| < k$.

Consider α_1 and α_2 such that $F_G(\alpha_1)(w) \neq F_G(\alpha_2)(w)$ and $|w| = k$. There are three cases with respect to the shape of α_1 and α_2 (we will often write just α, for $\alpha \in (N \cup \Omega \cup \bar{\Omega})^*$, as an abbreviation for $F_G(\alpha)$):

– α_1 or α_2 is of constant type.
 We report FAILURE in (2).
– $\alpha_1 = u_1 A \alpha_1'$ and $\alpha_2 = u_2 A \alpha_2'$.
 Let $\gamma = (x u_1 u_x)^{-1} u_2 A$, where $(x, u_x) = \min A$.
 Since $\alpha_1(w) \neq \alpha_2(w)$, there exists $a \in \Sigma$, $w_A, w' \in \Sigma^*$ such that $a w_A w' = w$, $a w_A \in L(A)$.
 If $\alpha_1'(w') \neq \gamma \alpha_2(w')$ then, by induction since $|w'| < k$ and the fact that $(\alpha_1', \gamma \alpha_2)$ is added to Q, the algorithm reports FAILURE.
 Otherwise, i.e., if $\alpha_1'(w') = \gamma \alpha_2(w')$ then we have:

$$u_1 A \alpha_1'(a w_A w') \neq u_2 A \alpha_2'(a w_A w')$$
$$u_1 A(a w_A) \alpha_1'(w') \neq u_2 A(a w_A) \alpha_2'(w')$$
$$u_1 A(a w_A) \gamma \neq u_2 A(a w_A)$$
$$u_1^{-1} u_2 A(a w_A) \neq A \gamma(a w_A)$$

By Lemma 2, the inequality holds if and only if $\mathrm{power}(u_1^{-1} u_2) \neq \mathrm{power}(\gamma)$ or $a^{-1} \mathrm{root}(u_1^{-1} u_2) A(w_A) \neq a^{-1} A \mathrm{root}(\gamma)(w_A)$.
 The inequality $\mathrm{power}(u_1^{-1} u_2) \neq \mathrm{power}(\gamma)$ is checked for in (5) and FAILURE is reported. Otherwise, $(a^{-1} \mathrm{root}(u_1^{-1} u_2) A, a^{-1} A \mathrm{root}(\gamma))$ is added to Q. Hence, by the induction hypothesis, the program eventually reports failure, since $|w_A| < k$.
– $\alpha_1 = u_1 A_1 \alpha_1'$ and $\alpha_2 = u_2 A_2 \alpha_2'$ with $A_1 < A_2$.
 Let $\gamma = (x u_1 u_x)^{-1} u_2 A_2$, where $(x, u_x) = \min A_1$.
 If $\gamma = \bot$ then in (3) we report FAILURE. Otherwise, we have two cases to consider:
  • One of A_1 and $L(A_2)$, but not both, is not defined for any prefix of w. Let $a w_1$ be the prefix of w such that $A_1(a w_1)$ or $A_2(a w_1)$ is defined. In such a case, $a^{-1} u_1 A_1 \gamma(w_1) \neq a^{-1} u_2 A_2(w_1)$, which is equivalent to $a^{-1} u_2^{-1} u_1 A_1 \gamma(w_1) \neq a^{-1} A_2(w_1)$.
  • There exist w_1 and w_γ such that $A_1(a w_1)$ and $A_2(a w_1 w_\gamma)$. Hence, $w = a w_1 w_\gamma w'$ with $a \in \Sigma$.
   If $\alpha_1'(w_\gamma w') \neq \gamma \alpha_2'(w_\gamma w')$ then, by induction hypothesis, the algorithm will report FAILURE.

  Otherwise, $\alpha_1'(w_\gamma w') = \gamma \alpha_2'(w_\gamma w')$, and therefore
$$u_1 A_1 \alpha_1'(a w_1 w_\gamma w') \neq u_2 A_2 \alpha_2'(a w_1 w_\gamma w') \text{ implies}$$
$$u_1 A_1(a w) \gamma(w_\gamma) \alpha_2'(w') \neq u_2 A_2(a w_1 w_\gamma) \alpha_2'(w'),$$
  i.e., $u_1 A_1(a w) \gamma(w_\gamma) \neq u_2 A_2(a w_1 w_\gamma)$.   □

3.2   Complexity

The efficiency of the algorithm follows from the fact that the number of insertions into $\mathcal{D}$ and $\mathcal{C}$ is polynomial.

Proposition 3. *The algorithm* EQUIVALENCE(A, B) *works in polynomial time with respect to* $|G| + v(G)$.

Proof. Let $k \overset{\text{def}}{=} \max\{|\alpha| \mid (A \to a\alpha) \in P\}$, i.e., the length of a longest rule in P. Therefore, for any $A \in N$, $(w, u) \in F_G(A)$ implies that $|u| \le k|w|$.

Since the number of insertions into $\mathcal{D}$ and $\mathcal{C}$ is $O(|N|)$, in the worst case Q can contain $O(|N||\Sigma|)$ targets (α, β). Notice that for all $(\alpha, \beta) \in Q$, $\min(||\alpha||, ||\beta||) \le k\, v(G)$. At that point no more targets can be added to Q. Therefore, the number of iterations of the while loop is $O(|N||\Sigma| k\, v(G))$.

Since the precomputing phase (calculating $\min F_G(A)$, for all $A \in N$, and calculating $\texttt{SingleOut}(G)$, Lemma 3) takes polynomial time in $|G| + v(G)$, and all operations in the algorithm are proportional to the size of the arguments (Lemma 1), the overall running time of the algorithm is polynomial.   □

3.3   Trace History of the Algorithm

Let $G = (\{0, 1\}, \{a, b\}, \{S, T, X, Y, Z\}, P)$ be a simple grammar with output, where P is given by productions:

$$S \to 1bZbaab, \quad T \to 1Ybaay, \quad X \to 0abba, \quad X \to 1aYba,$$
$$Y \to 0bbaXab, \quad Y \to 1bZbaab, \quad Z \to 0baXaY, \quad Z \to 1ZbaaY.$$

We show how $\text{Equivalence}(S, T)$ is computed.

We start by precomputing $\min F_G(A)$, for $A \in \{S, T, X, Y, Z\}$:
$(w_S, u_S) = (w_T, u_T) = (10000, bbaabbaabbaabbaabbaab)$,
$(w_X, u_X) = (0, abba)$, $(w_Y, u_Y) = (00, bbaabbaab)$, and
$(w_Z, u_Z) = (0000, baabbaabbaabbaab)$.

We set $X < Y < Z < S < T$ since $||X|| = 1$, $||Y|| = 2$, $||Z|| = 4$, $||S|| = ||T|| = 5$. We have also to precompute $\texttt{SingleOut}(G)$, which in our case is empty. The initialization step sets $Q = \{(S, T)\}$, $\mathcal{C} = \{\}$, and $\mathcal{D} = \{\}$.

The first iteration of the main while loop begins by retrieving the target $(\alpha_1, \alpha_2) = (S, T)$ from Q. The target is first simplified using $\mathcal{D}$ which does not change its state since $\mathcal{D}$ is empty. The longest common prefix of α_1 and α_2 is then removed, which again does not modify the target since S and T have no common prefix. Both α_1 and α_2 are of general type, therefore they are decomposed as $u_1.A.\alpha_1' = \varepsilon.S.\varepsilon$ and $u_2.B.\alpha_2' = \varepsilon.T.\varepsilon$. Then, since $(w_S, u_S) = (10000, bbaabbaabbaabbaabbaab)$ we compute

$$\gamma = (bbaabbaabbaabbaabbaab)^{-1}\texttt{Derived}(T, 10000) = \varepsilon,$$

and add $(\varepsilon, \varepsilon)$ to Q.

Finally, since the first nonterminal of α_1 and α_2 are different (i.e. $S \ne T$) and $T > S$, we add (T, S) to $\mathcal{D}$. We also compute $\texttt{Derived}(T, 0) = \bot$, $\texttt{Derived}(S, 0) = \bot$, $\texttt{Derived}(T, 1) = YbaaY$ and $\texttt{Derived}(S, 1) = bZbaab$, from which we set $Q = Q \cup \{(\bot, \bot), (YbaaY, bZbaab)\}$. This completes the iteration.

The full trace of the execution of the algorithm has been summarized in Fig. 2. The underlined elements in column Q are the targets (α_1, α_2) considered by the algorithm during the iteration. Column "*simplified (α_1, α_2)*" corresponds to the result of the simplification by $\mathcal{D}$ followed by the removal of the longest common prefix; the result is written in the form $(u_1.A.\alpha_1', u_2.B.\alpha_2')$. Column $\mathcal{C}$ contains the conjugation equations added (black ones) or checked for (gray ones) in the iteration. The last column, $\mathcal{D}$, contains the decomposition added in the iteration.

Q	simplified (α_1, α_2)	γ	$\mathcal{C}$	$\mathcal{D}$
(S,T)	$(\varepsilon.S.\varepsilon, \varepsilon.T.\varepsilon)$	ε		(T,S)
$(\varepsilon,\varepsilon),(\perp,\perp),$ $(Y\overline{baa}Y, bZbaab)$	$(\varepsilon,\varepsilon)$			
$(\perp,\perp),(Y\overline{baa}Y, bZbaab)$				
$(Y\overline{baa}Y, bZbaab)$	$(\varepsilon.Y.\overline{baa}Y, b.Z.baab)$	$\overline{b}Y$		$(Z, \overline{b}Y\overline{b}Y)$
$(\overline{baa}Y, \overline{b}Ybaab),$ $(\overline{ba}Xa Y, baXaY),$ $(Z\overline{baa}Y, ZbaaY)$	$(\overline{baa}.Y.\varepsilon, \overline{b}.Y.baab)$	$\overline{ba}\overline{ab}$	$(\overline{aa}bbY, Y\overline{ba}\overline{ab})$	
$(\overline{ba}Xa Y, baXaY),$ $(\overline{Z\overline{baa}Y}, ZbaaY),$ $(\varepsilon,\varepsilon),(\overline{a}Xab, bbaX\overline{ab}),$ $(\overline{aa\overline{b}}Zbaab, bZ)$	$(\varepsilon,\varepsilon)$			
$(\overline{Z\overline{baa}Y}, ZbaaY),$ $(\varepsilon,\varepsilon),(\overline{a}Xab, bbaX\overline{ab}),$ $(\overline{aa\overline{b}}Zbaab, bZ)$	$(\varepsilon,\varepsilon)$			
$(\overline{\varepsilon,\varepsilon}),(\overline{a}Xab, bbaX\overline{ab}),$ $(\overline{aa\overline{b}}Zbaab, bZ)$	$(\varepsilon,\varepsilon)$			
$(\overline{\overline{a}Xab, bbaX\overline{ab}}),$ $(\overline{aa\overline{b}}Zbaab, bZ)$	$(\overline{a}.X.ab, bba.X.\overline{ab})$	$abba$	$(abbaX, Xabba)$	
$(\overline{aa\overline{b}}Zbaab, bZ),(ab,ab),$ $(\overline{abbaabba, abbaabba}),$ $(abbaaYba, aYbaabba)$	$(\overline{aabb}.Y.\overline{b}Ybaab, \varepsilon.Y.\overline{b}Y)$	$baab$	$(bbaaY, Ybaab)$	
$(ab,ab),$ $(\overline{abbaabba, abbaabba}),$ $(abbaaYba, aYbaabba),$ $(\overline{b}Ybaab, baaY)$	$(\varepsilon,\varepsilon)$			
$(\overline{abbaabba, abbaabba}),$ $(\overline{abbaaYba, aYbaabba}),$ $(\overline{b}Ybaab, baaY)$	$(\varepsilon,\varepsilon)$			
$(\overline{abbaaYba, aYbaabba}),$ $(\overline{b}Ybaab, baaY)$	$(bbaa.Y.ba, \varepsilon.Y.baabba)$	$\overline{ba}\overline{ab}$	$(\overline{aabb}Y, Y\overline{ba}\overline{ab})$	
$(\overline{\overline{b}Ybaab, baaY}),$ (ba,ba)	$(\overline{b}.Y.baab, baa.Y.\varepsilon)$	$baab$	$(bbaaY, Ybaab)$	
$(ba,ba),(baab,baab)$	$(\varepsilon,\varepsilon)$			
$(\overline{baab,baab})$	$(\varepsilon,\varepsilon)$			
$empty$	Return SUCCESS			

Fig. 2. Execution of EQUIVALENCE proving $F_G(S) = F_G(T)$. In each iteration the *active* pair (α_1, α_2) is underlined.

4   Conclusions

We showed an algorithm which tests the equality of two simple functions for a grammar G in time polynomial with respect to $|G| + v(G)$ (the *shortest-word complexity* of G). In practical situations this algorithm works in polynomial time with respect to the size of G, since usually $v(G)$ is polynomial with respect to the size $|G|$ of the grammar. However it is theoretically possible that $v(G)$ is exponential with respect to $|G|$. Our algorithm is a first step towards a fully polynomial (with respect only to $|G|$) time algorithm.

References

1. Cédric Bastien, Jurek Czyzowicz, Wojciech Fraczak, and Wojciech Rytter. Prime normal form and equivalence of simple grammars. In *CIAA 2005*, volume 3845 of *LNCS*, pages 78–89. Springer, 2006.
2. Didier Caucal. A fast algorithm to decide on simple grammars equivalence. In *Optimal Algorithms*, volume 401 of *LNCS*, pages 66–85. Springer, 1989.
3. Bruno Courcelle. An axiomatic approach to the Korenjak-Hopcroft algorithms. *Mathematical Systems Theory*, 16(3):191–231, 1983.
4. Maxime Crochemore and Wojciech Rytter. *Text Algorithms*. Oxford University Press, New York, 1994.
5. Wojciech Debski and Wojciech Fraczak. Concatenation state machines and simple functions. In *CIAA 2004*, volume 3317 of *LNCS*, pages 113–124. Springer, 2004.
6. M.A. Harrison. *Introduction to formal language theory*. Addison Wesley, 1978.
7. Yoram Hirshfeld, Mark Jerrum, and Faron Moller. A polynomial algorithm for deciding bisimilarity of normed context-free processes. *Theoretical Computer Science*, 158(1–2):143–159, 1996.
8. Oscar H. Ibarra and Louis E. Rosier. On the decidability of equivalence of deterministic pushdown transducers. *Information Processing Letters*, 13(3):89–93, 1981.
9. A. J. Korenjak and J. E. Hopcroft. Simple deterministic languages. In *Proc. IEEE 7th Annual Symposium on Switching and Automata Theory*, IEEE Symposium on Foundations of Computer Science, pages 36–46, 1966.
10. Lothaire. *Combinatorics on words*. Cambridge University Press, United Kingdom, 1997.
11. Gérard Sénizergues. $T(A) = T(B)$?. In *ICALP'99*, volume 1644 of *LNCS*, pages 665–675. Springer, 1999.

On the Gap-Complexity of Simple RL-Automata*

F. Mráz[1], F. Otto[2], and M. Plátek[1]

[1] Charles University, Faculty of Mathematics and Physics
Department of Computer Science, Malostranské nám. 25
118 00 Praha 1, Czech Republic
mraz@ksvi.ms.mff.cuni.cz, Martin.Platek@mff.cuni.cz
[2] Fachbereich Mathematik/Informatik, Universität Kassel
34109 Kassel, Germany
otto@theory.informatik.uni-kassel.de

Abstract. *Analysis by reduction* is a method used in linguistics for checking the correctness of sentences of natural languages. This method is modelled by *restarting automata*. Here we introduce and study a new type of restarting automaton, the so-called t-sRL-*automaton*, which is an RL-automaton that is rather restricted in that it has a window of size 1 only, and that it works under a minimal acceptance condition. On the other hand, it is allowed to perform up to t rewrite (that is, delete) steps per cycle. Here we study the *gap-complexity* of these automata. The membership problem for a language that is accepted by a t-sRL-automaton with a bounded number of gaps can be solved in polynomial time. On the other hand, t-sRL-automata with an unbounded number of gaps accept NP-complete languages.

1 Introduction

The original motivation for introducing the restarting automaton was the desire to model the so-called *analysis by reduction* of natural languages. Analysis by reduction is usually presented by finite samples of sentences of a natural language and by sequences of their correct reductions (e.g., tree-banks) (see, e.g., [7]).

From a theoretical point of view the restarting automaton can be seen as a tool that yields a very flexible generalization of analytical grammars. On the other hand, the restarting automaton can be considered as a generalization and a refinement of the pushdown automaton (see, e.g., [9]) and the contraction automaton [13].

Up to now all models of restarting automata studied in the literature accept at least all deterministic context-free languages. Hence, they can be used to analyze

---

* The first and the third author were partially supported by the Grant Agency of the Czech Republic under Grant-No. 201/04/2102 and by the program 'Information Society' under project 1ET100300517. The first author was also partially supported by the Grant Agency of Charles University in Prague under Grant-No. 358/2006/A-INF/MFF.

O.H. Ibarra and Z. Dang (Eds.): DLT 2006, LNCS 4036, pp. 83–94, 2006.

only language classes above the level of deterministic context-free languages. In particular, regular languages or even finite languages are of too small a degree of complexity to be studied by restarting automata. However, in (corpus) linguistics essentially finite (though very large) approximations of infinite languages are often studied. As the motivation for restarting automata is derived from linguistic considerations, this is a shortcoming of the model.

Here we propose a way to remedy this situation. We introduce a new variant of the restarting automaton, the so-called *simple* RL-*automaton* (sRL - automaton), that is rather restricted in various aspects to ensure that its expressive power is limited. However, by admitting that t (≥ 1) delete operations may be performed in each cycle, the expressive power of the obtained model of the restarting automaton is parametrized by t, and hence, we obtain an infinite hierarchy of automata and therewith of language classes. Then we study the number of gaps generated during a reduction as a complexity measure for t-sRL-automata that is inspired by linguistic considerations (see, e.g., [2, 3]). This measure is related to the notion of j-monotonicity considered in [11], and it yields an infinite hierarchy of automata and language classes. In contrast to the situation for 2-monotone restarting automata, which accept NP-complete languages [5], it turns out that a bounded number of gaps implies that only feasible languages are accepted, that is, languages that are recognizable in polynomial time. However, with an unbounded number of gaps these automata still accept NP-complete languages.

The paper is structured as follows. After introducing the simple RL - automaton in Section 2, we will establish an infinite hierarchy based on the parameter t of delete operations that are permitted per cycle. In Section 3 various notions of monotonicity are presented for sRL-automata, and it is shown that sRL-automata can simulate standard RL-automata, preserving the type of monotonicity. In particular, this implies that t-right-left-monotone sRL-automata accept NP-complete languages. Then in Section 4 the gap-complexity is introduced, and it is shown that with bounded gap-complexity sRL-automata only accept feasible languages. The paper closes with a characterization of regular languages in terms of a special type of sRL-automata.

2   The t-sRL-Automaton

Here we describe in short the type of restarting automaton we will be dealing with. More details on restarting automata in general can be found in [9].

An sRL-*automaton* (*simple* RL-*automaton*) M is a (in general) nondeterministic machine with a finite-state control Q, a finite input alphabet Σ, and a head (window of size 1) that works on a flexible tape delimited by the left sentinel ¢ and the right sentinel \$. For an input $w \in \Sigma^*$, the initial tape inscription is ¢w\$. To process this input M starts in its initial state q_0 with its window over the left end of the tape, scanning the left sentinel ¢. According to its transition relation, M performs *move-right steps* which shift the window one position to the right, thereby possibly changing the state of M, *move-left steps* which shift the window one position to the left, thereby possibly changing the state of M, and *delete steps*, which delete the content of the window, thus shortening the tape, change the state, and shift the window to the right neighbour of the symbol deleted. Of course, neither the left sentinel ¢ nor the right sentinel $ must be deleted. At the right end of the tape M either halts and accepts, or it halts and rejects, or it *restarts*, that is, it places its window over the left end of the tape and reenters the initial state. It is required that before the first restart step and also between any two restart steps, M executes at least one delete operation.

A *configuration* of M is a string $\alpha q \beta$ where $q \in Q$, and either $\alpha = \lambda$ and $\beta \in \{ \text{¢} \} \cdot \Sigma^* \cdot \{ \$ \}$ or $\alpha \in \{ \text{¢} \} \cdot \Sigma^*$ and $\beta \in \Sigma^* \cdot \{ \$ \}$; here q represents the current state, $\alpha \beta$ is the current content of the tape, and it is understood that the window contains the first symbol of β. A *restarting configuration* is of the form $q_0 \text{¢} w \$$, where $w \in \Sigma^*$.

We observe that any finite computation of an sRL-automaton M consists of certain phases. A phase, called a *cycle*, starts in a restarting configuration, the window is moved along the tape by performing move-right, move-left, and (at least one) delete operations until a restart operation is performed and thus a new restarting configuration is reached. If no further restart operation is performed, each finite computation necessarily finishes in a halting configuration – such a phase is called a *tail*. We assume that no delete operation is executed in a tail computation.

We use the notation $u \vdash^c_M v$ to denote a cycle of M that begins with the restarting configuration $q_0 \text{¢} u \$$ and ends with the restarting configuration $q_0 \text{¢} v \$$; the relation $\vdash^{c*}_M$ is the reflexive and transitive closure of $\vdash^c_M$.

An input $w \in \Sigma^*$ is *accepted* by M, if there is an accepting computation which starts with the (initial) configuration $q_0 \text{¢} w \$$. By $L(M)$ we denote the language consisting of all words accepted by M; we say that M *recognizes (accepts) the language* $L(M)$. By $S(M)$ we denote the *simple language* accepted by M, which consists of all words that M accepts by tail computations. Obviously, $S(M)$ is a regular sublanguage of $L(M)$. By $RS(M)$ we denote the *reduction system* $RS(M) := (\Sigma^*, \vdash^c_M, S(M))$ that is induced by M. Observe that, for each $w \in \Sigma^*$, we have $w \in L(M)$ if and only if $w \vdash^{c*}_M v$ holds for some word $v \in S(M)$.

We say that M is an sRR-automaton if M does not use any move-left steps. By sRL (sRR) we denote the class of all sRL-automata (sRR-automata). A t-sRL-automaton ($t \geq 1$) is an sRL-automaton which uses at most t delete operations in a cycle, and similarly we obtain the t-sRR-automaton. By $\mathcal{L}(\mathsf{A})$ we denote the class of languages that are accepted by automata of type A (A-automata), and by $\mathcal{L}_{\leq n}(\mathsf{A})$ we denote the class of finite languages that are accepted by automata of type A and that do not contain any words of length exceeding the number n.

On the set of input words Σ^*, we define a partial ordering $\leq$ as follows:

$$u \leq v \text{ if and only if } u \text{ is a scattered subword of } v.$$

By $<$ we denote the proper part of $\leq$. Obviously, $\leq$ is well-founded, that is, there do not exist infinite descending sequences with respect to $<$.

For $L \subseteq \Sigma^*$, let $L_{\min} := \{ w \in L \mid u < w \text{ does not hold for any } u \in L \}$, that is, $L_{\min}$ is the set of minimal words of L. It is well-known that $L_{\min}$ is finite for each language L (see, e.g., [8]). We say that an sRL-automaton M accepting the language L works with *minimal acceptance* if it accepts in tail computations exactly the words of the language $L_{\min}$, that is, $S(M) = L_{\min}$. Thus, each word $w \in L \setminus L_{\min}$ is reduced to a word $w' \in L_{\min}$ by a sequence of cycles of M. We will use the prefix min- to denote sRL-automata that work with minimal acceptance.

Here we will mainly be interested in t-sRL-automata with minimal acceptance. In this way we will achieve similarities between certain classes of finite and infinite languages recognized by sRL-automata, as an sRL-automaton with minimal acceptance is forced to perform sequences of cycles even for accepting a regular language. In fact, this is even true for most finite languages.

Example 1. Let $t \geq 1$, and let $L^{<t>} := \{a^t, \lambda\}$. Then $L_{\min}^{<t>} = \{\lambda\}$. Hence, an sRL-automaton for the language $L^{<t>}$ that works with minimal acceptance must execute the cycle $a^t \vdash^c \lambda$, which means that it must execute t delete operations during this cycle. Hence, it is a t-sRL-automaton.

Concerning the relationship between sRR- and sRL-automata, we have the following important result which generalizes a corresponding result for RLWW-automata from [10].

Theorem 1. *For each integer $t \geq 1$ and each t-sRL-automaton M, there exists a t-sRR-automaton M' such that the reduction systems $RS(M)$ and $RS(M')$ coincide.*

Observe that, in each cycle, M' executes its up to t delete operations strictly from left to right, while M may execute them in arbitrary order.

Proof. Each cycle of a computation of M consists of (up to) $2t + 2$ phases. In phases $1, 3, \ldots, 2i + 1, \ldots, 2t + 1$, M shifts its window across the tape by executing move-left and move-right steps, in phases $2, 4, \ldots, 2i, \ldots, 2t$, M executes a delete operation, and in phase $2t + 2$, M performs a restart step. Thus, the sRR-automaton M' must guess the positions of the delete steps and the crossing sequences of M corresponding to the move-phases. As within a move-phase M need not visit the same tape cell twice while being in the same state, we see that the corresponding crossing sequence is bounded in length. Hence, M' can indeed guess the (up to) $t + 1$ crossing sequences and verify that they are consistent with each other and with the chosen delete operations.   □

Based on Theorem 1 we can describe a t-sRL-automaton by meta-instructions of the form

$$(\mathrm{c} \cdot E_0, a_1, E_1, a_2, E_2, \ldots, E_{s-1}, a_s, E_s \cdot \$),$$

where $1 \leq s \leq t$, $E_0, \ldots, E_s$ are regular languages (often represented by regular expressions), called the *regular constraints* of this instruction, and $a_1, \ldots, a_s \in \Sigma$ correspond to letters that are deleted by M during one cycle. On trying to execute this meta-instruction starting from a configuration $q_0 \mathrm{c} w \$$, M will get stuck (and so reject), if w does not admit a factorization of the form $w = v_0 a_1 v_1 a_2 \ldots v_{s-1} a_s v_s$ such that $v_i \in E_i$ for all $i = 0, \ldots, s$. On the other hand, if w admits factorizations of this form, then one of them is chosen nondeterministically, and $q_0 \mathord{\mathcal{c}} w\$$ is transformed into $q_0 \mathord{\mathcal{c}} v_0 v_1 \ldots v_{s-1} v_s \$$. In order to also describe the tails of accepting computations, we use accepting meta-instructions of the form $(\mathord{\mathcal{c}} \cdot E \cdot \$, \mathsf{Accept})$, where E is a regular language. Actually we can require that there is only a single accepting meta-instruction for M. If M works with minimal acceptance, then this accepting meta-instruction is of the form $(\mathord{\mathcal{c}} \cdot L(M)_{\min} \cdot \$, \mathsf{Accept})$.

Example 2. Let $t \geq 1$, and let $LR_t := \{\, c_0 w c_1 w c_2 \ldots c_{t-1} w \mid w \in \{a,b\}^* \,\}$, where $\Sigma_0 := \{a,b\}$ and $\Sigma_t := \{c_0, c_1, \ldots, c_{t-1}\} \cup \Sigma_0$. We obtain a t-sRR-automaton M_t for the language LR_t through the following sequence of meta-instructions:

$$(1)\ (\mathord{\mathcal{c}} c_0, a, \Sigma_0^* \cdot c_1, a, \Sigma_0^* \cdot c_2, \ldots, \Sigma_0^* \cdot c_{t-1}, a, \Sigma_0^* \cdot \$),$$
$$(2)\ (\mathord{\mathcal{c}} c_0, b, \Sigma_0^* \cdot c_1, b, \Sigma_0^* \cdot c_2, \ldots, \Sigma_0^* \cdot c_{t-1}, b, \Sigma_0^* \cdot \$),$$
$$(3)\ (\mathord{\mathcal{c}} c_0 c_1 \ldots c_{t-1} \$, \mathsf{Accept}).$$

It follows easily that $L(M_t) = LR_t$ holds, and that M_t works with minimal acceptance. Actually, the automaton M_t is even deterministic. Observe that the language LR_t cannot be accepted by an r-sRL-automaton for any $r < t$. First of all, LR_t cannot possibly be accepted by tail computations only, as it is not regular. However, by executing $r < t$ many delete steps a word $w \in LR_t$ will necessarily be transformed into a word $w' \notin LR_t$, which will then not lead to acceptance.

We emphasize the following properties of restarting automata, which are used implicitly in proofs. They play an important role in linguistic applications of restarting automata (e.g., for the analysis by reduction, grammar-checking, and morphological disambiguation).

Definition 1. (Correctness Preserving Property)
A t-sRL-automaton M is (strongly) correctness preserving if $u \in L(M)$ and $u \vdash_M^{c*} v$ imply that $v \in L(M)$.

Definition 2. (Error Preserving Property)
A t-sRL-automaton M is error preserving if $u \notin L(M)$ and $u \vdash_M^{c*} v$ imply that $v \notin L(M)$.

The following facts are easily verified.

Fact 3. *1. Each t-sRL-automaton is error preserving.*
2. Each deterministic t-sRL-automaton is correctness preserving.
3. There exist nondeterministic t-sRL-automata which are not correctness preserving.

From the witness languages $L^{\langle t \rangle}$ of Example 1 and LR_t of Example 2 we obtain the following hierarchy results. Here the prefix det- is used to denote deterministic automata.

Corollary 1. *For each suffix* $Y \in \{sRR, sRL\}$, *and each integer* $t \geq 2$,

(a) $\quad \mathcal{L}(\text{det-}(t-1)\text{-Y}) \subset \mathcal{L}(\text{det-}t\text{-Y})$ *and*
$\quad\quad\quad \mathcal{L}((t-1)\text{-Y}) \subset \mathcal{L}(t\text{-Y})$.

(b) $\quad \mathcal{L}(\text{min-det-}(t-1)\text{-Y}) \subset \mathcal{L}(\text{min-det-}t\text{-Y})$ *and*
$\quad\quad\quad \mathcal{L}(\text{min-}(t-1)\text{-Y}) \subset \mathcal{L}(\text{min-}t\text{-Y})$.

(c) $\mathcal{L}_{\leq n}(\text{min-det-}(t-1)\text{-Y}) \subset \mathcal{L}_{\leq n}(\text{min-det-}t\text{-Y})$ *and*
$\quad\quad \mathcal{L}_{\leq n}(\text{min-}(t-1)\text{-Y}) \subset \mathcal{L}_{\leq n}(\text{min-}t\text{-Y})$ *for each* $n \geq t$.

Example 3. There exists a 1-sRR-automaton M_{copy} that accepts the language

$$LR_2' := LR_2 \cdot \{a, b, \lambda\} = \{\, c_0 w c_1 w x \mid w \in \{a, b\}^*, x \in \{a, b, \lambda\} \,\}.$$

M_{copy} is given through the following sequence of meta-instructions:

(1) $(\mathcal{c}c_0c_1\$, \mathsf{Accept})$,
(2) $(\mathcal{c}c_0 \cdot (\{a,b\}^2)^*, a, c_1 \cdot (\{a,b\}^2)^* \cdot a\$)$,
(3) $(\mathcal{c}c_0 \cdot (\{a,b\}^2)^*, b, c_1 \cdot (\{a,b\}^2)^* \cdot b\$)$,
(4) $(\mathcal{c}c_0 \cdot (\{a,b\}^2)^* \cdot c_1 \cdot (\{a,b\}^2)^*, a, \$)$,
(5) $(\mathcal{c}c_0 \cdot (\{a,b\}^2)^* \cdot c_1 \cdot (\{a,b\}^2)^*, b, \$)$,
(6) $(\mathcal{c}c_0 \cdot (\{a,b\}^2)^* \cdot \{a,b\}, a, c_1 \cdot (\{a,b\}^2)^* \cdot \{a,b\} \cdot a\$)$,
(7) $(\mathcal{c}c_0 \cdot (\{a,b\}^2)^* \cdot \{a,b\}, b, c_1 \cdot (\{a,b\}^2)^* \cdot \{a,b\} \cdot b\$)$,
(8) $(\mathcal{c}c_0 \cdot (\{a,b\}^2)^* \cdot \{a,b\} \cdot c_1 \cdot (\{a,b\}^2)^* \cdot \{a,b\}, a, \$)$,
(9) $(\mathcal{c}c_0 \cdot (\{a,b\}^2)^* \cdot \{a,b\} \cdot c_1 \cdot (\{a,b\}^2)^* \cdot \{a,b\}, b, \$)$.

It is easily seen that M_{copy} accepts the language LR_2' working with minimal acceptance. On the other hand, the language LR_2' is not even growing context-sensitive. Assume to the contrary that this language is growing context-sensitive. Then there exists a nondeterministic shrinking two-pushdown automaton A that accepts this language [1]. Consider the language $L_{\text{copy}} := \{\, ww \mid w \in \{a, b\}^* \,\}$. Given an input $x \in \{a, b\}^*$, a shrinking two-pushdown automaton B can simply insert c_0 at the beginning and c_1 somewhere inside the word x, producing the word $c_0 x_1 c_1 x_2$, where the position for the latter insertion is chosen nondeterministically. While doing so the automaton verifies whether x is of even length. In the negative it will reject x, in the positive case it will simulate A for the input $c_0 x_1 c_1 x_2$. This will lead to acceptance if and only if $x_1 = x_2$. Thus, the shrinking two-pushdown automaton B accepts the language L_{copy}. This, however, contradicts the fact that L_{copy} is not growing context-sensitive (see, e.g., [1]).

As a consequence we obtain the following incomparability results.

Theorem 2. *The language classes* $\bigcup_{t \in \mathbb{N}_+} \mathcal{L}(\text{min-}t\text{-sRL})$ *and* $\bigcup_{t \in \mathbb{N}_+} \mathcal{L}(t\text{-sRL})$ *are incomparable under inclusion to the class* CFL *of context-free languages and to the class* GCSL *of growing context-sensitive languages.*

Proof. Already the class $\mathcal{L}(\text{min-1-sRR})$ contains a language that is not growing context-sensitive, as shown by the example above. On the other hand, it is shown in [4] that the context-free language $L_2 := \{\, a^n b^n \mid n \geq 0 \,\} \cup \{\, a^n b^m \mid m > 2n \geq 0 \,\}$ is not accepted by any RRW-automaton. The argument is based on the observation that in each cycle an RRW-automaton M for L_2 would have to guess whether to remove a factor of the form $a^i b^i$ or a factor of the form $a^i b^{2i}$ for some integer $i > 0$. Using pumping techniques it can then be shown that M violates the Error Preserving Property. The same argument also works for t-sRL-automata, that is, L_2 is not accepted by any t-sRL-automaton.   □

As we have seen above, t-sRL-automata accept some languages that are quite complicated in comparison to the context-free languages. However, we do not yet have a characterization for the expressive power of t-sRL-automata in general. The following result shows that this question has been solved at least for the special case of a single-letter alphabet.

Theorem 3. *Let M be a t-sRL-automaton. If $L(M) \subseteq \{a\}^*$, then $L(M)$ is regular.*

Proof. By Theorem 1 we can assume that M is a t-sRR-automaton. Further, as M does not accept any word containing a letter other than a, we can assume that the tape alphabet of M consists of the letter a only.

Assume that M is defined by the meta-instructions

$$I_i := (\text{¢} \cdot E_0^{(i)}, a, E_1^{(i)}, a, E_2^{(i)}, \ldots, E_{s_i-1}^{(i)}, a, E_{s_i}^{(i)} \cdot \$) \quad (1 \le i \le r)$$

and $I_0 := (\text{¢} \cdot S_0 \cdot \$, \mathsf{Accept})$, where all $E_j^{(i)}$ and S_0 are regular expressions. We now define another t-sRR-automaton M' through the meta-instruction I_0 and the meta-instructions I_i', $1 \le i \le r$, where I_i' is defined as

$$I_i' := (\text{¢} \cdot E_0^{(i)} \cdot E_1^{(i)} \cdot E_2^{(i)} \cdots E_{s_i-1}^{(i)} \cdot E_{s_i}^{(i)}, a^{s_i} \to \lambda, \$).$$

Here $a^{s_i} \to \lambda$ is used as a shorthand for the fact that s_i copies of the symbol a are deleted that are next to each other.

To complete the proof we establish the following two claims.

Claim 1. $L(M') = L(M)$.

Proof. Obviously, M' and M accept the same words in tail computations. Thus, it suffices to show that M' and M execute the same cycles. Assume that $u \vdash_M^c v$ is a possible cycle of M. Then there exists an index $i \in \{1, \ldots, r\}$ such that the tape content $\text{¢} \cdot u \cdot \$$ is transformed into $\text{¢} \cdot v \cdot \$$ by meta-instruction I_i. Hence, u can be factored as $u = u_0 a u_1 a u_2 \ldots u_{s_i-1} a u_{s_i}$ such that $u_j \in E_j^{(i)}$ for all $1 \le j \le s_i$ and $v = u_0 u_1 u_2 \ldots u_{s_i-1} u_{s_i}$. It is now immediate that by applying meta-instruction I_i', M' can execute the cycle $u \vdash_{M'}^c v$. Analogously, if $u \vdash_{M'}^c v$ by meta-instruction I_i', then also $u \vdash_M^c v$ by meta-instruction I_i. Thus, the languages $L(M)$ and $L(M')$ coincide.   □

Claim 2. $L(M') = L(A)$ for a nondeterministic finite-state acceptor A.

Proof. We present a nondeterministic finite-state acceptor A that, given a word $w = a^n$ as input, accepts if and only if there exists a sequence of cycles

$$w = w_m \vdash_{M'}^c w_{m-1} \vdash_{M'}^c \cdots \vdash_{M'}^c w_1 \vdash_{M'}^c w_0 \in S_0.$$

For each $i = 1, \ldots, m$, there exists a meta-instruction $I'_{j_i} = (\mathfrak{c} \cdot E'_{j_i}, a^{s_{j_i}} \to \lambda, \$)$ of M' such that $w_i = w_{i-1} \cdot a^{s_{j_i}}$ and $w_{i-1} \in E'_{j_i}$. Thus, scanning its input tape from left to right, A will simultaneously simulate finite-state acceptors for the languages E'_i, $1 \le i \le r$, and S_0. When it recognizes a prefix that belongs to S_0, then it decides nondeterministically whether this is the string w_0. In the affirmative it aborts the simulation of the finite-state acceptor for S_0, guesses an index $k \in \{1, \ldots, r\}$ such that the finite-state acceptor for E'_k is now in a final state, and continues with simulating the finite-state acceptors for the languages E'_i $(1 \le i \le r)$ for s_k steps. Then A again guesses an index $k' \in \{1, \ldots, r\}$ such that the finite-state acceptor for $E'_{k'}$ is now in a final state, and continues, otherwise, it halts and rejects. This process continues until A either rejects, or until the input has been read completely. If ℓ is the last index guessed, then A accepts if, since passing through a final state of the finite-state acceptor for the language E'_ℓ, exactly s_ℓ copies of the letter a have been read. It is now immediate that $L(A) = L(M')$.   □

Thus, the language $L(M)$ is regular.   □

Further, we have at least the following inclusion results.

Theorem 4. DCFL $\subset \bigcup_{t \in \mathbb{N}_+} \mathcal{L}(\text{min-det-}t\text{-sRL}) \subset \bigcup_{t \in \mathbb{N}_+} \mathcal{L}(\text{min-}t\text{-sRL})$.

Proof. Each deterministic context-free language L is accepted by some right-monotone deterministic RR-automaton M [4]. If M has a window of size k, then it can be simulated by a deterministic k-sRL-automaton M'. M' scans its tape from left to right until it detects a factor that is to be rewritten by M. Then it moves its window back by $k-1$ positions, and deletes the up to k symbols that M deletes in this cycle. By some additional cycles each word from the regular language $S(M)$ can then be reduced to a minimal word. Thus, DCFL $\subseteq \bigcup_{t \in \mathbb{N}_+} \mathcal{L}(\text{min-det-}t\text{-sRL})$.

The language $L := \{a^n b^n c, a^n b^{2n} d \mid n \ge 0\}$, which is not deterministic context-free, is easily seen to be accepted by a deterministic 3-sRL-automaton that works with minimal acceptance.

Thus, it follows that DCFL $\subset \bigcup_{t \in \mathbb{N}_+} \mathcal{L}(\text{min-det-}t\text{-sRL})$.

Using similar arguments as in the proof of Theorem 2 it can be shown that the language $L := \{a^n b^m \mid 0 \le n \le m \le 2n\}$ is not accepted by any deterministic t-sRL-automaton. However, it is accepted by the 3-sRL-automaton M that is given through the following meta-instructions:

$$(\mathfrak{c} \cdot a^*, a, \{\lambda\}, b, b^* \cdot \$), \quad (\mathfrak{c} \cdot a^*, a, \{\lambda\}, b, \{\lambda\}, b, b^* \cdot \$), \quad (\mathfrak{c} \cdot \$, \text{Accept}).$$

This shows that $\bigcup_{t \in \mathbb{N}_+} \mathcal{L}(\text{min-det-}t\text{-sRL}) \subset \bigcup_{t \in \mathbb{N}_+} \mathcal{L}(\text{min-}t\text{-sRL})$.   □

3   Monotonicity

Finally we turn to the notion of monotonicity for t-sRL-automata. Let M be a t-sRL-automaton. A configuration $C = \alpha q \beta$ of M in which a delete operation is to be applied is called a *delete configuration* of M. The number $|\beta|$ is called the *right distance* of C, denoted by $D_r(C)$, and the number $|\alpha|$ is the *left distance* of C, denoted by $D_l(C)$.

We say that a *sequence of delete configurations* $S = (C_1, C_2, \cdots, C_n)$ is *right-monotone* if $D_r(C_1) \geq D_r(C_2) \geq \ldots \geq D_r(C_n)$, that it is *left-monotone* if $D_l(C_1) \geq D_l(C_2) \geq \ldots \geq D_l(C_n)$, and that it is *right-left-monotone* if it is simultaneously right- and left-monotone. It is called *j-right-monotone, j-left-monotone* or *j-right-left-monotone* for some integer $j \geq 1$, if it can be partitioned into at most j interleaved subsequences $S_1, S_2, \ldots, S_j$ such that each of these subsequences is right-monotone, left-monotone or right-left-monotone, respectively. A computation of M is called *j-right-monotone, j-left-monotone* or *j-right-left-monotone* if the corresponding sequence of delete configurations is j-right-monotone, j-left-monotone or j-right-left-monotone, respectively.

An sRL-automaton M is called *t-right-monotone* (*t-left-monotone, t-right-left-monotone*) if it is a t-sRL-automaton and each of its computations is t-right-monotone (*t*-left-monotone, *t*-right-left-monotone). We will use the prefixes t-right-mon-, t-left-mon-, and t-right-left-mon- to denote the classes of t-right-, t-left-, and t-right-left-monotone sRL-automata, respectively.

For example, the deterministic t-sRR-automaton M_t of Example 2 is t-right-left-monotone. Concerning the j-right-, j-left-, and j-right-left-monotone (standard) RL-automata considered in [11] we have the following result.

Theorem 5. *For each prefix* Y $\in$ {right-left-mon, right-mon, left-mon} *and all integers* $j, k \geq 1$, *if* M *is a* j-Y-RL-*automaton with a window of size* k, *then there exists a* $(j \cdot k)$-Y-sRL-*automaton* M' *such that* $RS(M) = RS(M')$ *holds.*

Proof. Let M be a j-Y-RL-automaton over Σ with a window of size k. Then the sequence of cycles of each computation of M can be divided into j interleaved subsequences that are all Y-monotone. We describe an sRL-automaton M' that simulates the computations of M cycle by cycle. Within its finite-state control M' realizes a buffer of size k that it will use to store the content of the window of M. When M performs a rewrite step $u \rightarrow v$, then $|u| \leq k$, and v is obtained from u by deleting up to k symbols of u. Hence, M' can simulate this rewrite step by executing up to k delete steps, each deleting a single symbol. It follows that M' is a k-sRL-automaton, and that $RS(M') = RS(M)$. Further, if M is j-Y-monotone, then we consider M' as a $(j \cdot k)$-sRL-automaton, and it is obvious that as such M' is $(j \cdot k)$-Y-monotone. $\qquad\square$

From the proof above we see that a (standard) RL-automaton with a window of size k can be simulated by a k-sRL-automaton.

It is known that, for all three types of monotonicity, the language classes $(\mathcal{L}(j$-Y-RL$))_{j \geq 1}$ form a strict hierarchy (see [12] Theorem 7 (a)). In fact, this remains true when we restrict our attention to RL-automata with read/write windows of size two. Further, it is shown in [5] that already 2-right-monotone R-automata accept NP-complete languages. In fact, this result extends to right-left-monotonicity, as there are even 2-right-left-monotone R-automata that accept NP-complete languages [6]. However, as neither the proof in [5] nor the proof in

[6] gives the size of the window of the R-automaton constructed explicitly (it is rather large due to the encoding used), we only obtain the following complexity result for sRL-automata.

Corollary 2. *There exists an integer $t \geq 1$ such that $\mathcal{L}(t$-right-left-mon-sRL$)$ contains NP-complete languages.*

4   The Gap-Complexity

Let M be a t-sRL-automaton, let $w \in \Sigma^*$ be an input word, and let $w \vdash^c_M w_1 \vdash^c_M \cdots \vdash^c_M w_n$ be an initial part of a computation of M on input w. In each cycle M deletes up to t symbols from the tape content of the current restarting configuration. Instead of deleting symbols we can replace them by a special symbol ♦, thus obtaining a word w'_i for each $i = 1, \ldots, n$. The word w_i is obtained from w'_i by deleting all ♦-symbols. The *gap-number* of w_i denotes the number of factors from $♦^+$ that occur in w'_i. The *gap-number* of the above computation of M is the maximum of the gap-numbers of $w_1, \ldots, w_n$. The *gap-complexity* of M on input $w \in L(M)$ is the minimal gap-number of M over all accepting computations of M on input w. Finally, we say that M has *gap-complexity c*, if the gap-complexity of M on each input $w \in L(M)$ is at most c.

The 1-sRR-automaton M_{copy} for the language LR'_2 of Example 3 has gap-complexity 2, while the deterministic t-sRR-automaton M_t of Example 2 has gap-complexity t. Indeed, the accepting computation of M_t on the input $w = c_0abac_1abac_2\ldots c_{t-1}aba$ can be described by the following sequence of words with ♦-symbols:

$$c_0 ♦ bac_1 ♦ bac_2 \ldots c_{t-1} ♦ ba, c_0 ♦^2 ac_1 ♦^2 ac_2 \ldots c_{t-1} ♦^2 a, c_0 ♦^3 c_1 ♦^3 c_2 \ldots c_{t-1} ♦^3.$$

Observe that a part of a computation of a t-sRL-automaton M that has c gaps in each configuration is necessarily c-right-left-monotone. In this way the gap-complexity can be seen as a variation of the notion of j-right-left-monotonicity. From the examples above we obtain the following hierarchy results.

Proposition 1. *For all $c \in \mathbb{N}_+$, the class of languages accepted by sRL-automata with gap-complexity c is properly contained in the class of languages accepted by sRL-automata with gap-complexity $c + 1$.*

On the other hand, the t-sRL-automaton of Corollary 2, although being t-right-left-monotone, has unbounded gap-complexity. In fact, we have the following positive result on the complexity of languages that are accepted by t-sRL-automata with bounded gap-complexity.

Theorem 6. *If M is a t-sRL-automaton with gap-complexity c, then the membership problem for the language $L(M)$ can be solved in time $O(n^{2c+1+min(t,2c)})$.*

Proof. Let $w \in \Sigma^*$ be an input word of length n. With M, w, and the number c we associate a graph $G^{(c)}_M(w) = (V, E)$ as follows. The set V of vertices corresponds to the set of all words over $\Sigma \cup \{♦\}$ of length n that can be obtained from w by replacing symbols from Σ by $\blacklozenge$-symbols, and that have gap-number at most c. Within a word of length n there are $\binom{n+1}{2c}$ positions to place c gaps. Hence, we see that the number of vertices V is in the order of $O(n^{2c})$. The vertex w is called the *initial vertex* of $G_M^{(c)}(w)$, and a vertex $v' \in V$ is a *final vertex* if the word $v := \Pi_\Sigma(v')$ (that is, the projection onto Σ^*) belongs to the set $S(M)$.

Now a directed edge leads from a vertex $u' \in V$ to a vertex $v' \in V$, if M can execute the cycle $u \vdash_M^c v$, where u and v are obtained from u' and v', respectively, by deleting the $\blacklozenge$-symbols. As the graph $G_M^{(c)}(w)$ contains $O(n^{2c})$ many vertices, it is clear that $G_M^{(c)}(w)$ contains at most $O(n^{2c})$ many outgoing edges for any node. Moreover, as M can delete at most t symbols in a cycle, the number of outgoing edges for any node cannot exceed $O(n^t)$. Thus, the graph $G_M^{(c)}(w)$ can be computed from w and c in time $O(n^{2c+1+\min(t,2c)})$.

Now $w \in L(M)$ if and only if a final vertex v' can be reached from the initial vertex w in the graph $G_M^{(c)}(w)$. This can be checked in time $O(|V| + |E|)$, that is, in time $O(n^{2c+1+\min(t,2c)})$. $\square$

Thus, for t-sRL-automata, bounded gap-complexity separates those automata that accept feasible languages from those that accept NP-complete languages.

Corollary 3. *It is decidable in polynomial time whether a t-sRL-automaton M accepts a given word w with gap-complexity at most c.*

Proof. As in the proof of Theorem 6 we can associate with M, w, and the constant c a graph $G_M^{(c)}(w)$. Now M has an accepting computation for input w with gap-complexity at most c if and only if a final vertex v' can be reached from the initial vertex w in the graph $G_M^{(c)}(w)$. As this can be checked in polynomial time, this proves our result. $\square$

For a t-sRL-automaton M and an integer $c \geq 1$, we can thus define the language $L_{\mathrm{gap}}(M, c) := \{\, w \in L(M) \mid M \text{ accepts } w \text{ with gap-complexity at most } c \,\}$.

From Corollary 3 we see that the membership problem for languages of this form is decidable in polynomial time.

5   Analyzing Regular Languages

It is straightforward to see that each finite language is accepted by some t-sRL-automaton working with minimal acceptance. However, as seen in Example 1 the value of t depends on the particular language considered. Also all regular languages are accepted by t-sRL-automata. In fact, we have the following obvious result, where an sRL-automaton M is said to have the $\mathrm{mr}(k)$-*property* if it executes at most k right-move operations in any cycle or tail.

Theorem 7. *A language L is regular if and only if there exists an integer $t \geq 1$ and a deterministic t-sRL-automaton M with the $\mathrm{mr}(t)$-property such that M accepts L working with minimal acceptance.*

The above result implies in particular that all regular languages are accepted by sRL-automata with bounded gap-complexity. It should be interesting to classify regular languages with respect to the smallest value t for which they are accepted in this way, and with respect to the size of the description of these automata.

Acknowledgement. The authors wish to thank Hartmut Messerschmidt for many valuable discussions concerning the notions and results of this paper. They also want to thank an anonymous referee for a very thorough and helpful report.

References

1. G. Buntrock, F. Otto. Growing context-sensitive languages and Church-Rosser languages. *Information and Computation* 141 (1998) 1–36.
2. T. Holan. Dependency analyser configurable by measures. In: P. Sojka, I. Kopeček, K. Pala (eds.), *TSD 2002, Proc.*, *LNCS 2448*, Springer, Berlin, 2002, 81–88.
3. T. Holan, V. Kuboň, K. Oliva, M. Plátek. Two useful measures of word order complexity. In: A. Polguere, S. Kahane (eds.), *Workshop 'Processing of Dependency-Based Grammars,' COLING'98, Proc.*, University of Montreal, 1998, 21–28.
4. P. Jančar, F. Mráz, M. Plátek, J. Vogel. On monotonic automata with a restart operation. *Journal of Automata, Languages and Combinatorics* 4 (1999) 287–311.
5. T. Jurdziński, F. Otto, F. Mráz, M. Plátek. On the complexity of 2-monotone restarting automata. In: C.S. Calude, E. Calude, M.J. Dinneen (eds.), *DLT 2004, Proc.*, *LNCS 3340*, Springer, Berlin, 2004, 237–248.
6. T. Jurdziński, F. Otto, F. Mráz, M. Plátek. On the complexity of 2-monotone restarting automata. *Mathematische Schriften Kassel* 4/04, Universität Kassel, 2004.
7. M. Lopatková, M. Plátek, V. Kuboň. Modeling syntax of free word-order languages: Dependency analysis by reduction. In: V. Matoušek, P. Mautner, T. Pavelka (eds.), *TSD 2005, Proc.*, *LNCS 3658*, Springer, Berlin, 2005, 140–147.
8. M. Lothaire. *Combinatorics on Words*. Encyclopedia of Mathematics, Vol. 17, Addison-Wesley, Reading, 1983.
9. F. Otto. Restarting automata and their relations to the Chomsky hierarchy. In: Z. Esik, Z. Fülöp (eds.), *DLT 2003, Proc.*, *LNCS 2710*, Springer, Berlin, 2003, 55–74.
10. M. Plátek. Two-way restarting automata and j-monotonicity. In: L. Pacholski, P. Ružička (eds.), *SOFSEM 2001, Proc.*, *LNCS 2234*, Springer, Berlin, 2001, 316–325.
11. M. Plátek, F. Otto, F. Mráz. Restarting automata and variants of j-monotonicity. In: E. Csuhaj-Varjú, C. Kintala, D. Wotschke, G. Vaszil (eds.), *DCFS 2003, Proc.*, MTA SZTAKI, Budapest, 2003, 303–312.
12. M. Plátek, F. Otto, F. Mráz, T. Jurdziński. Restarting automata and variants of j-monotonicity. *Mathematische Schriften Kassel* 9/03, Universität Kassel, 2003.
13. S. H. von Solms. The characterization by automata of certain classes of languages in the context sensitive area. *Information and Control* 27 (1975) 262–271.

Noncanonical LALR(1) Parsing

Sylvain Schmitz

Laboratoire I3S, Université de Nice - Sophia Antipolis, France
schmitz@i3s.unice.fr

Abstract. This paper addresses the longstanding problem of the recognition limitations of classical LALR(1) parser generators by proposing the usage of noncanonical parsers. To this end, we present a definition of noncanonical LALR(1) parsers, NLALR(1). The class of grammars accepted by NLALR(1) parsers is a proper superclass of the NSLR(1) and LALR(1) grammar classes. Among the recognized languages are some nondeterministic languages. The proposed parsers retain many of the qualities of canonical LALR(1) parsers: they are deterministic, easy to construct, and run in linear time. We argue that they could provide the basis for a range of powerful noncanonical parsers.

1 Introduction

Testimonies abound on the shortcomings of classical LALR(1) parser generators like YACC [1]. The problem lies in the large *expressivity gap* between what can be specified using the context-free grammar they are fed with, and what can actually be parsed by the LALR(1) automaton they produce. Transforming a grammar until its LALR(1) parser becomes deterministic is arduous, and can obfuscate the attached semantics; moreover, some languages are simply not deterministic.

The expressivity gap vanishes when general parsers [2, 3] are preferred. Such a choice is however done at the expense of the detection of ambiguities. While this might seem acceptable for well established languages, for which the scrutiny of many implementors has pinpointed all ambiguous constructs, there always remains a risk of runtime problems if an unexpected ambiguity appears. The avoidance of such problems is clearly a desirable guarantee, thus motivating our option of restricting to some subclass of the unambiguous grammars.

This paper advocates an almost forgotten way of diminishing the expressivity gap: the usage of *noncanonical parsers*. We apply it to LALR(1) parsing by means of a generic construction. Therefore, we also allow immediate application to other LR-based parsing methods.

Noncanonical parsers have been thoroughly investigated on a theoretical level [4]. Surprisingly, there are very few practical noncanonical parsing methods, and their formal study remains largely unexplored. Indeed, the only one of clear practical interest is an extension to SLR(1) parsing [5]. Noncanonical parsers are however a powerful means of reducing the expressivity gap, while still rejecting any ambiguous syntax. In this they can be compared to LALR(k) parsers with $k > 1$ [6], or, to a larger extent, to parsers allowing unbounded regular

O.H. Ibarra and Z. Dang (Eds.): DLT 2006, LNCS 4036, pp. 95–107, 2006.
© Springer-Verlag Berlin Heidelberg 2006

lookaheads [7, 8, 9]. Like the latter, noncanonical parsers can recognize nondeterministic languages. The classes of grammars accepted by both methods are incomparable in general, but the class of languages accepted by noncanonical parsers is strictly wider than the one accepted by regular lookahead parsers [4]. And there is a winning argument in favor of noncanonical parsers: they can also increase the size of their lookahead window, possibly to an unbounded length [10]. This point motivates our study of noncanonical LALR(1) parsers, since NSLR(1) parsers are unfit for such extensions: their lookahead computation is not contextual.

Also in contrast with NSLR(1), our definitions rely on a prefix equivalence relation: we use the LR(0) equivalence so that the resulting parsers are LALR(1), but coarser equivalences could just as easily be used. Our specific choice of LALR(1) parsers can be explained by their wide adoption, their practical relevance, and the existence of efficient and broadly used algorithms for their generation [11]. We express our computations in the same framework and obtain a simple and efficient practical construction. The additional complexity of generating a NLALR(1) parser instead of a LALR(1) or a NSLR(1) one, as well as the increase of the parser size and the overhead on parsing performances are all quite small. Therefore, the improved parsing power comes at a fairly reasonable price.

The paper is organized as follows: Section 2 briefly introduces noncanonical parsing; Section 3 recalls the formal details of the canonical LALR(1) definition, which will be extended for its noncanonical counterpart in Section 4. We refer the interested reader to a separate research report [12] for a complete study, including grammar classes comparisons, alternative definitions for noncanonical LALR-based parsers, a concrete example of application, and omitted proofs.

Notation. The basic terminology, definitions, and notational conventions used in this paper are classical [13, 14]. Our context-free grammars are reduced and augmented to $\mathcal{G}' = \langle N', T', P', S' \rangle = \langle N \cup \{S'\}, T \cup \{\$\}, P \cup \{S' \rightarrow S\$\}, S' \rangle$. As usual, $A, B, C, \ldots$ denote nonterminals in N'; $a, b, c, \ldots$ denote terminals in T'; $u, v, w, \ldots$ denote strings in T'^*; X, Y, Z denote symbols in V'; $\alpha, \beta, \gamma, \ldots$ denote strings in V'^*; ε is the empty string or empty sequence; $k : \alpha$ is the prefix of length k of string α. Rightmost derivations are denoted by $\underset{\mathrm{rm}}{\Rightarrow}$, whereas leftmost derivations are denoted by $\underset{\mathrm{lm}}{\Rightarrow}$.

2   Noncanonical Parsing

A bottom-up parser reverses the derivation steps which lead to the terminal string it parses. For most bottom-up parsers, including LALR ones, these derivations are rightmost, and therefore the reduced phrase is the leftmost one, called the *handle* of the sentential form.

Noncanonical parsers allow the reduction of phrases which may not be handles [13]. A noncanonical parser is able to suspend a reduction decision where its canonical counterpart would not be deterministic, explore the remaining input, perform some reductions, resume to the conflict point and use nonterminals—resulting from the reduction of a possibly unbounded amount of input—in its lookahead window to infer its parsing decisions.

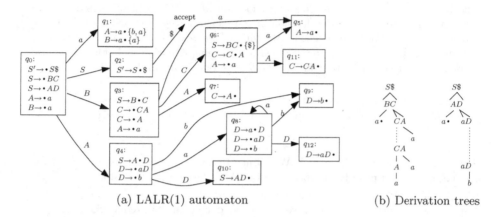

(a) LALR(1) automaton   (b) Derivation trees

Fig. 1. The conflict position in state q_1 for $\mathcal{G}_1$

2.1   Parsing Example

Consider for instance grammar $\mathcal{G}_1$ with rules $S \to BC \mid AD$, $A \to a$, $B \to a$, $C \to CA \mid A$, $D \to aD \mid b$, generating the language $\mathcal{L}_{\mathcal{G}_1} = aa^+ \mid aa^*b$.

The state q_1 in the automaton of Figure 1a is *inadequate*: the parser is unable to decide between reductions $A \to a$ and $B \to a$ when the lookahead is a. We see on the derivation trees of Figure 1b that, in order to choose between the two reductions, the parser has to know if there is a b at the very end of the input. This need for an unbounded lookahead makes $\mathcal{G}_1$ non-LR. A parser using a regular lookahead would solve the conflict by associating the distinct regular lookaheads a^*b and $a^+\$$ with the reductions to A and B respectively.

However, we notice that a single lookahead symbol (D or C) is enough: if the parser is able to explore the context on the right of the conflict, and to reduce some other phrases, then, it will reduce this context to a D or a C. When coming back to the conflict point, it will see a D or a C in the lookahead window.

Table 1 presents a noncanonical parse for a string in $\mathcal{L}_{\mathcal{G}_1}$. The noncanonical machine is not very different from the canonical one, except that it uses two stacks. The additional stack, the *input stack*, contains the (possibly reduced) right context, whereas the other stack is the classical *parsing stack*. Reductions push the reduced nonterminal on top of the input stack. There is no goto operation *per se*: the nonterminal on top of the input stack either allows a parsing decision which had been delayed, or is simply shifted.

We will now see how to transform and extend the canonical LALR(1) parser of Figure 1a to perform these parsing steps.

Table 1. The parse of the string aaa by the NLALR(1) parser for $\mathcal{G}_1$

parsing stack	input stack	actions
q_0	$aaa\$$	shift
$q_0 q_1$	$aa\$$	shift

The inadequate state q_1 is reached with lookahead a. The decision of reducing to A or B can be restated as the decision of reducing the right context to D or C. In order to perform the latter decision, we shift a and reach a state s_1 where we now expect a^*b and $a^*\$$. We are pretty much in the same situation as before: s_1 is also inadequate. But we know that in front of b or $\$$ a decision can be made:

$q_0 q_1 s_1$	$a\$$	shift

There is a new conflict between the reduction $A\to a$ and the shift of a to a position $D\to a\bullet D$. We also shift this a. The expected right contexts are still a^*b and $a^*\$$, so the shift brings us again to s_1:

$q_0 q_1 s_1 s_1$	$\$$	reduce using $A\to a$

The decision is made in front of $\$$. We reduce the a represented by s_1 on top of the parsing stack, and push the reduced symbol A *on top* of the input stack:

$q_0 q_1 s_1$	$A\$$	reduce using $A\to a$

Using this new lookahead, the parser is able to decide another reduction to A:

$q_0 q_1$	$AA\$$	reduce using $B\to a$

We are now back in state q_1. Clearly, there is no need to wait until we see a completely reduced symbol C in the lookahead window: A is already a symbol specific to the reduction to B:

q_0	$BAA\$$	shift
$q_0 q_3$	$AA\$$	shift
$q_0 q_3 q_7$	$A\$$	reduce using $C\to A$
$q_0 q_3$	$CA\$$	shift
$q_0 q_3 q_6$	$A\$$	shift
$q_0 q_3 q_6 q_{11}$	$\$$	reduce using $C\to CA$
$q_0 q_3$	$C\$$	shift
$q_0 q_3 q_6$	$\$$	reduce using $S\to BC$
q_0	$S\$$	shift, and then accept

2.2   Construction Principles

The LALR(1) construction relies heavily on the LR(0) automaton. This automaton provides a nice explanation for LALR lookahead sets: the symbols in the lookahead set for some reduction are the symbols expected next by the LR(0) parser, should it really perform this reduction.

Let us compute the lookahead set for the reduction $A\to a$ in state q_1. Should the LR(0) parser decide to reduce $A\to a$, it would pop q_1 from the parsing stack (thus be in state q_0), and then push q_4. We read directly on Figure 1a that three symbols are acceptable in q_4: D, a and b. Similarly, the reduction $B\to a$ in q_1 has $\{C, A, a\}$ for lookahead set, read directly from state q_3.

The intersection of the lookahead sets for the reductions in q_1 is not empty: a appears in both, which means a conflict. Luckily enough, a is not a *totally reduced symbol*: D and C are reduced symbols, read from kernel items in q_4 and

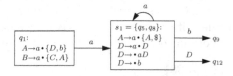

Fig. 2. State q_1 extended for noncanonical parsing

q_3. Conflicting lookahead a could be reduced, and later we might see a symbol on which we can make a decision instead. Thus, we shift the lookahead symbol a in order to reduce it and solve the conflict later. All the other symbols in the computed lookaheads allow to make a decision, so we leave them in the lookaheads sets, but we remove a from both sets.

Shifting a puts us in the same situation we would have been if we followed the transitions on a from both q_3 and q_4, since the noncanonical generation simulates both reductions in q_1. We create a noncanonical transition from q_1 on a to a noncanonical state $s_1 = \{q_5, q_8\}$, which will behave as the union of states q_5 and q_8. State s_1 will thus allow a reduction using $A{\rightarrow}a$ inherited from q_5, and the shifts of a, b and D inherited from q_8. We therefore need to compute the lookaheads for reduction using $A{\rightarrow}a$ in q_5. Using again the LR(0) simulation technique, we see on Figure 1a that this reduction would lead us to either q_7 or to q_{11}. In both cases, the LR(0) automaton would perform a reduction to C that would lead next to q_6. At this point, the LR(0) automaton expects either the end of file symbol \$, should a reduction to S occur, or an A or an a. The complete lookahead set for the reduction $A{\rightarrow}a$ in q_8 is thus $\{A, a, \$\}$.

The new state s_1 is also inadequate: with an a in the lookahead window, we cannot choose between the shift of a and the reduction $A{\rightarrow}a$. As before, we create a new transition on a from s_1 to a noncanonical state $s'_1 = \{q_5, q_8\}$. State q_5 is the state accessed on a from q_6. State q_8 is the state accessed from q_8 if we simulate a shift of symbol a.

State s'_1 is the same as state s_1, and we merge them. The noncanonical computation is now finished. Figure 2 sums up how state q_1 has been transformed and extended. Note that we just use the set $\{q_5, q_8\}$ in a noncanonical LALR(1) automaton; items represented in Figure 2 are only there to ease understanding.

3   LALR(1) Parsers

LALR parsers were introduced as practical parsers for deterministic languages. Rather than building an exponential number of LR(k) states, LALR(k) parsers add lookahead sets to the actions of the small LR(0) parser. We briefly recall some important definitions and results on LR(0) and LALR(1) parsers.

Valid Items and Prefixes. A dotted production $A{\rightarrow}\alpha{\bullet}\beta$ of $\mathcal{G}$ is a *valid LR(0) item* for string γ in V'^* if

$$S' \underset{\mathrm{rm}}{\Rightarrow}{}^* \delta A z \underset{\mathrm{rm}}{\Rightarrow} \delta \alpha \beta z = \gamma \beta z. \tag{1}$$

If such a derivation holds in $\mathcal{G}$, then γ in V'^* is a *valid prefix*.

The set of valid items for a given string γ in V'^* is denoted by $\text{Valid}(\gamma)$. Two strings δ and γ are equivalent if and only if they have the same valid items.

The valid item sets are obtained through the following computations:

$$\text{Kernel}(\varepsilon) = \{S' \to \bullet S\$\}, \tag{2}$$

$$\text{Kernel}(\gamma X) = \{A \to \alpha X \bullet \beta \mid A \to \alpha \bullet X\beta \in \text{Valid}(\gamma)\}, \tag{3}$$

$$\text{Valid}(\gamma) = \text{Kernel}(\gamma) \cup \{B \to \bullet\omega \mid A \to \alpha \bullet B\beta \in \text{Valid}(\gamma)\}. \tag{4}$$

LR(0) States. LR automata are pushdown automata that use equivalence classes on valid prefixes as their stack alphabet Q. We therefore denote explicitly states of a LR parser as $q = [\delta]$, where δ is some valid prefix in q the state reached upon reading this prefix. For instance, in the automaton of Figure 1a, state q_2 is the equivalence class $\{S\}$, while state q_8 is the equivalence class described by the regular language Aa^*a.

A pair $([\delta], X)$ in $Q \times V$ is a *transition* if and only if δX is a valid prefix. If this is the case, then $[\delta X]$ is the state accessed upon reading δX, thus the notation $[\delta X]$ also implies[1] a transition from $[\delta]$ on X, and $[\delta\alpha]$ a *path* on α.

LALR(1) Automata. The *LALR(1) lookahead set* of a reduction using $A \to \alpha$ in state q is

$$\text{LA}(q, A \to \alpha) = \{1{:}z \mid S' \underset{\text{rm}}{\Rightarrow}^* \delta Az \text{ and } q = [\delta\alpha]\}. \tag{5}$$

4   NLALR(1) Parsers

There is a number of differences between the LALR(1) and NLALR(1) definitions. The most visible one is that we accept nonterminals in our lookahead sets. We also want to know which lookahead symbols are totally reduced. Finally, we are adding new states, which are sets of LR(0) states. Therefore, the objects in most of our computations will be LR(0) states.

4.1   Valid Covers

We have recalled in the previous section that LR(0) states can be viewed as collections of valid prefixes. A similar definition for NLALR(1) states would be nice. However, due to the suspended parsing actions, the language of all prefixes accepted by a noncanonical parser is no longer a regular language. This means the parser will only have a regular approximation of the exact parsing stack language. The noncanonical states, being sets of LR(0) states (*i.e.*, sets of equivalence classes on valid prefixes), provide this approximation. We therefore define valid covers as valid prefixes covering the parsing stack language.

Definition 1. *String γ is a* valid cover *in $\mathcal{G}$ for string δ if and only if γ is a valid prefix and $\gamma \Rightarrow^* \delta$. We write $\hat{\delta}$ to denote some cover of δ and $\text{Cover}(L)$ to denote the set of all valid covers for the set of strings L.*

---

[1] We always assume when writing $[\delta X]$ that $\text{Valid}(\delta X)$ is not the empty set.

Remember for instance configuration $q_0q_1\|aa\$$ from Table 1. This configuration leads to pushing state $s_1 = \{q_5, q_8\}$, where both valid prefixes $(B|BC)a$ and Aa^*a of q_5 and q_8 are valid covers for the actual parsing stack prefix aa. Thus in s_1 we cover the parsing stack prefix by $(B \mid BC \mid Aa^*)a$.

4.2   Noncanonical Lookaheads

Noncanonical lookaheads are symbols in V'. Adapting the computation of the LALR(1) lookahead sets is simple, but a few points deserve some explanations.

First of all, noncanonical lookahead symbols have to be *non null, i.e.* X is non null if $X \Rightarrow^* ax$. Indeed, null symbols do not provide any additional right context information—worse, they can hide it. If we consider that we always perform a reduction at the earliest parsing stage possible, then they will never appear in a lookahead window.

Totally Reduced Lookaheads. Totally reduced lookaheads form a subset of the noncanonical lookahead set such that none of its elements can be further reduced. A conflict with a totally reduced symbol as lookahead of a reduction cannot be solved by a noncanonical exploration of the right context, since there is no hope of ever reducing it any further.

We define here totally reduced lookaheads as non null symbols which can follow the right part of the offending rule in a leftmost derivation.

Definition 2. *The set of* totally reduced lookaheads *for a reduction $A \rightarrow \alpha$ in LR(0) state q is defined by*

$$RLA(q, A \rightarrow \alpha) = \{X \mid S' \underset{lm}{\Rightarrow}^* zA\gamma X\omega, \gamma \Rightarrow^* \varepsilon, X \Rightarrow^* ax, \text{ and } q = [\hat{z}\alpha]\}.$$

Derived Lookaheads. The derived lookahead symbols are simply defined by extending (5) to the set of all non null symbols in V.

Definition 3. *The set of* derived lookaheads *for a reduction $A \rightarrow \alpha$ in LR(0) state q is defined by*

$$DLA(q, A \rightarrow \alpha) = \{X \mid S' \Rightarrow^* \delta AX\omega, X \Rightarrow^* ax, \text{ and } q = [\hat{\delta}\alpha]\}.$$

We obviously have that

$$LA(q, A \rightarrow \alpha) = DLA(q, A \rightarrow \alpha) \cap T'. \tag{6}$$

Conflicting Lookahead Symbols. Last, we need to compute which lookahead symbols would make the state inadequate. A noncanonical exploration of the right context is required for these symbols. They appear in the derived lookahead sets of several reductions and/or are transition labels. However, the totally reduced lookaheads of a reduction are not part of this lookahead set, for if they are involved in a conflict, then there is no hope of being able to solve it.

Definition 4. Conflicts lookahead set *for a reduction using $A{\to}\alpha$ in set s of LR(0) states is defined as*

$$CLA(s, A{\to}\alpha) = \{X \in DLA(q, A{\to}\alpha) \mid q \in s, X \notin RLA(q, A{\to}\alpha),$$
$$(q, X) \text{ or } (\exists p \in s, X \in DLA(p, B{\to}\beta))\}.$$

We then define the noncanonical lookahead set *for a reduction using $A{\to}\alpha$ in set s of LR(0) states as*

$$NLA(s, A{\to}\alpha) = (\bigcup_{q \in s} DLA(q, A{\to}\alpha)) - CLA(s, A{\to}\alpha).$$

We illustrate these definitions by computing the lookahead sets for the reduction using $A{\to}a$ in state $s_1 = \{q_5, q_8\}$ as in Section 2.2: $RLA(q_5, A{\to}a) = \{A, \$\}$, $DLA(q_5, A{\to}a) = \{A, a, \$\}$, $CLA(s_1, A{\to}a) = \{a\}$ and $NLA(s_1, A{\to}a) = \{A, \$\}$.

4.3   Noncanonical States

We said at the beginning of this section that states in the NLALR(1) automaton were in fact sets of LR(0) states. We denote by $[\![\delta]\!]$ the noncanonical state accessed upon reading string δ in V'^*.

Definition 5. Noncanonical state $[\![\delta]\!]$ *is the set of LR(0) states defined by*

$$[\![\varepsilon]\!] = \{[\varepsilon]\} \text{ and}$$
$$[\![\delta X]\!] = \{[\widehat{\hat{\gamma}AX}] \mid X \in CLA([\![\delta]\!], A{\to}\alpha), [\hat{\gamma}\alpha] \in [\![\delta]\!]\} \cup \{[\varphi X] \mid [\varphi] \in [\![\delta]\!]\}.$$

Noncanonical transition *from $[\![\delta]\!]$ to $[\![\delta X]\!]$ on symbol X, denoted by $([\![\delta]\!], X)$, exists if and only if $[\![\delta X]\!] \neq \emptyset$. Reduction $([\![\delta]\!], A{\to}\alpha)$ exists if and only if there exists a reduction $(q, A{\to}\alpha)$ and q is in $[\![\delta]\!]$.*

Note that these definitions remain valid for plain LALR(1) states since, in absence of a conflict, a noncanonical state is a singleton set containing the corresponding LR(0) state.

A simple induction on the length of δ shows that the LR(0) states considered in the noncanonical state $[\![\delta]\!]$ provide a valid cover for any accessing string of the noncanonical state. It basically means that the actions decided in a given noncanonical state make sense at least for a cover of the real sentential form prefix that is read.

The approximations done when covering the actual sentential form prefix are made on top of the previous approximations: with each new conflict, we need to find a new set of LR(0) states covering the parsing stack contents. This stacking is made obvious in the above definition when we write $\widehat{\hat{\gamma}AX}$. It means that NLALR(1) parsers are not prefix valid, but prefix cover valid.

Throughout this paper, we use the LR(0) automaton to approximate the prefix read so far. We could use more powerful methods—but it would not really be in the spirit of LALR parsing any longer; see [12] for alternative methods.

4.4 NLALR(1) Automata

Here we formalize noncanonical LALR(1) parsing machines. They are a special case of two-stack pushdown automata (2PDA). As said before, the additional stack serves as an input for the parser, and reductions push the reduced nonterminal on top of this stack. This behavior of reductions excepted, the definition of a NLALR(1) automaton is similar to the LALR(1) one.

Definition 6. *Let $M = (Q \cup V \cup \{\$, \|\}, R)$ be a rewriting system. A configuration of M is a string of the form*

$$[\![\varepsilon]\!][\![X_1]\!] \ldots [\![X_1 \ldots X_n]\!] \| \omega \$$$

where $X_1 \ldots X_n$ and ω are strings in V^*. We say that M is a NLALR(1) automaton if its initial configuration is $[\![\varepsilon]\!]\|w\$$ with w the input string in T^*, its final configuration is $[\![\varepsilon]\!][\![S]\!]\|\$$, and if each rewriting rule in R is of the form

- shift X *in state* $[\![\delta]\!]$, *defined if there is a transition* $([\![\delta]\!], X)$

$$[\![\delta]\!]\|X \vdash_{shift} [\![\delta]\!][\![\delta X]\!]\|,$$

- or reduce *by rule* $A {\to} X_1 \ldots X_n$ *of P in state* $[\![\delta X_1 \ldots X_n]\!]$ *with lookahead X, defined if $A {\to} X_1 \ldots X_n$ is a reduction in $[\![\delta X_1 \ldots X_n]\!]$ and lookahead X is in $NLA([\![\delta X_1 \ldots X_n]\!], A {\to} X_1 \ldots X_n)$*

$$[\![\delta X_1]\!] \ldots [\![\delta X_1 \ldots X_n]\!]\|X \vdash_{A \to X_1 \ldots X_n} \|AX.$$

The following rules illustrate Definition 6 on state s_1 of the NLALR(1) automaton for $\mathcal{G}_1$: $s_1\|a \vdash_{shift} s_1 s_1\|$, $s_1\|b \vdash_{shift} s_1\{q_9\}\|$, $s_1\|D \vdash_{shift} s_1\{q_{12}\}\|$, $s_1\|A \vdash_{A \to a}\|AA$ and $s_1\|\$ \vdash_{A \to a}\|A\$$.

According to Definition 6, NLALR(1) automata are able to backtrack by a limited amount, corresponding to the length of their window, at reduction time only. We know that noncanonical parsers using a bounded lookahead window operate in linear time [4]; the following theorem precisely shows that the total number of rules involved in the parsing of an input string is linear in respect with the number of reductions performed, which itself is linear with the input string length. This theorem uses an output effect τ which outputs the rules used for each reduction performed by M; we then call (M, τ) a NLALR(1) parser.

Theorem 1. *Let $\mathcal{G}$ be a grammar and (M, τ) its NLALR(1) parser. If π is a parse of w in M, then the number of parsing steps $|\pi|$ is related to the number $|\tau(\pi)|$ of derivations producing w in $\mathcal{G}$ and to the length $|w|$ of w by*

$$|\pi| = 2|\tau(\pi)| + |w|.$$

Since all the conflict lookahead symbols are removed from the noncanonical lookahead sets NLA, the only possibility for the noncanonical automaton to be nondeterministic would be to have a totally reduced symbol causing a conflict.

A context-free grammar $\mathcal{G}$ is NLALR(1) if its NLALR(1) automaton is deterministic, and thus if no totally reduced symbol can cause a conflict.

4.5 Computing the Lookaheads and Covers

The LALR(1) lookahead sets that are defined in Equation (5) can be expressed using the following definitions [11], where **lookback** is a relation between reductions and nonterminal LR(0) transitions, **includes** and **reads** are relations between nonterminal LR(0) transitions, and DR—standing for *directly reads*—is a function from nonterminal LR(0) transitions to sets of lookahead symbols.

$$([\delta\alpha], A\to\alpha) \textbf{ lookback } ([\delta], A), \tag{7}$$
$$([\delta\beta], A) \textbf{ includes } ([\delta], B) \text{ iff } B\to\beta A\gamma \text{ and } \gamma\Rightarrow^*\varepsilon, \tag{8}$$
$$([\delta], A) \textbf{ reads } ([\delta A], C) \text{ iff } ([\delta A], C) \text{ and } C\Rightarrow^*\varepsilon, \tag{9}$$
$$\text{DR}([\delta], A) = \{a \mid ([\delta A], a)\}. \tag{10}$$

Using the above definitions, we can rewrite Equation (5) as

$$\text{LA}(q, A\to\alpha) = \bigcup_{(q,A\to\alpha) \textbf{ lookback } \circ \textbf{ includes}^* \circ \textbf{ reads}^* (r,C)} \text{DR}(r, C). \tag{11}$$

This computation for LALR(1) lookahead sets is highly efficient. It can entirely be performed on the LR(0) automaton, and the union can be interleaved with a fast transitive closure algorithm [15] on the **includes** and **reads** relations.

Since we have a very efficient and widely adopted computation for the canonical LALR(1) lookahead sets, why not try to use it for the noncanonical ones?

Theorem 2

$$RLA(q, A\to\alpha) = \{X \mid X\Rightarrow^*ax, \psi\Rightarrow^*\varepsilon, C\Rightarrow\rho B\bullet\psi X\sigma \in Kernel(\delta\rho B) \text{ and}$$
$$(q, A\to\alpha) \textbf{\textit{ lookback }} \circ \textbf{\textit{ includes}}^*([\delta\rho], B)\}.$$

This theorem is consistent with the description of Section 2.2, where we said that C was a totally reduced lookahead for reduction $B\to a$ in q_1: item $S\to B\bullet C$ is in the kernel of state q_3 accessed by (q_0, B), and $(q_1, B\to a) \textbf{ lookback } (q_0, B)$.

Theorem 3. *Let us extend the* directly reads *function of (10) to*

$$DR([\delta], A) = \{X \mid ([\delta A], X) \text{ and } X\Rightarrow^*ax\}; \text{ then}$$
$$DLA(q, A\to\alpha) = \bigcup_{(q,A\to\alpha) \textbf{\textit{ lookback }} \circ \textbf{\textit{ includes}}^* \circ \textbf{\textit{ reads}}^* (r,C)} DR(r, C).$$

We are still consistent with the description of Section 2.2 since, using this new definition of the DR function, DR(q_0, B) is $\{a, C, A\}$.

To find the valid covers that approximate a sentential form prefix using the LR(0) automaton and to find the LALR lookahead sets wind up being very similar operations. This allows us to reuse our relational computations for the automaton construction itself, as illustrated by the following theorem.

Theorem 4. *Noncanonical state* $[\![\delta]\!]$ *is the set of LR(0) states defined by*

$$[\![\varepsilon]\!] = \{[\varepsilon]\} \ and$$
$$[\![\delta X]\!] = \quad \{[\gamma C X] \mid X \in CLA([\![\delta]\!], A {\rightarrow} \alpha), q \in [\![\delta]\!] \ and$$
$$(q, A {\rightarrow} \alpha) \ \textbf{\textit{lookback}} \circ \textbf{\textit{includes}}^* \circ \textbf{\textit{reads}}^*([\gamma], C)\}$$
$$\cup \ \{[\varphi X] \mid [\varphi] \in [\![\delta]\!]\}.$$

4.6 Practical Construction Steps

We present here a more informal construction, with the main steps leading to the construction of a NLALR(1) parser, given the LR(0) automaton.

1. Associate a noncanonical state $s = \{q\}$ with each LR(0) state q.
2. Iterate while there exists an inadequate[2] state s:
   (a) if it has not been done before, compute the RLA and DLA lookahead sets for the reductions involved in the conflict; save their values for the reduction and LR(0) state involved;
   (b) compute the CLA and NLA lookahead sets for s;
   (c) set the lookaheads to NLA for the reduction actions in s;
   (d) – if the NLA lookahead sets leave the state inadequate, meaning there is a conflict on a totally reduced lookahead, then report the conflict, and use a conflict resolution policy or terminate with an error;
       – if CLA is not empty, create transitions on its symbols and create new states if no fusion occurs. New states get new transition and reduction sets computed from the LR(0) states they contain. If these new states result from shift/reduce conflicts, the transitions from s on the conflicting lookahead symbol now lead to the new states.

This process always terminates since there is a bounded number of LR(0) states and thus a bounded number of noncanonical states.

Let us conclude this section with a few words on the size of the generated parsers. Since NLALR(1) states are sets of LR(0) states, we find an exponential function of the size of the LR(0) automaton as an upper bound on the size of the NLALR(1) automaton. This bound seems however pretty irrelevant in practice. The NLALR(1) parser generator needs to create a new state for each lookahead causing a conflict, which does not happen so often. All the grammars we studied created transitions to canonical states very quickly afterwards. Experimental results with NSLR(1) parsers show that the increase in size is negligible in practice [5].

---

[2] We mean here inadequate in the LR(0) sense, thus no lookaheads need to be computed yet.

5   Conclusion

We have presented a construction for noncanonical LALR(1) parsers. Such parsers are practical for some difficult syntax problems. They improve on both noncanonical SLR(1) parsers and canonical LALR(1) parsers, and their generation is only slightly more complex while their size and their performances are comparable.

For practical uses, we feel we would need an unbounded lookahead version of NLALR parsers. Though the cost to pay might be a quadratic parsing time in the worst case, the freedom offered to the grammar writer would probably be worth it. The ability to specify coarser equivalence relations instead of the LR(0) one would prove its usefulness in this setting where precision becomes critical.

In complement to previous theoretical work on noncanonical parsing [4], it would be interesting to formally study practical noncanonical parsers. To this end, we expect the concept of valid covers modulo an equivalence relation to be a good starting point.

Acknowledgements. The author is highly grateful to Jacques Farré and Ana Almeida Matos for their invaluable help in the preparation of this paper.

References

1. Johnson, S.C.: YACC — yet another compiler compiler. Computing science technical report 32, AT&T Bell Laboratories, Murray Hill, New Jersey (1975)
2. Earley, J.: An efficient context-free parsing algorithm. Communications of the ACM **13**(2) (1970) 94–102
3. Tomita, M.: Efficient Parsing for Natural Language. Kluwer Academic Publishers (1986)
4. Szymanski, T.G., Williams, J.H.: Noncanonical extensions of bottom-up parsing techniques. SIAM Journal on Computing **5**(2) (1976) 231–250
5. Tai, K.C.: Noncanonical SLR(1) grammars. ACM Transactions on Programming Languages and Systems **1**(2) (1979) 295–320
6. Charles, P.: A Practical method for Constructing Efficient LALR(k) Parsers with Automatic Error Recovery. PhD thesis, New York University (1991)
7. Čulik, K., Cohen, R.: LR-Regular grammars—an extension of LR(k) grammars. Journal of Computer and System Sciences **7** (1973) 66–96
8. Bermudez, M.E., Schimpf, K.M.: Practical arbitrary lookahead LR parsing. Journal of Computer and System Sciences **41**(2) (1990) 230–250
9. Farré, J., Fortes Gálvez, J.: A bounded-connect construction for LR-regular parsers. In Wilhelm, R., ed.: CC'01. Volume 2027 of Lecture Notes in Computer Science., Springer (2001) 244–258
10. Farré, J., Fortes Gálvez, J.: Bounded-connect noncanonical discriminating-reverse parsers. Theoretical Computer Science **313**(1) (2004) 73–91
11. DeRemer, F., Pennello, T.: Efficient computation of LALR(1) look-ahead sets. ACM Transactions on Programming Languages and Systems **4**(4) (1982) 615–649
12. Schmitz, S.: Noncanonical LALR(1) parsing. Technical Report I3S/RR-2005-21-FR, Laboratoire I3S (2005) URL `http://www.i3s.unice.fr/~mh/RR/2005/RR-05.21-S.SCHMITZ.pdf`.

13. Aho, A.V., Ullman, J.D.: The Theory of Parsing, Translation, and Compiling. Volume I: Parsing of Series in Automatic Computation. Prentice Hall, Englewood Cliffs, New Jersey (1972)
14. Sippu, S., Soisalon-Soininen, E.: Parsing Theory. Volume II: LR(k) and LL(k) Parsing of EATCS Monographs on Theoretical Computer Science. Springer (1990)
15. Tarjan, R.E.: Depth first search and linear graph algorithms. SIAM Journal on Computing **1**(2) (1972) 146–160

Context-Free Grammars and XML Languages*

Alberto Bertoni[1], Christian Choffrut[2], and Beatrice Palano[1]

[1] Dipartimento di Scienze dell'Informazione, Università degli Studi di Milano
Via Comelico 39/41, 20135 Milano – Italy
{bertoni, palano}@dsi.unimi.it
[2] L.I.A.F.A., Université Paris VII, 2 Place Jussieu, 75221 Paris – France
Christian.Choffrut@liafa.jussieu.fr

Abstract. We study the decision properties of XML languages. It was known that given a context-free language included in the Dyck language with sufficiently many pairs of parentheses, it is undecidable whether or not it is an XML language. We improve on this result by showing that the problem remains undecidable when the language is written on a unique pair of parentheses. We also prove that if the given language is deterministic, then the problem is decidable; while establishing whether its surfaces are regular turns out to be undecidable whenever the deterministic language is contained in the Dyck language with two pairs of parentheses. Our results are based on a "pumping property" of what Boasson and Berstel call the surface of a context-free language.

1 Introduction

The World Wide Web Consortium has adopted XML (Extended Markup Language) as standard format for data exchanging on the Web. XML is a markup language [8]: words are documents composed by text and markups called tags. Data are represented by text and tags are used to give information about text blocks.

Tags act as parenthesis: they can be open or closed. Each opening tag has a unique associated closing tag, and conversely. Informally, an XML document is "correct" if the sequence w of tags in the document is a well-formed prime parenthesized word, that is, w belongs to the Dyck language [3] D_A defined on the set A of different tags and w cannot be decomposed as product of words in D_A. A Document Type Definition (DTD) is used to set the different types of tags deserving to be admitted in an XML documents; a DTD also states how tags can be nested in the document. By considering only the syntactic part, a DTD can be view as a particular type of context-free grammar G. A XML document d is said to be "valid" for a given DTD G if the sequence w of tags in d belongs to the language generated by G.

The aim of [1] is the analysis of this particular class of grammars and the corresponding class of languages, called XML-languages. In that paper the notion

---

* Partially supported by MURST, under the project "COFIN: Automi e linguaggi formali: aspetti matematici ed applicativi".

O.H. Ibarra and Z. Dang (Eds.): DLT 2006, LNCS 4036, pp. 108–119, 2006.

of surfaces of a language is introduced, and it is shown that this notion is a key-concept for XML-languages. Informally, the surface of an opening tag a is the set of sequences of opening tags that are sons of a (i.e., the tags immediately under a that may follow a in a document before the closing tag $\bar{a}$ is reached). A characterization of XML-languages, based on surfaces is given in [1]: $L \subseteq D_A$ is a XML-language iff, for every $a \in A$, the surface S_a of L is regular and L is the maximum among the languages having S_a's as surfaces. Moreover, some decision problems are studied. In particular, it is shown that it is decidable whether a context-free language is contained in D_A, but it is undecidable whether a context-free language is an XML-language, for sufficiently large A. They also prove that it is undecidable whether the surfaces of a context-free language $L \subseteq D_A$ are regular, for sufficiently large A, but it is decidable whether the surfaces of L are finite.

In this paper we prove that every surface of a context-free language $L \subseteq D_A$ satisfies the pumping lemma. This explains why it is decidable whether the surfaces of a context-free language are finite; a furthermore consequence is that the surfaces of context-free language $L \subseteq D_A$ are regular if $|A| = 1$. So, the problem to establishing whether the surfaces of a context-free language $L \subseteq D_A$ are regular is trivially decidable if $|A| = 1$. However, enough surprisingly, we prove that the problem to establishing whether a context-free language $L \subseteq D_A$ is an XML-language is undecidable even if $|A| = 1$. For $|A| > 1$, we prove that it is undecidable whether a deterministic context-free language contained in D_A admits regular surfaces; however, enough surprisingly, we prove that the problem to establishing whether a deterministic context-free language is an XML-language becomes decidable.

2   Preliminaries

In this Section we introduce concepts and notations useful through all the paper.

XML Languages. Let A be a finite set of opening tags and let $\overline{A}$ be an isomorphic copy of the corresponding closing tags. Set $T = A \cup \overline{A}$; a word $x \in T^*$ is correctly parenthesized if any opening tags is followed by a corresponding closing tag $\bar{a}$ and, if tag a' follows a, then a' should be closed before a. The Dyck language D_A is the set of correctly parenthesized word $x \in T^*$. Let $P_a = aT^*\bar{a} \cap D_A$; the language $P_A = \bigcup_{a \in A} P_a$ is the set of Dyck *primes* over A. P_A is a bifix code, i.e., no word in P_A is suffix or prefix of another word in P_A and every word in D_A can be univocally factorized by Dyck primes. An *XML-document* is a word d over T; it is *well-formed* if $d \in P_A$. An *XML-grammar* is a system composed by a terminal alphabet $T = A \cup \overline{A}$, a set of variables $V_A = \{X_a \mid a \in A\}$, a distinguished variable X_s called *axiom* and, for each $a \in A$, a regular language $R_a \subset V_A^*$ defining the (possible infinite) set of productions

$$\{X_a \rightarrow aw\bar{a} \mid w \in R_a\} \quad \text{(in short } X_a \rightarrow aR_a\bar{a}\text{).}$$

An *XML-language* is a language generated by an XML-grammar. It can be proved that every XML language is a deterministic context-free language [3]. The converse does not hold.

Example 1. P_s is generated by the XML-grammar $\langle A \cup \overline{A}, V_A, X_s, \mathcal{P} \rangle$ where the production rules in $\mathcal{P}$ are $X_a \rightarrow a \left(\sum_{b \in A} X_b \right)^* \overline{a}$; hence P_s is an XML-language; on the contrary, if $|A| > 1$, then P_A is deterministic contex free language that is not an XML language.

Any word $w \in D_A$ satisfying $w \neq \varepsilon$, can be univocally factorized as $w = w_1 \cdots w_m$, where w_i's belong to P_A. The *type* is the morphism of the sub-monoid D_A into A^* defined by $\mathtt{type}(aw\overline{a}) = a$ for all prime $aw\overline{a} \in D_A$. E.g., $\mathtt{type}(aa\overline{a}\overline{a}bb\overline{b}\overline{b}\overline{b}\overline{b}a\overline{a}) = abba$. The *trace* of a prime $q = aw\overline{a}$ is the type of w: $\mathtt{trace}(q) = \mathtt{trace}(aw\overline{a}) = \mathtt{type}(w)$. Let $\mathrm{Fact}(L)$ be the set of factors of words in L, i.e., $\mathrm{Fact}(L) = \{z \mid yzy' \in L, \text{ for some words } y, y'\}$ and let $F_a(L) = \mathrm{Fact}(L) \cap P_a$. Let $L \subseteq D_A$; then we call a-*surface* of L $S_a(L) = \{\mathtt{trace}(w) \mid w \in F_a(L)\}$.

Example 2. For the language $L = \{ab^n\overline{b}^n(c\overline{c})^m\overline{a} \mid n \geq 1, m \geq 0\}$, the surfaces are easily seen to be $S_c(L) = \{\varepsilon\}$, $S_b(L) = \{b, \varepsilon\}$ and $S_a(L) = \{bc^m \mid m \geq 0\}$.

Let ϕ be the function that associates with every language $L \subseteq D_A$ the class of its surfaces, i.e., $\phi(L) = \{S_a(L) \mid a \in A\}$.

In [1, Theorem 4.2.] is given the following characterization of XML-languages in terms of their surfaces. It says that given a collection of regular languages indexed by the letters of the alphabet, there exists an XML language having this collection as set of surfaces and that this language is the largest with this set of surfaces.

Theorem 1. $L \subseteq D_A$ *is an XML-language if and only if* $\phi(L)$ *is a class of regular languages and* $L = \bigcup_{\{X \subseteq D_A \mid \phi(X) = \phi(L)\}} X$.

Reduced Words. It is well-known that a word belongs to the Dyck language if and only if it can be reduced to the empty word through productions erasing factors of the form $a\overline{a}$ where $a \in A$. Formally, the *Dyck reduction* on the alphabet A is the *semi-Thue system* $\mathcal{S}$ defined by the *rules* $a\overline{a} \rightarrow \varepsilon$, for all $a \in A$. A word z is said *irreducible with respect to* $\mathcal{S}$ or simply *irreducible* when $\mathcal{S}$ is understood, if it has no factor equal to $a\overline{a}$, for every $a \in A$. Every word $w \in T^*$ can be reduced to an irreducible word obtained by iteratively the above rules on w. Since this reduction is terminating and confluent, a word possesses a unique reduced form denoted by $\rho(w)$ and called *the reduced of* w. Notice that for any factor x of words in D_A, we have $\rho(x) \in \overline{A}^* A^*$. The following lemma holds for Dyck reductions:

Lemma 1. *If* x *and* y *are irreducible words, then there exist two unique factor-izations* $x = x_1 x_2$, $y = y_1 y_2$ *such that* $\rho(xy) = x_1 y_2$, $|x_2| = |y_1|$ *and* $\rho(x_2 y_1) = \varepsilon$.

Lemma 2. *Let* s *be a suffix (resp.* p *a prefix) of a Dyck word and let* x *and* y *be words in* A^*. *Then* $\rho(sp) = \overline{x}y$ *implies* $\rho(s) = \overline{x}$ *and* $\rho(p) = y$.

We extend in natural way ρ to languages $L \subseteq T^*$: $\rho(L) = \{\rho(w) \mid w \in L\}$. The crucial observation is that $\emptyset \neq L \subseteq D_A$ holds if and only if $\rho(L) = \{\varepsilon\}$.

3 Pumping Lemma for Surfaces of Context-Free Languages

In this section we prove that, if L is a context-free language and $L \subseteq D_A$ then, for every $a \in A$, the surface $S_a(L)$ satisfies the pumping lemma as defined for the context-free languages. As a consequence, if $|A| = 1$ then $S_a(L)$ is a regular language.

We first recall the following technical Lemma.

Lemma 3. [1, Lemma 6.3] *If there exists the derivation $Y \Rightarrow^+ sYd$, with variable Y and words $s, d \in (A \cup \overline{A})^*$, then there exist words $x, y, q, p \in A^*$ such that p, q are conjugate (i.e., p and q have same length and p is a factor of q^2) and*

$$\rho(s) = \overline{x}px \quad and \quad \rho(d) = \overline{y}\,\overline{q}y.$$

The pair (s, d) of a derivation $Y \overset{+}{\Rightarrow} sYd$ is called *flat* if $\rho(s) = \overline{x}x$ and $\rho(d) = \overline{y}y$, i.e., the word q of Lemma 3 is the empty word. The following property holds for a flat pair (the proof is omitted)

Lemma 4. *Let (s, d) a flat pair and d (resp. s) a proper factor of a Dyck prime w, i.e., $w = w'dw''$ (resp. $w = w'sw''$). Then, for all integers $k \geq 0$, the word $w'd^kw''$ (resp. $w = w's^kw''$) is a Dyck prime, and $\mathtt{type}(w'd^kw'') = \mathtt{type}(w)$ (resp. $\mathtt{type}(w's^kw'') = \mathtt{type}(w)$).*

Now consider a reduced context-free grammar $G = \langle T = A \cup \overline{A}, \{X_1, \ldots, X_M\}, \mathcal{P}, X_1 \rangle$, where $\mathcal{P}$ are production rules in Chomsky Normal Form. Let $L_k = \{w \in T^* \mid X_k \Rightarrow^* w\}$ and $\mathrm{Irr}_k = \rho(L_k)$, so that L_1 is the language generated by G. We suppose that $L_1 \subseteq D_A$, or equivalently $\rho(L_1) = \{\varepsilon\}$; in [1] it is proved that $|\mathrm{Irr}_k| < \infty$, for $1 \leq k \leq M$. The following lemma is a refinement (the proof is omitted)

Lemma 5. *For $1 \leq k \leq M$, there are $a, b, c \in A^*$ with $|a| + |b| + 2|c| \leq 4^M$, ε the maximum common prefix of a and b, and $\mathrm{Irr}_k \subseteq \{\overline{a}\,\overline{w}wb \mid w \text{ suffix of } c\}$. In particular, the words in Irr_k have length at most 4^M.*

The upper bound given in the previous lemma has a consequence on the existence of flat pairs in a given sequence of derivations. In fact, it holds the following lemma

Lemma 6. *Let Y be a nonterminal of G and*

$$Y \overset{+}{\Rightarrow} s_1 u_1 p_1 Y d_1, \quad \cdots, \quad Y \overset{+}{\Rightarrow} s_f u_f p_f Y d_f$$

be a sequence of derivations in G such that each s_j (resp. p_j) is a suffix (resp. a prefix) of a suitable Dyck prime, u_j is a Dyck word, for $1 \leq j \leq f$, and $p_j s_{j+1}$ is a Dyck prime, for $1 \leq j < f$. If $f \geq 4^M + k$, at least k pairs $(s_j u_j p_j, d_j)$'s in the sequence of derivation are flat.

Proof. By applying Lemma 3, we get

$$(a) \quad \rho(s_j u_j p_j) = \overline{x_j} \gamma_j x_j \quad \text{and} \quad \rho(d_j) = \overline{y_j} \, \overline{\delta_j} y_j \quad (b). \tag{1}$$

where γ_j and δ_j are conjugate. Since $\rho(s_j u_j p_j) = \rho(\rho(s_j)\rho(u_j)\rho(p_j)) = \rho(s_j)\rho(p_j)$ holds, the equation (1.a) yields

$$(a) \quad \rho(s_j) = \overline{x_j} \quad \text{and} \quad \rho(p_j) = \gamma_j x_j \quad (b). \tag{2}$$

Hence we have the following equalities

$$\begin{aligned}
\rho(s_{j+1}) &= \overline{\rho(p_j)} & &\text{since } p_j s_{j+1} \text{ is a Dyck prime} \\
&= \overline{x_j} \, \overline{\gamma_j} & &\text{by Equation (2.b)} \\
&= \rho(s_j)\overline{\gamma_j} & &\text{by Equation (2.a).}
\end{aligned}$$

The last equality applied recursively implies $\rho(s_f) = \rho(s_1)\overline{\gamma_1} \ldots \overline{\gamma_f}$. Let $t \in (A \cup \overline{A})^*$ be a word such that, $Y \Rightarrow^* t$, so that $Y \Rightarrow^* s_f u_f p_f t d_f$. Since $\rho(u_f) = \varepsilon$, it holds

$$\rho(s_f u_f p_f t d_f) = \overline{x_1} \, \overline{\gamma_1} \ldots \overline{\gamma_f} \rho(p_f t d_f).$$

As a consequence of Lemma 5, we have

$$4^M \geq |\rho(s_f u_f p_f t d_f)| \geq |\gamma_1| + \cdots + |\gamma_f|.$$

If $f \geq 4^M + k$, at least k words in $\gamma_1, \ldots, \gamma_f$ must be equal to ε. Hence the result. $\qquad \square$

By definition of XML languages, the surface of such a language relative to an arbitrary letter a of the alphabet is regular. In fact, a weaker result holds for arbitrary context-free languages included in the Dyck language, as they all satisfy a pumping property.

Theorem 2. (PUMPING LEMMA FOR SURFACES)
Let $L \subseteq D_A$ be a context-free language and $a \in A$. Then the surface language $S_a(L)$ satisfy the pumping property, i.e., there exists a positive constant H such that every word $z \in S_a(L)$, with $|z| \geq H$, can be factorized as $z = uvwxy$ such that the following holds:
- *$vx \neq \varepsilon$,*
- *$|vwx| \leq H$,*
- *$uv^k wx^k y \in S_a(L)$, for every $k \geq 0$.*

Proof. Let G be a grammar in Chomsky Normal Form generating L, with M non terminal symbols. Let H be an integer whose value will be fixed later and consider a word $z = z_1 z_2 \cdots z_n \in S_a(L)$ of length $n \geq H$. By definition there exists a word $w = \alpha a w_1 w_2 \ldots w_n \overline{a} \beta \in L$, with $w_1, \ldots, w_n$ primes such that **type**$(w_j) = z_j$, for $1 \leq j \leq n$. Let W be the sequence of words $w_1, \ldots, w_n$.

Fix a derivation-tree of w in G and consider a suitable path in the tree starting from the root and stopping the first time it reaches a vertex from which there hangs a tree generating a proper factor of an occurrence in W. Given a vertex internal in this path, next vertex is the child which is the root of the subtree generating a factor of w with most occurrences of words in W (either one of the two children in case of a tie). We are interested in the terminal part of this path starting from the first vertex under which there hangs a tree generating a factor of w containing at most H and at least $\frac{H}{2}$ occurrences of W: let $X_0 \ldots X_i \ldots X_N$ be the sequence of nonterminals that are the labels of the nodes in the terminal part. This leads to the sequence

$$
\begin{aligned}
X_0 \quad &\overset{+}{\Rightarrow} \ell_1 X_1 r_1 \\
&\;\;\vdots \\
X_i \quad &\overset{+}{\Rightarrow} \ell_{i+1} X_{i+1} r_{i+1} \\
&\;\;\vdots \\
X_{N-1} &\overset{+}{\Rightarrow} \ell_N X_N r_N \\
X_N \quad &\overset{+}{\Rightarrow} e \in T^*
\end{aligned}
\tag{3}
$$

where ℓ_i and r_i are terminal words and where one of them is empty. In particular, $\ell_1 \cdots \ell_N e r_N \cdots r_1$ is a factor of w.

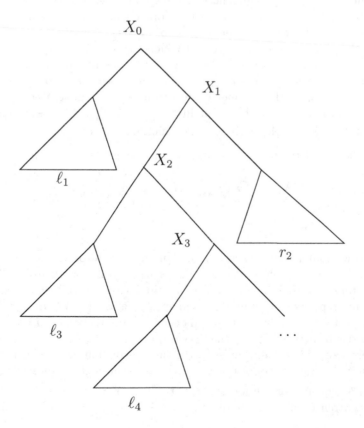

In order to obtain new derivation sequences satisfying certain desirable properties, we shall use a general method consisting of grouping successive derivation steps. The nonterminals occurring on the left handsides define a subsequence of $X_0, \ldots, X_i, \ldots$ The general form will be as follows.

$$
\begin{aligned}
X_0 &\overset{+}{\Rightarrow} \ell'_1 X_{j_1} r'_1 = \ell_1 \cdots \ell_{j_1} X_{j_1} r_{j_1} \cdots r_1 \\
X_{j_1} &\overset{+}{\Rightarrow} \ell'_{j_2} X_{j_2} r'_{j_2} = \ell_{j_1+1} \cdots \ell_{j_2} X_{j_1} r_{j_2} \cdots r_{j_1+1} \\
&\vdots \\
X_{j_k} &\overset{+}{\Rightarrow} \ell'_{j_{k+1}} X_{j_{k+1}} r'_{j_{k+1}} = \ell_{j_k+1} \cdots \ell_{j_{k+1}} X_{j_{k+1}} r_{j_{k+1}} \cdots r_{j_k+1} \\
&\vdots
\end{aligned}
\tag{4}
$$

In a derivation of the kind $X \Rightarrow^+ \ell Y r$, we call ℓ (r) respectively prefix (suffix) of the derivation.

$$
\text{Set } h = \lfloor \tfrac{\log H}{2} \rfloor - 1.
$$

Claim 1. By grouping the sequence (3) we may find a sequence of length $m \geq 2h$ such that each derivation has a prefix or a suffix containing at least an occurrence of an element in W.

Indeed, the sequence is obtained following our general method. Starting from the sequence (3) we group as few successive steps as necessary so that at each step $k \geq 0$, either $\ell'_{j_{k+1}}$ or $r'_{j_{k+1}}$ contains at least one occurrence of a word in W. I.e., j_1 is the first index for which either $\ell_1 \cdots \ell_{j_1}$ or $r_{j_1} \cdots r_1$ contains an occurrence of a word in W, $j_2 > j_1$ is the first index for which either $\ell_{j_1+1} \cdots \ell_{j_2}$ or $r_{j_2} \cdots r_{j_1+1}$ contains an occurrence of a word in W etc ... Let m be the length of the sequence just defined. We now show that $m \geq 2h$ holds. Because of the choice of the initial sequence (3), if c_{j_k} is the number of occurrences of W generated by the non terminal X_{j_k}, then we have the inequalities: $\lfloor \tfrac{c_{j_k}}{2} \rfloor \leq c_{j_{k+1}} \leq c_{j_k} - 1$. Therefore, we have $m \geq \lfloor \tfrac{\log H}{2} \rfloor = \lfloor \log H - 1 \rfloor \geq 2h$ which proves the claim.

Consider the tail of the previous sequence consisting of the last $2h$ derivations.

$$
\begin{aligned}
X_{j_{m-2h}} &\overset{+}{\Rightarrow} \ell'_{j_{m-2h+1}} X_{j_{m-2h+1}} r'_{j_{m-2h+1}} \\
&\vdots \\
X_{j_{m-1}} &\overset{+}{\Rightarrow} \ell'_{j_m} X_{j_m} r'_{j_m}
\end{aligned}
\tag{5}
$$

Assume without loss of generality that in the derivation sequence (5), there exists h steps whose prefixes contain an occurrence of a word in W. Then by grouping steps in this sequence, we may obtain a sequence of length h such that each step has a prefix containing an occurrence of a word in W. Indeed, if a step of (5) is such that its prefix contains an occurrence of W but the i next steps do not, then group these i steps to the previous one.

As there exist M nonterminals, in the last derivation sequence there exists a nonterminal which occurs at least $f = \lfloor \tfrac{h}{M} \rfloor$ times, therefore by grouping we obtain a sequence of the form $Y \overset{+}{\Rightarrow} \ell''_1 Y r''_1$ ... $Y \overset{+}{\Rightarrow} \ell''_f Y r''_f$, where $\ell''_1, \ldots, \ell''_f$ contain occurrences of W. Consequently, $\ell''_2 \cdots \ell''_{f-1}$ is a factor of $w_1 \cdots w_n$. Thus we have a sequence of the form $Y \overset{+}{\Rightarrow} s_1 u_1 p_1 Y d_1 \; \ldots \; Y \overset{+}{\Rightarrow} s_{f-2} u_{f-2} p_{f-2} Y d_{f-2}$, where Y is a nonterminal and each s_j (resp. p_j) is a suffix (resp. a prefix) of a suitable prime in W, u_j is a product of consecutive elements in W, $1 \leq j \leq f - 2$, and $p_j s_{j+1}$ is the occurrence of W which follows u_j, for $1 \leq j < f - 2$. Furthermore, because the factor $s_j u_j p_j$ contains an occurrence of W, we may adopt the convention that s_j is not the empty word and that p_j is not an occurrence of W.

By applying Lemma 6, if $f - 2 \geq 4^M + 2$, at least two pairs $(s_i u_i p_i, d_i)$ and $(s_j u_j p_j, d_j)$ are flat. The factors d_i and d_j are disjoint in $w_1 \cdots w_n \bar{a} \beta$, therefore one among d_i and d_j, say d_j to fix ideas, does not contain the occurrence $\bar{a}$ which precedes the suffix β. Indeed, we have two cases. In case 1, d_j is a factor of $w_1 \cdots w_n$ and in case 2 d_j is a factor of β. In both cases for some $r < l$ we have $s_j = w_r''$, $p_j = w_l'$ and $u_j = w_{r+1} \cdots w_{l-1}$ where $w_J = w_J' w_J'' \in W$, for $J = r, l$, i.e., $s_j u_j p_j = w_r'' w_{r+1} \cdots w_{l-1} w_l'$.

Since the pair $(s_j u_j p_j, d_j)$ is flat, equality (2) ensures that there exist words $x, y \in A^*$ such that $\rho(s_j u_j p_j) = \bar{x} x$ and $\rho(d_j) = \bar{y} y$ hold. This preliminary observation allows us to establish the following two facts.

Claim 2. $w_l' w_r''$ is a Dyck prime and $\texttt{type}(w_r) = \texttt{type}(w_l' w_r'')$ (the proof is omitted).

Claim 3. $\rho((s_j u_j p_j)^k) = \rho(s_j u_j p_j)$ and $\rho((d_j)^k) = \rho(d_j)$ for all integers $k \geq 0$. Trivial.

CASE 1: There are two subcases according to whether or not d_j is a proper factor of one of the occurrences $w_1, \ldots, w_n$.

SUBCASE 1.1: The factor d_j is not a proper factor of one of the occurrences of W. Reasoning as for $s_j u_j p_j$ in Claim 2, we obtain that d_j is of the form $w_t'' w_{t+1} \cdots w_{s-1} w_s'$ for some $t < s$, some $w_t = w_t' w_t''$ and some $w_s = w_s' w_s''$ where $w_s' w_t''$ is a Dyck prime satisfying $\texttt{type}(w_t) = \texttt{type}(w_s' w_t'')$. Hence the word w is of the form

$$\alpha a w_1 \cdots w_r' s_j u_j p_j w_l'' \cdots w_t' d_j w_s'' \cdots w_n \bar{a} \beta$$

Since $(s_j u_j p_j, d_j)$ is an iterative pair, we get for all $k \geq 0$

$$\alpha a w_1 \cdots w_{r-1} w_r' (s_j u_j p_j)^k w_l'' w_{l+1} \cdots w_{t-1} w_t' (d_j)^k w_s'' w_{s+1} \cdots w_n \bar{a} \beta \in L.$$

Using Claim 3, we obtain $\rho(a w_1 \cdots w_r' (s_j u_j p_j)^k w_l'' \cdots w_t' (d_j)^k w_s'' \cdots w_n \bar{a}) = \rho(a w_1 \cdots w_r' s_j u_j p_j w_l'' \cdots w_t' d_j w_s'' \cdots w_n \bar{a}) = \varepsilon$, which implies $a w_1 \cdots w_r' (s_j u_j p_j)^k w_l'' \cdots w_t' (d_j)^k w_s'' \cdots w_n \bar{a} \in F_a(L)$. We compute

$$\begin{aligned}
&\texttt{trace}(a w_1 \cdots w_{r-1} w_r' (s_j u_j p_j)^k w_l'' \cdots w_t' (d_j)^k w_s'' \cdots w_n \bar{a}) \\
&= \texttt{type}(w_1 \cdots w_{r-1} w_r' (w_r'' w_{r+1} \cdots w_{l-1} w_l')^k w_l'' \cdots w_t' (w_t'' \cdots w_{s-1} w_s')^k w_s'' \cdots w_n) \\
&= \texttt{type}(w_1 \cdots w_{r-1} (w_r w_{r+1} \cdots w_{l-1})^k w_l \cdots (w_t w_{t+1} \cdots w_{s-1})^k w_s \cdots w_n) \\
&= z_1 \cdots z_{r-1} (z_r \cdots z_{l-1})^k z_l \cdots z_{t-1} (z_t \cdots z_{s-1})^k z_s \cdots z_n \in S_a(L).
\end{aligned}$$

SUBCASE 1.2: The factor d_j is a proper factor of one of the occurrences of W. (Omitted).

CASE 2: d_j is a factor of β. (Omitted).   □

Since each language over a unary alphabet is regular if and only if is satisfies the pumping lemma, an immediate consequence of Theorem 2 is the following

Corollary 1. *Let $L \subseteq D_{\{a\}}$ be a context-free language on $\{a, \bar{a}\}$. The surface $S_a(L)$ is a regular language.*

Proof. Since $S = S_a(L)$ is a unary language that verifies the pumping lemma, there exists a constant H such that, for a word $a^n \in S$ such that $n \geq H$, there exists a constant $c \leq H$ satisfying $a^{n+kc} \in S$, for every $k \geq -1$.

Let $P = H!$: if $a^n \in S$, then $a^{n+kP} \in S$, for every $k \geq 0$. For $0 \leq r < P$ let Π_r be $\Pi_r = \{a^n \in S \mid n \geq H, \langle n \rangle_P = r\}$. If $S \cap \Pi_r \neq \emptyset$, call $n_r = \min\{n \mid a^n \in S \cap \Pi_r\}$, so that $S \cap \Pi_r = a^{n_r}(a^P)^*$. Since $S = (S \cap a^{\leq H}) \cup \bigcup_{S \cap \Pi_r \neq \emptyset} a^{n_r}(a^P)^*$, then S is regular.   □

4   Deciding Whether or Not a Language Is XML

In this section we study the problem of testing whether or not a context-free language is an XML-language. Formally, we consider

– CONTEXT-FREE XML-LANGUAGE   (CFX$_A$)
  INSTANCE: A context-free grammar G with alphabet $A \cup \overline{A}$.
  QUESTION: Is the language L_G generated by G an XML-language?

Notice that it is undecidable the regularity of surfaces of context-free languages contained in Dyck languages on A, if $|A|$ is sufficiently large (see, [1]). However, if $|A| = 1$, then the problem is trivially decidable (Corollary 1). In [1], the undecidability of CFX$_A$ is shown for $|A|$ sufficiently large. The presented argument does not work if $|A| = 1$. Surprisingly, we show that CFX$_A$ is undecidable, even if $|A| = 1$, by using a reduction to the following undecidable problem: is a context-free language on the input alphabet $\Sigma = \{0, 1\}$ equal to Σ^* (see, e.g., [3, Theorem 14.4])?

Theorem 3. *CFX$_A$ is undecidable for any input alphabet $A \neq \emptyset$.*

Proof. (Outline). Consider a context-free language $L \subseteq \{0, 1\}^*$ and the two Dyck primes $p = a\bar{a}$ and $q = aa\bar{a}\bar{a}$ on the alphabet $\{a, \bar{a}\}$. Consider the morphism $\psi : \{0, 1\}^* \to D_a$ satisfying $\psi(0) = p$ and $\psi(1) = q$, and extend it to languages in the natural way. The language $T(L) = \left(P_a \setminus a\{p, q\}^*\bar{a}\right) \cup a\psi(L)\bar{a}$ is clearly context-free if L is.

Moreover we have $L = \{0, 1\}^* \Leftrightarrow \psi(L) = \{p, q\}^* \Leftrightarrow a\psi(L)\bar{a} = a\{p, q\}^*\bar{a} \Leftrightarrow T(L) = P_a$.

We are left to prove that $T(L)$ is XML if and only if it is equal to P_a. Since the language P_a is XML, it suffices to check that if there exists a word in P_a not belonging to $T(L)$, then $T(L)$ is not XML. Observe that the set $T(L)$ contains the subset $a(a^3\bar{a}^3)^*\bar{a}$ which shows that its surface is a^*. But there exists a unique XML with a given surface, in this case it is P_a which completes the proof.   □

Now, we consider deterministic context-free languages. We are interesting in the following problem

- DETERMINISTIC CONTEXT-FREE XML-LANGUAGE   (DCFX$_A$)
  INSTANCE: A deterministic context-free grammar G with alphabet $A \cup \overline{A}$.
  QUESTION: Is the language L_G generated by G an XML-language?

As before, we first analyze the problem of testing the regularity of surfaces of this class of languages. We show that, if $|A| > 1$, the problem of testing the regularity of surfaces of deterministic context-free languages, contained in Dyck languages on A, remains undecidable. In what follows, we denote by L^c the complement of the language L.

Theorem 4. *Given $A = \{a, b\}$, it is undecidable whether the surfaces of a context-free deterministic language contained in D_A are regular.*

Proof. Using standard tools, we consider a reduction to Post Correspondence Problem (PCP). An instance of PCP is given by two alphabets $\Sigma = \{a_1, \ldots, a_n\}$, $B = \{b_1, b_2\}$ and by two morphisms $\varphi, \psi : \Sigma^* \to B^*$. Let $L_\varphi = \{x\varphi(x) \mid x \in \Sigma^*\}$, $L_\psi = \{x\psi(x) \mid x \in \Sigma^*\}$: PCP has solution on (φ, ψ) if and only if $L_\varphi \cap L_\psi \neq \emptyset$. We recall that $L_\varphi \cap L_\psi$ are deterministic context-free, moreover $L_\varphi \cap L_\psi = \emptyset$ if and only if $L_\varphi \cap L_\psi$ is regular.

Let $\xi : (\Sigma \cup B)^* \to \{a, b\}^*$ be a morphism such that $\xi((\Sigma \cup B)^*)$ is a prefix code[1] and let $\mathcal{L}_\varphi = \xi(L_\varphi)^c$. $\mathcal{L}_\varphi$ is deterministic context free, moreover

$$\mathcal{L}_\varphi \cup \mathcal{L}_\psi \text{ is regular } \Leftrightarrow \xi(L_\varphi) \cap \xi(L_\psi) \text{ is regular } \Leftrightarrow L_\varphi \cap L_\psi \text{ is regular.}$$

Finally, consider the morphism $\chi : \{a, b\}^* \to \{a, \bar{a}, b, \bar{b}\}^*$ given by $\chi(a) = a\bar{a}$, $\chi(b) = b\bar{b}$, and let $\mathcal{L}_{\varphi, \psi} = a^2 \chi(\mathcal{L}_\varphi)\bar{a}^2 \cup a\chi(\mathcal{L}_\psi)\bar{a}$. It holds that $\mathcal{L}_{\varphi, \psi}$ is a deterministic context-free language contained in the Dyck language on $\{a, \bar{a}, b, \bar{b}\}$. Moreover, the surface $S_a(\mathcal{L}_{\varphi, \psi}) = \{\varepsilon, a\} \cup \mathcal{L}_\varphi \cup \mathcal{L}_\psi$, hence $S_a(\mathcal{L}_{\varphi, \psi})$ is regular if and only if $\mathcal{L}_\varphi \cup \mathcal{L}_\psi$ is regular if and only if PCP does not admit solutions on (φ, ψ).   □

Surprisingly, DCFX$_A$ is decidable for every alphabet A. In [1, Theorem 7.1.] it was proved that it is decidable, given a regular language, whether or not it is an XML language. So, we improve this result by showing that it is still decidable when given a deterministic context-free language. We assume that the input grammar is reduced in the sense that all terminal symbols of the rules have an occurrence in a word of the language generated by the grammar.

---

[1] We recall that $L \subseteq \Sigma^*$ is a *code* whenever each word in L^+ can be univocally decomposed as product of words in L. In addition:
  - if, for any $v, w \in \Sigma^+$, we have $vw \in L \Rightarrow v \notin L$, then L is a *prefix code*;
  - if, for any $v, w \in \Sigma^+$, we have $vw \in L \Rightarrow w \notin L$, then L is a *suffix code*.
  L is a *bifix code* whenever it is both a prefix and a suffix code.

Procedure IsDetXML
INPUT: a deterministic context free grammar G on alphabet $A \cup \overline{A}$
1. if, for all $a \in A$, $L_G \nsubseteq aD_A\overline{a}$ then REJECT
        else $s =$ element $a \in A$ such that $L_G \subseteq aD_A\overline{a}$;
2. for all $a \in A$ find α_a, w_a, β_a so that α_a contains no occurrence
        of a or $\overline{a}$ and that $\alpha_a w_a \beta_a \in L_G$ and $w_a \in F_a(L_G)$ holds;
3. if there exists $a \in A$ such that

$$L_a = L_G \cap \alpha_a a \left(\bigcup_{\sigma \in A} w_\sigma \right)^* \overline{a} \beta_a$$

    not regular then REJECT;
4. construct the XML grammar

$$X = \langle A \cup \overline{A}, \{X_a \mid a \in A\}, \mathcal{P} = \{X_a \to aR_a\overline{a} \mid a \in A\}, X_s \rangle, \text{ where}$$

    R_a is a regular expression for $\{X_{z_1} \cdots X_{z_k} | z_1 \cdots z_k \in S_a(L_a)\}$;
5. if $L_X = L_G$ then ACCEPT else REJECT.

Theorem 5. *The procedure* **IsDetXML** *correctly determines whether or not a given deterministic grammar generated an XML language.*

Proof. We first justify our claim that **IsDetXML** is a procedure, i.e., that all instructions are effective. This is clearly the case for instruction 1 by [1, Corollary 5.4]. Concerning instruction 2, the preliminary hypothesis guarantees that there exists a word in the language which contains an occurrence of the letter a. Using the pumping lemma, such a word may be assumed of length $2^{\mathcal{O}(M)}$ where M is the number of nonterminals of a grammar in Chomsky Normal Form generating L_G. Then this word has a factorization of the form $\alpha_a w_a \beta_a \in L_G$ and $w_a \in F_a(L_G)$ where the word α_a contains no occurrence of a or of $\overline{a}$. Notice that L_a as defined in instruction 3 is the intersection of the deterministic context-free L_G and the regular $\alpha_a a(\cup_{\sigma \in A} w_\sigma)^* \overline{a} \beta_a$. Therefore, by [7] it is decidable whether L_a is regular and furthermore, it is also possible to construct an automaton which recognizes it, and therefore an XML grammar as in instruction 4. Finally, L_G and L_X are two deterministic context-free languages and the problem to determine whether or not $L_G = L_X$ holds is decidable by [6].

It is clear, that if the procedure returns ACCEPT then the language is XML since the language L_X is by construction XML. So we only need to prove the converse. Assume the language is XML. The control passes to instruction 3 for which two assertions must be verified. First we show that L_a and L_G have the same surface relative to the letter a which implies in particular that $S_a(L_a)$ is regular. Second, we show that the language $S_a(L_a)$ is regular if and only if so is the language L_a.

Concerning the first assertion, let $W = \{w_\sigma\}_{\sigma \in A}$ be the set of Dyck primes computed at instruction 2. Consider a letter $b \in \Sigma$ and let $ubw\overline{b}v \in L_G$ be an arbitrary word, where $bw\overline{b}$ is a Dyck prime. The word w decomposes into $w_1 w_2 \ldots w_n$ where $w_i = a_i u_i \overline{a_i}$ are Dyck primes, for all $i = 1, \ldots, n$. Because of [1, Theorem 4.4.], the Dyck primes $w_i = a_i u_i \overline{a_i}$ and $w_{a_i} \in W$ have the same contexts and therefore $u b w_{a_1} \ldots w_{a_n} \overline{b} v$ is also in the language. We use the same argument to the Dyck primes $b w_{a_1} \ldots w_{a_n} \overline{b}$ and w_b which shows that $\alpha_b b w_{a_1} \ldots w_{a_n} \overline{b} \beta_b$ is also in the language L_G, i.e., that $a_1 \ldots a_n$ belongs to $S_b(L_b)$.

Concerning the second assertion, write $\beta_a = v_a \beta_a'$ where v_a is the longest prefix of β_a belonging to D_a. Since $\{w_a \mid a \in A\}$ is a bifix code, L_a is regular if and only if $B_a = \{b_1 \ldots b_n \in A^* \mid \alpha_a a w_{b_1} \ldots w_{b_n} \overline{a} \beta_a \in L_G\}$ is regular. Now, observe that we have

$$S_a(L_a) - (B_a \bigcup S_a(\{v_a\})) \subseteq \bigcup_{b \in \Sigma} S_a(\{w_b\})$$

Since the surface of a finite language is finite, we observe that $S_a(L_a)$ is regular if and only if so is L_a. Finally, instruction 5 is successful since an XML language characterized by its surfaces which completes the proof.   □

5 Conclusion

We have proved that establishing whether the surface of a context-free language $L \subseteq D_A$ is regular is trivially decidable for $|A| = 1$. However, the problem to establishing whether L is an XML-language is undecidable even for $|A| = 1$. For a deterministic context-free language $L \subseteq D_A$, the opposite holds: the problem to establishing whether L is an XML-language is decidable; instead, it is undecidable whether L admits regular surfaces for $|A| > 1$. Hence, our results show that the known characterization of XML languages recalled in Theorem 1 does not represent a valid tool for investigating XML property.

References

1. J. Berstel and L. Boasson. Formal properties of XML grammars and languages. *Acta Inf.*, 38:649-671, 2002.
2. R. Book, S. Even, S. Greibach and G. Ott. Ambiguity in graphs and expressions. *IEEE Trans. Comput.*, C-20:149-153, 1971.
3. J. Hopcroft and J. Ullman. *Formal Languages and their relation to automata.* Addison-Wesley, 1969.
4. D.E. Knuth. A characterization of parenthesis languages. *Inf. Cont.*, 11:269-289, 1967.
5. M. Lohrey. Word problems and membership problems on compressed words. Preprint, 2004. http://www.informatik.uni-stuttgart.de/fmi/ti/mitarbeiter/Lohrey.
6. G. Senizergues. L(A)=L(B)? decidability results from complete formal systems. *Theor. Comp. Sci.*, 251:1-166, 2001.
7. R.E. Stearns. A regularity test for pushdown machines. *Inf. Cont.*, 11:323-340, 1967.
8. W3C Recommendation REC-xml-19980210. *Extensible Markup Language (XML) 1.0*, 10 February 1998. http://www.w3.org/TR/REC-XML.

Synchronization of Pushdown Automata

Didier Caucal

IRISA–CNRS, Campus de Beaulieu, 35042 Rennes, France
caucal@irisa.fr

Abstract. We introduce the synchronization of a pushdown automaton by a sequential transducer associating an integer to each input word. The visibly pushdown automata are the automata synchronized by an one state transducer whose output labels are $-1, 0, 1$. For each transducer, we can decide whether a pushdown automaton is synchronized. The pushdown automata synchronized by a given transducer accept languages which form an effective boolean algebra containing the regular languages and included in the deterministic real-time context-free languages.

1   Introduction

It is well-known that the context-free languages are not closed under intersection and complementation, and that the deterministic context-free languages are not closed under intersection and union. Alur and Madhusudan have shown that the languages accepted by the visibly pushdown automata form a boolean algebra included in the deterministic real-time context-free languages [AM 04]. The notion of visibly pushdown automaton is based on the synchronization between the input symbols and the actions performed on the stack: this enforces that the variation of the stack height is entirely characterized by the input word.

It appears that the closure results for the languages accepted by the visibly pushdown automata are based on a geometrical property of their graphs with regard to the stack height. This geometrical property which holds for every pushdown graph (not only visibly) was discovered by Muller and Schupp [MS 85]. A simple adaptation of their result shows that the graph of every pushdown automaton is regularly generated by increasing stack height [Ca 95]. This regularity is described by a finite deterministic graph grammar which in n steps of parallel rewritings, generates the graph restricted to the configurations with stack height at most n.

In this article, we generalize the notion of synchronization to abstract from the stack height. Towards this goal, we introduce a sequential transducer associating an integer to each input word. Provided that this transducer defines a norm for the vertices of the pushdown graph, we show that we can decide whether the graph can be generated regularly with regard to that norm. This is the notion of synchronization by a transducer. For any fixed transducer, the languages accepted by the pushdown automata synchronized by this transducer,

O.H. Ibarra and Z. Dang (Eds.): DLT 2006, LNCS 4036, pp. 120–132, 2006.

are deterministic real-time context-free languages and form an effective boolean algebra containing the regular languages.

2   Graphs and Finite Automata

By allowing labels not only on arcs but also on vertices, we define graphs as a simple extension of automata: the vertex labelling is not restricted to indicate initial and final vertices but it also permits to add information on vertices (e.g. the vertices accessible from a given colour, more generally the vertices defined by a μ-formula, ...). First we give some notations.

Let $\mathbb{N}$ be the set of non-negative integers and $\mathbb{Z}$ be the set of integers. For any set E, we denote $|E|$ its cardinality. For every $n \geq 0$, E^n is the set of n tuples of elements of E, and $E^* = \bigcup_{n \geq 0} E^n$ is the free monoid generated by E for the *concatenation*: $(e_1, \ldots, e_n) \cdot (e'_1, \ldots, e'_n) = (e_1, \ldots, e_n, e'_1, \ldots, e'_n)$. A finite set E of symbols is an *alphabet* of *letters*, and E^* is the set of *words* over E. Any word $u \in E^n$ is of *length* $|u| = n$ and is represented by the juxtaposition of its letters: $u = u(1) \ldots u(|u|)$. The word of length 0 is the *empty word* ε. We denote $|u|_P := |\{ 1 \leq i \leq |u| \mid u(i) \in P \}|$ the *number of occurrences* of $P \subseteq E$ in $u \in E^*$. For any binary relation $R \subseteq E \times F$ from E into a set F, we write also $e\,R\,f$ for $(e, f) \in R$, and we denote $Dom(R) := \{ e \mid \exists\, f,\ e\,R\,f \}$ the *domain* of R, and $Im(R) := \{ f \mid \exists\, e,\ e\,R\,f \}$ the *image* (or the range) of R.

Now we present our notion of graph which generalizes the notion of automaton. Let L and C be disjoint countable sets of symbols for respectively labelling arcs and labelling vertices. Here a graph is simple, oriented, arc labelled in a finite subset of L and vertex labelled in a finite subset of C. Precisely, a *graph* G is a subset of $V \times L \times V \cup C \times V$ where V is an arbitrary set such that its *vertex* set

$$V_G := \{ p \mid \exists\, a, q,\ (p, a, q) \in G \ \vee\ (q, a, p) \in G \} \cup \{ p \mid \exists\, c,\ (c, p) \in G \}$$

is finite or countable, with its *vertex label* set or *colour* set

$$C_G := \{ c \in C \mid \exists\, p,\ (c, p) \in G \} \quad \text{is finite,}$$

and its *arc label* set or *label* set

$$L_G := \{ a \in L \mid \exists\, p, q,\ (p, a, q) \in G \} \quad \text{is finite.}$$

Any (p, a, q) of G is a *labelled arc* of *source* p, of *goal* q, with label a, and is identified with the labelled transition $p \xrightarrow{a}_G q$ or directly $p \xrightarrow{a} q$ if G is understood. Any (c, p) of G is a vertex p labelled by c and is also written $c\,p$ if G is understood. We denote $V_{G,i} := \{ p \mid i\,p \in G \}$ the set of vertices of G labelled by the colour $i \in C$.

A graph is *deterministic* if distinct arcs with the same source have distinct labels:

$$r \xrightarrow{a} p \ \wedge\ r \xrightarrow{a} q \ \implies\ p = q.$$

Note that a graph G is finite if and only if it has a finite vertex set V_G. For instance $\{ r \xrightarrow{b} p,\, p \xrightarrow{a} s,\, p \xrightarrow{b} q,\, q \xrightarrow{a} p,\, q \xrightarrow{b} s,\, i\,r,\, g\,p,\, h\,p,\, f\,s,\, f\,t \}$ is a finite graph of vertices p, q, r, s, t, of colours i, f, g, h, and of (arc) labels a, b. It is represented below.

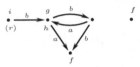

Note that a vertex r is depicted by a dot named by (r) where parentheses are used to differentiate a vertex name with a vertex label (a colour).

For any $p \in V_G$, $d^+(p) := |\{\,(a, q)\mid p \xrightarrow{a} q\,\}|$ and $d^-(p) := |\{\,(q, a)\mid q \xrightarrow{a} p\,\}|$ are respectively the *out-degree* and the *in-degree* of p; $d(p) := d^+(p) + d^-(p)$ is the *degree* of p and $d_G := sup\{\,d(p)\mid p \in V_G\,\}$ is the *degree* of G.

A graph is of *finite degree* (or *locally finite*) if $d(p) < \omega$ for any vertex p; a graph G is of *bounded degree* (or *locally bounded*) if $d_G < \omega$.

A graph G without vertex label *i.e.* $C_G = \emptyset$ is called an *uncoloured graph*.

The *restriction* (or the *induced subgraph*) of a graph G to a vertex subset P is

$$G_{|P} := \{\,p \xrightarrow{a}_{G} q \mid p, q \in P\,\} \cup \{\,cp \in G \mid p \in P\,\}.$$

Any tuple $(p_0, a_1, p_1, \ldots, a_n, p_n)$ such that $n \geq 0$ and $p_0 \xrightarrow{a_1} p_1 \ldots p_{n-1} \xrightarrow{a_n} p_n$, denoted also by the word $p_0 a_1 p_1 \ldots a_n p_n$ (which have a sense if $V_G \cap L_G^* = \emptyset$), is a *path* from p_0 to p_n labelled by $u = a_1 \ldots a_n$, and we write $p_0 \xRightarrow{u}_{G} p_n$ or directly $p_0 \xRightarrow{u} p_n$ if G is understood; for $n = 0$, the path $p_0 \xRightarrow{\varepsilon}_{G} p_0$ is reduced to $p_0 \in V_G$. For any $U \subseteq L^*$, we write $p \xRightarrow{U} q$ if $p \xRightarrow{u} q$ for some $u \in U$. We also write $p \xRightarrow{}_{G}{}^* q$ if $p \xRightarrow{L^*}_{G} q$.

We say that a vertex r is a *root* of G if every vertex p is accessible from r: $r \Longrightarrow^* p$. The *accessible subgraph* $G_{/p} := G_{|\{q \mid p \Rightarrow^* q\}}$ of a graph G from a vertex p is the restriction of G to the vertices accessible from p.

Given a graph G and vertex sets $P, Q \subseteq V_G$, we denote $L(G, P, Q)$ the language of path labels from vertices in P to vertices in Q:

$$L(G, P, Q) := \{\,u \mid \exists\, p \in P\; \exists\, q \in Q\; p \xRightarrow{u}_{G} q\,\}.$$

Given colours $i, f \in C$, we define $L(G, i, f) := L(G, V_{G,i}, V_{G,f})$ the path labels from the set $V_{G,i}$ of vertices labelled by i to the set $V_{G,f}$ of vertices labelled by f.

For instance taking the previous graph, its path labels from i to f is $b(ba)^* (a + bb)$.

So a finite graph G with two colours i and f is a *finite automaton* recognizing the language $L(G, i, f)$. The family

$$Rat(T^*) := \{\,L(G, i, f)\mid |G| < \omega \,\wedge\, i, f \in C\,\}$$

of languages over T recognized by the finite automata coincides with the family of *rational languages* (or *regular languages*). So the finite graphs describe the structures of the rational languages and permit to derive properties on these languages. For the context-free languages which are the languages recognized by the pushdown automata, their graphs are generated by the deterministic graph grammars which are also really powerful to get properties on these languages. Our purpose is to use the deterministic graph grammars in order to describe geometrically the notion of visibly pushdown automata and to get in this way a natural generalization.

3   Graph Grammars and Pushdown Automata

A pushdown automaton is a particular case of a word rewriting system where the rules are only applied by prefix. By restriction to rational vertex sets, the pushdown automata and the word rewriting systems define the same (prefix) graphs which are the graphs of bounded degree and regular in the sense that they can be generated by a deterministic grammar [Ca 90]. We extend this result to the pushdown automata which are in a weak form used by the visibly pushdown automata (cf. Theorem 3.1). We recall that the graphs of the pushdown automata are regular by increasing length (cf. Proposition 3.2).

We fix an alphabet T of *terminals*. Recall that a *labelled word rewriting system* R is a finite subset of $N^* \times T \times N^*$ for some alphabet N of *non-terminals* *i.e.* is a finite uncoloured graph of (arc) labels in T and whose vertices are words over N. The graph:

$$G \cdot P := \{ uw \xrightarrow{a} vw \mid u \xrightarrow[G]{a} v \wedge w \in P \}$$

is the *right concatenation* of any graph $G \subseteq N^* \times T \times N^*$ by any language $P \subseteq N^*$. Rewritings of a system are generally defined as applications of rewriting rules in every context. On the contrary, we are here concerned with prefix rewriting [Bü 64]. The *prefix transition graph* of R is the uncoloured graph $R \cdot N^*$ which is of bounded degree and has a finite number of non isomorphic connected components.

A subclass of labelled word rewriting systems is the standard model of real-time pushdown automata. A *pushdown automaton* R (without ε-rule) is a finite set of rules of the form:

$$pA \xrightarrow{a} qU \quad \text{with} \quad p, q \in Q, \ A \in P, \ U \in P^*, \ a \in T$$

where P and Q are disjoint alphabets of respectively *pushdown letters* and *states*.

The *transition graph* of R is $R \cdot P^* = \{ pAV \xrightarrow{a} qUV \mid pA \xrightarrow[R]{a} qU \wedge V \in P^* \}$ the restriction of the prefix transition graph $R \cdot (P \cup Q)^*$ of R to the rational set QP^* of *configurations*.

A strong way to normalize the rules of pushdown automata is given by a *weak pushdown automaton* R which is a finite set of rules of the following form:

$$p \xrightarrow{a} q \quad \text{or} \quad p \xrightarrow{a} qA \quad \text{or} \quad pA \xrightarrow{a} q \quad \text{with} \quad p, q \in Q, \ A \in P, \ a \in T.$$

Its transition graph $R \cdot P^*$ is isomorphic to $S \cdot P^* \bot$ where $\bot$ is a new pushdown letter (the bottom of the stack) and S is the following pushdown automaton:

$$S = \{ pA \xrightarrow{a} qA \mid p \xrightarrow[R]{a} q \wedge A \in P \cup \{\bot\} \}$$
$$\cup \ \{ pB \xrightarrow{a} qAB \mid p \xrightarrow[R]{a} qA \wedge B \in P \cup \{\bot\} \} \cup \{ pA \xrightarrow{a} q \mid pA \xrightarrow[R]{a} q \}.$$

The labelled word rewriting systems and the weak pushdown automata define the same prefix transition graphs, hence also for the pushdown automata which are intermediate devices.

Theorem 3.1 *The transition graphs of weak pushdown automata,*
               the transition graphs of pushdown automata,
               the prefix transition graphs of labelled word rewriting systems,
have up to isomorphism the same
               accessible subgraphs: the rooted regular graphs of bounded degree,
               connected components: the connected regular graphs of bounded degree,
               rational restrictions: the regular graphs of bounded degree.

This theorem has been first established in [Ca 90] and completed in [Ca 95] but without considering the weak pushdown automata.

It remains to recall what is a regular graph and more exactly to reintroduce the notion of a deterministic graph grammar to generate a graph. Such a generation needs to use non-terminal arcs linking several vertices and called hyperarcs.

Let F be a set of symbols called *functions*, graded by a mapping $\varrho : F \longrightarrow \mathbb{N}$ associating to each function f its *arity* $\varrho(f)$, and such that

$$F_n := \{\, f \in F \mid \varrho(f) = n \,\} \ \text{ is countable for every } n \geq 0,$$

with $F_1 = C$ and $F_2 = L$.

A *hypergraph* G is a subset of $\bigcup_{n \geq 0} F_n V^n$ where V is an arbitrary set such that

   its *vertex* set $V_G := \{\, p \in V \mid FV^*pV^* \cap G \neq \emptyset \,\}$ is finite or countable,
   its *label* set $F_G := \{\, f \in F \mid fV^* \cap G \neq \emptyset \,\}$ is finite.

Any $fv_1\ldots v_{\varrho(f)} \in G$ is a *hyperarc* labelled by f and of successive vertices $v_1, \ldots, v_{\varrho(f)}$; it is depicted for $\varrho(f) \geq 2$ as an arrow labelled f and successively linking $v_1, \ldots, v_{\varrho(f)}$:

The transformation of a hypergraph G by a function h from V_G into any set V is the graph $h(G) := \{\, fh(v_1)\ldots h(v_{\varrho(f)}) \mid fv_1\ldots v_{\varrho(f)} \in G \,\}$. Note that the graphs are the hypergraphs whose any label is of arity 1 or 2: any arc $p \xrightarrow{a} q$ corresponds to the hyperarc apq.

A *graph grammar* R is a finite set of rules of the form $fx_1\ldots x_{\varrho(f)} \longrightarrow H$ where $fx_1\ldots x_{\varrho(f)}$ is a hyperarc joining pairwise distinct vertices $x_1 \neq \ldots \neq x_{\varrho(f)}$ and H is a finite hypergraph. The labels of the left hand sides form the set N_R of *non-terminals* of R:

$$N_R := \{\, X(1) \mid X \in Dom(R) \,\},$$

and the labels of R which are not non-terminals form the set T_R of *terminals*:

$$T_R := \{\, X(1) \notin N_R \mid \exists\, P \in Im(R),\, X \in P \,\}.$$

Any graph grammar R is used to generate graphs of arc labels in T hence we assume that $T_R \subset T \cup C$. We will use capital letters for the non-terminals and small letters for the terminals. Starting from any non-terminal hyperarc, we want to generate by a graph grammar a unique graph up to isomorphism. So we restrict any graph grammar to be *deterministic*: there is only one rule per non-terminal. For instance taking $A \in F_0$, $B \in F_3$, $a, b, c \in T$ and $i, f \in C$, the following two rules:

$$A \longrightarrow \{ip, fr, Bpqr\} \quad ; \quad Bxyz \longrightarrow \{axp, bxy, cqy, byz, crz, Bpqr\}$$

constitute a deterministic graph grammar which is represented below:

For any (deterministic graph) grammar R, the *rewriting* $\underset{R}{\longrightarrow}$ is the binary relation between hypergraphs defined by $M \underset{R}{\longrightarrow} N$ if we can choose a non-terminal hyperarc $X = As_1 \ldots s_p$ in M and a rule $Ax_1 \ldots x_p \longrightarrow H$ in R to replace X by H in M:

$$N = (M - X) + h(H)$$

for some function h mapping x_i to s_i, and the other vertices of H injectively to vertices outside of M; this rewriting is denoted by $M \underset{R, X}{\longrightarrow} N$. The rewriting $\underset{R, X}{\longrightarrow}$ of a hyperarc X is extended in an obvious way to the rewriting $\underset{R, E}{\longrightarrow}$ of any set E of non-terminal hyperarcs. A *complete parallel rewriting* $\underset{R}{\Longrightarrow}$ is the rewriting according to the set of all non-terminal hyperarcs: $M \underset{R}{\Longrightarrow} N$ if $M \underset{R, E}{\longrightarrow} N$ where E is the set of all non-terminal hyperarcs of M.

For instance, the first three steps of the parallel derivation from the hypergraph $\{A\}$ according to the above grammar are depicted in the figure below.

Let $[H] := H \cap (CV_H \cup TV_H V_H)$ be the set of terminal arcs and of coloured vertices of any hypergraph H.

A *regular graph*, also called a *hyperedge replacement equational graph* [Co 90], is a graph G *generated* by a hypergraph grammar R from a non-terminal hyperarc X. More formally, G is isomorphic to a graph in the following set $R^\omega(X)$ of isomorphic graphs:

$$R^\omega(X) := \{ \, \bigcup_{n \geq 0} [H_n] \mid H_0 = X \wedge \forall \, n \geq 0, \, H_n \underset{R}{\Longrightarrow} H_{n+1} \, \}.$$

For instance by continuing infinitely the previous derivation, we get the infinite graph:

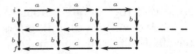

In particular any regular graph of finite degree is of bounded degree.

The regular graphs trace the context-free languages: for any regular graph G (not necessarily bounded) and for any colours $i, f \in C$, $L(G, i, f)$ is a context-free language and by Theorem 3.1, the converse is true. Graph grammars are suitable to deduce the pumping lemma, or to prove the Parikh lemma. Here we will use graph grammars to describe geometrically the notion of visibly pushdown automaton and to extend it.

A regular graph can be generated by several grammars. For instance instead of generating the previous graph by 'vertical slides', we can generate it by 'diagonal slides' using the following grammar:

$$A \longrightarrow \underset{B}{\overset{i}{\underset{b}{\overset{a}{\nearrow}}}} \; ; \; \underset{(y)}{\overset{(x)}{\downarrow}} B \longrightarrow \underset{(y)}{\overset{(x)}{\underset{b}{\overset{a}{\lessgtr}}}} \Bigg\} C \; ; \; \underset{(z)}{\overset{(x)}{\underset{(y)}{\Bigg\}}}} C \longrightarrow \underset{(z)}{\overset{(x)}{\underset{(y)}{\overset{a}{\lessgtr}}}} \Bigg\} C$$

We specify the regularity of a graph G according to a mapping g from V_G into $\mathbb{N}$. Precisely for every $n \geq 0$, we define the graph $G_{g,n}$ of the first n *levels* of G according to g by

$$G_{g,n} := \{ \, p \xrightarrow[G]{a} q \mid g(p) < n \vee g(q) < n \, \} \cup \{ \, cp \in G \mid g(p) < n \, \}.$$

We say that a graph G is *regular by* g if there exists a grammar R and a non-terminal hyperarc I such that for any parallel derivation $I \underset{R}{\Longrightarrow^n} H$, the set of terminal arcs of H is $[H] = G_{g,n}$ and its vertex set of its non-terminal hyperarcs is $V_{H-[H]}$ which is included in

$$\{ \, p \in V_G \mid g(p) = n \, \} \cup \{ \, p \in V_G \mid g(p) > n \wedge \exists \, q \, (p \longleftrightarrow q \wedge g(q) < n) \, \}$$

with the notation $p \longleftrightarrow q$ for $\exists \, a, \; p \xrightarrow{a} q \vee q \xrightarrow{a} p$.

So any graph regular by some mapping is of bounded degree.

We consider the regularity of the transition graph $R \cdot P^*$ of any pushdown automaton R according to the *stack height* $|U|$ of any configuration pU where $p \in Q$ is a state and $U \in P^*$ is a pushdown word. When R is weak then $R \cdot P^*$ is regular by stack height with the grammar reduced to this unique rule:

$$Z q_1 \ldots q_n \longrightarrow R \cup \{ \, Z(q_1 A) \ldots (q_n A) \mid A \in P \, \} \quad \text{for} \quad \{q_1, \ldots, q_n\} = Q.$$

By synchronisation product of this rule with any finite automaton, we deduce that any rational restriction of the transition graph of any weak pushdown automaton is regular by stack height (or by length). This result is extended to any pushdown automaton.

Proposition 3.2 *The rational restrictions of the prefix transition graphs of labelled word rewriting systems are regular by length.*

This proposition has been established for any morphism [Ca 95].

We can now present a geometrical description of the visibly pushdown automata, and its extension by synchronization with a sequential transducer with integer output.

4   Visibly Pushdown Automata

We present the visibly pushdown automata defined in [AM 04] with the main result (cf. Theorem 4.1), and we consider the regularity of their transition graphs by stack height.

The visibly pushdown automata are given according to a splitting of the alphabet T of terminals into three disjoint alphabets T_{-1}, T_0, T_1 to indicate respectively the letters allowed to pop the topmost stack symbol, to unchange the stack, and to push a symbol on the stack. A *visibly pushdown automaton* R is a finite set of rules of the following form:

$$pA \xrightarrow{a} q \quad \text{or} \quad p \xrightarrow{b} q \quad \text{or} \quad p \xrightarrow{c} qA \quad \text{or} \quad p\bot \xrightarrow{a} q\bot$$

with $p, q \in Q$, $A \in P$, $a \in T_{-1}, b \in T_0, c \in T_1$, where $P, Q, \{\bot\}$ are disjoint alphabets of respectively *pushdown letters*, of *states* and of the bottom of the stack. The transition graph of R is $R \cdot P^* \bot$ and the language $L(R \cdot P^* \bot, I\bot, FP^*\bot)$ recognized from a set $I \subseteq Q$ of *initial states* to a set $F \subseteq Q$ of *final states* is a *visibly pushdown language*.

For instance taking $a \in T_1$, $c \in T_0$, $b \in T_{-1}$, the language $\{ a^n c b^n \mid n \geq 0 \}$ is a visibly pushdown language and the Lukasiewicz language *i.e.* the language $L(G, A)$ generated by the context-free grammar $G = \{A \longrightarrow aAA, A \longrightarrow b\}$, is also a visibly pushdown language. But the language $\{ a^n b a^n \mid n \geq 0 \}$ and the language $L(G, A)$ generated by the grammar $G = \{A \longrightarrow aAAA, A \longrightarrow b\}$ are not visibly pushdown languages for any partition of T in T_{-1}, T_0 and T_1. So the visibly pushdown languages are not preserved in general by morphism and inverse morphism.

Any rational language over T is a visibly pushdown language according to any partition $T = T_{-1} \cup T_0 \cup T_1$: for any finite T-graph H and any $I, F \subseteq V_H$, the rational language $L(H, I, F) = L(R \cdot P^* \bot, I\bot, FP^*\bot)$ for $P = \{A\}$ reduced to a unique pushdown letter A and for the following visibly pushdown automaton:

$$R = \{ p \xrightarrow{a} qA \mid p \xrightarrow[H]{a} q \wedge a \in T_1 \} \cup \{ p \xrightarrow{a} q \mid p \xrightarrow[H]{a} q \wedge a \in T_0 \}$$
$$\cup \{ pA \xrightarrow{a} q \mid p \xrightarrow[H]{a} q \wedge a \in T_{-1} \} \cup \{ p\bot \xrightarrow{a} q\bot \mid p \xrightarrow[H]{a} q \wedge a \in T_{-1} \}.$$

The family of visibly pushdown languages is an extension of the regular languages with same basic closure properties.

Theorem 4.1 [AM 04] *For any partition of the input letters, the class of visibly pushdown languages is a subfamily of deterministic real-time context-free languages, and is an effective boolean algebra closed by concatenation and its transitive closure.*

In particular the universality problem and the inclusion problem are decidable for the visibly pushdown languages. For any visibly pushdown automaton R and by discarding the rules $p\bot \xrightarrow{a} q\bot$ (for $a \in T_{_1}$), note that

$$pU \underset{R \cdot P^*}{\overset{u}{\Longrightarrow}} qV \quad \Longrightarrow \quad |V| - |U| = |u|_{T_1} - |u|_{T_{-1}}$$

which implies the following first key point:

$$|u|_{T_1} - |u|_{T_{-1}} = |v|_{T_1} - |v|_{T_{-1}} = |U| \quad \text{for any } u, v \in L(R \cdot P^*, I, pU).$$

A second key point is given by Proposition 3.2: any rational restriction of $R \cdot P^*$ can be generated by a graph grammar S by stack height. We will see that these two key points are sufficient to establish Theorem 4.1 (without the closure by concatenation and its transitive closure). These two key points indicate that the weak form of a visibly pushdown automaton is inessential. We need that the transition graph G restricted to the configurations accessible from a given set I, satisfies the property that for every vertex s, any label u of a path from I to s has the same value $|u|_{T_1} - |u|_{T_{-1}}$ (first key point) called the norm of s, and that G is regular by norm (second key point). Note that the norm of a vertex can be negative which allows to discard the rules of the form $p\bot \xrightarrow{a} q\bot$. Finally we will generalize the visibility defined by the partition $T = T_{-1} \cup T_0 \cup T_1$ to any sequential transducer from T^* into $\mathbb{Z}$.

5   Synchronized Pushdown Automata

The synchronization of pushdown automata over T is done according to a sequential transducer A from T^* into $\mathbb{Z}$. The synchronization by A is defined for the regular graphs of bounded degree which are by Theorem 3.1 the rational restrictions of the transition graphs of pushdown automata. It is decidable whether a regular graph is synchronized by A (cf. Proposition 5.4), and the traces of the graphs synchronized by A form an effective boolean algebra of deterministic real-time context-free languages containing the rational languages (cf. Theorem 5.8).

We fix a colour $i \in C$ to indicate initial vertices.

A sequential *transducer* (or generalized sequential machine) from the free monoid T^* into the additive monoid $\mathbb{Z}$ is a finite graph A of label set $L_A \subset T \times \mathbb{Z}$ and of colour set $C_A = \{i\}$ such that A is input deterministic:

$$p \xrightarrow[A]{(a,x)} q \wedge p \xrightarrow[A]{(a,y)} r \quad \Longrightarrow \quad x = y \wedge q = r$$
$$i\,p\,,\,i\,q \in A \quad \Longrightarrow \quad p = q \qquad \text{(a unique state is coloured by } i\text{)}.$$

A (sequential) transducer A realizes the transduction

$$L(A, i, V_A) = \{ (u, m) \mid \exists\, s, t,\ is \in A \wedge s \xRightarrow[A]{(u,m)} t \}$$

of the label set of the paths from the vertex coloured by i to any vertex, for the operation

$$(u, m).(v, n) := (uv, m + n) \quad \text{for every } u, v \in T^* \text{ and } m, n \in \mathbb{Z}.$$

For instance taking a unique state r, the transducer $\{i\,r,\ r \xrightarrow{(a,1)} r,\ r \xrightarrow{(b,-1)} r\}$
represented in the next figure, realizes $\{\ (u,\ |u|_a - |u|_b) \mid u \in \{a,b\}^*\ \}$.
For any (sequential) transducer A, we denote $L(A):=Dom(L(A,i,V_A))$ its first
projection, and we say that A is *complete* if $L(A) = T^*$. For any word $u \in L(A)$,
there is a unique integer $\| u \|_A$ called the *norm* of u in A such that $(u, \| u \|_A) \in$
$L(A,i,V_A)$.

A transducer A is *visible* if it has a unique state, is complete and the value of
any arc can be only $-1,0,1$: $|V_A| = 1$, $L_A \subseteq T \times \{-1,0,1\}$ and $Dom(L_A) = T$;
in that case for any $i \in \{-1,0,1\}$, $T_i = \{\ a \mid (a,i) \in L_A\ \}$.

We say that a graph G is *compatible with* a transducer A if for any vertex s of
G, there is a path from (a vertex coloured by) i to s and the labels of the paths
from i to s are in $L(A)$ and have the same norm:

$$\emptyset \neq L(G,i,s) \subseteq L(A) \quad \wedge \quad u,v \in L(G,i,s) \implies \| u \|_A = \| v \|_A;$$

in that case we denote $\| s \|_A := \| u \|_A$ for any $u \in L(G,i,s)$. In the next figure,

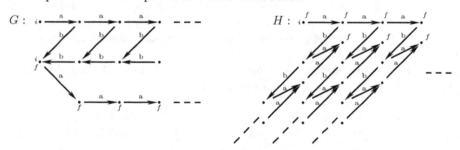

we have a graph G compatible with a visible transducer for $T = \{a,b\}$; note
that $L(G,i,f) = \{\ u \in T^* \mid |u|_a = |u|_b\ \}$ is not a visibly pushdown language.
For G compatible with A and $H \subseteq G$, H is compatible with A. Let us give
another fact.

Lemma 5.1 *For any regular graph G and any transducer A,*
$$G^A := \{\ s \xrightarrow{(a,x)} t \mid s \xrightarrow{a}_G t \wedge \exists\, p,q,\ p \xrightarrow{(a,x)}_A q \wedge i \xRightarrow{Dom(L(A,i,p))}_G s\ \} \cup (G \cap CV_G)$$
is a regular graph, and we can decide whether G is compatible with A.

Here are represented by increasing norm two regular graphs of finite degree which
are compatible with the previous visible transducer.

G :

H :

Their languages $L(G,i,f) = \{\ a^n b^n \mid n \geq 0\ \} a^*$ and $L(H,i,f) = a^* \{\ b^n a^n \mid n \geq 0\ \}$ give by intersection the language $\{\ a^n b^n a^n \mid n \geq 0\ \}$ which is not context-free, hence is not the language between colours of a regular graph. We now
discard the graph H because we cannot generate it by increasing norm: we
would need non-terminal hyperarcs having an infinite number of vertices.

We say that a graph is *synchronized* by a transducer A if it is compatible with
A and regular by the absolute value of the norm $\| \ \|_A$.

The graph above Lemma 5.1 is generated by increasing norm with the following grammar:

Let us give a graph synchronized by A with the same path labels.

Lemma 5.2 *For any transducer A, the following graph:*
$$\overrightarrow{A} := \{ (p,n) \xrightarrow{a} (q, n+x) \mid p \xrightarrow{(a,x)}_{A} q \ \wedge \ n \in \mathbb{Z} \} \cup \{ i\,(p,0) \mid i\,p \in A \}$$
is synchronized by A and $L(G, i, V_G) = L(A)$.

Let us give a simple characterization of the regular graphs which are synchronized. We say that any graph G compatible with a transducer A is *finitely compatible* with A if for every integer $n \in \mathbb{Z}$, the vertex set $\{ s \in V_G \mid \|s\|_A = n \}$ is finite. By definition, any synchronized graph by A is regular and finitely compatible with A; the converse is true.

Proposition 5.3 *For any transducer A,*
G *is synchronized by* A $\iff$ G *is regular and finitely compatible with* A.

This permits to extend the decidability of Lemma 5.1 to the synchronization problem.

Proposition 5.4 *For any transducer A, we can decide whether a regular graph G is synchronized by A, and in the affirmative, we can construct a graph grammar generating G by increasing norm $\| \ \|_A$.*

In particular we can decide whether a regular graph is visibly synchronized (we have only a finite number of visible transducers). We fix another colour $f \in C$ to indicate final vertices. The visibly pushdown languages are extended to any transducer A: a *synchronized language* by A is $L(G, i, f)$ for some graph G synchronized by A. Let us give basic examples of synchronized languages.

Example 5.5 The languages synchronized by a transducer A such that $L_A \subsetneq T \times \{0\}$ are all the rational languages included in $L(A)$.

Example 5.6 Taking $m \geq 0$, the language $L_m := L(G, X)$ generated by the context-free grammar $G = \{X \longrightarrow aX^m, X \longrightarrow b\}$ is synchronized by the transducer $A = \{i\,p, p \xrightarrow{(a, m-1)} p, p \xrightarrow{(b, -1)} p\}$. This transducer has a unique state, and it is visible for $m = 0$ ($L_0 = \{a, b\}$), for $m = 1$ ($L_1 = a^*b$) and for $m = 2$ (L_2 is the Lukasiewicz language). For $m > 2$, L_m is not a visibly pushdown language. The language $\{ u \in \{a, b\}^* \mid |u|_b = (m-1)|u|_a \}$ is also synchronized by A. More generally for $m, n \geq 0$, $L_{m,n} := \{ u \in \{a, b\}^* \mid m\,|u|_a = n\,|u|_b \}$ is a language synchronized by $\{i\,p, p \xrightarrow{(a, m)} p, p \xrightarrow{(b, -n)} p\}$.
For $m, n > 0$, $L_{m,n}$ is not a visibly pushdown language in the sense of [AM 04].

Example 5.7 The linear language $\{\, uc\widetilde{u} \mid u \in \{a,b\}^* \,\}$ for $\widetilde{u}$ the mirror of u, is synchronized by $\{\, i\,p\,,\; p \xrightarrow{(a,1)} p\,,\; p \xrightarrow{(b,1)} p\,,\; ,p \xrightarrow{(c,0)} q\,,\; q \xrightarrow{(a,-1)} q\,,\; q \xrightarrow{(b,-1)} q \,\}$. Such a language cannot be synchronized by an one state transducer: we need several integers for the labels a and b (here 1 and -1).

We establish effective closure properties of the synchronized languages by A as for the rational languages: we apply the classical constructions on finite automata to the graph grammars generating by $\| \ \|_A$. By synchronization product of $\overrightarrow{A}$ by any finite automaton, we deduce that any rational language included in $L(A)$ is synchronized by A. By disjoint union of two graph grammars generating by $\| \ \|_A$, we deduce that the intersection of two synchronized languages by A remains synchronized. By synchronization product of two graph grammars generating by $\| \ \|_A$, we obtain that the intersection of two synchronized languages by A remain synchronized. Finally by a determinization of any graph grammar generating by $\| \ \|_A$, we show that any synchronized language is deterministic context-free, and its complement with respect to $L(A)$ remains synchronized.

Theorem 5.8 *For any transducer A, the class of synchronized languages contains all the rational languages in $L(A)$, is a subfamily of deterministic real-time context-free languages, and is an effective boolean algebra with respect to $L(A)$.*

This generalization of the visibly pushdown automata has permitted to work with unrestricted pushdown automata (the synchronization is independent of the length of the words in the rules), and by allowing any integer and several states (instead of $-1, 0, 1$ and a unique state). This paper also indicates that the deterministic graph grammars can be a powerful tool to investigate properties of context-free languages.

Thanks to Christof Löding for a survey on the visibly pushdown automata which has been at the origin of this paper. The first half part of this paper has been done during a stay in Udine; many thanks to Angelo Montanari for his invitation. The second half part of this paper has been done during a stay in Aachen; many thanks to Wolfgang Thomas for his support. Thanks to Arnaud Carayol for his help in the drafting of this paper.

References

[AM 04]   R. ALUR and P. MADHUSUDAN *Visibly pushdown languages*, 36th STOC, ACM Proceedings, L. Babai (Ed.), 202–211 (2004).

[Bü 64]   R. BÜCHI *Regular canonical systems*, Archiv für Mathematische Logik und Grundlagenforschung 6, 91–111 (1964) [also in *The collected works of J. Richard Büchi*, Springer-Verlag, New York, S. Mac Lane, D. Siefkes (Eds.), 317–337 (1990)].

[Ca 90]   D. CAUCAL *On the regular structure of prefix rewriting*, 15th CAAP, LNCS 431, A. Arnold (Ed.), 87–102 (1990) [a full version is in Theoretical Computer Science 106, 61–86 (1992)].

[Ca 95]   D. CAUCAL *Bisimulation of context-free grammars and of pushdown automata*, CSLI volume 53, Stanford, 85–106 (1995).

[Co 90]   B. COURCELLE *Graph rewriting: an algebraic and logic approach*, Handbook of Theoretical Computer Science Vol. B, J. Leeuwen (Ed.), Elsevier, 193–242 (1990).

[MS 85]   D. MULLER and P. SCHUPP *The theory of ends, pushdown automata, and second-order logic*, Theoretical Computer Science 37, 51–75 (1985).

Context-Dependent Nondeterminism for Pushdown Automata

Martin Kutrib[1] and Andreas Malcher[2]

[1] Institut für Informatik, Universität Giessen,
Arndtstr. 2, D-35392 Giessen, Germany
kutrib@informatik.uni-giessen.de
[2] Institut für Informatik, Johann Wolfgang Goethe-Universität,
D-60054 Frankfurt am Main, Germany
a.malcher@em.uni-frankfurt.de

Abstract. Pushdown automata using a limited and unlimited amount of nondeterminism are investigated. Moreover, nondeterministic steps are allowed only within certain contexts, i.e., in configurations that meet particular conditions. The relationships of the accepted language families with closures of the deterministic context-free languages (DCFL) under regular operations are studied. For example, automata with unbounded nondeterminism that have to empty their pushdown store up to the initial symbol in order to make a guess are characterized by the regular closure of DCFL. Automata that additionally have to reenter the initial state are (almost) characterized by the Kleene star closure of the union closure of the prefix-free deterministic context-free languages. Pushdown automata with bounded nondeterminism are characterized by the union closure of DCFL in any of the considered contexts. Proper inclusions between all language classes discussed are shown. Finally, closure properties of these families under AFL operations are investigated.

1 Introduction

One of the central questions in automata theory asks for the power of non-determinism in bounded-resource computations. Traditionally, nondeterministic devices have been viewed as having as many nondeterministic guesses as time steps. The studies of this concept of unlimited nondeterminism led, for example, to the famous open LBA-problem or the unsolved question whether or not P equals NP. In order to gain further understanding of the nature of nondeterminism, in [3, 12] it has been viewed as an additional limited resource at the disposal of time or space bounded computations. The well-known proper inclusion between the deterministic and nondeterministic real-time multitape Turing machine languages is refined by showing an infinite hierarchy between the deterministic real-time Turing machine languages and the languages acceptable by real-time Turing machines whose number of nondeterministic steps is logarithmically bounded. In [14] this result is further generalized to arbitrary dimensions, and extended to time complexities in the range between real time and linear time.

O.H. Ibarra and Z. Dang (Eds.): DLT 2006, LNCS 4036, pp. 133–144, 2006.

In [2] limited nondeterminism is added to deterministic complexity classes independent of the computational model for the class. For these Guess-and-Check models the nondeterministically chosen bits are appended to the input. If for some choice this extended input belongs to the deterministic complexity class, then the original input is accepted. A good survey of limited nondeterminism reflecting the state-of-the-art at its time is [4].

Extensive investigations are also done on limited nondeterminism in the context of finite automata and pushdown automata. In [13] the nondeterminism is restricted depending on the size of finite automata. The authors prove an infinite nondeterministic hierarchy below a logarithmic bound, and relate the amount of nondeterminism to the number of states necessary for deterministic finite automata to accept the same language. An automata independent quantification of the inherent nondeterminism in regular languages is dealt with in [5]. Recently, measures of nondeterminism in finite automata have been investigated in [11].

Two measures for the nondeterminism in pushdown automata are proposed in [17]. By bounding the number of nondeterministic steps depending on the length of the input, a hierarchy of three classes is obtained. A modification of that measure can be found in [15]. The second measure depends on the depth of the directed acyclic graph that represents a given pushdown automaton. The corresponding proof of an infinite nondeterministic hierarchy of properly included classes is completed in [16].

Measuring the nondeterminism by branching has been introduced for finite automata in [5]. In [6, 8] it is studied in connection with pushdown automata. In [8] infinite hierarchies in between the deterministic context-free (DCFL) and context-free languages (CFL) depending on the amount of nondeterminism or on the amount of ambiguity are shown. In [6] lower bounds for the minimum amount of nondeterminism to accept certain context-free languages are established.

The main goal of this paper is to investigate pushdown automata with limited and unlimited context-dependent nondeterminism measured by branching. The branching of a transition step is defined to be the number of choices the automaton has, and the branching of a computation is the product of the branchings of all steps. Context-dependence means that it is additionally required that nondeterministic transition steps are only allowed within certain contexts, i.e., in configurations that meet particular conditions. The relationships of the accepted language families with closures of the deterministic context-free languages under regular operations are studied. This language class is particularly interesting, because deterministic context-free languages are not closed under the regular operations union, concatenation, and Kleene star and thus the regular closure increases the computational capacity. In fact, the regular closure contains, e.g., inherently ambiguous languages such as $\{a^m b^m c^n\} \cup \{a^m b^n c^n\}$ [7]. Moreover, the time complexity is still as optimal as for deterministic context-free languages, namely of order $O(n)$ [1].

The main result of this paper is that the language families accepted by pushdown automata with context-dependent nondeterminism can be characterized as subsets of the regular closure of the deterministic context-free languages.

Furthermore, one language family is shown to be equivalent to the regular closure and thus a nice automata characterization of the regular closure is obtained.

The paper is organized as follows. In the following section we present some basic notions and definitions. Section 3 deals with the computational power of pushdown automata with finite branching. This restriction yields the characterization by the union closure of the deterministic context-free languages in any of the considered contexts. Section 4 is devoted to the computational power of pushdown automata with unbounded branching. For example, it is shown that automata with unbounded branching that have to empty their pushdown store up to the initial symbol are characterized by the regular closure of the deterministic context-free languages. Furthermore, it is shown that all language classes discussed form a proper hierarchy. In Section 5, basically, the closure properties of the families in question under the AFL operations are exhibited. It turns out that the regular closure of the deterministic context-free languages is closed under all AFL operations except for homomorphism whereas the language class accepted by pushdown automata with unbounded branching which have to empty their pushdown store and to reenter the initial state is an anti-AFL, i.e., not closed under any AFL operation. Finally, the open question [9] whether or not the union closure of the deterministic context-free languages is closed under concatenation is answered negatively.

2   Preliminaries

Let Σ^* denote the set of all words over the finite alphabet Σ. The empty word is denoted by λ and $\Sigma^+ = \Sigma^* \setminus \{\lambda\}$. The reversal of a word w is denoted by w^R and for the length of w we write $|w|$. The number of occurrences of an alphabet symbol $a \in \Sigma$ in $w \in \Sigma^*$ is denoted by $|w|_a$. Set inclusion and strict set inclusion are denoted by $\subseteq$ and $\subset$, respectively. The complement of a language $L \subseteq \Sigma^*$ is denoted by $\overline{L}$. The set of mappings from some set M to some set N is denoted by N^M. We write REG for the family of regular languages.

2.1   Closures, Pushdown Automata

In general, a *family of languages* is a collection of languages containing at least one non-empty language. Let $\mathscr{L}$ be a family of languages and $op_1, \dots, op_k, k \in \mathbb{N}$, be a finite number of operations defined on $\mathscr{L}$. Then $\Gamma_{op_1,\dots,op_k}(\mathscr{L})$ denotes the *least family of languages which contains all members of $\mathscr{L}$ and is closed under* $op_1, \dots, op_k$. In particular, we consider the operations union ($\cup$), concatenation ($\bullet$), and Kleene star ($*$), which are called *regular operations*. Accordingly, we write Γ_{REG} for the *regular closure*, i.e. $\Gamma_{\cup,\bullet,*}$.

Considering a computation of a pushdown automaton we call a single step nondeterministic if the automaton has more than one choice for its next move. The *branching* of the step is defined to be the number of choices. The branching of a computation is the product of the branchings of all steps of the computation. This measure of nondeterminism has been introduced for finite automata in [5]. In [6,8] it is studied in connection with pushdown automata.

To be more precise, we continue with the stepwise formalization of pushdown automata with bounded branching. For convenience, throughout the paper we use Σ_λ for $\Sigma \cup \{\lambda\}$. Intuitively, a nondeterministic pushdown automaton has unbounded branching.

A *pushdown automaton* (PDA) is a system $\mathcal{M} = \langle Q, \Sigma, \Gamma, \delta, q_0, Z_0, F \rangle$, where Q is a finite set of states, Σ is the finite input alphabet, Γ is a finite pushdown alphabet, δ is a mapping from $Q \times \Sigma_\lambda \times \Gamma$ to finite subsets of $Q \times \Gamma^*$ called the transition function, $q_0 \in Q$ is the initial state, $Z_0 \in \Gamma$ is a particular pushdown symbol, called the bottom-of-pushdown symbol, which initially appears on the pushdown store, and $F \subseteq Q$ is the set of accepting states.

A *configuration* of a pushdown automaton is a triple (q, w, γ), where q is the current state, w the unread part of the input, and γ the current content of the pushdown store, the leftmost symbol of γ being the top symbol. If p, q are in Q, a is in Σ_λ, w is in Σ^*, γ and β are in Γ^*, and Z is in Γ, then we write $(q, aw, Z\gamma) \vdash_\mathcal{M} (p, w, \beta\gamma)$, if the pair (p, β) is in $\delta(q, a, Z)$. In order to simplify matters, we require that during any computation the bottom-of-pushdown symbol appears only at the bottom of the pushdown store. Formally, we require that if (p, β) is in $\delta(q, a, Z)$, then either β does not contain Z_0 or $\beta = \beta' Z_0$, where β' does not contain Z_0, and $Z = Z_0$. As usual, the reflexive transitive closure of $\vdash_\mathcal{M}$ is denoted by $\vdash_\mathcal{M}^*$. The subscript $\mathcal{M}$ will be dropped whenever the meaning remains clear. Furthermore, the meaning of Γ will never conflict with the closure operator.

The *language accepted* by $\mathcal{M}$ with accepting states is

$$T(\mathcal{M}) = \{w \in \Sigma^* \mid (q_0, w, Z_0) \vdash^* (q, \lambda, \gamma), \text{ for some } q \in F \text{ and } \gamma \in \Gamma^*\}.$$

The *language accepted* by $\mathcal{M}$ by empty pushdown store is

$$N(\mathcal{M}) = \{w \in \Sigma^* \mid (q_0, w, Z_0) \vdash^* (q, \lambda, \lambda), \text{ for some } q \in Q\}.$$

Intuitively, the branching of a deterministic pushdown automaton is bounded as much as possible, i.e., bounded to one. A PDA is a deterministic pushdown automaton (DPDA), if there is at most one choice of action for any possible configuration. In particular, there must never be a choice of using an input symbol or of using λ input. Formally, a pushdown automaton $\mathcal{M} = \langle Q, \Sigma, \Gamma, \delta, q_0, Z_0, F \rangle$ is *deterministic* if (i) $\delta(q, a, Z)$ contains at most one element, for all a in Σ_λ, q in Q, and Z in Γ, and (ii) for all q in Q and Z in Γ: if $\delta(q, \lambda, Z)$ is not empty, then $\delta(q, a, Z)$ is empty for all a in Σ.

The family of deterministic context-free languages (DCFL) is closed, e.g., under MIN, MAX, inverse homomorphism and intersection with regular sets, but not under homomorphism. Moreover, it is closed under complementation, right quotient with regular sets, and left quotient with a fixed string, but it is not closed under union, intersection, concatenation, or Kleene star [7]. Let DCFL$_e$ denote the set of all languages which are accepted by DPDAs that accept by empty pushdown store. It is known that DCFL$_e \subset$ DCFL and that DCFL$_e$ is equivalent to the set of all prefix-free deterministic context-free languages [10]. In the sequel a subscript e indicates that the corresponding PDA accepts by empty pushdown.

2.2   Pushdown Automata with Context-Dependent Nondeterminism

Now we turn to branching in more detail. Let $\mathcal{M} = \langle Q, \Sigma, \Gamma, \delta, q_0, Z_0, F \rangle$ be a pushdown automaton, $q \in Q$, $a \in \Sigma_\lambda$, $w \in \Sigma^*$, $\gamma \in \Gamma^*$, and $Z \in \Gamma$.

(i) The branching of a single step $\beta_\mathcal{M}(q, a, Z)$ is defined to be $|\delta(q, a, Z)|$.

(ii) The branching of a configuration c with $c \vdash c'$ is $\beta_\mathcal{M}(c) = |\{c'' \mid c \vdash c''\}|$.

(iii) A sequence of configurations (computation) $C = c_0 \vdash \cdots \vdash c_k$ has branching $\prod_{i=0}^{k-1} \beta_\mathcal{M}(c_i)$.

(iv) For words $w \in T(\mathcal{M})$ we define the branching as

$$\beta_\mathcal{M}(w) = \min\{\beta_\mathcal{M}(C) \mid C \text{ is an accepting computation of } \mathcal{M} \text{ on } w\}.$$

(v) Finally, let the branching of $\mathcal{M}$ be $\beta_\mathcal{M} = \sup\{\beta_\mathcal{M}(w) \mid w \in T(\mathcal{M})\}$.

Next we put several restrictions on PDAs and call the resulting devices *pushdown automata with context-dependent nondeterminism* (nPDA).

In particular, we will bound the branching by constants k (k-nPDA), or allow unbounded branching (∞-nPDA). We write *fin*-nPDA to indicate finite branching, that is, there exists some k such that the device is a k-nPDA.

The next step is to put restrictions on configurations. That is, conditions that have to be met in order to make a new guess with branching greater than one. In particular, we will consider devices with no further condition, devices that have to empty their pushdown store up to the initial symbol, devices that have to return to the initial state q_0, and devices that have to reinitialize completely, which means both to empty their pushdown store and to return to the initial state.

(i) If the steps $\delta(q, a, Z_0)$ are the only ones with branching greater than one, then the resulting device is a (k, Z_0)-nPDA or a (∞, Z_0)-nPDA.

(ii) If the steps $\delta(q_0, a, Z)$ are the only ones with branching greater than one, then the resulting device is a (k, q_0)-nPDA or a (∞, q_0)-nPDA.

(iii) If the steps $\delta(q_0, a, Z_0)$ are the only ones with branching greater than one, then the resulting device is a (k, q_0, Z_0)-nPDA or a (∞, q_0, Z_0)-nPDA.

Thus, (k, Z_0)-nPDAs are allowed to make a new guess only if the pushdown store is empty up to the initial symbol, whereas k-nPDAs can make new guesses not depending on the pushdown store height.

In general, we denote the *family of languages accepted* by devices of type X by $\mathscr{L}(X)$.

3   Characterization of Finite Context-Dependent Nondeterminism

In order to prove the main result of this section, the characterization of nPDA with finite branching, we need the fact that we may assume without loss of generality that there is never a choice between a λ- and a non-λ step. The result is obvious for nondeterministic pushdown automata. Here, the amount of nondeterminism has to be kept finite. A proof of the following lemma may be found in [8].

Lemma 1. *Let $\mathcal{M}$ be an nPDA with finite branching. Then we can efficiently construct an equivalent nPDA $\mathcal{M}'$ with finite branching of the same type, such that for any $q \in Q$ and $Z \in \Gamma$, whenever $\delta(q, \lambda, Z)$ is defined then $\delta(q, a, Z)$ is undefined for all $a \in \Sigma$.*

From the following lemma the desired characterization can be derived.

Lemma 2. *Let $\mathcal{M}$ be an nPDA with finite branching. Then we can efficiently construct an equivalent nPDA $\mathcal{M}'$ with finite branching of the same type whose only nondeterministic step is the very first one. The first nondeterministic step is a λ-step that preserves the content of the pushdown store.*

Proof. Without loss of generality we may assume that $\mathcal{M} = \langle Q, \Sigma, \Gamma, \delta, q_0, Z_0, F \rangle$ has the property provided by Lemma 1. Moreover, there is a positive integer k such that $\mathcal{M}$ is a k-nPDA. Let $\ell = \lceil \log_2(k) \rceil$, i.e. roughly an upper bound for the number of nondeterministic steps of $\mathcal{M}$.

We define a set $N \subseteq Q \times \Sigma_\lambda \times \Gamma$ such that $(q, a, Z) \in N$ if $|\delta(q, a, Z)| > 1$, that is the set of contexts in which nondeterminism appears. We observe that due to the properties of $\mathcal{M}$, for any configuration the set N contains at most one triple such that δ is applicable.

Now the construction of $\mathcal{M}' = \langle Q', \Sigma, \Gamma, \delta', q_0', Z_0, F' \rangle$ is as follows. Set

$$Q' = \{q_0'\} \cup \left(Q \times ((Q \times \Gamma^*)^N)^\ell \times \{0, \ldots, \ell\} \right) \text{ and}$$
$$F' = F \times ((Q \times \Gamma^*)^N)^\ell \times \{0, \ldots, \ell\}.$$

The first step of $\mathcal{M}'$ is to guess $\mathcal{M}$'s guesses (at most ℓ) in advance. It does so by guessing ℓ mappings from the nondeterministic contexts N to the possible actions. For all $a \in \Sigma_\lambda$,

$$\delta'(q_0', \lambda, Z_0) = \{((q_0, f_1, \ldots, f_\ell, 0), Z_0) \mid f_i \in (Q \times \Gamma^*)^N \text{ such that}$$
$$f_i(q, a, Z) = (p, \beta) \text{ if and only if } (p, \beta) \in \delta(q, a, Z), 1 \leq i \leq \ell\}.$$

It should be noted that there are at most finitely many such mappings $f \in (Q \times \Gamma^*)^N$. Thus, Q' and F' are finite. To simulate deterministic steps of $\mathcal{M}$, we define for all $1 \leq i \leq \ell$,

$$\delta'((q, f_1, \ldots, f_\ell, i), a, Z) = \{((p, f_1, \ldots, f_\ell, i), \beta)\}$$

if $(q, a, Z) \notin N$ and $\delta(q, a, Z) = (p, \beta)$.

In order to simulate a nondeterministic step of $\mathcal{M}$, we apply the previously guessed mapping f_i to the current situation and increase the index i. Define for all $1 \leq i \leq \ell$,

$$\delta'((q, f_1, \ldots, f_\ell, i), a, Z) = \{((p, f_1, \ldots, f_\ell, i+1), \beta)\}$$

if $(q, a, Z) \in N$ and $f_{i+1}(q, a, Z) = (p, \beta)$. $\square$

Theorem 1. *A language L is accepted by a fin-nPDA ((fin, q_0)-nPDA, (fin, Z_0)--nPDA, (fin, q_0, Z_0)-nPDA, respectively), if and only if L belongs to the union closure of the deterministic context-free languages $\Gamma_\cup(DCFL)$.*

Proof. Let $\mathcal{M} = \langle Q, \Sigma, \Gamma, \delta, q_0, Z_0, F \rangle$ be a *fin*-nPDA. By Lemma 2 we may assume that $\mathcal{M}$ makes only one nondeterministic step in every computation which is the very first one, a λ-step, and preserves the content of the pushdown store. We conclude that the sole nondeterministic step requires the initial state q_0 which is never reentered during the whole computation. So, let $S = \{q \in Q \mid (q, Z_0) \in \delta(q_0, \lambda, Z_0)\}$. Then we construct a deterministic pushdown automaton $\mathcal{M}_q$ for each $q \in S$.

$\mathcal{M}_q = \langle Q \setminus \{q_0\}, \Sigma, \Gamma, \delta', q_0, Z_0, F \rangle$, where $\delta'(q, a, Z) = \delta(q, a, Z)$ for all $q \in Q \setminus \{q_0\}$, $a \in \Sigma$, and $Z \in \Gamma$. Clearly, all $\mathcal{M}_q$ are deterministic, and $T(\mathcal{M}) = \bigcup_{q \in S} T(\mathcal{M}_q)$.

Conversely, let $L = L_1 \cup \cdots \cup L_k$ and $\mathcal{M}_i = \langle Q_i, \Sigma, \Gamma_i, \delta_i, q_{0,i}, Z_0, F_i \rangle$ be deterministic pushdown automata with $T(\mathcal{M}_i) = L_i$, for $1 \leq i \leq k$. Without loss of generality we assume that the Q_i are disjoint and q_0 is a new state. Then we construct

$$\mathcal{M} = \langle \{q_0\} \cup \bigcup_{i=1}^{k} Q_i, \Sigma, \bigcup_{i=1}^{k} \Gamma_i, \delta, q_0, Z_0, \bigcup_{i=1}^{k} F_i \rangle,$$

where $\delta(q_0, \lambda, Z_0) = \{(q_{0,i}, Z_0) \mid 1 \leq i \leq k\}$ and $\delta(q, a, Z) = \delta_i(q, a, Z)$ if and only if $q \in Q_i$ and $Z \in \Gamma_i$. It follows immediately from the construction that $\mathcal{M}$ is a *(fin, q_0, Z_0)*-nPDA, and that $T(\mathcal{M}) = L$. $\qquad\square$

4   Characterization of Unlimited Context-Dependent Nondeterminism

If the nondeterminism allowed is unbounded and may be used within any context or only when the PDA is in its initial state, then the computational capacity is not reduced.

Theorem 2. $\mathscr{L}(\infty\text{-}nPDA) = CFL$ *and* $\mathscr{L}((\infty, q_0)\text{-}nPDA) = CFL$

Proof. Obviously, every ∞-nPDA or (∞, q_0)-nPDA accepts a context-free language. Since every PDA can be considered as a ∞-nPDA, we obtain the first equation. To show the second equation we use the fact that every context-free language can be accepted by a one-state PDA (cf. [10]). $\qquad\square$

We now want to characterize the language family accepted by (∞, Z_0)-nPDAs by the regular closure of the deterministic context-free languages. As a first step we show that each language from the regular closure of the prefix-free deterministic context-free languages is accepted by some (∞, Z_0)-nPDA$_e$ and vice versa.

Theorem 3. *A language L is accepted by a (∞, Z_0)-nPDA$_e$ if and only if L belongs to the regular closure of the prefix-free deterministic context-free languages $\Gamma_{\mathrm{REG}}(DCFL_e)$.*

Proof. Since any prefix-free deterministic context-free language is accepted by some (∞, Z_0)-nPDA$_e$, and the language family $\mathscr{L}((\infty, Z_0)$-nPDA$_e)$ is closed under the regular operations (cf. Lemma 3), any language $L \in \Gamma_{\mathrm{REG}}(\mathrm{DCFL}_e)$ is accepted by some (∞, Z_0)-nPDA$_e$.

To show the converse let $\mathcal{M} = \langle Q, \Sigma, \Gamma, \delta, q_0, Z_0, \emptyset \rangle$ be a (∞, Z_0)-nPDA$_e$. Due to space considerations we can only sketch the proof. Roughly, the idea is as follows. We first transform $\mathcal{M}$ into an equivalent context-free grammar $\mathcal{G}$. The transformation follows the standard technique [10]. Then we observe that $\mathcal{G}$ is of a certain form which allows us to consider $\mathcal{G}$ as a right-linear grammar $\mathcal{G}'$ whose terminal symbols represent prefix-free deterministic context-free languages generated by non-terminals of $\mathcal{G}$. Next, $\mathcal{G}'$ is transformed into an equivalent regular expression $\mathcal{E}$, where each symbol in $\mathcal{E}$ still represents a prefix-free deterministic context-free language generated by non-terminals of $\mathcal{G}$. Therefore, $N(\mathcal{M})$ can be described as a regular expression whose symbols are prefix-free deterministic context-free languages. Thus, $N(\mathcal{M}) \in \Gamma_{\mathrm{REG}}(\mathrm{DCFL}_e)$.   □

Now, we can prove the desired machine characterization of the regular closure. Moreover, the next result reveals the interesting fact that for (∞, Z_0)-nPDA it does not matter whether they accept by empty pushdown store or by accepting states whereas between DCFL$_e$ and DCFL a proper inclusion is known.

Theorem 4. *The following language classes are equivalent.*

$$\Gamma_{\mathrm{REG}}(DCFL_e) = \mathscr{L}((\infty, Z_0)\text{-}nPDA_e) = \mathscr{L}((\infty, Z_0)\text{-}nPDA) = \Gamma_{\mathrm{REG}}(DCFL)$$

In particular, a language L is accepted by a (∞, Z_0)-nPDA if and only if L belongs to the regular closure of the deterministic context-free languages.

Proof. It is shown in [1] that $\Gamma_{\mathrm{REG}}(\mathrm{DCFL}_e) = \Gamma_{\mathrm{REG}}(\mathrm{DCFL})$. The above theorem shows that $\Gamma_{\mathrm{REG}}(\mathrm{DCFL}_e) = \mathscr{L}((\infty, Z_0)\text{-nPDA}_e)$. Since it is not difficult to convert a (∞, Z_0)-nPDA$_e$ to an equivalent (∞, Z_0)-nPDA, we obtain $\mathscr{L}((\infty, Z_0)\text{-nPDA}_e) \subseteq \mathscr{L}((\infty, Z_0)\text{-nPDA})$. Thus, we know that $\Gamma_{\mathrm{REG}}(\mathrm{DCFL}) \subseteq \mathscr{L}((\infty, Z_0)\text{-nPDA})$. In order to show $\mathscr{L}((\infty, Z_0)\text{-nPDA}) \subseteq \Gamma_{\mathrm{REG}}(\mathrm{DCFL})$ let $\mathcal{M}$ be a (∞, Z_0)-nPDA. $\mathcal{M}$ can be easily transformed into a (∞, Z_0)-nPDA$_e$ accepting $T(\mathcal{M})\$$ where $\$$ is a new alphabet symbol. Thus, $T(\mathcal{M})\$ \in \Gamma_{\mathrm{REG}}(\mathrm{DCFL}_e)$ and hence $T(\mathcal{M})\$$ can be represented as a regular expression with prefix-free deterministic context-free atoms $D_1, \dots, D_n$. Since DCFL is closed under right quotient with regular sets, we conclude that $D_1\{\$\}^{-1}, \dots, D_n\{\$\}^{-1}$ are deterministic context-free languages. Hence, $T(\mathcal{M})$ can be represented as a regular expression with deterministic context-free atoms and thus is in Γ_{REG} (DCFL).   □

We next characterize those pushdown automata which are allowed to make a new guess only when the pushdown store is empty up to the initial symbol and, in addition, the initial state is attained. The proof is omitted due to space considerations.

Theorem 5. *A language L is accepted by a (∞, q_0, Z_0)-nPDA if and only if L admits a factorization $L_1^* L_2$, where $L_1 \in \Gamma_\cup(DCFL_e)$ and $L_2 \in \Gamma_\cup(DCFL)$. This language family is denoted by $\mathscr{L}_*$.*

Table 1. Characterizations of context-dependent pushdown automata

Restriction	Characterization
fin, $(\textit{fin}, q_0)$, $(\textit{fin}, Z_0)$, $(\textit{fin}, q_0, Z_0)$	$\Gamma_\cup(\text{DCFL})$
∞, (∞, q_0)	CFL
(∞, Z_0)	$\Gamma_{\text{REG}}(\text{DCFL})$
(∞, q_0, Z_0)	$\mathcal{L}_*$

Table 1 summarizes the characterization results. The relations between the language families characterized are summarized in the next theorem.

Theorem 6. $\Gamma_\cup(\textit{DCFL}) \subset \mathcal{L}_* \subset \Gamma_{\text{REG}}(\textit{DCFL}) \subset \textit{CFL}$

Proof. We consider the language $L_0 = (\{a^n b^n c \mid n \geq 0\} \cup \{a^n b^{2n} c \mid n \geq 0\})^*$ which is not in $\Gamma_\cup(\text{DCFL})$ due to [6] and Theorem 1. Obviously, $L_0 \in \mathcal{L}_*$ which shows the first proper inclusion.

Since $\Gamma_{\text{REG}}(\text{DCFL})$ is closed under union and $\mathcal{L}_*$ is not (Lemma 6), we obtain that the second inclusion is a proper one.

For the last inclusion we consider $L = \{c^n w w^R c^n \mid n \geq 0, w \in \{a, b\}^+\} \in$ CFL. We have to show that $L \notin \Gamma_{\text{REG}}(\text{DCFL})$. In contrast to the assertion assume that L is accepted by a (∞, Z_0)-nPDA $\mathcal{M} = \langle Q, \{a, b, c\}, \Gamma, \delta, q_0, Z_0, F \rangle$.

For each state $q \in Q$ we define the set W_q of prefixes of accepted words which reinitialize the pushdown store in state q:

$$W_q = \{w \in \{a, b, c\}^+ \mid (q_0, w, Z_0) \vdash^+ (q, \lambda, Z_0)$$
$$\text{and there exists } v \in \{a, b, c\}^* \text{ such that } (q, v, Z_0) \vdash^* (q_f, \lambda, \gamma),$$
$$\text{for some } q_f \in F, \gamma \in \Gamma^*\}$$

In [6,7] it is shown that the language $L_w = \{w w^R \mid w \in \{a, b\}^*\}$ does not belong to $\Gamma_\cup(\text{DCFL})$. By Theorem 1 we obtain that L_w is not accepted by any $(\textit{fin}, Z_0)$-nPDA. Now we consider words of the form $c^* \{a, b\}^+$ in the sets W_q. Assume that in each of the finitely many sets there are only words of this form with the same number of leading c's, respectively. Let n_0 be the maximal number appearing. Then all words $c^{n_0+1} w w^R c^{n_0+1}$, $w \in \{a, b\}^+$, are accepted with finitely many reinitializations of the pushdown store, i.e., with finite branching. This implies by an immediate construction that the language L_w is accepted by a $(\textit{fin}, Z_0)$-nPDA. From the contradiction we obtain that there exists a set W_q that contains at least two words, say z_1 and z_2, that have different numbers of leading c's, say n_1 and n_2.

Now we derive a contradiction to our first assumption as follows. Let $i \in \{1, 2\}$. Since z_i belongs to W_q there is a word v_i such that $z_i v_i$ is accepted. The word z_i is of the form $c^{n_i} \{a, b\}^+$. Therefore, v_i has the form $\{a, b\}^* c^{n_i}$. Since for $\mathcal{M}$ it makes no difference whether state q with empty pushdown store is reached by processing input z_1 or z_2, it accepts the words $z_1 v_2$ and $z_2 v_1$ as well. But $z_1 v_2$ has the form $c^{n_1} \{a, b\}^+ c^{n_2}$ and, thus, does not belong to L. $\qquad\square$

5 Closure Properties

In this section we consider closure properties of context-dependent nondeterministic pushdown automata languages. By the characterization results obtained in the previous sections the properties of some families are known. For other families the properties are also interesting for their own. For example, it is natural to ask for the closure properties of closures. We start with some straightforward constructions whose details are omitted.

Lemma 3. *The language family $\mathscr{L}((\infty, Z_0)\text{-}nPDA)$ is closed under intersection with regular sets, union, concatenation, and Kleene star. $\mathscr{L}(fin\text{-}nPDA)$ is closed under intersection with regular sets and union. Both families are closed under inverse homomorphism.*

We now turn to non-closure results.

Lemma 4. *$\mathscr{L}(fin\text{-}nPDA)$, $\mathscr{L}((\infty, q_0, Z_0)\text{-}nPDA)$, and $\mathscr{L}((\infty, Z_0)\text{-}nPDA)$ are not closed under homomorphism and complementation.*

Proof. First, we observe that all families contain DCFL. Since all families are proper subsets of CFL and every context-free language can be represented as the homomorphic image of a deterministic context-free language (Chomsky-Schützenberger Theorem, see, e.g., [7]), all families are not closed under homomorphism.

Next, consider the language $L = \{a^i b^j c^k \mid i, j, k \geq 0, (i \neq j \text{ or } j \neq k)\} \cup \overline{a^* b^* c^*}$ which is contained in all of the above families. The assumption that one of the above families is closed under complementation implies that it contains the non-context-free language $\overline{L} = \{a^n b^n c^n \mid n \geq 0\}$ which is a contradiction. $\square$

The next lemma answers an open question raised in [9].

Lemma 5. *The language family $\mathscr{L}(fin\text{-}nPDA) = \Gamma_\cup(DCFL)$ is not closed under concatenation and Kleene star.*

Proof. Let

$$L_a = \{a^i b^j c^k \mid i, j, k \geq 1 \text{ and } k \leq i\} \text{ and}$$
$$L_b = \{a^i b^j c^k \mid i, j, k \geq 1 \text{ and } k \leq j\}$$

be two languages. We observe that L_a^* and L_b are deterministic context-free languages and, therefore, belong to $\Gamma_\cup(DCFL)$.

It suffices to show that the concatenation $L_a^* L_b$ does not belong to $\Gamma_\cup(DCFL)$. Assume contrarily it would. Then there are languages $L_1, \ldots, L_m \in DCFL$, for some $m \geq 1$, such that $L_a^* L_b = \bigcup_{\ell=1}^m L_\ell$.

We consider the language operation MIN, that is

$$\text{MIN}(L) = \{w \mid w \in L \text{ and no } v \in L \text{ is a proper prefix of } w\}.$$

Since the family DCFL is closed under MIN all languages $L_\ell' = \text{MIN}(L_\ell)$, $1 \leq \ell \leq m$ are deterministic context free languages. It is easily seen that all words of L'_ℓ are of the form $\{a^i b^j c^k \mid i,j,k \geq 1$ and $k \leq i$ and $k > j\}^* \{a^i b^j c^k \mid i,j,k \geq 1$ and $k \leq j\}$.

Trivially, all languages L'_ℓ are context free. The context-free languages are closed under (nondeterministic) gsm-mappings. We can easily construct a gsm that chooses one but not the last of the subwords $a^i b^j c^k$ and maps all other symbols to λ. We conclude that the resulting languages are context-free languages. Additionally, their words are of the form $\{a^i b^j c^k \mid i,j,k \geq 1$ and $k \leq i$ and $k > j\}$. Since $\mathrm{MIN}(L_a^* L_b)$ is infinite, at least one of the resulting languages contains infinitely many words that differ at least in the number of b's, let us say language L_r.

It remains to be shown that L_r is not context free. To this end, let n be the constant of Ogden's lemma and consider some word $w = a^i b^j c^k \in L_r$ such that $j \geq n$. Let the positions of the b's be distinguished. Then w admits a factorization $uvwxy$ such that v and x together have at least one distinguished position, vwx has at most n distinguished positions, and $uv^s wx^s y \in L_r$ for all $s \geq 1$.

If either v or x contains two distinct symbols, then $uv^2 wx^2 y$ does not belong to L_r. Now at least one of v and x must contain b's since only b's are in distinguished positions. Thus, either vx does not contain a's or does not contain c's. In both cases we can find some constant s such that $uv^s wx^s y$ does not belong to L_r. Thus L_r is not a context-free language.

To show non-closure under Kleene star we consider the language L_0 from Theorem 6. Clearly, $L = \{a^n b^n c \mid n \geq 0\} \cup \{a^n b^{2n} c \mid n \geq 0\} \in \Gamma_\cup(\mathrm{DCFL})$, but $L_0 = L^* \notin \Gamma_\cup(\mathrm{DCFL})$ due to [6] and Theorem 1. $\square$

Table 2. Closure properties of context-dependent pushdown automata languages

Restriction	Characterization	$\cup$	$\bullet$	$*$	h	h^{-1}	$\cap R$	$\overline{}$
$fin, (fin, q_0), (fin, Z_0), (fin, q_0, Z_0)$	$\Gamma_\cup(\mathrm{DCFL})$	+	$-$	$-$	$-$	+	+	$-$
$\infty, (\infty, q_0)$	CFL	+	+	+	+	+	+	$-$
(∞, Z_0)	$\Gamma_{\mathrm{REG}}(\mathrm{DCFL})$	+	+	+	$-$	+	+	$-$
(∞, q_0, Z_0)	$\mathscr{L}_*$	$-$	$-$	$-$	$-$	$-$	$-$	$-$

Finally, we obtain that $\mathscr{L}((\infty, q_0, Z_0)\text{-nPDA})$ is an anti-AFL, i.e., a language class not closed under union, concatenation, Kleene star, homomorphism, inverse homomorphism, and intersection with regular sets. This is particularly interesting, since it provides an example of an anti-AFL which is a proper subset of the context-free languages. Due to space considerations the proof is omitted.

Lemma 6. *The language family* $\mathscr{L}((\infty, q_0, Z_0)\text{-nPDA}) = \mathscr{L}_*$ *is an anti-AFL.*

Putting together the characterization results from the previous sections and the closure properties shown in this section, we obtain Table 2.

References

1. Bertsch, E., Nederhof, M.J.: Regular closure of deterministic languages. SIAM J. Comput. **29** (1999) 81–102
2. Cai, L., Chen, J.: On the amount of nondeterminism and the power of verifying. SIAM J. Comput. **26** (1997) 733–750
3. Fischer, P.C., Kintala, C.M.R.: Real-time computations with restricted nondeterminism. Math. Systems Theory **12** (1979) 219–231
4. Goldsmith, J., Levy, M.A., Mundhenk, M.: Limited nondeterminism. SIGACT News **27** (1996) 20–29
5. Goldstine, J., Kintala, C.M., Wotschke, D.: On measuring nondeterminism in regular languages. Inform. Comput. **86** (1990) 179–194
6. Goldstine, J., Leung, H., Wotschke, D.: Measuring nondeterminism in pushdown automata. J. Comput. System Sci. **71** (2005) 440–466
7. Harrison, M.A.: Introduction to Formal Language Theory. Addison-Wesley, Reading (1978)
8. Herzog, C.: Pushdown automata with bounded nondeterminism and bounded ambiguity. Theoret. Comput. Sci. **181** (1997) 141–157
9. Herzog, C.: Nondeterminism in context-free languages (in German). PhD thesis, University of Frankfurt, Germany (1999)
10. Hopcroft, J.E., Ullman, J.D.: Introduction to Automata Theory, Language, and Computation. Addison-Wesley, Reading, Massachusetts (1979)
11. Hromkovič, J., Seibert, S., Karhumäki, J., Klauck, H., Schnitger, G., : Communication complexity method for measuring nondeterminism in finite automata. Inform. Comput. **172** (2002) 202–217
12. Kintala, C.M.: Computations with a Restricted Number of Nondeterministic Steps. PhD thesis, Pennsylvania State University (1977)
13. Kintala, C.M., Wotschke, D.: Amounts of nondeterminism in finite automata. Acta Inform. **13** (1980) 199–204
14. Kutrib, M.: Refining nondeterminism below linear time. J. Autom., Lang. Comb. **7** (2002) 533–547
15. Salomaa, K., Yu, S.: Limited nondeterminism for pushdown automata. Bull. EATCS **50** (1993) 186–193
16. Salomaa, K., Yu, S.: Measures of nondeterminism for pushdown automata. J. Comput. System Sci. **49** (1994) 362–374
17. Vermeir, D., Savitch, W.: On the amount of nondeterminism in pushdown automata. Fund. Inform. **4** (1981) 401–418

Prime Decompositions of Regular Languages

Yo-Sub Han[1], Kai Salomaa[2,*], and Derick Wood[3,**]

[1] System Technology Division, Korea Institute of Science and Technology,
P.O. Box 131, Cheongryang, Seoul, Korea
emmous@kist.re.kr
[2] School of Computing, Queen's University, Kingston, Ontario K7L 3N6, Canada
ksalomaa@cs.queensu.ca
[3] Department of Computer Science, The Hong Kong University of Science and
Technology, Clear Water Bay, Kowloon, Hong Kong, SAR
dwood@cs.ust.hk

Abstract. We investigate factorizations of regular languages in terms of prime languages. A language is said to be strongly prime decomposable if any way of factorizing the language yields a prime decomposition in a finite number of steps. We give a characterization of the strongly prime decomposable regular languages and using the characterization we show that every regular language over a unary alphabet has a prime decomposition. We show that there exist co-context-free languages that do not have prime decompositions.

1 Introduction

A language is said to be prime [12, 14] if it cannot be written as a catenation of two languages neither one of which is the singleton language consisting of the empty word. A prime decomposition of a language is a factorization where all the components are prime languages. The original work on prime decompositions concentrated mainly on finite languages [12]. Factorizations of prefix-free or infix-free regular languages into prime components that in turn are required to be prefix-free or infix-free, respectively, are considered in [3, 6]. Decompositions of factorial languages are investigated in [1].

Any finite language always has a prime decomposition, although it need not be unique [12, 14]. Work on factorizations of finite languages leads to nontrivial questions concerning commutativity. Recent work in this direction and more references can be found e.g. in [10].

Generally the decomposition of a language can be chosen in very different ways and it turns out to be somewhat difficult to find languages without any prime decompositions. We give a construction of a nonregular language that provably does not have any prime decomposition.

* The author was supported by the Natural Sciences and Engineering Research Council of Canada Grant OGP0147224.
** The author was supported under the Research Grants Council of Hong Kong Competitive Earmarked Research Grant HKUST6197/01E.

O.H. Ibarra and Z. Dang (Eds.): DLT 2006, LNCS 4036, pp. 145–155, 2006.
© Springer-Verlag Berlin Heidelberg 2006

We consider also a stronger factorization property that requires that any refinement of a decomposition of the language leads to a prime decomposition in a finite number of steps. We call such languages strongly prime decomposable. We give necessary and sufficient conditions for a regular language to be strongly prime decomposable. The characterization establishes that the property is decidable for regular languages.

Using the characterization of the strongly prime decomposable languages we show that every regular language over a unary alphabet has a prime decomposition. As a by-product of the proof we establish the existence of prime decompositions for context-free languages over arbitrary alphabets where, roughly speaking, the set of "short words" of the corresponding length set is not closed under any multiple of the cycle of the length set.

The main open question remaining is whether all regular languages have prime decompositions.

2   Language Decompositions

Let Σ be a finite alphabet. A language is any subset of Σ^*. The length of a word $w \in \Sigma^*$ is $|w|$. The catenation of languages L_1 and L_2 over Σ is $L_1 \cdot L_2 = \{w \in \Sigma^* \mid (\exists u_i \in L_i, i = 1, 2)\ w = u_1 u_2\}$. For all unexplained notions in language theory we refer the reader e.g. to [9, 16, 17].

We say that a language L has a non-trivial decomposition if we can write $L = A \cdot B$ where $A, B \neq \{\varepsilon\}$. In the following, unless otherwise mentioned, by a decomposition or a factorization of a language we always mean a non-trivial decomposition.

A nonempty language $L \neq \{\varepsilon\}$ is said to be *prime* if L has no decompositions. For a given regular language L it is decidable whether or not L has a decomposition [11, 12], i.e., whether or not L is prime. More generally, the regular language decomposition problem is decidable for all operations defined by letter-bounded regular sets of trajectories [5].

Definition 2.1. [12] *A prime decomposition of a language L is a factorization*

$$L = L_1 \cdot \ldots \cdot L_m, \tag{1}$$

where each of the languages L_i, $i = 1, \ldots, m$, is prime.

A language, unlike an integer, can have also infinite factorizations, that is, decompositions into an infinite product of nontrivial factors. Here we restrict consideration to decompositions having finitely many components. Infinite factorizations obviously would involve interesting and different types of questions.

A finite language (distinct from $\emptyset$, $\{\varepsilon\}$) clearly always has a prime decomposition. On the other hand, a prime decomposition need not be unique even for finite languages [12]. Any prefix-free regular language has a unique decomposition in terms of prime languages if it is additionally required that the components are regular and prefix-free [3, 7]. Interestingly, the analogous property does not hold for decompositions of infix-free regular languages [6]. A factorial language is a language that is closed under the subword operation. In [1] it is shown that a factorial language has a unique canonical decomposition, where the components satisfy certain minimality conditions, into indecomposable factorial components.

Example 2.1. Let $H \subseteq \Sigma^n$, $n \geq 1$, be a set of words of length n. We show that H^* has the following prime decomposition

$$H^* = (\{\varepsilon\} \cup H) \cdot (\bigcup_{i=1}^{\infty} H^{2i-1} \cup \{\varepsilon\}) \tag{2}$$

Since the equality obviously holds, it is sufficient to verify that the two factors on the right side are prime.

In any decomposition $\{\varepsilon\} \cup H = AB$ both of the sets A and B must contain ε. Then the equality can only hold if one of A and B contains all words of H and the other set is $\{\varepsilon\}$, that is, $\{\varepsilon\} \cup H$ has only trivial decompositions.

In order to see that the second language on the right side of (2) is prime, assume that we can write

$$\bigcup_{i=1}^{\infty} H^{2i-1} \cup \{\varepsilon\} = AB \tag{3}$$

for some $A, B \subseteq \Sigma^*$. Again ε has to be in both A and B. Thus A or B cannot contain any nonempty words shorter than n and all words of H must be in A or B. If both A and B contain words of H then AB would have some word of length $2n$. We assume that $H \subseteq A$, the other possibility being symmetric. Again all words of H^3 must be in A or B, and similarly as above we see that the only possibility is that $H^3 \subseteq A$ since otherwise the catenation of A and B would have some word of length $4n$. By induction it follows that $A = \bigcup_{i=1}^{\infty} H^{2i-1} \cup \{\varepsilon\}$ and $B = \{\varepsilon\}$.

It seems that earlier work [12] did not expect that the Kleene-star of languages as in Example 2.1 could have prime decompositions. In fact, we do not know any regular language L such that L provably has no prime decompositions. In Section 4 we show that every regular language over a unary alphabet has a prime decomposition.

Next we show that there exist nonregular languages without any prime decompositions. Let $\Sigma = \{a, b\}$. We define $H_0 \subseteq \Sigma^*$ as follows:

$$H_0 = \{a^{i_1} b^{i_1} a^{i_2} b^{i_2} \cdots a^{i_k} b^{i_k} \mid k \geq 0, 1 \leq i_1 < i_2 < \ldots < i_k\}.$$

Lemma 2.1. *The language H_0 does not have any prime decomposition.*

Proof. Consider an arbitrary decomposition of H_0,

$$H_0 = L_1 \cdot \ldots \cdot L_m, \tag{4}$$

$m \geq 1$. For the sake of contradiction assume that (4) is a prime decomposition.

By the maximal ab-prefix, mab-prefix, (respectively, mab-suffix) of a word w we mean the longest prefix (respectively, longest suffix) of w that is in a^*b^*.

Consider a fixed $i \in \{1, \ldots, m\}$. We claim that if the mab-prefix of some word in L_i is of a form

$$a^j b^k, \ j \neq k, \ j, k \geq 0, \tag{5}$$

then all words in L_i must have the same mab-prefix $a^j b^k$. This follows from the observation that if L_i has two words u_1, u_2 where the mab-prefix u_{mp} of u_1 is as in (5) and the mab-prefix of u_2 is distinct from u_{mp}, then for any fixed $v \in L_1 \cdots L_{i-1}$ and $w \in L_{i+1} \cdots L_m$ only one of the words vu_1w and vu_2w can be in H_0.

Now if all words in L_i have the same mab-prefix as in (5) (which is not the empty word since $j \neq k$) we get a decomposition for L_i by factoring out the common prefix.

Since L_i is prime, the above means that we need to consider only the case where the mab-prefix of all words in L_i, $i = 1, \ldots, m$, is of a form $a^j b^j$, $j \geq 1$. (Note that in this case the mab-prefixes need not be identical, e.g., it is possible that $L_i = \{a^j b^j a^{j+1} b^{j+1}, a^j b^j, a^{j+1} b^{j+1}, \varepsilon\}$.) With a completely symmetric argument we see that the same property holds for mab-suffixes.

By a balanced word we mean a word of the form $a^j b^j$, $j \geq 0$. From the above we can conclude that for all $i \in \{1, \ldots, m\}$,

the mab-prefix and the mab-suffix of any word in L_i is balanced. (6)

Thus all words occurring in L_i, $1 \leq i \leq m$, are of the form

$$w_i = a^{k_{1,i}} b^{k_{1,i}} \cdots a^{k_{r,i}} b^{k_{r,i}}, \ 0 < k_{1,i} < \ldots < k_{r,i}, \ r \geq 0. \tag{7}$$

Now if we consider an arbitrary word $w_{i+1} = a^{k_{1,i+1}} b^{k_{1,i+1}} \cdots a^{k_{s,i+1}} b^{k_{s,i+1}} \in L_{i+1}$, the equation $k_{r,i} < k_{1,i+1}$ has to hold since otherwise $w_i w_{i+1}$ cannot occur as a subword of a word in H_0.

Now the equation (4) implies that, for all $i = 1, \ldots m - 1$, there exist integers M_i and N_i ($M_1 = 1$, $N_i = M_{i+1} - 1$) such that L_i consists of exactly all the words as in (7) where $M_i \leq k_{1,i}$ and $k_{r,i} \leq N_i$, and L_m consists of all words as in (7) where $k_{1,i} > N_{m-1}$.

It follows that (4) is not a prime decomposition since, for example,

$$L_m = \{\varepsilon, a^{N_{m-1}+1} b^{N_{m-1}+1}\} \cdot A,$$

where A consists of all words as in (7) where $k_{1,i} > N_{m-1} + 1$. This concludes the proof. $\qquad \square$

The language H_0 used in Lemma 2.1 is not context-free but its complement is context-free. It should be noted that Lemma 2.1 does not require any assumptions concerning the component languages, that is, H_0 doesn't have a prime decomposition even if the components could be non-recursively enumerable languages.

We conclude with the following question.

Open Problem. *Does there exist a context-free (or even a regular) language L such that L has no prime decomposition.*

3   Strong Prime Decomposition Property

In the previous section we saw (in Example 2.1) that regular languages can have artificial prime decompositions even if the natural way of decomposing the language does not result in a prime decomposition, i.e., the components could always be factorized further.

Example 3.1. Let $L = \varepsilon + a^2 a^*$. We note that $L = L \cdot L$ or $L = (\varepsilon + a^2) \cdot L$ so obviously L has many different factorizations with arbitrarily many components. However, L has also the following prime decomposition

$$(\varepsilon + a^2)(\varepsilon + a^3)(\varepsilon + \bigcup_{i=1}^{\infty}(a^2)^{2i-1}).$$

Note that the last component is an instance of the left side of (3) that was shown to be prime in Example 2.1.

Here we consider a stronger version of the prime decomposition property that prevents situations as in Example 3.1.

Definition 3.1. *Let $L \subseteq \Sigma^*$. The* index *of a non-trivial decomposition of L,*

$$L = L_1 \cdot \ldots \cdot L_m \tag{8}$$

is m. The decomposition index *of L is the maximum index of any non-trivial decomposition of L if the maximum exists. Otherwise, we say that the decomposition index of L is infinite.*

If a language L has a finite decomposition index, we say that L is *strongly prime decomposable*. When L is strongly prime decomposable, any way of iteratively decomposing L has to stop after a finite number of steps, i.e., the refinement of any decomposition results in a prime decomposition in a finite number of steps.

Clearly all finite languages are strongly prime decomposable since the decomposition index of a finite language L is at most the length of the longest word in L. The language L considered in Example 3.1 has a prime decomposition but it is not strongly prime decomposable. An example of a strongly prime decomposable infinite language is $a^* + b^*$. This follows from Theorem 3.1 below.

For presenting a characterization of the strongly prime decomposable regular languages we recall some notation and a result from [12, 14]. Let $A = (Q, \Sigma, \delta, q_0, Q_F)$ be a deterministic finite automaton (DFA). For a subset $P \subseteq Q$ we define the languages

$$R_1^P = \{w \in \Sigma^* \mid \delta(q_0, w) \in P\},$$
$$R_2^P = \bigcap_{p \in P} \{w \in \Sigma^* \mid \delta(p, w) \in Q_F\}.$$

Proposition 3.1. [12] *Let $A = (Q, \Sigma, \delta, q_0, Q_F)$ be the minimal DFA for a language L and assume that we can write $L = L_1 L_2$. Then*

$$L = R_1^P R_2^P,$$

where $P \subseteq Q$ is defined by

$$P = \{p \in Q \mid (\exists w \in L_1)\, \delta(q_0, w) = p\}.$$

Furthermore, we know that $L_i \subseteq R_i^P$, $i = 1, 2$.

Theorem 3.1. *A regular language L is not strongly prime decomposable if and only if there exist regular languages H_1, H_2, H_3, where H_2 contains some nonempty word such that*

$$L = H_1(H_2)^* H_3. \tag{9}$$

Proof. The "if"-direction follows from the observation that, for any $k \geq 1$, the equation (9) gives for L a decomposition of index at least k:

$$L = H_1(H_2 \cup \{\varepsilon\})^{k-1}(H_2)^* H_3. \tag{10}$$

(The index of the decomposition (10) is between k and $k + 2$ depending on whether or not H_1 or H_3 is the trivial language $\{\varepsilon\}$.)

Next we prove the "only-if"-direction. Let $A = (Q, \Sigma, \delta, q_0, Q_F)$ be the minimal DFA for L. Since L is not strongly prime decomposable we can write

$$L = L_1 L_2 \cdot \ldots \cdot L_m,$$

where $m = 2^{|Q|} + 1$ and $L_i \neq \{\varepsilon\}$, $i = 1, \ldots, m$. Furthermore, by [12] (Proposition 3.1 above) we know that the languages L_i can be chosen to be regular.

Define $P_i = \{p \in Q \mid (\exists w \in L_1 \cdot \ldots \cdot L_i)\, \delta(q_0, w) = p\}$, $i = 1, \ldots m - 1$. By Proposition 3.1,

$$L = R_1^{P_i} R_2^{P_i}, \quad i = 1, \ldots, m - 1. \tag{11}$$

Here $R_j^{P_i}$, $j = 1, 2$, is as defined in Proposition 3.1.

Since $m - 1 \geq 2^{|Q|}$ and $P_i \neq \emptyset$, $i = 1, \ldots, m-1$, there exist $j, k \in \{1, \ldots, m-1\}$, $j < k$, such that $P_j = P_k$. This means that for all $p \in P_j$ and $w \in L_{j+1} \cdot \ldots \cdot L_k$ we have

$$\delta(p, w) \in P_j \ (= P_k).$$

Thus (11) implies that for all $r \geq 1$,

$$R_1^{P_j}(L_{j+1} \cdot \ldots \cdot L_k)^r R_2^{P_j} \subseteq L.$$

Consequently, $L = R_1^{P_j}(L_{j+1} \cdot \ldots \cdot L_k)^* R_2^{P_j}$ and $L_{j+1} \cdot \ldots \cdot L_k$ is not empty or $\{\varepsilon\}$ since $j < k$. $\qquad\square$

It is known that primality is decidable for regular languages [12]. As a corollary of the proof of Theorem 3.1 we see that also the strong prime decomposition property is decidable for regular languages.

Corollary 3.1. *Given a regular language L it is decidable whether or not L is strongly prime decomposable.*

Proof. Let $A = (Q, \Sigma, \delta, q_0, Q_F)$ be the minimal DFA for L. In the "only if" part of the proof of Theorem 3.1 it is established that if L is not strongly prime decomposable, there exist $P \subseteq Q$ and a nonempty language $K_P \neq \{\varepsilon\}$ such that $L = R_1^P R_2^P$ (where R_j^P, $j = 1, 2$, is defined as in Proposition 3.1) and $\delta(p, w) \in P$ for all $p \in P$ and $w \in K_P$. Conversely, the existence of P and K_P as above implies that $L = R_1^P (K_P)^* R_2^P$ and hence, by the first part of Theorem 3.1, L is not strongly prime decomposable.

Given $P \subseteq Q$, a language K_P as above exists if and only if some nonempty word of length at most $s = |Q|^{|P|}$ takes each state of P to a state in P. Note that if this property holds for some word of length greater than s, using a pumping argument it follows that the property has to hold for a word of length at most s. Hence we can determine whether P and K_P as above exist by testing the required property for all subsets of Q. □

The algorithm given by Corollary 3.1 is extremely inefficient since it relies on an exhaustive search of subsets of the state set of the minimal DFA for L. It is probable that an efficient (e.g. a polynomial time) algorithm cannot be found since there is no known polynomial time algorithm even to test primality of a regular language [12].

4   Unary Regular Languages

We want to show that every regular language over a unary alphabet has a prime decomposition. First we recall some terminology concerning regular languages over a unary alphabet. A standard reference is [2], and references to more recent work on unary regular languages can be found e.g. in [4, 8].

A DFA A with a unary input alphabet can be divided into a *tail* which has the states that are not reachable from themselves with any non-empty word, and the *cycle* consisting of the remaining states of A. Naturally, A has no accepting states in the cycle if the language recognized by it is finite. If A is minimal, it is additionally required that all states are pairwise inequivalent. If the tail of A accepts words $a^{j_1}, \ldots a^{j_{r-1}}$ and the length of the cycle of A is m, the language accepted by A is denoted by a regular expression

$$a^{j_1} + \ldots + a^{j_{r-1}} + a^{j_r}(a^{i_1} + \ldots a^{i_{s-1}})(a^m)^*, \tag{12}$$

$0 \leq j_1 < \ldots j_{r-1} < j_r$, $0 \leq i_1 < \ldots < i_{s-1} < m$, $r, s \geq 0$. We use the names "tail" and "cycle" also when referring to the corresponding parts of a regular expression as in (12).

Lemma 4.1. *Let $L \subset \{a\}^*$ be any unary language. Then L^* is the union of a finite language and a linear language, that is, $L^* = F \cup \{a^{i \cdot p} \mid i \geq 0\}$ where $p \geq 0$ and $F \subseteq \{a\}^*$ is finite. Furthermore, p divides the length of any word in F.*

Proof. If L is empty or $L = \{\varepsilon\}$, the property holds by choosing $F = \emptyset$ and $p = 0$. Otherwise, if p is the greatest common divisor of the lengths of all words in L, there exists $M_p \geq 1$ such that for all $n > M_p$, $a^n \in L$ if and only if n is a multiple of p. We can choose F as the set of all words in L of length at most M_p. The length of any word in F is divided by p.   □

Lemma 4.2. *Let $L \subseteq \{a\}^*$ be a regular language such that*

$$L = LR^* \tag{13}$$

where R contains a nonempty word. Then L has a prime decomposition.

Proof. Let L be denoted by a regular expression as in (12). By factoring out the shortest word we can assume without loss of generality that $\varepsilon \in L$, that is, $j_1 = 0$. We assume that m (using the notations of (12)) is the cycle length of the minimal DFA for L and all words ε, a^{j_2}, $\ldots a^{j_{r-1}}$, $a^{j_r+i_1}$, $\ldots$, $a^{j_r+i_{s-1}}$ are pairwise inequivalent. These properties hold if the tail and cycle of (12) are as in the minimal DFA for L. Note that (13) implies that L is infinite and hence the minimal DFA has a cycle containing an accepting state, that is, $m \geq 1$.

By Lemma 4.1 we can write

$$R^* = \varepsilon + a^{k_1} + \ldots + a^{k_{t-1}} + a^{k_t}(a^n)^*, \tag{14}$$

where $0 < k_1 < \ldots < k_t$, $t \geq 1$, are all multiples of n. Here we require that $k_t \geq 1$ and as the word a^{k_t} we can choose the first nonempty word that is in the cycle of R^*. (The expression (14) does not need to correspond to the minimal DFA for R^*. This would be the case, for example, if the minimal DFA is cyclic, i.e., it has no tail.) Since R contains a nonempty word, it follows that $n \geq 1$.

By (13), $uv \in L$ for all $u \in L$ and $v \in R^*$. Since m is the cycle length of the minimal DFA for L, this implies that m divides n, and consequently the length of any word in R^* is a multiple of m. Write

$$a^{k_t} = c \cdot m, \quad c \geq 1.$$

Then

$$L = (\varepsilon + a^{j_2} + \ldots + a^{j_{r-1}} + a^{j_r}(a^{i_1} + \ldots + a^{i_{s-1}} + a^{i_1+m} + \ldots \tag{15}$$
$$+ a^{i_{s-1}+m} + \ldots + a^{i_1+(c-1)m} + \ldots + a^{i_{s-1}+(c-1)m}))(a^{k_t})^*.$$

In (15) the inclusion from right to left follows by (13) since all words in the first factor are in L and $(a^{k_t})^* \subseteq R^*$ because k_t is a multiple of n. The inclusion from left to right follows using the simple observation that the right side of (15) is obtained from the regular expression (12) for L with cycle length m by repeating the original cycle c times and taking $c \cdot m$ to be the new cycle length.

In the right side of (15) the first component has a prime decomposition since it is a finite language. The second component has a prime decomposition by Example 2.1.   □

The construction of Lemma 4.2 is illustrated in the next example. In particular, the example shows that in the factorization (15) we could not use $(a^n)^*$ as a factor for L where n is the cycle length of the minimal DFA for R^*.

Example 4.1. Let

$$L = \varepsilon + a^5 + a^{12} + a^{17}(a^3)^* + a^{18}(a^3)^*,$$

and let $R = (a^{12} + a^{18})^*$. Now $L = LR^*$ and the the construction from the proof of Lemma 4.2 gives for L the factorization

$$L = (\varepsilon + a^5 + a^{12} + a^{17} + a^{18} + a^{20} + a^{21} + a^{23} + a^{24} + a^{26} + a^{27})(a^{12})^*.$$

It can be noted that the cycle length of R^* is 6. However, $(a^6)^*$ is not a factor of L since $\varepsilon, a^5 \in L$ and $a^6, a^{11} \notin L$.

Theorem 4.1. *Every regular language over a unary alphabet has a prime decomposition.*

Proof. Let $L \subseteq \{a\}^*$ be regular. If we can write $L = L_1(L_2)^*$ for regular languages L_1 and L_2, where L_2 contains a nonempty word, then also $L = L(L_2)^*$ holds and, by Lemma 4.2, L has a prime decomposition.

If there exist no regular languages L_i, $i = 1, 2$, $L_2 \neq \{\varepsilon\}$, $L_2 \neq \emptyset$, such that $L = L_1(L_2)^*$, then using the commutativity of catenation of unary languages and Theorem 3.1 we get that L is strongly prime decomposable. $\qquad \square$

Let Σ be an arbitrary finite alphabet and $L \subseteq \Sigma^*$. The *length set* of L is the language over the unary alphabet $\{a\}$ defined by

$$\text{length}(L) = \{a^k \mid (\exists w \in L) \, |w| = k\}.$$

A language L over a non-unary alphabet may have more structure than the corresponding length set and decompositions of the length set of L do not necessarily yield a factorization of L. For example, the language $\{bc, cb\}$ is prime but its length set has the factorization $\{aa\} = \{a\} \cdot \{a\}$. Conversely, however, corresponding to any decomposition of L there exists a decomposition of the length set of L. This gives the following lemma.

Lemma 4.3. *Let Σ be a finite alphabet and $L \subseteq \Sigma^*$. If $\text{length}(L)$ is strongly prime decomposable, then the same holds for L.*

Proof. If L has a non-trivial decomposition $L = L_1 \cdot L_2$, then $\text{length}(L_1) \cdot \text{length}(L_2)$ is a non-trivial decomposition of $\text{length}(L)$. Hence, if L has an infinite decomposition index, the same holds for $\text{length}(L)$. In other words, if $\text{length}(L)$ is strongly prime decomposable, so is L. $\qquad \square$

The result of Lemma 4.3 can be used to show the existence of prime decompositions for context-free languages where the tail of the length set is "not closed" under any multiple of the cycle length of the minimal DFA for the length set. Note that the length set of a context-free language is always regular [9, 16].

Theorem 4.2. *Let L be a context-free language and let m be the cycle length of the minimal DFA for* length(L). *If for some $d \geq 0$ and $M_d \geq 1$, $a^d \in$ length(L) and, for all $i \geq M_d$, $a^{d+i \cdot m} \notin$ length(L), then L has a prime decomposition.*

Proof. Assume that length(L) has a decomposition length$(L) = MR^*$ in terms of regular languages M and R, where R contains a nonempty word. Then length$(L) =$ length$(L)R^*$ and, by the proof of Lemma 4.2, we know that there is a constant c such that $a^d \in$ length(L) implies that, for all $i \geq 1$, $a^{d+i \cdot c \cdot m}$ is in length(L). This contradicts the assumptions for length(L).

Hence there do not exist regular languages M and R, $R \neq \emptyset$, $R \neq \{\varepsilon\}$, such that length$(L) = MR^*$. By Theorem 3.1, length(L) is strongly prime decomposable and Lemma 4.3 implies that also L is strongly prime decomposable. $\qquad\square$

The conditions of Theorem 4.2 apply, for example, to any context-free language L such that L has a word of odd length and there exists a constant $M_L \geq 1$ such that all words of L of length greater than M_L have even length. The assumption that L is context-free is needed to guarantee that the length set of the language is regular.

Finally, we note that Theorem 4.1 cannot be extended for arbitrary unary languages. Recently, Rampersad and Shallit [13], and A. Salomaa and Yu [15], have independently given examples of non-regular unary languages that provably do not have a prime decomposition. The first mentioned language consists of all words over a unary alphabet whose length when represented in ternary notation does not contain a 2. The second mentioned language consists of all words whose length when represented in binary has no 1's in odd positions from the right. These languages are higher in the complexity hierarchy than the language of Lemma 2.1 in the sense that they are not co-context-free.

5   Conclusions

We have established an effective characterization of the strongly prime decomposable regular languages. Using the characterization it is easy to construct regular languages (over a unary or a non-unary alphabet) that are not strongly prime decomposable, i.e., that have an infinite decomposition index. We have shown that every regular language over a unary alphabet has a prime decomposition. The main open problem remaining is whether all regular languages over arbitrary alphabets have at least one prime decomposition. We conjecture a positive answer to this question.

References

1. Avgustinovich, S.V., Frid, A.E.: A unique decomposition theorem for factorial languages. Intern. J. of Algebra and Computation **15** (2005) 149–160
2. Chrobak, M.: Finite automata and unary languages. Theoret. Comput. Sci. **47** (1986) 149–158

3. Czyzowicz, J., Fraczak, W., Pelc, A., Rytter, W.: Linear-time prime decomposition of regular prefix codes. Internat. J. Foundations of Computer Science **14** (2003) 1019–1031
4. Domaratzki, M., Ellul, K., Shallit, J., Wang, M.-W.: Non-uniqueness and radius of cyclic unary NFAs. Internat. J. Foundations of Computer Science **16** (2005) 883–896
5. Domaratzki, M., Salomaa, K.: Decidability of trajectory based equations. Theoret. Comput. Sci. **345** (2005) 304–330
6. Han, Y.-S., Wang, Y., Wood, D.: Infix-free regular expressions and languages, Internat. J. Foundations of Computer Science, to appear.
7. Han, Y.-S., Wood, D.: The generalization of generalized automata: Expression automata. In: Domaratzki, M., Okhotin, A., Salomaa, K., Yu, S. (eds.): Implementation and Application of Automata, CIAA'04. Lecture Notes in Computer Science, Vol. 3317. Springer-Verlag (2005) 156–166
8. Holzer, M., Kutrib, M.: Unary language operations and their nondeterministic state complexity. In: Ito, M., Toyama, M. (eds.): Developments in Language Theory, DLT'02. Lecture Notes in Computer Science, Vol. 2450. Springer-Verlag (2003) 162–172
9. Hopcroft, J.E., Ullman, J.D.: Introduction to Automata Theory, Languages, and Computation. Addison-Wesley Publishing Company (1979)
10. Karhumäki, J.: Finite sets of words and computing. In: Margenstern, M. (ed.): Machines, Computations and Universality, MCU04. Lecture Notes in Computer Science, Vol. 3354. Springer-Verlag (2005) 36–49
11. Kari, L., Thierrin, G.: Maximal and minimal solutions to language equations, J. Comput. System Sci. **53** (1996) 487–496
12. Mateescu, A., Salomaa, A., Yu, S.: Factorizations of languages and commutativity conditions. Acta Cybernetica **15** (2002) 339–351
13. Rampersad, N., Shallit, J.: private communication.
14. Salomaa, A., Yu, S.: On the decomposition of finite languages. In: Rozenberg, G., Thomas, W. (eds.): Developments in Language Theory, DLT'99. World Scientific (2000) 22–31
15. Salomaa, A., Yu, S.: private communication.
16. Wood, D.: Theory of Computation. John Wiley & Sons, New York, NY, (1987)
17. Yu, S.: Regular languages. In: Rozenberg, G., Salomaa, A. (eds.): Handbook of Formal Languages, Vol. I. Springer-Verlag (1997) 41–110

On Weakly Ambiguous Finite Transducers

Nicolae Santean and Sheng Yu

Department of Computer Science,
University of Western Ontario, London, ON N6A 5B8, Canada

Abstract. By weakly ambiguous (finite) transducers we mean those transducers that, although being ambiguous, may be viewed to be at arm's length from unambiguity. We define input-unambiguous (IU) and input-deterministic (ID) transducers, and transducers with finite codomain (FC). IU transductions are characterized by nondeterministic bimachines and ID transductions can be represented as a composition of sequential functions and finite substitutions. FC transductions are recognizable and can be expressed as finite unions of subsequential functions. We place these families along with uniformly ambiguous (UA) and finitely ambiguous (FA) transductions in a hierarchy of ambiguity. Finally, we show that restricted nondeterministic bimachines characterize FA transductions. Perhaps the most important aspect of this work consists in defining nondeterministic bimachines and describing their power by linking them with weakly ambiguous finite transducers (IU and FA).

1 Overview

Arguably one of the most intriguing machines that realize rational transductions are the bimachine, designed by Schutzenberger ([11]) and studied, among others, by Eilenberg who stated their importance in [3, §11.7, Theorem 7.1, p. 321]. A bimachine is a compact representation of the composition of a left and a right sequential transducer, and it characterizes the family of rational functions. A few variations of the original design have been studied in [9], where it has been shown that the scanning direction of its two reading heads does not matter. A natural question which has not been addressed so far is "what family of transductions are realized by bimachines that operate nondeterministically?". We show that these machines characterize the family of transductions that can be written as a composition of a rational function and a finite substitution. They are equivalent to the so-called input-unambiguous transducers (IU), which are close relatives of the classical unambiguous transducers. We also show that nondeterministic bimachines can "simulate" (i.e., they give a representation of) rational relations with finite codomain (FC). Surprisingly, we prove that FC transductions belong strictly to the family of recognizable relations and that they can be written as a finite union of subsequential functions. We notice that nondeterministic bimachines are a compact representation of the composition of a left sequential transducer and a right input-deterministic (ID) transducer - which is a close relative of the classical right sequential transducer. Finally, we define restricted

O.H. Ibarra and Z. Dang (Eds.): DLT 2006, LNCS 4036, pp. 156–167, 2006.

nondeterministic bimachines to be those which do not reset themselves at each computation step. Surprisingly, we observe that this restriction increases their representation power, allowing them to characterize the entire family of finitely ambiguous (FA) rational relations. Basically, the reset/no-reset dichotomy reveals the difference between IU and FA families. Thus, by investigating the computational power of nondeterministic bimachines, we have been led to a study of various degrees of weak ambiguity in finite transducers.

The paper is structured as follows. In Section 2 we introduce transducers and ambiguity. We give a normalized form for IU transducers and we characterize ID transductions. Theorem 1 states the connection between IU and ID transductions by means of right sequential functions. In Section 3 we build a hierarchy of ambiguity by introducing FA, UA and FC transductions and by establishing their mutual relations. Since FC is a newly introduced family, we give a Mezei-like characterization of FC transductions, thus proving their recognizability (Theorem 2) and leading to a representation as a finite union of subsequential functions. Section 4 holds the most important results of the paper: theorems 3, 4 and 5. We define several types of nondeterministic bimachines, show that some types are equivalent and characterize the family of IU transductions, and reveal that restricted nondeterministic bimachines characterize FA transductions. All the proofs are omitted and can be found in [10].

2   Input-Unambiguous and Input-Deterministic Finite Transducers

In the following we assume known basic notions of automata theory ([5], [8], [13]). By DFA and NFA we understand deterministic and nondeterministic finite automata, and by ϵ-free NFA we understand NFA with no ϵ-transitions, where ϵ denotes the empty word.

By a finite transducer over the alphabets X and Y we understand a finite automaton over the product of free monoids $X^* \times Y^*$. In other words, a transducer is a finite automaton whose transition labels are elements of $X^* \times Y^*$, with the meaning that the first component of the label is an input word and the second component is an output word. It is well known that finite transducers realize **rational word relations** (see for example [8, §IV.1.2, p. 566]), denoted by $\mathrm{Rat}(X^* \times Y^*)$, or simply Rat when the alphabets are understood. By RatF we understand the family of rational functions.

Formally, a transducer is a tuple $T = (Q, X, Y, E, q_0, F)$, where Q is a set of states, X, Y are alphabets, q_0 is an initial state, F is a set of final states and E is a **finite** set of transitions which are elements of $Q \times X^* \times Y^* \times Q$. The transduction (binary word relation) realized by T will be denoted by $|T| \colon X^* \to Y^*$ and is defined similarly to the language accepted by an NFA. The transducer T is **normalized** if the following conditions hold:

1. $E \subseteq Q \times (X \cup \{\epsilon\}) \times (Y \cup \{\epsilon\}) \times Q$;
2. $F = \{q_f\}$, $q_f \neq q_0$;
3. $(p, x, \alpha, q) \in E \Rightarrow p \neq q_f$, $q \neq q_0$.

It is known that any rational transduction is realized by a normalized finite transducer and that any transducer can algorithmically be normalized.

By a **useful** state (or transition, path, loop, etc.) in a transducer we understand a state (or transition, path, loop, etc.) which is used in at least one successful computation. By an ϵ-**input loop** we understand a loop (in the transition graph) whose transitions have only ϵ-input labels.

The notion of ambiguity for automata (and transducers) relates to the number of possible successful computations performed by an automaton for a given input. For example, a DFA is unambiguous, whereas an NFA can have various degrees of ambiguity.

Definition 1. *An ϵ-NFA A is unambiguous (UNFA) if each word is the label of at most one successful computation in A.*

Let $T = (Q, X, Y, E, q_0, F)$ be a finite transducer. The **input automaton** of T is the finite automaton $A = (Q, X, \delta, q_0, F)$, where δ is given by

$$\forall x \in X^*: \quad q \in \delta(p, x) \Leftrightarrow \exists \alpha \in Y^* : (p, x, \alpha, q) \in E, \quad \text{where } p, q \in Q .$$

If the transducer is normalized, then its input automaton is an ϵ-NFA, otherwise it may be a lazy NFA (its transitions are labelled with words rather than letters or ϵ.).

Definition 2. *A finite transducer T is called* **input-unambiguous** *(IU, for short) if its input automaton is unambiguous (i.e., an UNFA).*

Notice that a transducer can still have different successful paths with same input labels and nevertheless be input-unambiguous. One such situation is depicted in Figure 1. Notice also the difference between this definition and the classical

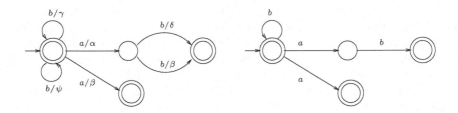

Fig. 1. An IU transducer and its input automaton

definition of **unambiguous transducers** ([1, p. 114]).

Remark 1. In our formalism, we imply that an IU transducer cannot have useful ϵ-input loops, in the same way as an unambiguous automaton cannot have useful ϵ-loops.

An IU transduction is a transduction realized by an IU transducer. Given an arbitrary IU transducer, there exists an equivalent IU transducer in normal form, in the sense mentioned at the begining of Section 2. Indeed, the standard normalization algorithm (see for example [1, §III.6, Corollary 6, p. 79]) does not change the degree of ambiguity of a transducer.

We recall that a **trim transducer** has only useful states. Without loss of generality, we follow the convention that if the initial state of a transducer is also final then the pair (ϵ, ϵ) is realized by the transducer. This convention has a theoretical explanation which we choose to ignore here, due to its interference with the definition of ambiguity and normalization.

Lemma 1. *Any IU transduction $\tau : X^* \to Y^*$ with $\tau(\epsilon) = \epsilon$ or $\tau(\epsilon) = \emptyset$ is realized by a trim IU transducer $T = (Q, X, Y, E, q_0, F)$ which satisfy the following conditions:*

(i) $E \subset Q \times X \times Y^* \times Q$;
(ii) if $\tau(\epsilon) = \epsilon$ then $F = \{q_0, q_f\}$, else $F = \{q_f\}$, and $q_f \neq q_0$;
(iii) $(p, x, \alpha, q) \in E \Rightarrow q \neq q_f, p \neq q_0$.

One can notice that it is decidable whether a finite transducer is IU or not. The decision can be reduced to whether an ϵ-NFA is UNFA or not.

In the following we recall sequential transducers and functions in order to draw a parallel with ID transducers which will be defined in the following. A (**left**) **sequential transducer** is a tuple $T' = (Q, X, Y, \delta, \lambda, q_0)$, where Q, X and Y are as usual and $\delta : Q \times X \to Q$ and $\lambda : Q \times X \to Y^*$ are partial functions (transition and output functions) with a same domain $(dom(\delta) = dom(\lambda))$, that are extended in the usual way. This transducer is a particular finite transducer that has all its states final and has the transition set given by $E = \{(q, x, \lambda(q, x), \delta(q, x))/(q, x) \in dom(\delta)\}$. This type of transducers represents a subfamily of rational functions: sequential functions. A **right sequential transducer** is a sequential transducer that reads its input and writes its output from right to left. It is known that any rational function can be written as a composition of a left and a right sequential function ([1]).

Definition 3. *An **input-deterministic** (ID) transducer is a tuple $T = (Q, X, Y, \delta, \omega, q_0)$ where X, Y are alphabets, Q is a finite set of states, and*

$$\delta : Q \times X \to Q, \; and \; \omega : Q \times X \to \mathcal{FP}(Y^*)$$

are partial functions with the same domain, denoting the transition and the output function. ($\mathcal{FP}(Y^*)$ denotes all finite parts of Y^*)

In other words, an ID transducer is similar to a sequential transducer, with the exception that reading an input letter leads to a finite number of output choices. Notice that a transducer is ID if and only if its input automaton is deterministic – hence justifying its name. As usual we define the family of ID transductions to be the family of all transductions that are realized by ID transducers.

Lemma 2. *A transduction is ID if and only if it is the composition of a sequential transduction and a finite substitution.*

Theorem 1. *Let $\tau : X^* \to Y^*$ be a transduction with $\tau(\epsilon) = \epsilon$. Then τ is an IU transduction if and only if there exist a right sequential function $\mu : X^* \to Z^*$ and an ID transduction $\nu : Z^* \to Y^*$ such that $\tau = \nu \circ \mu$. Moreover, μ can be chosen to be total and length preserving.*

Intuitively, in the above decomposition the sequential transducer represents the set of unique successful paths of the unambiguous transducer, whereas the ID transducer represents the nondeterminism of the output process.

It is also worth mentioning that a transduction is IU if and only if it is the composition of a left sequential function and a "right" ID transducer, fact that can be proven similar to Theorem 1. Here, by a right ID transducer we understand a transducer that scans the input from right to left and writes the output from right to left as well. It is apparent by now the similarity between this characterization and the characterization of rational functions by right and left sequential functions.

3   A Hierarchy of Ambiguity

In order to place IU and ID transductions into a proper context, in the following we recall two known families of rational transductions: finitely and uniformly ambiguous.

Definition 4. *A rational transduction $\tau : X^* \to Y^*$ is* finitely ambiguous *(FA) if $| \tau(u) | < \aleph_0, \forall u \in X^*$. We say that τ is* uniformly ambiguous *(UA) if there is a constant N such that $| \tau(u) | < N, \forall u \in X^*$.*

These families of transductions have been studied and used in various application in the past ([4], [6]). For example, it is known that an UA rational transduction can be written as a finite union of rational functions ([6]), and one can easily decide whether a rational transduction is in FA (this is equivalent to detecting non-trivial ϵ-input loops in a finite transducer). However, we are not aware of whether it is decidable if a rational transduction is in UA or not. Next we aim at finding the relationship between all these families of rational word relations.

Corollary 1
$$IU \subset FA \ .$$

This is a direct consequence of Remark 1: since an IU transducer has no ϵ-input loops, any input word can trigger a finite number of words to be written on the output tape. It affirms that the transductions realized by IU transducers are in FA, however they are not necessarily in UA. Indeed, the following example shows an IU transducer which realizes a transduction that is not uniformly ambiguous.

Example 1. The transducer in Figure 2 realizes the transduction τ given by:

$$\forall n \geq 1 : \quad \tau(a^n) = \begin{cases} \bigcup_{i=1}^{n} \{x^i\}, & \text{if } n \text{ is even} \\ \bigcup_{i=1}^{n} \{y^i\}, & \text{otherwise} \end{cases} ,$$

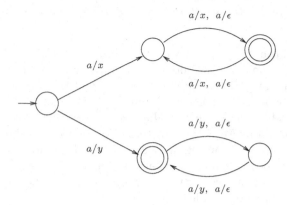

Fig. 2. An IU transducer whose transduction is not UA

which clearly is not UA, however it is IU. On the other hand, not all rational transductions which are UA are necessarily IU. The transduction

$$\tau = \{(a^n, x^n)/n \geq 1\} \cup \{(a^n, y^n)/n \geq 1\} \tag{1}$$

(with a, x, y different letters) is UA (notice that it is written as a union of two rational functions), however it is not IU. Indeed, a transducer T realizing τ must have two successful computations for each input word a^n: one outputting x^n and the other y^n, for all integers n. If these two successful computations coincide in the input automaton of T, then in T must exist a successful computation which "shuffles" x and y on the output tape, hence T cannot be IU.

Definition 5. *A rational transduction* $\tau : X^* \to Y^*$ *is with* `finite codomain` *(FC) if* $| \tau(X^*) | < \aleph_0$.

Obviously, it is decidable whether a rational transduction is in FC or not (it is equivalent to deciding whether the output automaton of a transducer accepts a finite language or not).

Lemma 3. *A rational transduction* $\tau : X^* \to Y^*$ *is in FC if and only if it can be written as*

$$\tau = \bigcup_{i \in I} [L_i \times R_i] \ ,$$

where I is finite, $\{L_i\}_{i \in I}$ *are disjoint regular languages and* $\{R_i\}_{i \in I}$ *are finite languages.*

One consequence of this lemma is the connection between transductions with finite codomain and subsequential transductions. Recall that a **(left) subsequential transducer** T' is a sequential transducer $T = (Q, X, Y, \delta, \lambda, q_0)$ (as defined in Section 2) together with a terminal output function $\rho : Q \to Y^*$, that realizes the rational function $| T' | (w) = | T | (w)\rho(\delta(q_0, w))$. It is known that there exist rational functions that can not be realized by either sequential or subsequential transducers. For more on the topic consult [1, §IV.2].

Corollary 2. *Any FC transduction can be written as a finite union of subsequential functions.*

In order to reveal the recognizability of FC transductions we recall that a recognizable set in a monoid is a set defined by an action over that monoid (see, for example, [8, p. 252]). Recall also that a subset of the direct product of two monoids (also a monoid) is recognizable if and only if it can be written as a finite union of blocks (a block is a direct product of two recognizable sets). This characterization is known as Mezei's characterization of recognizable sets in direct product monoids (see [3, Proposition 12.2, p. 68, and the note at p. 75]). Then, the recognizability of FC is a consequence of Lema 3. In the following, by Rec we understand the set of recognizable transductions over the alphabets X and Y, i.e., the family of recognizable subsets of $X^* \times Y^*$.

Theorem 2
$$FC \subset Rec \cap IU \ .$$

Notice that obviously $FC \subset UA$. Notice also that FC and the family of rational functions overlap, but are incomparable.

Remark 2. Although both FC and ID are included in IU, there is no relation of inclusion between FC and ID. For example, the transduction $\mu : \{a\}^* \to \{a\}^*$ given by

$$\forall n \geq 1: \ \mu(a^n) = \bigcup_{i=1}^{n} \{a^i\}$$

is in ID but not in FC; whereas the transduction $\nu : \{a\}^* \to \{a, b\}^*$ given by

$$\forall n \geq 1: \ \nu(a^n) = \begin{cases} a, & \textit{if } n \textit{ is even} \\ b, & \textit{otherwise} \end{cases}$$

is in FC (and in RatF, incidentally) but not in ID. Consequently, we may also infer that both FC and ID are `strictly` included in IU.

In Figure 3 we present a hierarchy describing different levels of ambiguity, where by dots we denote the areas where we have provided examples, including the following three:

$$FC \setminus (RatF \cup ID): \ \tau_1(a^n) = \begin{cases} \{x, y\}, & \textit{if } n \textit{ is even} \\ z, & \textit{otherwise} \end{cases} \ ,$$

$$FA \setminus (UA \cup IU): \ \tau_2(a^n) = \{\epsilon\} \cup \bigcup_{i=1}^{n} x^i \cup \bigcup_{i=1}^{n} y^i \ ,$$

$$(UA \cap IU) \setminus (ID \cup RatF \cup FC): \ \tau_3(a^n) = \begin{cases} \{x, y\}, & \textit{if } n \textit{ is even} \\ z^n, & \textit{otherwise} \end{cases} \ .$$

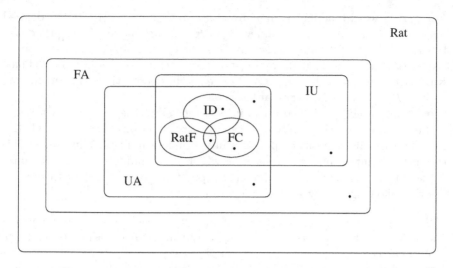

Fig. 3. Different degrees of ambiguity (dots represent examples)

4   Nondeterministic Bimachines

In the following we consider all input-unambiguous transducers to be trim and normalized according to Lemma 1. We are now aiming at giving a bimachine-characterization of IU.

Definition 6. *A bimachine $B = (Q, P, X, Y, \delta_Q, \delta_P, q_0, p_0, \omega)$ over X and Y is composed of*

two finite sets of states Q and P,
a finite input alphabet X and a finite output alphabet Y,
two partial next state functions

$$\delta_Q : Q \times X \to Q \text{ and } \delta_P : X \times P \to P \ ,$$

two initial states $q_0 \in Q$ and $p_0 \in P$,
and a partial output function $\omega : Q \times X \times P \to Y^*$.

The next-state functions are extended to operate on words as follows:

- $\forall q \in Q$ and $p \in P : \delta_Q(q, \epsilon) = q$ and $\delta_P(\epsilon, p) = p$;
- $\forall q \in Q, p \in P, a \in X$ and $w \in X^+$:

$$\delta_Q(q, wa) = \delta_Q(\delta_Q(q, w), a) \text{ and } \delta_P(aw, p) = \delta_P(a, \delta_P(w, p)).$$

Notice that function δ_P "reads" its argument word in reverse. We consider a similar extension of the output function:

- $\forall q \in Q$ and $p \in P : \omega(q, \epsilon, p) = \epsilon$;
- $\forall q \in Q, p \in P, a \in X$ and $w \in X^+$:

$$\omega(q, wa, p) = \omega(q, w, \delta_P(a, p))\omega(\delta_Q(q, w), a, p).$$

The partial word function realized by B is a function $f_B : X^* \to Y^*$, defined by $f_B(w) = \omega(q_o, w, p_0)$ if ω is defined in (q_0, w, p_0) and is undefined otherwise. Notice that $f_B(\epsilon) = \epsilon$ for any bimachine B. In essence, a bimachine is composed of two partial automata without final states (more precisely, all states act as final) and an output function. Indeed, (Q, X, δ_Q, q_0) will denote the `left automaton` of B and (P, X, δ_P, p_0) its `right automaton`.

Bimachines are of great theoretical importance since they are specifically designed to characterize the family of rational word functions. To our knowledge, so far there has been no attempts to study nondeterministic bimachines. We distinguish 3 components of a bimachine which are candidate to nondeterminism: the left and right automata and the output function. According to this, we define the following new types of bimachines:

1. `FNObm` : with finitely nondeterministic output (at each "step" the bimachine nondeterministically writes a word on the output tape, choosing from a finite set of choices);
2. `NTbm` : with nondeterministic transitions (the two underlying automata are nondeterministic: ϵ-NFA);
3. `LNTbm` : with left nondeterministic transitions (only the "left automaton" is nondeterministic);
4. `RNTbm` : with right nondeterministic transitions (only the "right automaton" is nondeterministic);
5. `NTObm` : with both nondeterministic transitions and finitely nondeterministic output;

and we denote by FNO, NT, LNT, etc. the families of transductions realized by these types of bimachines.

It is important to observe that at each computation step of an NTbm B, both the left and the right automata of B are "reset" to their initial state. This point is made clear in Figure 4. While reading w_1, the left automaton reaches the state q, through the computation(path) labelled w_1. However, in the next computation step, the left automaton reads $w_1 a$ and performs the computation labelled $w_1 a$ that may not overlap with the previous computation (more precisely, the computation labelled $w_1 a$ is not necessarily prefixed by the computation labelled w_1). This is due to the fact that the left automaton is reset to the initial state before reading $w_1 a$ (it does not continue the computation from q while reading a).

Theorem 3

$$FNO = NT = LNT = RNT = NTO \ .$$

In other words, it does not matter which component of the bimachine is nondeterministic. For this reason, we are allowed to employ the term `nondeterministic bimachine` in a generic sense.

It has been shown in [9] that the scanning direction of the reading heads of a (deterministic) bimachine does not matter. It is natural to question whether this property still holds for nondeterministic bimachines.

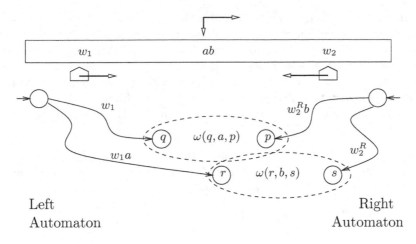

Fig. 4. NTbm behavior: each computation step involves a "reset"

Corollary 3. *The parsing direction of the reading heads of a nondeterministic bimachine does not matter.*

The same statement also applies to restricted nondeterministic bimachines - defined later. It tells that convergent, left sequential, right sequential, and divergent nondeterministic bimachines all have equal power. This is a consequence of Theorem 3: one may use FNO bimachines and adapt the proof in [9, T.16, p. 135] to the nondeterministic case.

We are now ready to state one of the main results of this paper, namely a bimachine characterization of IU rational transductions.

Theorem 4. *A transduction τ with $\tau(\epsilon) = \epsilon$ is IU rational* **if and only if** *it is realized by a nondeterministic bimachine.*

Consequence of Lemma 2 and Theorem 1 we obtain another characterization of IU transductions, that by Theorem 4 becomes a characterization of nondeterministic bimachines as well:

Corollary 4. *A transduction $\tau : X^* \to Y^*$ is IU if and only if there exists a rational function $\mu : X^* \to Z^*$ and a finite substitution $\sigma : Z^* \to \mathcal{FP}(Y^*)$ such that $\tau = \sigma \circ \mu$.*

Notice that it is decidable whether a nondeterministic bimachine is single-valued (realizes a rational function). Indeed, one can first construct an equivalent IU transducer whose functionality can be decided ([12], [2]). Notice also that the number of outputs for a given input of an IU transduction is a linear function of the length of the input and the length of any output is also a linear function of the length of the input. The converse does not hold, as the transduction (1) in Section 3, Example 1 is not IU, however it verifies these conditions. Finally, a surprising consequence of Corollary 4 and Theorem 2 is that any FC transduction can be represented by a composition of a rational function and a finite substitution as well.

So far we have introduced nondeterministic bimachines with a special behavior: at each computation step, these bimachines perform a "reset", i.e., they set their underlying automata to be in initial state. Then a natural question occurs, that is, "what would happen if we inhibit the reset?". This leads to the definition of another type of nondeterministic bimachine: a **restricted nondeterministic bimachine**. At each step, these bimachines are forced to continue their computation from the states reached at the previous step (nevertheless, they remain nondeterministic).

Definition 7. *A restricted nondeterministic bimachine (RNTbm) is a bimachine with nondeterministic transitions (NTbm) and multiple initial states $B = (Q, P, X, Y, \delta_Q, \delta_P, I_Q, I_P, \omega)$, where the output function is extended as follows:*

-   $\forall q \in Q, p \in P: \ \omega(q, \epsilon, p) = \{\epsilon\};$

-   $\forall w = a_1...a_n \in X^+$ *(where* $\forall i \in \{1, ..., n\} : a_i \in X$*),*
    $\forall q_0 \in I_Q, p_0 \in I_P, \ \omega(q_0, w, p_0)$ *is given by:*

$$\{ \ \omega(q_0, a_1, p_{n-1})\omega(q_1, a_2, p_{n-2})...\omega(q_{n-2}, a_{n-1}, p_1)\omega(q_{n-1}, a_n, p_0) \ /$$
$$q_1 \in \delta_Q^*(q_0, a_1), ..., q_{n-1} \in \delta_Q^*(q_{n-2}, a_{n-1}),$$
$$p_1 \in \delta_P^*(a_n, p_0), ..., p_{n-1} \in \delta_P^*(a_2, p_{n-2}) \ \}$$

Notice that by this behavior, the bimachine still operates nondeterministically. However, the current states of its automata depend on the previous current states. Surprisingly, although this seems like a restriction, RNTbm's have a greater power than NTbm's. Notice also that we allow multiple initial states - for improving the formalism. At the beginning of the operation, a RNT bimachine sets itself nondeterministically into two initial states corresponding to its left and right automata.

Theorem 5. *A transduction τ with $\tau(\epsilon) = \epsilon$ is in FA if and only if it is realized by a RNTbm.*

This theorem together with Theorem 4 completes the characterization of nondeterministic bimachines: they realize either IU or FA rational transductions, with respect to whether a reset is or not in place. Notice in Figure 3 the gap between deterministic bimachines (RatF) and nondeterministic ones (IU, FA).

5   Conclusion and Further Work

The goal of this paper has been twofold: to introduce nondeterministic bimachines and to study weakly ambiguous finite transducers. Nondeterministic bimachines can realize FC relations; however, they can do better than that: they exactly represent the family of transductions that are the composition of rational functions and finite substitutions. The transducer counterpart of these machines is the input-unambiguous transducer, which is a slight variation of the classical notion of unambiguous transducer. FC relations are recognizable and they have a particular "Mezei representation", as a finite union of blocks with certain properties: their left components are disjoint and their right ones are finite. This leads in a natural way to the representation of FC relations as a finite union of subsequential functions - notice the parallel with the uniformly ambiguous rational relations, that are finite unions of rational functions. Nondeterministic bimachines can work in two "modes": with or without reset. We have proven that suppressing the reset in between computation steps increases their power: they now characterize the family of finitely ambiguous transductions. Finally, we believe that all major rational families of transductions have a "bimachine" counterpart. In particular, we leave for immediate work the study of "ϵ-RNT" bimachines (i.e., RNT bimachines with ϵ-advancement) that we believe characterize the entire family of rational relations.

References

1. Jean Berstel. *Transductions and Context-Free Languages*. B. G. Teubner, Stuttgart, 1979.
2. Meera Blattner and Tom Head. Single Valued a-Transducers. *Journal of Computer and System Sciences*, 15(3):310–327, 1977.
3. Samuel Eilenberg. *Automata, Languages and Machines*, volume A. Academic Press, Inc., Orlando, FL, 1974.
4. Eitan M. Gurari and Oscar H. Ibarra. A Note on Finite-Valued and Finitely Ambiguous Transducers. *Mathematical Systems Theory*, 16(1):61–66, 1983.
5. John E. Hopcroft and Jeffrey D. Ullman. *Introduction to Automata Theory, Languages, and Computation - 1st edition*. Addison-Wesley Longman Publishing Co. Inc., Boston, MA, 1979.
6. Emmanuel Roche and Yves Schabes. Introduction to Finite-State Devices in Natural Language Processing. Technical Report TR-96-13, Mitsubishi Electric Research Laboratories, 1996.
7. Grzegorz Rozenberg and Arto Salomaa. *Handbook of Formal Languages*. Springer Verlag, Berlin Heidelberg New York, 1997.
8. Jacques Sakarovitch. *Éléments de Théorie des Automates*. Vuibert Informatique, Paris, 2003.
9. Nicolae Santean. Bimachines and Structurally-Reversed Automata. *Journal of Automata, Languages and Combinatorics*, 9(1):121–146, 2004.
10. Nicolae Santean and Sheng Yu. Nondeterministic Bimachines and Rational Relations with Finite Codomain. Technical Report 649, ISBN-10: 0-7714-2552-X, ISBN-13: 978-0-7714-2552-X, University of Western Ontario, 2005.
11. Marcel-Paul Schützenberger. A Remark on Finite Transducers. *Information and Control*, 4(2-3):185–196, 1961.
12. Marcel-Paul Schützenberger. Sur les Relations Rationelles. *Lecture Notes in Computer Science*, 33:209–213, 1975.
13. Sheng Yu. Regular Languages. *In [7]*, 1:41–110, 1997.

Ciliate Bio-operations on Finite
String Multisets*

Jürgen Dassow[1] and György Vaszil[2,**]

[1] Otto-von-Guericke-Universität Magdeburg,
Fakultät für Informatik
PSF 4120, D–39016 Magdeburg, Germany
`dassow@iws.cs.uni-magdeburg.de`
[2] Computer and Automation Research Institute,
Hungarian Academy of Sciences
Kende u. 13–17, H-1111 Budapest, Hungary
`vaszil@sztaki.hu`

Abstract. We study properties of string operations motivated by the
gene assembly process of ciliates. We examine the effect of these operations applied to finite multisets of words which make it possible to study
not only sequential, but also different types of parallel derivation strategies. We compare the classes of finite languages which can be obtained
by the different strategies, and show that although the string operations
we consider are reversible, their parallel application can produce effects
resembling the irreversibility of the biological process of gene assembly
in ciliates.

1   Introduction

Ciliates are a diverse group of very simple single-celled organisms. One of the
unifying characteristics of the group is the presence of two kinds of function-
ally different nuclei in the same cell, a micro-nucleus and a macro-nucleus. The
macro-nucleus contains the genetic information for producing proteins, while the
micro-nucleus contributes only to sexual reproduction where it is transformed
into the macro-nucleus. This transformation process of gene assembly is one
of the most involved DNA processing known in living organisms because the
structure of the micro-nuclear genome is very much different from the structure
of the macro-nuclear genome. While genes in the macro-nucleus are contiguous
sequences, in the micro-nucleus they are broken into pieces which occur individ-
ually or in groups, separated by long sequences of non-coding DNA. During gene

---

* Research supported in part by the Intergovernmental S&T Cooperation Programme
of the Office of Research and Development Division of the Hungarian Ministry
of Education and the Federal Ministry of Education and Research of Germany
(BMBF) under grant no. D-35/2000 and by the Hungarian Scientific Research Fund,
"OTKA", under grant no. F 037567.
** The author's stay in Magdeburg was also supported by a research fellowship from
the Alexander von Humboldt Foundation, Germany.

O.H. Ibarra and Z. Dang (Eds.): DLT 2006, LNCS 4036, pp. 168–179, 2006.

assembly, the non-coding parts are eliminated the coding parts are transferred and rearranged in their proper order.

Since single or double stranded DNA molecules can be thought of as a sequence of nucleotides or pairs of nucleotides, it is natural to represent them as strings of symbols and to use the tools and techniques provided by formal language theory to study the behavior of the molecules represented by the strings.

Several models for the gene de-scrambling process of ciliates were also established. In [6], [7], the gene assembly is modeled by three unary operations corresponding to intra-molecular recombinations, while in [10], [11], binary operations corresponding to intermolecular recombinations are proposed. The relationship of the two types of models is investigated in [9] where it is shown that if the binary operations of the intermolecular model are defined to be irreversible, as the operations of the intra-molecular model are, then the two different gene assembly strategies are equivalent, they are able to assemble the same sets of strings.

In the present paper we also investigate formal language theoretic string operations proposed to model the DNA sequence rearrangements observed during the gene assembly of ciliates. We consider both inter- and intra-molecular recombinations, so our approach resembles both types of models mentioned above, but the operations we propose are all reversible, in which sense we more closely follow [11].

The basic idea of our model is the following. There is a collection of strings and a finite set of operations modeling the recombinant behavior of the genetic material. The language generated by the system consists of all words which can be obtained from the elements of the initial collection by (iterative) applications of these operations. This idea is not new, see for example [2], [3], [4], but our model is essentially different from those considered so far. Similarly to [5], the string collections we consider are finite multisets, thus, in each derivation step, only a finite number of strings is available in a finite number of copies. This makes our model uninteresting from the point of view of the generative power since only finite languages can be produced. On the other hand, however, it enables the study of the effect of the parallel application of the operations, in contrast to the case when the generated strings are available in arbitrary many copies making the generation process essentially sequential.

We consider a sequential and the two different types of parallel derivation modes (see also [5]), and investigate the effect these have on the behavior of the system. The possible parallelism of the gene assembly process of ciliates was also considered in [8] in the framework of the intra-molecular model, but our approach is not only different because of the different model we use, but also because of our different intention. We do not only consider parallelism as a way to speed-up the process by executing more operations simultaneously, but also as a way to control it, to eliminate some of the many derivation paths that otherwise would be made available by the reversible string operations. In this respect our research is related to [1] and [12] where a similar control is achieved by the use of so called template strings, words which themselves are not changed during a recombination operation but only guide and control the process.

In what follows, we first define the model called multiset systems with ciliate operations, together with the sequential and the different types of parallel modes of application of the recombination operations, then examine the effect that these different modes have on the multisets or on the words which can be produced by the system.

By reasons of space limitations, some proofs are omitted.

2   Definitions

Let V be a finite set of symbols (or letters) called an alphabet, and let the finite sequences of its elements be called words (or strings) over V. Let V^* and V^+ be the set of all words over V (including the empty word λ) and the set of all non-empty words over V, respectively. For a word $w \in V^*$ and a letter $a \in V$, we denote the length of w by $|w|$ and the number of occurrences of a in w by $\#_a(w)$. For any set of words $L \subseteq V^*$, let $alph(L) \subseteq V$ denote the set of symbols appearing in the words, $alph(L) = \{a \in V \mid \#(w)_a > 0 \text{ for some } w \in L\}$. Let V^i denote the set of all words of length at most i for some $i \geq 0$, and let the set of all subwords x of a string $w = \alpha x \beta$ where $w, x, \alpha, \beta \in V^*$ be denoted by $sub(w)$. The mirror image (or reverse) of a string $w \in V^*$ is denoted by w^R.

The circular word $\langle x \rangle$, $x \in V^*$ over V is defined as $\langle x \rangle = \{x_2 x_1 \mid x = x_1 x_2,$ where $x_1, x_2 \in V^*\}$. The set of subwords of a circular word $\langle x \rangle$ is defined as $sub(\langle x \rangle) = \bigcup_{x' \in \langle x \rangle} sub(x')$. The length of the circular word $\langle x \rangle$ is the length of x, $|\langle x \rangle| = |x|$, the number of occurrences of a symbol a in $\langle w \rangle$ is $\#_a(\langle w \rangle) = \#_a(w)$. The set of all circular words over the alphabet V is denoted by $V^{\langle * \rangle}$, and the set of all circular words of length at most i, $i \geq 0$, is denoted by $V^{\langle i \rangle}$.

Formally, a multiset M over a set A is a mapping of A into the set $\mathbb{N}$ of non-negative integers. $M(x)$ is called the multiplicity of $x \in A$ in M. The support of multiset M is the set $supp(M) = \{x \in A \mid M(x) > 0\}$. We say that $x \in A$ is an element of M, denoted by $x \in M$, if $x \in supp(M)$. M is called finite or empty, if its support, $supp(M)$ is finite or empty, respectively. We denote empty multisets in the same way as empty sets, by the symbol $\emptyset$. The union of two multisets M_1 and M_2 over the same set, A, is denoted as $M_1 \cup M_2$ with $(M_1 \cup M_2)(x) = M_1(x) + M_2(x)$ for all $x \in A$.

A finite multiset M can be represented as a collection off elements containing $M(x)$ occurrences of x. For example, the multiset M over $\{a, b\}^*$ with $M(a) = M(b) = M(aba) = 1$, $M(ab) = M(ba) = 2$ and $M(x) = 0$ in all other cases can, be represented as $M = [a, b, ab, ab, ba, ba, aba]^1$. Obviously, as for sets, the order of the elements in this representation is not fixed and can be changed without changing the multiset.

---

[1] We use the brackets [and] instead of { and } in order to distinguish multisets from sets.

Let M be a multiset over $V^* \cup V^{\langle * \rangle}$. The cardinality, the ordinary cardinality, and the length of M is defined as

$$\#(M) = \sum_{x \in V^* \cup V^{\langle * \rangle}} M(x), \quad \#_o(M) = \sum_{x \in V^*} M(x), \quad \text{and } l(M) = \sum_{x \in V^* \cup V^{\langle * \rangle}} M(x)|x|,$$

respectively.

For a multiset over $V^* \cup V^{\langle * \rangle}$, $M = [w_1, w_2, \dots, w_n]$ (in the representation above) we have $l(M) = |w_1' w_2' \dots w_n'|$ where $w_i' = w_i$ for $w_i \in V^*$ and $w_i' \in \langle w_i \rangle$ if $w_i \in V^{\langle * \rangle}$, $1 \le i \le n$. Moreover, for M over $V^* \cup V^{\langle * \rangle}$ and $a \in V$, we set $\#_a(M) = \#_a(w_1' w_2' \dots w_n')$ with $w_1' w_2' \dots w_n'$ as above.

Definition 1. A *scheme with ciliate operations* (a CO scheme in short) is a pair (V, P), where V is an alphabet and $P \subseteq V^+$ is a finite set of pointers such that $\alpha \in P$ implies $\alpha^R \in P$.

Definition 2. Let (V, P) be a CO scheme as above. We define the following operations with respect to the given CO scheme.

A *loop direct repeat excision* on $w \in V^*$ is defined as

$$lde(w) = \{(x\alpha z, \langle y\alpha \rangle) \mid w = x\alpha y\alpha z, \ x, z \in V^*, y \in V^+, \alpha \in P, sub(y) \cap P = \emptyset \}.$$

A *loop direct repeat insertion* of the circular word $\langle y\alpha \rangle$, $y \in V^+, \alpha \in P$, $sub(y) \cap P = \emptyset$, into the linear word $w \in V^*$ is defined as

$$ldi(w, \langle y\alpha \rangle) = \{x\alpha y\alpha z \mid w = x\alpha z, \ x, z \in V^*\}.$$

A *hairpin inverted repeat excision/reinsertion* and a *double loop alternating direct repeat excision/reinsertion* on the linear word $w \in V^*$ is defined as

$$hi(w) = \{x\alpha y^R \alpha^R z \mid w = x\alpha y\alpha^R z, \ x, z \in V^*, y \in V^+, \alpha \in P\},$$
$$dlad(w) = \{x\alpha u\beta z\alpha y\beta v \mid w = x\alpha y\beta z\alpha u\beta v, \ x, z, v \in V^*, y, u \in V^+, \alpha, \beta \in P\}.$$

We say that $[w] \Rightarrow_{hi} [v]$ if $v \in hi(w)$, $[w] \Rightarrow_{dlad} [v]$ if $v \in dlad(w)$ for some $v, w \in V^*$, $[w] \Rightarrow_{lde} [v_1, v_2]$ if $(v_1, v_2) \in lde(w)$ for some $w, v_1 \in V^*$ and $v_2 \in V^{\langle * \rangle}$, and $[w_1, w_2] \Rightarrow_{ldi} [v]$ if $v \in ldi(w_1, w_2)$ for some $v, w_1 \in V^*$ and $w_2 \in V^{\langle * \rangle}$.

Definition 3. A *multiset system with ciliate operations* (an MCO system in short) is a triple $G = (V, P, I)$ where (V, P) is a CO scheme and I is a finite multiset over V containing only linear words, that is, $w \in I$ implies $w \in V^*$.

Definition 4. Let $G = (V, P, I)$ be an MCO system as above. For two multisets M and M', we define the following three types of the derivation relation.

- M' can be obtained from M by a *sequential derivation step*, denoted as $M \Rightarrow_s M'$ if $M = [v_1, \dots, v_n]$, $M' = [w_1, \dots, w_m]$ and either
  1. $m = n$ and $[v_1] \Rightarrow_X [w_1]$ for $X \in \{hi, dlad\}$, while $w_i = v_i, 2 \le i \le n$, for some appropriate indexing of the elements, or
  2. $m = n + 1$ and $[v_1] \Rightarrow_{lde} [w_1, w_{n+1}]$, while $w_i = v_i, 2 \le i \le n$, for some appropriate indexing of the elements, or

3. $m = n - 1$ and $[v_1, v_n] \Rightarrow_{ldi} [w_1]$, while $w_i = v_i, 2 \leq i \leq m$ for some appropriate indexing of the elements.

- M' can be obtained from M by a *maximally parallel derivation step*, denoted as $M \Rightarrow_{mp} M'$ if M and M' can be partitioned into four sub-multisets as $M = M_1 \cup M_2 \cup M_3 \cup M_4$ and $M' = M_1' \cup M_2' \cup M_3' \cup M_4'$ where $\#(M_1) + \#(M_2) + \#(M_3) > 0$ (thus, at least one of these partitions is not empty) and the following conditions hold.

  1. Either $M_1 = \emptyset$, in which case $M_1' = \emptyset$, or if $M_1 = [s_1, \ldots, s_k]$ for some $k \geq 1$, then $M_1' = [w_1, \ldots, w_k]$ where $[s_i] \Rightarrow_X [w_i]$, $1 \leq i \leq k$, $X \in \{hi, dlad\}$.
  2. Either $M_2 = \emptyset$, in which case $M_2' = \emptyset$, or if $M_2 = [t_1, \ldots, t_l]$ for some $l \geq 1$, then $M_1' = [x_1, y_1, \ldots, x_l, y_l]$ where $[t_i] \Rightarrow_{lde} [x_i, y_i]$, $1 \leq i \leq l$.
  3. Either $M_3 = \emptyset$, in which case $M_3' = \emptyset$, otherwise M_3 consists of an even number of elements $M_3 = [u_1, v_1, \ldots, u_m, v_m]$ for some $m \geq 1$, and $M_3' = [z_1, \ldots, z_m]$ where $[u_i, v_i] \Rightarrow_{ldi} [z_i]$, $1 \leq i \leq m$.
  4. Finally, $M_4' = M_4$ where if $M_4 \neq \emptyset$, then $lde(p) = hi(p) = dlad(p) = ldi(p, q) = \emptyset$ for all $p, q \in M_4$.

- We define a *strongly maximally parallel derivation step*, $M \Rightarrow_{smp} M'$, to be a maximally parallel derivation step with the four partitions of $M = M_1 \cup M_2 \cup M_3 \cup M_4$ where M_i, $1 \leq i \leq 4$ is exactly as above, chosen in such a way that considering any other maximally parallel derivation step with $M = \bar{M}_1 \cup \bar{M}_2 \cup \bar{M}_3 \cup \bar{M}_4$, the condition $\#(M_4) \leq \#(\bar{M}_4)$ holds.

Thus, in the sequential mode, one operation is applied to one string (or two strings in the case of ldi) in one derivation step. In the maximally parallel mode, a multiset of strings is chosen to which operations can be applied, in such a way that no other string or string pair can be added to the chosen multiset to which further operations are applicable. A multiset of strings is chosen also in the strongly maximally parallel mode, but this time in such a way, that it contains the maximal number of strings from the possible choices that are available. In other words, a strongly maximally parallel derivation step is a maximal parallel derivation step where the number of words not involved in any operation is the least possible.

Let us denote the reflexive and transitive closure of $\Rightarrow_s, \Rightarrow_{mp}$, and $\Rightarrow_{smp}$ by $\Rightarrow_s^*, \Rightarrow_{mp}^*$, and $\Rightarrow_{smp}^*$, respectively.

Definition 5. Let $G = (V, P, I)$ be an *MCO* system as above. The *multiset language*, the *word language*, and the *strong word language* generated by G in the sequential, maximal parallel, and strongly maximal parallel derivation modes are defined as follows.

$$mL(G, X) = \{M \mid I \Rightarrow_X^* M\},$$
$$wL(G, X) = \{w \in V^* \mid w \in M \text{ for some } M \in mL(G, X)\},$$
$$swL(G, X) = wL(G, X) \cup \{w \in u \mid u \in V^{\langle * \rangle}, u \in M$$
$$\text{for some } M \in mL(G, X)\},$$

where $X \in \{s, mp, smp\}$.

Thus, the multiset language contains all multisets which can be obtained from the initial one, the word language contains all linear words that can be obtained as elements of a multiset of the multiset language, and the strong word language contains all linear words plus all possible linearizations of the circular words that can be obtained as elements of a multiset of the multiset language.

For $X \in \{m, w, sw\}$ and $Y \in \{s, mp, smp\}$, we denote the classes of all languages $XL(G, Y)$ which can be generated by an MCO system G, by $X\mathcal{L}(Y)$. If we restrict to MCO systems $G = (V, P, I)$ with $\#(I) = n$, we use the notation $X\mathcal{L}_n(Y)$.

3 Multiset Languages

In this section we investigate the properties of multiset languages, the sets of multisets which can be obtained by various types of MCO systems from the initial configuration.

Lemma 1. *For an MCO system $G = (V, P, I)$, $a \in V$, $X \in \{s, mp, smp\}$, and any $M \in mL(G, X)$,*

$$l(M) = l(I), \ \#_a(M) = \#_a(I), \ and \ \#_o(M) = \#_o(I).$$

Remark. By Lemma 1, $m\mathcal{L}_n(X)$ and $m\mathcal{L}_{n'}(X')$ for $X, X' \in \{s, mp, smp\}$, and $n, n' \in \mathbb{N}$ are disjoint classes of multisets if $n \neq n'$, since the elements of $m\mathcal{L}_n(X)$ and those of $m\mathcal{L}_{n'}(X')$ have different ordinary cardinalities, n and n', respectively.

Thus, $m\mathcal{L}(X) = m\mathcal{L}(X')$ $(m\mathcal{L}(X) \subseteq m\mathcal{L}(X'))$, where $X, X' \in \{s, mp, smp\}$, if and only if $m\mathcal{L}_n(X) = m\mathcal{L}_n(X')$ $(m\mathcal{L}_n(X) \subseteq m\mathcal{L}_n(X'))$ for all $n \in \mathbb{N}$, and furthermore, $m\mathcal{L}(X)$ and $m\mathcal{L}(X')$ are incomparable if and only if $m\mathcal{L}_n(X)$ and $m\mathcal{L}_n(X')$ are incomparable for at least one n.

Therefore, in order to study the relationship between $m\mathcal{L}(X)$ and $m\mathcal{L}(X')$, it is sufficient to investigate the relationship between $m\mathcal{L}_n(X)$ and $m\mathcal{L}_n(X')$ for each $n \in \mathbb{N}$.

Theorem 2

1. $m\mathcal{L}_1(s) = m\mathcal{L}_1(mp)$,
2. $m\mathcal{L}_1(smp)$ and $m\mathcal{L}_1(X)$ are incomparable for $X \in \{s, mp\}$.
3. $m\mathcal{L}_n(s)$, $m\mathcal{L}_n(mp)$, and $m\mathcal{L}_n(smp)$ are pairwise incomparable for $n \geq 2$.

Proof. 1. The equality is obvious because there is always at most one non-circular word present, and this means that there is at most one operation which is applied in any derivation step.

2. Consider the MCO system $G = (\{a, b\}, \{b\}, [babab])$. Let

$$M_1 = [babab], \ M_2 = [bab, \langle ab \rangle], \ M_3 = [b, \langle ab \rangle, \langle ab \rangle].$$

Then we have

$$L_1 = mL(G, X) = \{M_1, M_2, M_3\}, \ X \in \{s, mp\} \ and$$
$$L_2 = mL(G, smp) = \{M_1, M_2\}.$$

The language L_1 cannot be generated in the strongly maximal parallel mode, because after starting with M_1 and obtaining M_2, a strongly maximal derivation step can only lead to M_1 again, and since the initial multisets cannot contain circular words, to start with M_1 would be the only possibility.

To generate L_2 in the sequential or in the maximal parallel mode with a system $G' = (\{a,b\}, P', M_1)$, either $a \in P'$ or $b \in P'$ because one of these pointers is necessary to obtain M_2. Now, if any of these pointers is present, then M_3 is also obtained which means that L_2 cannot be generated by any G' in the sequential or in the maximal parallel derivation mode.

3. We start by considering the case when $n = 2$, and first prove the incomparability of $mL_2(s)$ to both $mL_2(mp)$ and $mL_2(smp)$. Consider the MCO system

$$G = (\{a,b\}, \{a\}, [aba, aba]),$$

and let

$$M_1 = [aba, aba], \quad M_2 = [aba, a, \langle ba \rangle], \quad M_3 = [a, a, \langle ba \rangle, \langle ba \rangle], \quad M_4 = [a, ababa].$$

Then

$$L_1 = mL(G, s) = \{M_1, M_2, M_3, M_4\},$$

$$L_2 = mL(G, mp) = mL(G, smp) = \{M_1, M_3\}.$$

In order to prove the incomparability of $mL_2(s)$ to $mL_2(mp)$ and $mL_2(smp)$, it is sufficient to show that $L_1 \notin mL_2(mp)$, $L_1 \notin mL_2(smp)$, and $L_2 \notin mL_2(s)$.

Assume that $L_1 \in mL_2(X)$, $X \in \{mp, smp\}$. Then there is an MCO system $G' = (\{a,b\}, P', I')$ such that $L_1 = mL(G', mp)$ or $L_1 = mL(G', smp)$, and $I' \in \{M_1, M_4\}$.

If $I' = M_1$, then any generation of $M \neq I'$ from I' requires the application of the lde operation with respect to the pointer a. But if $a \in P'$, then $b \notin P'$, and we have $M_1 \Rightarrow_X M_3$ and $M_3 \Rightarrow_X M_1$, $X \in \{mp, smp\}$.

If $I' = M_4$, then in order to obtain any $M \neq I'$ from I', either $a \in P'$ or $b \in P'$. If $b \in P'$, then we only get $M_4 \Rightarrow_X M_2 \Rightarrow_X M_4$, $X \in \{mp, smp\}$, thus $a \in P'$ and $b \notin P'$ must hold. But then $M_4 \Rightarrow_X M_2$, and from M_2 we only obtain one of M_2 and M_4 in the maximal parallel mode, or only M_2 in the strongly maximal parallel mode.

Now assume that $L_2 \in mL_2(s)$. Then there is an MCO system $G'' = (\{a,b\}, P'', I'')$ with $L_2 = mL(G'', s)$. Again, $I'' = M_1$ because M_3 contains circular words. Any operation applied to M_1 in the sequential mode yields a multiset containing one occurrence of aba. Thus, we produce M_1 or a set different from both M_1 and M_3. Therefore, $mL(G'', s) = \{M_1\}$ or $M \in mL(G'', s)$ for some $M \notin \{M_1, M_3\}$. Thus, $mL(G'', s) \neq L_2$ in contrast to the choice of G''.

The incomparability of $mL_2(mp)$ and $mL_2(smp)$ can be similarly shown, we omit the proof due to space limitations.

The proof for $n \geq 3$ can be given by augmenting the languages used in the arguments with $n - 2$ such words to which no operations can be applied.   □

Now by Lemma 1, the remark above, and Theorem 2, we obtain the following corollary.

Corollary 3. *The classes of multiset languages* $m\mathcal{L}(s)$, $m\mathcal{L}(mp)$, *and* $m\mathcal{L}(smp)$ *are pairwise incomparable.*

4   Word Languages and Strong Word Languages

Now we continue by considering the sets of words, linear or circular, which can be generated by the various types of MCO systems.

Lemma 4. *For any MCO system* $G = (V, P, I)$,

$$XL(G,Y) \subseteq \bigcup_{i=0}^{l(I)} V^i, \tag{1}$$

where $X \in \{w, sw\}$, *and* $Y \in \{s, mp, smp\}$.

In the next two lemmas, we prove that equality cannot hold in (1).

Lemma 5. *Let* $X \in \{w, sw\}$ *and let* $Y \in \{s, mp, smp\}$. *If* $G = (V, P, I)$ *is an MCO system such that* $|V| \geq 2$ *and* $l(I) > 1$, *then*

$$\bigcup_{i=j}^{l(I)} V^i \setminus XL(G,Y) \neq \emptyset \text{ for any } 0 \leq j < l(I).$$

Lemma 6. *For any MCO system* $G = (V, P, I)$ *over a one-letter alphabet* $V = \{a\}$, *and any* $X \in \{w, sw\}$, $Y \in \{s, mp, smp\}$, *if* $\{a^i, a^{i+1}\} \subseteq XL(G,Y)$, *for some* $i \geq 2$, *then* $l(I) \geq i + 3$.

Proof. Let us assume that $G = (\{a\}, P, I)$, and $a^i, a^{i+1} \in XL(G,Y)$ for some X, Y as above. If $a^i, a^{i+1} \in I$, then $l(I) \geq 2i + 1 \geq i + 3$ (since $i \geq 2$), thus our statement holds. If at least one of a^i or a^{i+1} is not in I, then this means that starting the derivation from I, we can obtain two not necessarily disjoint pairs of a linear word and a possibly empty set of circular words (w_1, C_1) and (w_2, C_2), $w_i \in V^*$, $C_i \subset V^{\langle * \rangle}$, $1 \leq i \leq 2$, such that the elements of the first pair can be combined through ldi operations to obtain a^i, and the elements of the second pair can be combined to obtain a^{i+1}.

Now, since a^i and a^{i+1} have different length, at least one of the elements of one pair has to be different from another element of the other pair, thus, either w_1 and w_2 are different, or there is a circular word $c \in C_2$, $c \notin C_1$. This gives us a lower bound on the length of any configuration (which equals the length of the initial multiset) as $l(I) \geq |w_1| + |w_2| + l(C_1)$ in the first case, or $l(I) \geq |w_1| + |c| + l(C_1)$ in the second.

Since the shortest possible pointer contains two symbols (if $a \in P$ then no lde operations can be applied), the minimal length of a linear word which can participate in an ldi operation cannot be less then *two*, and the minimal length of a circular word (the result of an lde operation) is at least *three*. Thus, considering the first case, $|w_1| + l(C_1) = i$ and $|w_2| + l(C_1) = i + 1$ implies $|w_2| = |w_1| + 1$, and since the minimal length of any linear word which can participate in the *ldi* operation is *two*, we obtain $|w_2| \geq 3$, that is, $l(I) \geq i + 3$. If we consider the second case, we have $l(I) \geq i + |c|$, where c is a circular string that can be inserted into a linear word through an *ldi* operation. The minimal length of such words is *three*, so we have $l(I) \geq i + 3$ also in this case.   □

Let $L \subseteq V^*$ be an arbitrary finite language and let $a \in V$. The *MCO* system $G = (V, \{a^n\}, L)$ with $n > max\{|x| \mid x \in L\}$ satisfies $XL(G, Y) = L$ for all $X \in \{w, sw\}$, $Y \in \{s, mp, smp\}$ since no application of any operation is possible. This fact together with Lemma 4 implies that the languages $XL(Y)$ where X, Y as above, coincide with the class of finite languages, thus, we have the following result.

Theorem 7. $XL(s) = X'L(mp) = X''L(smp)$, $X, X', X'' \in \{w, sw\}$.

More interesting questions arise if we consider the class of languages which can be generated from initial multisets with a certain number of elements, that is, if we consider the families $XL_n(Y)$, $X \in \{w, sw\}$, $Y \in \{s, mp, smp\}$.

Lemma 8. *For any* $n \geq 1$, $X, X' \in \{w, sw\}$, *and* $Y, Y' \in \{s, mp, smp\}$,

$$XL_{n+1}(Y) \setminus X'L_n(Y') \neq \emptyset.$$

Let us now continue by examining the relationship of language classes which can be obtained from initial multisets of the same cardinality with different types of derivation modes.

Lemma 9. *For any* $n \geq 1$ *and* $X \in \{w, sw\}$,

$$XL_n(s) \cap XL_n(mp) \cap XL_n(smp) \neq \emptyset.$$

Proof. Let, for all $j \geq 1$,

$$L_j = \{a^{3^i} \mid 1 \leq i \leq j\}. \tag{2}$$

The MCO system

$$G = (\{a\}, \{a\}, [a^3, a^9, a^{27}, \ldots, a^{3^n}])$$

where all elements of the axiom multiset have the multiplicity 1, thus $\#(I) = n$, generates L_n as a word language and as a strong word language in all modes, because the application of the *hi* and *dlad* operations do not change the words and the *ldi*, *lde* operations cannot be applied.   □

Lemma 10. *For* $X \in \{w, sw\}$ *and* $n \geq 2$,

$$(XL_n(mp) \cap XL_n(smp)) \setminus XL_n(s) \neq \emptyset.$$

Proof. We first prove the statement for $n = 2$. Consider the MCO system

$$G = (\{a, b, c, d, e\}, \{ab, ba, cd, dc\}, [abcdba, cdecd])$$

having an initial multiset with two elements. Then

$$[abcdba, cdecd] \Rightarrow_Y [abdcba, cd, \langle ecd \rangle] \Rightarrow_Y [abcdba, cdecd]$$

for $Y \in \{mp, smp\}$, thus, the languages generated by G are

$$L_w = wL(G, Y) = \{abcdba, cdecd, abdcba, cd\} \text{ and}$$

$$L_{sw} = swL(G, Y) = L_w \cup \langle ecd \rangle.$$

Suppose now that $XL(G, Y) = XL(G', s)$, $X \in \{w, sw\}$, Y as above, for some

$$G' = (\{a, b, c, d, e\}, P', I'),$$

where $\#(I') = 2$. We show that this is not possible.

If $I' \in \{[abcdba, cdecd], [abdcba, cdecd]\}$ then G' needs to generate the word cd. If cd is somehow obtained from $abcdba$, then a word not containing cd as a subword is also produced, and since there is no such word in L_w or L_{sw}, cd must be cut out from $cdecd$. This means that either $c \in P'$, $d \in P'$, or $cd \in P'$ and both cd and $\langle ecd \rangle$ are obtained by an lde operation. But then $abcdecdba \in ldi(abcdba, \langle ecd \rangle)$ would also be produced which is a contradiction.

If $I' \in \{[w_1, w_2] \mid w_1, w_2 \in \{abcdba, abdcba, cd\} \}$, then no word containing the symbol e, if $I' = [cdecd, cd]$, then no word containing the symbols a, b can be produced.

Thus, we can conclude that $wL(G', s) = L_w$ is a contradiction.

To see that $swL(G', s) = L_{sw}$ is also impossible, we need to check in addition the word pairs containing elements of $\langle ecd \rangle$.

If $I' \in \{[w_1, w_2] \mid w_1 \in \{abcdba, abdcba\}, w_2 \in \langle ecd \rangle\}$, then the generation of the word $cdecd$ requires all the available copies of c, d. This means that words containing only a, b must also be generated which is a contradiction. On the other hand, if $I' \in \{[w_1, w_2] \mid w_1 \in \{cdecd, cd\}, w_2 \in \langle ecd \rangle\}$, then no word containing a or b can be generated, thus there is no G' with $\#(I') = 2$ and $XL(G, Y) = XL(G', s)$, $X \in \{w, sw\}$, $Y \in \{mp, smp\}$, which completes the proof for the case of $n = 2$.

To see that the statement also holds for $n \geq 3$, consider the languages $L'_n = L_X \cup L_{n-2}$ where $X \in \{w, sw\}$, and L_{n-2} is the unary language for $j = n - 2$ defined at (2) in the proof of Lemma 9 above, with the assumption that $alph(L_X) \cap alph(L_{n-2}) = \emptyset$.

Since L_X and L_{n-2} are languages over disjunct alphabets, their words have to be produced independently in any MCO system generating L'_n. As shown in Lemma 8, the production of L_{n-2} requires $n - 2$ different words in the initial multiset, and as shown above, the remaining two words cannot produce L_X in the sequential mode, so we can conclude that the statement of our lemma holds for any $n \geq 2$.   □

There are also languages which can be generated by systems working in the sequential and maximal parallel derivation modes, but not in the strongly maximal parallel one.

Lemma 11. *For* $X \in \{w, sw\}$ *and* $n \geq 2$,

$$(X\mathcal{L}_n(s) \cap X\mathcal{L}_n(mp)) \setminus X\mathcal{L}_n(smp) \neq \emptyset.$$

And finally, there are languages generated by systems working in the strongly maximal parallel derivation mode which cannot be generated in the sequential or maximal parallel modes.

Lemma 12. *For* $X \in \{w, sw\}$ *and* $n \geq 2$,

$$X\mathcal{L}_n(smp) \setminus (X\mathcal{L}_n(mp) \cup X\mathcal{L}_n(s)) \neq \emptyset.$$

Based on Lemma 9, Lemma 10, Lemma 11, and Lemma 12, we can formulate the following statement.

Theorem 13. *For any* $X \in \{w, sw\}$ *and* $n \geq 2$, *the language classes* $X\mathcal{L}_n(smp)$ *and* $X\mathcal{L}_n(Y)$, *for* $Y \in \{s, mp\}$, *are incomparable.*

5   Conclusion

We have studied the effect of applying reversible recombination operations on finite string collections in the sequential, the maximally parallel, and the strongly maximally parallel manner.

We have seen that the class of sets of multisets generated by the three modes are pairwise incomparable.

The classes of word and strong word languages, in general, coincide with the class of finite languages, but considering systems with initial multisets of a given cardinality we have shown the incomparability of the language classes produced in the strongly maximally parallel mode and those produced in the maximally parallel or the sequential mode. About the relationship of the language classes generated by the maximally parallel and the sequential mode, we know that they cannot be equal since by Lemma 10, there are word languages generated in maximally parallel mode which cannot be produced sequentially.

References

1. Daley, M., McQuillan, I.: Template-guided DNA recombination. Theoretical Computer Science **330** (2005) 237–250
2. Daley, M.: Kari, L.: Some properties of ciliate bio-operations. In: Developments in Language Theory, 6th International Conference, DLT 2002, Kyoto, Japan, September 18-21, 2002, revised papers. Edited by M. Ito, M. Toyama. Volume 2450 of Lecture Notes in Computer Science, Springer-Verlag, 2003, 117–127

3. Daley, M., Ibarra, O., Kari, L.: Closure properties and decision questions of some language classes under ciliate bio-operations. Theoretical Computer Science **306** (2003) 19–38
4. Dassow, J., Holzer, M.: Language families defined by a ciliate bio-operation: Hierarchies and decision problems. International Journal of Foundations of Computer Science **16** (2005) 645–662
5. Dassow, J., Vaszil, Gy.: Multiset splicing systems. BioSystems **74** (2004) 1–7
6. Ehrenfeucht, A., Harju, T., Petre, I., Prescott, D. M., Rozenberg, G.: Computation in Living Cells. Gene Assembly in Ciliates. Springer-Verlag, 2003
7. Ehrenfeucht, A., Prescott, D. M., Rozenberg, G.: Computational aspects of gene (un)scrambling in ciliates. In: Evolution as Computation. Edited by L. Landweber, E. Winfree. Springer-Verlag, 2001, 216–256
8. Harju, T., Li, Ch., Petre, I., Rozenberg, G.: Parallelism in gene assembly. In: DNA Computing. 10th International Workshop on DNA Computing, DNA 10, Milan, Italy, June 7-10, 2004. Revised, selected papers. Edited by C. Ferretti, G. Mauri, C. Zandron. Volume 3384 of Lecture Notes in Computer Science, Springer-Verlag, 2005, 140–150
9. Harju, T., Petre, I., Rozenberg, G.: Two models for gene assembly in ciliates. In: Theory is Forever. Edited by J. Karhumäki, Gh. Păun, G. Rozenberg. Volume 3113 of Lecture Notes in Computer Science, Springer-Verlag, 2004, 89–101
10. Landweber, L. F., Kari, L.: The evolution of cellular computing: Nature's solution to a computational problem. BioSystems, **52** (1999) 3–13 (Special issue: Proceedings of the 4th DIMACS meeting on DNA based computers, guest editors: L. Kari, H. Rubin, D. Wood.)
11. Kari, L., Kari, J., Landweber, L. F.: Reversible molecular computation in ciliates. In: Jewels are Forever. Edited by J. Karhumäki, H. Maurer, Gh. Păun, G. Rozenberg. Springer-Verlag 1999, 353–363
12. Prescott, D. M., Ehrenfeucht, A., Rozenberg, G.: Template-guided recombination for IES elimination and unscrambling of genes in stichotrichous ciliates. Journal of Theoretical Biology **222** (2003) 323–330

Characterizing DNA Bond Shapes
Using Trajectories

Michael Domaratzki[*]

Jodrey School of Computer Science, Acadia University,
Wolfville, NS B4P 2R6, Canada
mike.domaratzki@acadiau.ca

Abstract. We consider the use of DNA trajectories to characterize
DNA bond shapes. This is an extension of recent work on the bond-
free properties of a language. Using an new definition of bond-freeness,
we show that we can increase the types of DNA bond shapes which are
expressible. This is motivated by the types of bond shapes frequently
found in DNA computing.

We examine the algebraic properties of sets of trajectories. In particular, we consider rotation of trajectories and weakening of the bonding
conditions expressed by a set of DNA trajectories. We also consider decidability results for bond-freeness with respect to our definition.

1 Introduction

DNA computing is the process of translating computational problems into
strands of DNA such that reactions on DNA strands indicate solutions to the
problems [2, 12]. When constructing a set of DNA strands for use in DNA computing, it is imperative that the types of bonds which can and cannot appear
between strands in this set are well-understood to ensure the desired outcome.
Because of this, the study of the description and characterization of bond shapes
in DNA has received a great deal of attention in the literature [4, 5, 6, 7, 8, 9, 10].
One of the most promising ways of interpreting and assessing bonding between
strands is via the use of trajectories.

The use of *DNA trajectories* was proposed by Kari *et al.* [10] as a means of
formalizing the visual description of the bonding of two strands. The inclusion of
the term *trajectories* in DNA trajectories reflects the underlying use of shuffle on
trajectories [11] to interpret the bonding described by a set of DNA trajectories.
The strength of DNA trajectories is that they clearly represent the bonding
which occurs between two single-stranded DNA molecules, and does so with the
use of the well-understood concept of shuffle on trajectories. Further, trajectories
are robust with respect to changes in the description such as the lengths of bonds.

However, as we demonstrate in this paper, the interpretation of DNA trajectories proposed by Kari *et al.* [10] is restricted somewhat in its descriptional power.
To resolve this, we propose a more natural interpretation of DNA trajectories,

---

[*] Research supported in part by a grant from NSERC.

O.H. Ibarra and Z. Dang (Eds.): DLT 2006, LNCS 4036, pp. 180–191, 2006.

whose capacity to describe DNA bonding patterns is greater than the original definition. With this definition, more complex interactions between strands of a set of DNA are describable. This change in interpretation is motivated by the types of bonds which are seen in experimental DNA computing.

We consider algebraic properties of our new definition related to redundancy. In particular, we can characterize the effect of weakening of bonding and rotation of DNA trajectories and their effect on sets of trajectories. We have also shown that determining whether a regular language is bond-free with respect to a set of DNA trajectories is decidable even if the set of DNA trajectories is context-free, and is decidable in quadratic time if the set of DNA trajectories is regular.

2 Definitions

For additional background in formal languages and automata theory, please see Rozenberg and Salomaa [13]. For an introduction to DNA computing, see Păun *et al.* [12] or Amos [2]. Let Σ be a finite set of symbols, called *letters*. Then Σ^* is the set of all finite sequences of letters from Σ, which are called *words*. The empty word ϵ is the empty sequence of letters. The *length* of a word $w = w_1 w_2 \cdots w_n \in \Sigma^*$, where $w_i \in \Sigma$, is n, and is denoted $|w|$. Note that ϵ is the unique word of length 0. Given a word $w \in \Sigma^*$ and $a \in \Sigma$, $|w|_a$ is the number of occurrences of a in w. By extension, for any $S \subseteq \Sigma$, we let $|w|_S = \sum_{a \in S} |w|_a$.

A *language* L is any subset of Σ^*. By $\overline{L}$, we mean $\Sigma^* - L$, the complement of L. We denote singleton languages $\{w\}$ simply by w.

Let Σ, Δ be alphabets and $h : \Sigma \to \Delta$ be any function. Then h can be extended to a *morphism* $h : \Sigma^* \to \Delta^*$ via the condition that $h(uv) = h(u)h(v)$ for all $u, v \in \Sigma^*$. Similarly, h can be extended to an *anti-morphism* via the condition that condition that $h(uv) = h(v)h(u)$ for all $u, v \in \Sigma^*$. An *involution* θ is any function $\theta : \Sigma \to \Sigma$ such that θ^2 is the identity mapping on Σ. Note that an involution satisfies $\theta^{-1} = \theta$ and every involution is a bijection. If θ is extended to a morphism, we say that it is a *morphic involution*, while we use the term *anti-morphic involution* for the case when θ is extended to an anti-morphism. We denote by $\iota : \Sigma^* \to \Sigma^*$ the identity morphism, and $\mu : \Sigma^* \to \Sigma^*$ denotes the identity anti-morphism (or mirror involution).

When dealing with DNA as words over the alphabet $\{A, C, G, T\}$, we adopt the convention that words are read from the 5'-end to the 3'-end. When illustrating this fact, we draw an arrow from the 5'-end to the 3'-end.

A *deterministic finite automaton* (DFA) is a five-tuple $M = (Q, \Sigma, \delta, q_0, F)$ where Q is the finite set of states, Σ is the alphabet, $\delta : Q \times \Sigma \to Q$ is the transition function, $q_0 \in Q$ is the distinguished start state, and $F \subseteq Q$ is the set of final states. We extend δ to $Q \times \Sigma^*$ in the usual way. A word $w \in \Sigma^*$ is accepted by M if $\delta(q_0, w) \in F$. The *language accepted* by M, denoted $L(M)$, is the set of all words accepted by M. A language is called *regular* if it is accepted by some DFA.

A *context-free grammar* (CFG) is a four-tuple $G = (V, \Sigma, P, S)$, where V is a finite set of non-terminals, Σ is a finite alphabet, $P \subseteq V \times (V \cup \Sigma)^*$ is a finite set of productions and $S \in V$ is a distinguished start non-terminal. If $(\alpha, \beta) \in P$, we usually denote this by $\alpha \to \beta$. If $G = (V, \Sigma, P, S)$ is a CFG, then given two words $\alpha, \beta \in (V \cup \Sigma)^*$, we denote $\alpha \Rightarrow_G \beta$ if $\alpha = \alpha_1 \alpha_2 \alpha_3$, $\beta = \alpha_1 \beta_2 \alpha_3$ for $\alpha_1, \alpha_2, \alpha_3, \beta_2 \in (V \cup \Sigma)^*$ and $\alpha_2 \to \beta_2 \in P$. Let $\Rightarrow_G^*$ denote the reflexive, transitive closure of $\Rightarrow_G$. Then the language generated by a grammar $G = (V, \Sigma, P, S)$ is given by $L(G) = \{x \in \Sigma^* : S \Rightarrow_G^* x\}$. If a language is generated by a CFG, then it is a context-free language (CFL).

2.1 Trajectory-Based Operations

We now define shuffle on trajectories, the main tools for examining bond-free properties in this paper.

The shuffle on trajectories operation is a method for specifying the ways in which two input words may be merged, while preserving the order of symbols in each word. Each trajectory $t \in \{0, 1\}^*$ with $|t|_0 = n$ and $|t|_1 = m$ specifies one particular way in which we can form the shuffle on trajectories of two words of length n (as the left operand) and m (as the right operand). The word resulting from the shuffle along t will have length $n + m$, with a letter from the left input word in position i if the i-th symbol of t is 0, and a letter from the right input word in position i if the i-th symbol of t is 1.

Formally [11], let x and y be words over an alphabet Σ and t, the *trajectory*, be a word over $\{0, 1\}$. The shuffle of x and y on trajectory t is denoted by $x \, \sqcup\!\sqcup_t \, y$. If $x = ax'$, $y = by'$ (with $a, b \in \Sigma$) and $et \in \{0, 1\}^*$ (with $e \in \{0, 1\}$), then

$$x \, \sqcup\!\sqcup_{et} \, y = \begin{cases} a(x' \, \sqcup\!\sqcup_t \, by') & \text{if } e = 0; \\ b(ax' \, \sqcup\!\sqcup_t \, y') & \text{if } e = 1. \end{cases}$$

If $x = ax'$ ($a \in \Sigma$), $y = \epsilon$ and $et \in \{0, 1\}^*$ ($e \in \{0, 1\}$), then

$$x \, \sqcup\!\sqcup_{et} \, \epsilon = \begin{cases} a(x' \, \sqcup\!\sqcup_t \, \epsilon) & \text{if } e = 0; \\ \emptyset & \text{otherwise.} \end{cases}$$

If $x = \epsilon$, $y = by'$ ($b \in \Sigma$) and $et \in \{0, 1\}^*$ ($e \in \{0, 1\}$), then

$$\epsilon \, \sqcup\!\sqcup_{et} \, y = \begin{cases} b(\epsilon \, \sqcup\!\sqcup_{t'} \, y') & \text{if } e = 1; \\ \emptyset & \text{otherwise.} \end{cases}$$

We let $x \, \sqcup\!\sqcup_\epsilon \, y = \emptyset$ if $\{x, y\} \neq \{\epsilon\}$. Finally, if $x = y = \epsilon$, then $\epsilon \, \sqcup\!\sqcup_t \, \epsilon = \epsilon$ if $t = \epsilon$ and $\emptyset$ otherwise.

It is not difficult to see that if $t = \prod_{i=1}^n 0^{j_i} 1^{k_i}$ for some $n \geq 0$ and $j_i, k_i \geq 0$ for all $1 \leq i \leq n$, then we have that

$$x \, \sqcup\!\sqcup_t \, y = \{\prod_{i=1}^n x_i y_i \; : \; x = \prod_{i=1}^n x_i, y = \prod_{i=1}^n y_i,$$
$$\text{with } |x_i| = j_i, |y_i| = k_i \text{ for all } 1 \leq i \leq n\}$$

if $|x| = |t|_0$ and $|y| = |t|_1$, and $x \, \sqcup\!\sqcup_t \, y = \emptyset$ if $|x| \neq |t|_0$ or $|y| \neq |t|_1$.

We extend shuffle on trajectories to *sets* $T \subseteq \{0,1\}^*$ *of trajectories* as follows:

$$x \sqcup_T y = \bigcup_{t \in T} x \sqcup_t y.$$

Further, for $L_1, L_2 \subseteq \Sigma^*$, we define

$$L_1 \sqcup_T L_2 = \bigcup_{\substack{x \in L_1 \\ y \in L_2}} x \sqcup_T y.$$

Thus, for example, it is not hard to see that if $T = 0^*1^*$, then $L_1 \sqcup_T L_2 = L_1 L_2$ (the usual concatenation operation) while if $T = 0^*1^*0^*$, $L_1 \sqcup_T L_2 = L_1 \leftarrow L_2$, the insertion operation, defined by $x \leftarrow y = \{x_1 y x_2 : x = x_1 x_2\}$. When $T = \{0,1\}^*$, we get the shuffle operation, denoted simply by $\sqcup$.

2.2   *S*-Bond-Free Properties

We consider DNA trajectories, defined by Kari *et al.* [10]. A DNA trajectory is a word over the alphabet

$$V_D = \left\{ \binom{b}{b}, \binom{f}{f}, \binom{f}{\epsilon}, \binom{\epsilon}{f} \right\}.$$

Intuitively, a set of DNA trajectories defines a bonding between DNA. The occurrence of $\binom{b}{b}$ implies a bond at a certain position, while $\binom{f}{f}$ (resp., $\binom{f}{\epsilon}$, $\binom{\epsilon}{f}$) denotes two nucleotides which are not bonded (resp., an extra unbonded nucleotide on the top strand, an extra unbonded nucleotide on the bottom strand). Kari *et al.* [10] use DNA trajectories to denote undesirable bonding properties between strands in DNA codeword design, as we explain below in Section 4. For example, the DNA trajectory $\binom{f}{\epsilon}\binom{f}{\epsilon}\binom{b}{b}\binom{b}{b}$ represents the bonding of a strand of length two with a strand of length four, creating a so-called *sticky end* of length two on the 3'-end of the strand of length 4.

We require the following morphisms. Let $\varphi_u, \varphi_d : V_D^* \to \{0,1\}^*$ be morphisms defined by

$$\varphi_u(\binom{b}{b}) = 0,\ \varphi_u(\binom{f}{y}) = 1,\ \text{for } y \in \{f, \epsilon\},\ \varphi_u(\binom{\epsilon}{f}) = \epsilon$$
$$\varphi_d(\binom{b}{b}) = 0,\ \varphi_d(\binom{y}{f}) = 1,\ \text{for } y \in \{f, \epsilon\},\ \varphi_d(\binom{f}{\epsilon}) = \epsilon$$

We now turn to our main definition, which expresses bond-freeness of a set of words using DNA trajectories. Our definition is an extension of the previous use of the term bond-freeness by Kari *et al.* [10]. Let Σ be an alphabet and $\theta : \Sigma \to \Sigma$ be an involution, extended to a morphism or anti-morphism. Let $L \subseteq \Sigma^*$ be a language and $S \subseteq V_D^*$ be a set of DNA trajectories. Then L is S-bond-free with respect to θ if

$$\forall w \in \Sigma^+, x, y \in \Sigma^*, s \in S,$$
$$(w \sqcup_{\varphi_u(s)} x \cap L \neq \emptyset, w \sqcup_{\varphi_d(s)} y \cap \theta(L) \neq \emptyset) \Rightarrow xy = \epsilon. \tag{1}$$

Note that $s \in S$ is used for both the upper and lower shuffles. We omit the phrase "with respect to θ" if θ is understood or unimportant. We illustrate the definition with two examples.

Example 1. Let $S = \binom{f}{\epsilon}\binom{b}{b}^+ \cup \binom{b}{b}^+\binom{\epsilon}{f}$. A graphical representation of S is given in Figure 1. From (1), for any language L, if there exist $w \in \Sigma^+$ and $a \in \Sigma$ such that $aw, \theta(w) \in L$ or $w, \theta(wa) \in L$, then L is not S-bond free. For example, if $\theta = \mu$ (the identity anti-morphism), then $\{ab, cba\}$ is not S-bond-free (with $w = ab$). However, $\{aba, aab\}$ is S-bond-free.

If $\theta = \iota$ (the identity morphism), then $\{ab, b\}$ is not S-bond-free (with $w = b$), but $\{ab, ba\}$ is S-bond-free (as well as the language $\{aab, b\}$).

Fig. 1. A graphical representation of S

Example 2. This example is motivated by the well-known DNA computing experiment of Adleman [1]. In this experiment, vertices and edges of a directed graph were encoded as single-strands of DNA, 20 nucleotides in length. Except for the start and end vertices, bonding is only allowed between 10 nucleotides of a vertex and 10 nucleotides of an edge. Let τ be the Watson-Crick anti-morphism. Define $S_{\mathrm{ok}} \subseteq V_D^*$ as

$$S_{\mathrm{ok}} = \left\{ \binom{f}{\epsilon}^{10}\binom{b}{b}^{10}\binom{\epsilon}{f}^{10}, \binom{\epsilon}{f}^{10}\binom{b}{b}^{10}\binom{f}{\epsilon}^{10} \right\}.$$

Intuitively, S_{ok} represents those bonds that are allowed in the DNA computation. The set S_{ok} is represented in Figure 2. Let $S = \{s \in V_D^* - S_{\mathrm{ok}} : |\varphi_u(s)| = |\varphi_d(s)| = 20\}$. Consider any $L \subseteq \{A, C, G, T\}^{20}$ with $L \cap \tau(L) = \emptyset$ (such L are called τ-*non-overlapping* by Kari *et al.* [6]). If L is S-bond-free, consider $\alpha, \beta \in L$ such that α can partially bond with β. Then there exist $w \in \Sigma^+$, $x, y \in \Sigma^*$ and $s \in S$ such that $\alpha \in w \amalg_{\varphi_u(s)} x$ and $\theta(\beta) \in w \amalg_{\varphi_d(s)} y$. Note however, that $|\varphi_u(s)| = |\alpha| = 20$ and $|\varphi_d(s)| = |\beta| = 20$. Thus, since L is S-bond-free, we must have $s \in S_{\mathrm{ok}}$, i.e., $\alpha = \alpha_1\alpha_2$ and $\beta = \beta_1\beta_2$ where $|\alpha_i| = |\beta_j| = 10$ for $1 \le i, j \le 2$ and further either $\alpha_1 = \tau(\beta_2)$ or $\alpha_2 = \tau(\beta_1)$. That is, the bonding is only allowed in the cases specified by S_{ok} and illustrated by Figure 2.

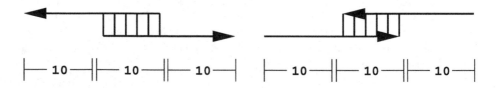

Fig. 2. A graphical representation of S_{ok}

We note that another example of the application of the current definition is given by Kari *et al.* [8], who define θ-overhang-free languages, which are defined precisely by $S = \binom{f}{\epsilon}^* \binom{b}{b}^* \binom{\epsilon}{f}^* \cup \binom{\epsilon}{f}^* \binom{b}{b}^* \binom{f}{\epsilon}^*$.

3   Preliminary Results

We first present some preliminary results which will be employed throughout the paper.

Lemma 1. *Let* $S_1 \subseteq S_2 \subseteq V_D^*$, $L \in \Sigma^*$ *and* θ *be a morphic or anti-morphic involution. If* L *is* S_2-*bond-free then* L *is* S_1-*bond-free.*

Lemma 2. *Let* $S_1, S_2 \subseteq V_D^*$. *For all languages* L, L *is* S_1-*bond-free and* S_2-*bond-free if and only if* L *is* $S_1 \cup S_2$-*bond-free.*

The union-closure property defined by Lemma 2 is an important observation because it is useful in many important design properties: to design a set of DNA strands satisfying two properties described by S_1 and S_2, we simply take the union $S_1 \cup S_2$ of these two properties.

We can also consider the effect of concatenation on sets of DNA trajectories.

Lemma 3. *Let* θ *be a morphic involution, and* $S_1, S_2 \subseteq V_D^*$. *If* $L_1 L_2$ *is* $S_1 S_2$-*bond-free then* L_1 *is* S_1-*bond-free or* L_2 *is* S_2-*bond-free.*

We note that it is not true that if L_1 is S_1-bond-free and L_2 is S_2-bond-free, then $L_1 L_2$ is $S_1 S_2$-bond-free. Indeed, let $S_1 = \{\binom{\epsilon}{f}\binom{b}{b}\binom{f}{\epsilon}\}$, $S_2 = \{\binom{b}{b}\}$, $L_1 = \{a\}$, $L_2 = \{aa\}$ and $\theta = \iota$, the identity morphism. Then note that L_i is S_i-bond-free for $1 \le i \le 2$. However, $L_1 L_2 = \{aaa\}$ is not $S_1 S_2$-bond-free.

Further, Lemma 3 does not hold for anti-morphic involutions. To see this, let $L_1 = \{a, ab\}$, $L_2 = \{c, ac\}$, $S_1 = \{\binom{\epsilon}{f}\binom{b}{b}\}$ and $S_2 = \{\binom{f}{\epsilon}\binom{f}{f}\}$, and $\theta = \mu$, the identity anti-morphism. Then we can verify that $L_1 L_2 = \{aac, ac, abac, abc\}$ is $S_1 S_2$-bond-free. However, L_1 is not S_1-bond-free and L_2 is not S_2-bond-free.

4   *S*-Bond-Free and Bond-Free Properties

We discuss the relationship between S-bond-freeness, which we have defined in Section 2.2, and the existing definition of bond-free properties with respect to a pair of sets of trajectories, previously given by Kari *et al.* [10]. We begin with the definition of bond-free properties using shuffle on trajectories (Kari *et al.* actually give a definition of bond-free properties which are more general than what we present here, but we, like them, restrict our analysis to the case where the operations are defined by shuffle on trajectories).

Let Σ be an alphabet and $\theta : \Sigma \to \Sigma$ be an involution, extended to a morphism or anti-morphism. Let $T_1, T_2 \subseteq \{0, 1\}^*$ be sets of trajectories. We say that $L \subseteq \Sigma^*$ is bond-free with respect to T_1, T_2 if the following condition holds:

$$\forall w \in \Sigma^+, x, y \in \Sigma^* (w \sqcup_{T_1} x \cap L \ne \emptyset, w \sqcup_{T_2} y \cap \theta(L) \ne \emptyset) \Rightarrow xy = \epsilon. \qquad (2)$$

Thus, intuitively, the condition is similar to S-bond-freeness, except that separate sets of trajectories T_1 and T_2 are specified to indicate separately the bonding of the two single-strands together.

As we show in the following lemma, this definition does not allow us to specify as much as S-bond-freeness:

Lemma 4. *There exist an alphabet Σ, an involution $\theta : \Sigma \to \Sigma$, extended to either a morphism or anti-morphism, and $S \subseteq V_D^*$ such that, for all $T_1, T_2 \subseteq \{0, 1\}^*$, the S-bond-free property and the bond-free property with respect to T_1, T_2 do not coincide.*

Note that Lemma 4 can be established using the set S from Example 2. This emphasizes the importance of S-bond-freeness, since its ability to describe more complex DNA bonding properties is clearly desirable.

Bond-free properties with respect to a pair of sets of trajectories were previously interpreted using DNA trajectories by Kari *et al.* [10]. We now relate this usage to ours. For any $T_1, T_2 \subseteq \{0, 1\}^*$, let $S(T_1, T_2) \subseteq V_D^*$ be defined by

$$S(T_1, T_2) = \{s \in V_D^* \ : \ \varphi_u(s) \in T_1, \varphi_d(s) \in T_2\}.$$

As an example, if $T_1 = 0^*1^*$ and $T_2 = 1^*0^*$, then $S(T_1, T_2) = \binom{\epsilon}{f}^* \binom{b}{b}^* \binom{f}{\epsilon}^*$. Under this definition, given a set S of DNA trajectories, L is bond-free with respect to S if the following condition holds:

$$\forall w \in \Sigma^+, x, y \in \Sigma^*, s_1, s_2 \in S \tag{3}$$
$$(w \sqcup_{\varphi_u(s_1)} x \cap L \neq \emptyset, w \sqcup_{\varphi_d(s_2)} y \cap \theta(L) \neq \emptyset) \Rightarrow xy = \epsilon.$$

Note that the DNA trajectories s_1, s_2 are not required to be the same in this definition. Thus, given $T_1, T_2 \subseteq \{0, 1\}^*$, the condition in (3) with $S = S(T_1, T_2)$ is exactly the same as the condition in (2). We also note briefly that the fact that s_1, s_2 are not required to be the same leads to the situation as Lemma 4: S-bond-freeness can express conditions not expressible using bond-free properties with respect to any S' (in the sense of (3)).

5 Properties of S-Bond-Freeness

We now investigate properties of S-bond-free languages and sets of DNA trajectories. We say that a set $S \subseteq V_D^*$ is universal if, for all languages L, L is S-bond-free. As a trivial example, $S = \emptyset$ is universal. In the following theorem, we characterize universal sets of DNA trajectories.

Theorem 1. *A set $S \subseteq V_D^*$ is universal if and only if $S \subseteq \binom{b}{b}^* \cup \{\binom{f}{f}, \binom{f}{\epsilon}, \binom{\epsilon}{f}\}^*$.*

For equivalence between sets of DNA trajectories, we have the following result (in the following, $\triangle$ denotes symmetric difference for languages: $L_1 \triangle L_2 = (L_1 - L_2) \cup (L_2 - L_1)$):

Lemma 5. *If $S_1, S_2 \subseteq V_D^*$ with $S_1 \triangle S_2 \subseteq \binom{b}{b}^* \cup \{\binom{f}{f}, \binom{f}{\epsilon}, \binom{\epsilon}{f}\}^*$, then S_1-bond-freeness and S_2-bond-freeness coincide.*

We now investigate additional algebraic properties related to equivalence and simplification of sets of trajectories.

5.1   Symmetry of DNA Trajectories

Our first simplification comes from the fact that, when considering anti-morphic involutions, the meaning of a DNA trajectory is preserved under half-rotations. In particular, let $\rho : V_D \to V_D$ be the involution defined by

$$\rho\left(\binom{b}{b}\right) = \binom{b}{b} \quad \rho\left(\binom{f}{f}\right) = \binom{f}{f} \quad \rho\left(\binom{f}{\epsilon}\right) = \binom{\epsilon}{f} \quad \rho\left(\binom{\epsilon}{f}\right) = \binom{f}{\epsilon}.$$

In what follows, we may extend ρ to a morphism or an anti-morphism, depending on whether we are examining bond-freeness of a set of DNA trajectories with respect to a morphic or anti-morphic involution θ.

We note that if ρ is extended to an anti-morphism, it has the same effect as a half-rotation on a DNA trajectory. For instance, $\rho\left(\binom{b}{b}\binom{b}{b}\binom{f}{f}\binom{f}{\epsilon}\binom{f}{\epsilon}\right) = \binom{\epsilon}{f}\binom{\epsilon}{f}\binom{f}{f}\binom{b}{b}\binom{b}{b}$. This is represented graphically in Figure 3.

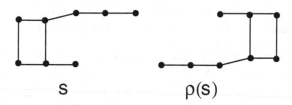

S   ρ(S)

Fig. 3. A graphical representation of the action of ρ

We begin with two technical results which will help us. The first formally states that ρ inverts the role of 'top' and 'bottom' in a DNA trajectory:

Proposition 1. *Let $\rho : V_D \to V_D$ be extended to an anti-morphism (resp., to a morphism). For all $s \in V_D^*$, the following equalities hold: $\varphi_u(s) = \varphi_d(\rho(s))^R$ and $\varphi_u(s) = \varphi_d(\rho(s))^R$. (resp., $\varphi_u(s) = \varphi_d(\rho(s))$ and $\varphi_u(s) = \varphi_d(\rho(s))$).*

Proposition 2. *Let $\rho : V_D \to V_D$ be extended to an anti-morphism (resp., to a morphism). Let $s \in V_D^*$, $u, v, x, y, z \in \Sigma^*$ and θ be an anti-morphism (resp., morphism). The memberships $u \in x \sqcup_{\varphi_u(s)} y$ and $\theta(v) \in x \sqcup_{\varphi_d(s)} z$ hold if and only if the memberships $v \in \theta(x) \sqcup_{\varphi_u(\rho(s))} \theta(z)$ and $\theta(u) \in \theta(x) \sqcup_{\varphi_d(\rho(s))} \theta(y)$ hold.*

We now show that adding the ρ-image of DNA trajectories in S does not change the set of languages which are bond-free:

Theorem 2. *Let θ be an anti-morphic involution (resp., morphic involution), $\rho : V_D \to V_D$ be extended to a anti-morphism (resp., morphism) and $S \subseteq V_D^*$. Let $S' = S \cup \rho(S)$. Then S-bond-freeness and S'-bond-freeness with respect to θ coincide.*

We can generalize Theorem 2 as follows: if S' is any set of DNA trajectories with $S \subseteq S' \subseteq S \cup \rho(S)$, then S-bond-freeness and S'-bond-freeness coincide. The increased number of DNA trajectories in S' are redundant, which leads us to consider the reverse operation of *removing* DNA trajectories to eliminate some redundancy from a set of trajectories.

To consider these issues, we first fix an arbitrarily chosen total order on V_D, say $\binom{b}{b} < \binom{f}{f} < \binom{\epsilon}{f} < \binom{f}{\epsilon}$. Let $<$ also denote the lexicographic order on V_D^* and $x \leq y$ for $x, y \in V_D^*$ denote that either $x < y$ or $x = y$.

If $S \subseteq V_D^*$, let $\mu(S)$ be defined by $\mu(S) = \{s \in S \; : \; \rho(s) \notin S \text{ or } s \leq \rho(s)\}$. We can see that $\mu(S) \cup \rho(\mu(S)) \supseteq S \supseteq \mu(S)$. Thus, $\mu(S)$-bond-freeness coincides with S-bond-freeness. (Note that this is true for *all* fixed total orders on V_D, and is also true if $\leq$ is replaced by $\geq$ in the definition of $\mu(S)$.) For finite sets of DNA trajectories S, moving from S to $\mu(S)$ reduces the number of trajectories, and hence reduces the complexity of S. However, in the following example, we see that μ may nontrivially increase the complexity of a set of trajectories:

Example 3. Let ρ be extended to an anti-morphism. Consider $S = \binom{b}{b}^* \binom{f}{f} \binom{b}{b}^*$. Then under the ordering we have fixed, $\mu(S) = \{\binom{b}{b}^n \binom{f}{f} \binom{b}{b}^m \; : \; n \geq m \geq 0\}$. Thus, S is a regular set of DNA trajectories, while $\mu(S)$ is context-free.

Example 3 also shows that sometimes, introducing redundancy may reduce the complexity for infinite sets of DNA trajectories: if we take $S = \{\binom{b}{b}^n \binom{f}{f} \binom{b}{b}^m \; : \; n \geq m \geq 0\}$, then $S \cup \rho(S) = \binom{b}{b}^* \binom{f}{f} \binom{b}{b}^*$ (again, let ρ be anti-morphic). However, this is not always the case: note that if $S_1 = \{\binom{b}{b}^n \binom{f}{f} \binom{b}{b}^n \; : \; n \geq 0\}$, then $S_1 \cup \rho(S_1) = S_1 = \mu(S_1)$. Further work is necessary to examine the effect of simplifying sets of DNA trajectories by introducing or removing redundancy.

5.2   Equivalence Via Weakening Bonds

We now consider another way for distinct sets of trajectories to be equivalent. The important observation is that if bonds in a DNA trajectory are replaced with non-bonded regions, the resulting trajectory represents all the bonding of the original trajectory, plus some additional bonding situations. Thus, a set of trajectories which includes both the original and modified trajectories contains some redundancy. To formalize this, we define a partial order $\prec$ on words over V_D^*. Let $s_1, s_2 \in V_D^*$ with

$$\varphi_u(s_1) = \prod_{i=1}^{n} 1^{j_i} 0^{k_i}, \text{ and } \varphi_d(s_1) = \prod_{i=1}^{n} 1^{\ell_i} 0^{k_i},$$

for $n \geq 0$ and $j_i, k_i, \ell_i \geq 0$ for all $1 \leq i \leq n$. Then $s_2 \prec s_1$ if there exist $\alpha_1, \ldots, \alpha_n \in \{0, 1\}^*$ such that the following three conditions hold:

(i) $\varphi_u(s_2) = \prod_{i=1}^{n} 1^{j_i} \alpha_i$ and $\varphi_d(s_2) = \prod_{i=1}^{n} 1^{\ell_i} \alpha_i$;
(ii) $|\alpha_i| = k_i$ for all $1 \leq i \leq n$; and
(iii) $\prod_{i=1}^{n} \alpha_i \notin 1^*$.

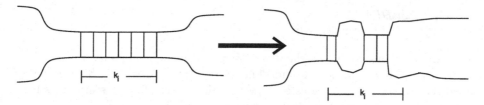

Fig. 4. A portion of s_1 is shown on the left, and a portion of s_2 is shown on the right

The situation is illustrated in Figure 4, which demonstrates that if $s_2 \prec s_1$, then we can modify s_1 to get s_2 by replacing a bonded region of length k_i in s_1 with a region in s_2 which possibly contains some non-bonded regions. The relation $\prec$ has also been employed in the study of hairpin conditions defined by DNA trajectories [3], where it is important in studying closure properties.

Example 4. Consider $s_1, s_2 \in V_D^*$ given by

$$s_1 = \binom{b}{b}\binom{b}{b}\binom{f}{\epsilon}\binom{b}{b}\binom{f}{f}, \quad s_2 = \binom{b}{b}\binom{f}{f}\binom{f}{\epsilon}\binom{f}{\epsilon}\binom{f}{f}\binom{\epsilon}{f}.$$

Note that $\varphi_u(s_1) = 00101, \varphi_u(s_2) = 01111, \varphi_d(s_1) = 0001$, and $\varphi_d(s_2) = 0111$. Thus, $s_2 \prec s_1$ holds with $\alpha_1 = 01$ and $\alpha_2 = 1$.

Note that Example 4 demonstrates that the relation $\prec$ is not simply defined by the idea "possibly replace $\binom{b}{b}$ with $\binom{f}{f}$" (an alternate characterization of $\prec$ of a similar form can be given via a rewriting system [3]). We now define the minimal set of DNA trajectories with respect to $\prec$. Let $S \subseteq V_D^*$. Then $\min(S) = \{s \in S \; : \; \forall t (\neq s) \in S, t \nprec s\}$.

Theorem 3. *Let $S \subseteq V_D^*$. Then for all Σ, all $L \subseteq \Sigma^*$ and all morphic or antimorphic involutions θ, L is S-bond-free if and only if L is $\min(S)$-bond-free.*

6   Decidability

We now turn to decidability problems. We first prove a result which allows us to give positive decidability results:

Theorem 4. *For all alphabets Σ, there exist an alphabet $\Delta \supseteq \Sigma$, an operation $\sigma : 2^{\Sigma^*} \times 2^{V_D^*} \to 2^{\Delta^*}$ and regular languages $R_1, R_2 \subseteq \Delta^*$ such that for all $L \subseteq \Sigma^*$ and $S \subseteq V_D^*$, L is S-bond-free if and only if $\sigma(L, S) \cap R_1 \subseteq R_2$.*

Corollary 1. *Given a context-free set $S \subseteq V_D^*$ of DNA trajectories and a regular language $L \subseteq \Sigma^*$, it is decidable if L is S-bond-free.*

We can also modify the proof of Theorem 4 to give the following result:

Theorem 5. *Let $S \subseteq V_D^*$ be a regular set of DNA trajectories. The following problem is decidable in quadratic time:*

Input: an DFA M.
Output: Is L(M) S-bond-free?

We can also consider undecidability. We prove a undecidability result which states that S-bond-freeness is undecidable for context-free L with respect to a fixed regular set of trajectories S. However, the power of the result comes from L rather than from any interaction between L and S.

Lemma 6. *There exists a fixed regular set of DNA trajectories S such that the following problem is undecidable: "Given an alphabet Σ, an anti-morphic (resp., morphic) involution $\theta : \Sigma^* \to \Sigma^*$ and a CFL $L \subseteq \Sigma^*$, is L S-bond-free?"*

7   Conclusions

In this paper, we have given a new definition of bond-freeness with respect to sets of DNA trajectories. The new definition, called S-bond-freeness, is able to express more complex properties than previous definitions which used either binary trajectories or DNA trajectories. We have illustrated some significant bonding conditions expressible through S-bond-freeness, and also proven that it is more expressive than previous definitions.

We have investigated properties of DNA trajectories related to equivalence, rotations and weakening of bonding. These characterizations are clearly expressed in terms of morphisms and partial orders. We have also examined the decidability properties of S-bond-free languages: it is decidable whether a regular language L is S-bond-free for any context-free set of DNA trajectories S. We have proven it is undecidable to determine if a context-free language L is S-bond-free with respect to regular sets of DNA trajectories S.

Acknowledgments

I am very grateful for discussions with Petr Sosík on the topics of this paper, and his insight.

References

1. Adleman, L.: Molecular computation of solutions to combinatorial problems. Science **266** (1994) 1021–1024
2. Amos, M.: Theoretical and Experimental DNA Computation. Springer (2005)
3. Domaratzki, M.: Hairpin structures defined by DNA trajectories. Technical Report TR-2006-001, Jodrey School of Computer Science, Acadia University (2006) Submitted for publication.
4. Hussini, S., Kari, L., Konstantinidis, S.: Coding properties of DNA languages. Theor. Comp. Sci. **290**(3) (2002) 1557–1579
5. Jonoska, N., Kephart, D., Mahalingam, K.: Generating DNA code words. Congressus Numerantium **156** (2002) 99–110

6. Kari, L., Kitto, R., Thierrin, G.: Codes, involutions and DNA encoding. Volume 2300 of LNCS. (2003) 376–393
7. Kari, L., Konstantinidis, S., Losseva, E., Sosík, P., Thierrin, G.: Hairpin structures in DNA words. In Carbone, A., Daley, M., Kari, L., McQuillan, I., Pierce, N., eds.: The 11th International Meeting on DNA Computing: DNA 11, Preliminary Proceedings. (2005) 267–277
8. Kari, L., Konstantinidis, S., Losseva, E., Wozniak, G.: Sticky-free and overhang-free DNA languages. Acta Inform. **40** (2003) 119–157
9. Kari, L., Konstantinidis, S., Sosík, P.: Bond-free languages: Formalisms, maximality and construction methods. Int. J. Found. Comp. Sci. **16** (2005) 1039–1070
10. Kari, L., Konstantinidis, S., Sosík, P.: On properties of bond-free DNA languages. Theor. Comp. Sci. **334** (2005) 131–159
11. Mateescu, A., Rozenberg, G., Salomaa, A.: Shuffle on trajectories: Syntactic constraints. Theor. Comp. Sci. **197** (1998) 1–56
12. Păun, G., Rozenberg, G., Salomaa, A.: DNA Computing: New Computing Paradigms. Springer (1998)
13. Rozenberg, G., Salomaa, A., eds.: Handbook of Formal Languages. Springer (1997)

Involution Solid and Join Codes

Nataša Jonoska[1], Lila Kari[2], and Kalpana Mahalingam[2]

[1] University of South Florida, Department of Mathematics, Tampa, FL-33620
jonoska@math.usf.edu
[2] University of Western Ontario, Department of Computer Science,
London, Ontario, Canada-N6A 5B7
{lila, kalpana}@csd.uwo.ca

Abstract. In this paper we study a generalization of the classical notions of solid codes and comma-free codes: involution solid codes (θ-solid) and involution join codes (θ-join). These notions are motivated by DNA strand design where Watson-Crick complementarity can be formalized as an antimorphic involution. We investigate closure properties of these codes, as well as necessary conditions for θ-solid codes to be maximal. We show how the concept of θ-join can be utilized such that codes that are not themselves θ-comma free can be split into a union of subcodes that are θ-comma free.

1 Introduction

When using single stranded DNA molecules in DNA nanotechnology and DNA computing it is important to minimize the errors that are due to unwanted cross-hybridization. Such errors occur if two different bits of information are encoded as single stranded DNA molecules that are totally or partially complementary. This complementarity induces unintentional hybridizations and such encodings should be avoided (See Fig. 1).

Several attempts have been made to address this issue and many authors have proposed various solutions. Such approaches to the design of DNA encodings without undesirable bonds and secondary structures were summarized in [20] and [24]. For more details we refer the reader to [1, 3, 4, 5, 6, 8, 23]. One approach to this issue of "good encodings" is theoretical study of the algebraic and code-theoretic properties of DNA encodings through formal language theory. In [11], Kari et al. introduced such theoretical approach to the problem of designing code words. Properties of languages that avoid certain undesirable bonds were discussed in several follow-up papers [13, 18, 19, 21, 22].

This paper follows the approach introduced in [11] and investigate the formal language and coding theoretic notions inspired and motivated by DNA encoded information. In order to model the characteristics of the biologically encoded information, we replace the identity function by a composition of the complement function with the mirror-image function (such a function is a correct mathematical model of the Watson-Crick complement of DNA strands) or, more generally, by an arbitrary involution (a function θ with the property that θ^2 equals identity)

O.H. Ibarra and Z. Dang (Eds.): DLT 2006, LNCS 4036, pp. 192–202, 2006.

[20]. Moreover, as needed, we replace regular homomorphisms by antimorphic functions, to allow for the property of reversal in complementary DNA strands. These lead to natural, as well as theoretically elegant, generalizations of classical notions such as prefix codes, suffix codes, infix codes, bifix codes, intercodes, comma-free codes, etc [13, 18, 19, 22]. These are but some examples of ways in which classical notions and results in formal language theory and algebraic informatics can be meaningfully generalized. This paper continues this line of research by introducing and studying the notions of θ-solid and θ-join codes. If θ is the identity function, the θ-solid codes and θ-join codes respectively become the well known solid and join codes respectively.

Section 2 includes the definitions and we introduce the new concept of θ-solid codes and include some properties of θ-overlap-free codes. Section 3 contains the closure properties of θ-solid codes. Note that the results obtained for θ-solid codes hold true for solid codes when θ is the identity function. We show that the property of being a θ-solid code is decidable for regular languages and provide results about maximal θ-solid codes. In Section 4 we generalize the concept of join codes to θ-join codes and develop a method to extract a sequence of subsets which are θ-comma-free from a set that is not θ-comma-free. Due to space restrictions, some of the proofs are omitted.

2   Definitions

An alphabet Σ is a finite non-empty set of symbols. A word u over Σ is a finite sequence of symbols in Σ. We denote by Σ^* the set of all words over Σ, including the empty word 1 and, by Σ^+, the set of all non-empty words over Σ. For a word $w \in \Sigma^*$, the length of w is the number of non empty symbols in w and is denoted by $|w|$. Throughout the rest of the paper, we concentrate on sets $X \subseteq \Sigma^+$ that are *codes* such that every word in X^+ can be written uniquely as a product of words in X, or equivalently, X^+ is a free semigroup generated by X. For the background on codes we refer the reader to [2, 26]. For a language $X \subseteq \Sigma^*$, let

$$\mathrm{PPref}(X) = \{u \in \Sigma^+ \mid \exists v \in \Sigma^+, uv \in X\}$$
$$\mathrm{PSuff}(X) = \{u \in \Sigma^+ \mid \exists v \in \Sigma^+, vu \in X\}$$
$$\mathrm{PSub}(X) = \{u \in \Sigma^+ \mid \exists v_1, v_2 \in \Sigma^*, v_1 v_2 \neq 1, v_1 u v_2 \in X\}$$

We recall the definitions initiated in [11, 18] and used in [12, 19].

An involution $\theta : \Sigma \to \Sigma$ of a set Σ is a mapping such that θ^2 equals the identity mapping.

Definition 1. *Let $\theta : \Sigma^* \to \Sigma^*$ be a morphic or antimorphic involution and $X \subseteq \Sigma^+$ be a code.*

1. *The set X is called θ-strict if $X \cap \theta(X) = \emptyset$.*
2. *The set X is called θ-infix if $\Sigma^* \theta(X) \Sigma^+ \cap X = \emptyset$ and $\Sigma^+ \theta(X) \Sigma^* \cap X = \emptyset$.*
3. *The set X is called θ-comma-free if $X^2 \cap \Sigma^+ \theta(X) \Sigma^+ = \emptyset$.*

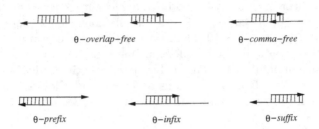

Fig. 1. Schematic representation for forbidden inter molecular DNA hybridizations. The 3′ ends are indicated with an arrow.

Note that when θ is identity θ-infix code and θ-comma-free code are just infix and comma-free codes [2, 26]. In [12, 13] θ-strict codes are called strictly θ and in [18] they are called θ-non overlapping.

Solid codes were introduced in [25]. Certain combinatorial and closure properties of solid codes were discussed in [15]. We recall the definition of solid codes used in [17] defined by using a characterization given in [15].

Definition 2. *A set $X \subseteq \Sigma^+$ is a* solid-code *if*

1. *X is an infix code*
2. *$\mathrm{PPref}(X) \cap \mathrm{PSuff}(X) = \emptyset$.*

The notion of solid codes was extended to involution solid codes in [22]. Note that when the involution map denotes the Watson-Crick complement, the set of involution-solid codes comprises of DNA strands that do not overlap with the complement of any other DNA strand (see Fig.1).

Definition 3. *Let $X \subseteq \Sigma^+$.*

1. *The set X is called θ-overlap free if $\mathrm{PPref}(X) \cap \mathrm{PSuff}(\theta(X)) = \emptyset$ and $\mathrm{PSuff}(X) \cap \mathrm{PPref}(\theta(X)) = \emptyset$.*
2. *X is a θ-solid code if X is θ-infix and θ-overlap free.*
3. *X is a maximal θ-solid code iff for no word $u \in \Sigma^+ \setminus X$, the language $X \cup \{u\}$ is a θ-solid code.*

Note 1. Let X be such that X^n is θ-overlap free for some $n \geq 1$ then X^i, $1 \leq i \leq n$ is also θ-overlap free.

Throughout the rest of the paper we use θ to be either a morphic or antimorphic involution unless specified otherwise. Note that X is θ-overlap free (θ-solid) iff $\theta(X)$ is θ-overlap free (θ-solid).

Proposition 1. *If X is a θ-strict-solid code then X^+ is θ-overlap free.*

Proof. We need to show that $\text{PPref}(X^+) \cap \text{PSuff}(\theta(X^+)) = \emptyset$ and $\text{PSuff}(X^+) \cap \text{PPref}(\theta(X^+)) = \emptyset$. Suppose there exists $x \in \text{PPref}(X^+) \cap \text{PSuff}(\theta(X^+))$ such that $x = x_1 x_2 .. x_i a = \theta(b)\theta(y_1)...\theta(y_j)$ for $x_i, y_j \in X$ for all i, j and $a \in \text{PPref}(X)$ and $\theta(b) \in \text{PSuff}(\theta(X))$. Then x_1 is not a subword of $\theta(b)$ as this contradicts the assumption that X is θ-infix. Also, $\theta(b)$ being in $\text{PPref}(x_1)$) contradicts the assumption that X is θ-solid. Hence $\text{PPref}(X^+) \cap \text{PSuff}(\theta(X^+)) = \emptyset$. Similarly, $\text{PSuff}(X^+) \cap \text{PPref}(\theta(X^+)) = \emptyset$.   □

Corollary 1. *Let $X, Y \subseteq \Sigma^+$ be such that $X \cup Y$ is θ-strict-solid. Then XY is θ-overlap free.*

Proposition 2. *Let X be a regular language. It is decidable whether or not X is θ-overlap free.*

3   Properties of Involution Solid Codes

In this section we consider the closure properties of the class of involution solid codes. It turns out that involution solid codes are closed under a restricted kind of product, arbitrary intersections and catenation closure while not closed under union, complement, product and homomorphisms. The first two properties are immediate consequences of the definitions.

Proposition 3. *The class of θ-solid codes is closed under arbitrary intersection and θ-solid codes is not closed under union, complement, concatenation and homomorphism.*

Example 1. Consider the θ-solid codes $\{a\}$ and $\{ab\}$ over the alphabet set $\Sigma = \{a, b\}$ and with θ being an antimorphic involution that maps $a \mapsto b$ and $b \mapsto a$. The sets $\{a, ab\} = \{a\} \cup \{ab\}$ and $\{aba\} = \{ab\}\{a\}$ are not θ-solid. Let $h : \Sigma^* \mapsto \Sigma^*$ be homomorphism such that $h(a) = aba$ and $h(b) = bab$. Note that $\{a\}$ is θ-solid but $h(a) = aba$ is not θ-solid.

Proposition 4. *If X is a θ-solid code then X is a θ-comma-free code.*

Proof. According to proposition 3.5 in [14], X is θ-comma-free if and only if X is θ-infix and $X_{ip}X_{is} \cap \theta(X) = \emptyset$ where $X_{ip} = \text{PPref}(\theta(X)) \cap \text{PSuff}(X)$ and $X_{is} = \text{PSuff}(\theta(X)) \cap \text{PPref}(X)$. Since X is θ-solid, X is θ-infix and $X_{ip} = X_{is} = \emptyset$ and hence $X_{ip}X_{is} \cap \theta(X) = \emptyset$.   □

Note that the converse of the above proposition does not hold in general. For example let $X = \{aa, baa\}$ and let θ be an antimorphic involution such that $\theta : a \to b, b \to a$, then $\theta(X) = \{bb, bba\}$. It is easy to check that X is θ-comma-free. But $ba \in \text{PPref}(X) \cap \text{PSuff}(\theta(X))$ which contradicts condition 2 of definition 3.

Proposition 5. *Let $X, Y \subseteq \Sigma^+$ be such that X and Y are θ-strict and $X \cap \theta(Y) = \emptyset$. If $X \cup Y$ is θ-solid then XY is θ-solid.*

Proof. Suppose XY is not θ-infix, then there exists $x_1, x_2 \in X$ and $y_1, y_2 \in Y$ such that $x_1y_1 = a\theta(x_2y_2)b$ for some $a, b \in \Sigma^*$ not both empty. When θ is morphic, $x_1y_1 = a\theta(x_2)\theta(y_2)b$. Then either $\theta(x_2)$ is a subword of x_1 or $\theta(y_2)$ is a subword of y_1 which is a contradiction with $X \cup Y$ being θ-infix. Similar contradiction arises when θ is antimorphic. From Corollary 1, XY is θ-overlap free and hence XY is θ-solid. □

Corollary 2. *X is a θ-strict-solid code if and only if X^+ is a θ-strict-solid code.*

Proposition 6. *Let X be a regular language. It is decidable whether or not X is a θ-solid code.*

Proof. It has been proved in [11] that it is decidable whether X is θ-infix or not. From Proposition 2 it is decidable whether X is θ-overlap free or not. Hence for a regular X, one can decide whether X is θ-solid or not.

The following proposition gives a method for constructing θ-strict-solid codes.

Proposition 7. *Let θ be a morphic or antimorphic involution and $X \subseteq \Sigma^+$ be θ-strict-solid code. Then $Y = \{u_1vu_2 : u_1u_2, v \in X, u_1, u_2 \in \Sigma^*\}$ is a θ-solid code.*

The next proposition provides a general method for constructing not just θ-solid codes, but maximal θ-solid codes.

Proposition 8. *Let θ be an antimorphic involution. Let $\Sigma = A \cup B \cup C$ such that A, B, C are disjoint sets such that A and C are θ-strict and $A \cap \theta(B) = \emptyset$ and $C \cap \theta(B) = \emptyset$. Then $X = AB^*C$ is a maximal θ-solid code.*

Example 2. Let $\Sigma = \{a, b, c, d\}$ and θ be an antimorphic involution such that θ maps $a \mapsto c$ and $b \mapsto d$. Then $X = \{a\}\{b, d\}^*\{c\}$ is a maximal θ-solid code.

From the above definitions and propositions we can deduce the following.

Lemma 1. *Let θ be an antimorphic involution.*

1. *Let $\Sigma_1, ..., \Sigma_n$ be a partition of Σ such that Σ_i is θ-strict for all i. Then every language $\Sigma_i\Sigma_j$ is θ-solid.*
2. *If Σ_1, Σ_2 is a partition of Σ such that Σ_i is θ-strict for $i = 1, 2$, then $\Sigma_1\Sigma_2$ is maximal θ-solid code.*
3. *Let $A \subseteq \Sigma$ such that $A = \theta(A)$ and $X \subseteq A^+$. Then X is maximal θ-solid code over A if and only if $X \cup (\Sigma \setminus A)$ is maximal θ-solid code over Σ.*
4. *Let $B \subseteq \Sigma$ such that $B \cap \theta(B) = \emptyset$. Then $X = B^+$ is θ-solid code.*

The next proposition provides conditions under which the involution solid codes are preserved under a morphic or antimorphic mapping.

Proposition 9. *Let Σ_1 and Σ_2 be finite alphabet sets and let f be an injective morphism or antimorphism from Σ_1 to Σ_2^*. Let $\theta_1 : \Sigma_1^* \mapsto \Sigma_1^*$ and $\theta_2 : \Sigma_2^* \mapsto \Sigma_2^*$ be both morphic or both antimorphic involutions such that $f(\theta_1(x)) = \theta_2(f(x))$. Define $P = Pref(\theta_2(f(X)), S = Suff(\theta_2(f(X))$ and $A = \Sigma_2^* \setminus f(\Sigma_1^*)$.*

Suppose $(A^+P \cap SA^+) \cap f(\Sigma_1^+) = \emptyset$ and $A^+PA^+ \cap f(\Sigma_1) = \emptyset$. Then

1. *If X is θ_1-strict-infix (comma-free) then $f(X)$ is θ_2-strict-infix (comma-free).*
2. *If X is a θ_1-solid code then $f(X)$ is a θ_2-solid code.*

Proof. The first statement was proved in [14]. We consider the case of θ-solid codes. Let X be θ_1-solid code. Note that $f(X)$ is θ_2-infix by the first part of the proposition. We need to show that $\text{PPref}(f(X)) \cap \text{PSuff}(\theta_2(f(X))) = \emptyset$ and $\text{PSuff}(f(X)) \cap \text{PPref}(\theta_2(f(X))) = \emptyset$. Let θ_1 and θ_2 be morphic involutions and let f be an injective antimorphism. Suppose there exists $a \in \Sigma^+$ such that $a \in \text{PPref}(f(x_1x_2))$ and $a \in \text{PSuff}(\theta_2(f(y_1y_2)))$ for some $x_1x_2, y_1y_2 \in X$. Note that $f(x_1x_2) = f(x_2)f(x_1)$ and $\theta_2(f(y_1y_2)) = f(\theta_1(y_1y_2)) = f(\theta_1(y_1)\theta_1(y_2)) = f(\theta_1(y_2))f(\theta_1(y_1))$. Hence if $a = f(x_2) = f(\theta_1(y_1))$ then $x_2 = \theta_1(y_1)$ since f is injective which is a contradiction to $\text{PPref}(X) \cap \text{PSuff}(\theta_1(X)) = \emptyset$. The other case follows similarly. Hence $f(X)$ is θ_2-solid. □

4   Involution Join Codes

In [10], Head, by using the coding properties relative to a language [9], showed how a sequence of subsets which are comma-free ([2, 16, 26]) from a set that is not comma-free can be obtained. The codes of this sequence are called join codes Similarly a sequence of subsets which are θ-comma-free codes from a given set that is not θ-comma-free can be obtained ([11, 12, 13, 18, 14]). The ith element in this sequence of codes θ-join codes is called θ-join code of level i. In this section we have several observations about these codes.

Definition 4. *Let $X \subseteq \Sigma^*$ and $w \in \Sigma^*$. Then the context of the word w in X is defined as the set $C_X(w) = \{(u, v) : uwv \in X, u, v \in \Sigma^*\}$.*

The following was defined in [10] and used in [7].

Definition 5. *A word w in X is a join for X if $(u, v) \in C_{X^*}(w)$ then both u and v are in X^*. The set of all joins for X is denoted $J(X)$.*

Recall that when X is a code, $J(X)$ is comma-free subset of X (see [10]), but $J(X)$ is not necessarily the maximal comma-free subset of X.s

Example 3. Let $X = \{aab, aba, bab\}$ over the alphabet set $\Sigma = \{a, b\}$. Note that $J(X) = \{aab\}$ but $Y = \{aab, bab\} \subseteq X$ is the maximal comma-free subset of X since $\Sigma^+ Y \Sigma^+ \cap Y^2 = \emptyset$.

Similar to the definition for $J(X)$, we define $J_\theta(X)$ such that $J_\theta(X)$ is θ-comma-free. The authors in [7] define $J_\theta(X)$ as $J(X) \setminus \theta(J(X))$, but such defined $J_\theta(X)$ is not necessarily θ-comma-free in general. For example, consider $X = \{aab, bab, abbb\}$. For an antimorphic involution θ with $a \to b$ and $b \to a$, $\theta(X) = \{abb, aba, aaab\}$. Note that $\Sigma^+ \{aab, abbb\} \Sigma^+ \cap X^2 = \emptyset$ but $\Sigma^+ X \Sigma^+ \cap X^2 \neq \emptyset$. Hence $J(X) = \{aab, abbb\}$ is comma-free subset of X and $J_\theta(X) = J(X)$ since $\theta(J(X)) \cap J(X) = \emptyset$. But $J_\theta(X)$ is not θ-comma-free since, $aab\theta(aab)b = aab.abbb \in (J_\theta(X))^2$. We alter the definition for $J_\theta(X)$ such that $J_\theta(X)$ becomes θ-comma-free for all involutions θ.

Definition 6. *A word $w \in X$ is a θ-join for X if $(u, v) \in C_{X^*}(\theta(w))$, then both u and v are also in X^*. The set of all θ-joins for X is denoted with $J_\theta(X)$.*

Lemma 2. *Let $X \subseteq \Sigma^+$ and θ be a morphic or antimorphic involution. Then the following hold true.*

1. *$J_\theta(X) = J(X)$ when θ is identity.*
2. *Let $X \cup \theta(X)$ be a code. When X is θ-comma-free, $J_\theta(X) = X$.*
3. *$X \cup \theta(X)$ is θ-comma-free if and only if $X \cup \theta(X)$ is comma-free and hence $X \cup \theta(X)$ is an infix code.*

Note that $X \cup \theta(X)$ being θ-infix does not necessarily imply that $X \cup \theta(X)$ is θ-comma-free. For example let $X = \{aa, aba\}$ with θ being antimorphic involution mapping $a \to b$ and $b \to a$. Then we have $\theta(X) = \{bb, bab\}$. It is easy to check that $X \cup \theta(X)$ is θ-infix but not θ-comma-free since $ababb = a(bab)b = a\theta(aba)b$.

Lemma 3. *If $X \cup \theta(X)$ is a code then $J_\theta(X)$ is θ-comma-free.*

Proof. Suppose $J_\theta(X)$ is not θ-comma-free, then there are $x, y, z \in J_\theta(X)$ such that $a\theta(z)b = xy$ for some $a, b \in \Sigma^*$. Since $z \in J_\theta(X)$ with $a\theta(z)b \in X^*$, for $a, b \in X$, xy has two distinct factorizations in $X \cup \theta(X)$. This contradicts that $X \cup \theta(X)$ is a code. Hence $J_\theta(X)$ is θ-comma-free.   □

Corollary 3. *If $X \cup \theta(X)$ is a code then $\Sigma^+\theta(J_\theta(X))\Sigma^+ \cap X^2 = \emptyset$.*

Note that $J_\theta(X)$ is not necessarily the maximal θ-comma-free subset of X.

Example 4. Let $X = \{abb, aab, aba\}$ over the alphabet set $\Sigma = \{a, b\}$ and for an antimorphic θ with $a \to b$ and $b \to a$, $\theta(X) = \{abb, bab, aab\}$. Note that $Y = \{aab, abb\}$ is the maximal θ-comma-free subset of X. But $J_\theta(X) = \{aab\}$.

Lemma 4. *Let $X \cup \theta(X)$ be a code and let Y be the maximal subset of X such that $\Sigma^+\theta(Y)\Sigma^+ \cap X^2 = \emptyset$. Then $Y = J_\theta(X)$.*

Proof. Since $J_\theta(X)$ is θ-comma-free, $J_\theta(X) \subseteq Y$. Note that $\Sigma^+\theta(Y)\Sigma^+ \cap X^2 = \emptyset$ if and only if $\Sigma^+\theta(Y)\Sigma^+ \cap X^* = \emptyset$. Then for $w \in Y$ and for all $(u, v) \in C_{X^*}(\theta(w))$, $u = v = 1$ which implies $w \in J_\theta(X)$. Hence $J_\theta(X) = Y$.

A possible communication by transferring single stranded DNA molecules can be done in the following way. To derive the meaning represented by a single stranded DNA molecule, by allowing attachments of complementary pieces of single stranded molecules. The meaning conveyed by the message molecule is expressed by the sequence of complementary code word molecules that attach. In such cases, each code word that is part of a θ-comma-free code has a unique place for hybridization, and the whole "message" molecule can be recovered. However, if the set of code words is not θ-comma-free, then we can extract a subset of code words that form a θ-comma-free set, i.e., the θ-joins for the code words. These code words would have unique places for annealing to the "message" DNA (see Fig. 2) leaving positions complementary to other words "empty". From the

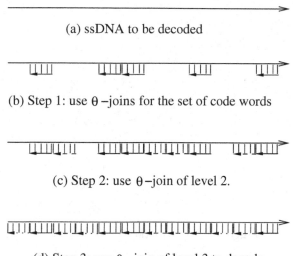

(a) ssDNA to be decoded

(b) Step 1: use θ–joins for the set of code words

(c) Step 2: use θ–join of level 2.

(d) Step 3: use θ–join of level 3 to decode.

Fig. 2. Step-by-step recovery of a "message" encoded within a single stranded DNA

remaining code words another θ-comma-free set can be extracted, i.e., the θ-joins for the remaining code words. Now these words can anneal in a unique way to the the empty places that were not occupied by the first set of words. If this process ends, the "message" can be uniquely decoded. This process is schematically represented in Fig. 2.

Formally, let $X = X_0$ and $X_1 = X \setminus J_\theta(X)$. We define X_i, $i \geq 0$, a chain of descending subsets of X such that $X_{i+1} = X_i \setminus J_\theta(X_i)$ where $J_\theta(X_i)$ is a θ-join of X_i (see Fig. 3). We call $J_\theta(X_i)$ the θ-join at level i. When θ is identity the θ-join code at level 1 is precisely $J(X)$.

Definition 7. *The set X is called as θ-split code if $X = \bigcup_{k=0}^{\infty} J_\theta(X_k)$.*

If there exists m such that $X_{m+1} = \emptyset$ then X is called as θ-k-split where $k = \min\{m : X_{m+1} = \emptyset\}$.

Example 5. Let $X = \{abb, aab, aba\}$ over the alphabet set $\Sigma = \{a, b\}$ and for an antimorphic θ with $a \to b$ and $b \to a$, $\theta(X) = \{abb, bab, aab\}$. Note that $J_\theta(X) = \{aab\}$ and hence $X_1 = X \setminus J_\theta(X) = \{abb, aba\}$. But $J_\theta(X_1) = \emptyset$. Hence X is not a θ-split code.

We assume that for a set X, $X \cup \theta(X)$ is a code throughout the rest of this section.

Proposition 10. *X is a θ-split code if and only if $\theta(X)$ is θ-split code.*

Note that it is possible to find an X such that X is θ-infix but not θ-split. For example let $X = \{aba, bab\}$ and let θ be an antimorphic involution such that θ maps $a \mapsto b$. Then X is θ-infix and $J_\theta(X) = \emptyset$.

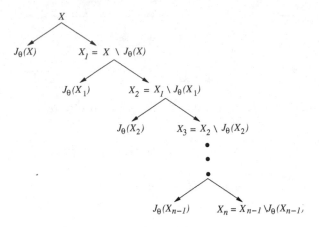

Fig. 3. The construction of θ-split code

Corollary 4. *If X and Y are such that $X \cup Y$ is a θ-split code, then XY is θ-split code.*

Corollary 5. *If X is a θ-split code, than X^n is a θ-split code for all $n \geq 1$.*

5   Concluding Remarks

The theory of codes, born in the context of information theory, has been developed as an independent subject using both combinatorial and algebraic methods. The objective of the theory of codes, from an elementary point of view, is the study of properties concerning factorizations of words into a sequence of words taken from a given set. Solid codes were introduced in [25] in the context of the study of disjunctive domains. Certain combinatorial properties of solid codes have been investigated in [15] and results concerning maximal solid codes of variable length were presented in [17]. In the hierarchy of codes, solid codes lie below the class of comma-free codes. By using code properties relative to a language [9], a sequence of generalizations of the concept of a comma-free code was developed in [10]. In other words, Head showed a way to partition a non-necessarily comma-free code into a sequence of subsets all of which are comma-free. The codes of this sequence are the join codes of various levels. The comma-free codes are precisely the join codes of the first level. The split codes of level k allow the segmentation of messages to be made in a sequence of k steps for which each step has the simplicity of a comma-free segmentation.

In this paper we extended the concepts of solid codes and join codes to incorporate the notion of an involution function replacing the identity function (An involution function θ is such that θ^2 equals identity). An involution code refers to any of the generalization of classical notion of codes ([2, 16, 26]) that replace the identity function with the involution morphic or antimorphic function in a way explained in Definition 1. Involution codes were introduced in [11] in the process of designing DNA strands suitable for computation. Along these lines properties of θ-comma-free codes, θ-infix codes, θ-prefix codes, θ-outfix codes, θ-intercodes were introduced and studied in [11, 13, 14, 19, 21]. This paper completes this line of research by investigating the notions of θ-solid codes and θ-join codes. Several closure properties of θ-overlap free and θ-solid codes were discussed and we introduced θ-split codes as codes that can be "split", i.e., partition into a sequence of θ-comma-free codes. Properties of a code X that is a θ-split code of finite or infinite level remain to be determined. Note that if θ is the identity function, these notions become the well known notions of solid respectively join codes and the results obtained hold true for solid respectively join codes. Generalizations where θ is substituted with an arbitrary morphism seem as a natural next step.

Acknowledgment. Research supported by NSERC and Canada Research Chair grants for Lila Kari and NSF grants CCF–0523928 and CCF–0432009 for Natasha Jonoska.

References

1. E.B. Baum, *DNA sequences useful for computation*, unpublished article, available at: http://www.neci.nj.nec.com/homepages/eric/seq.ps (1996).
2. J. Berstel, D. Perrin, *Theory of Codes*, Academis Press, Inc. Orlando Florida, 1985.
3. R.Deaton, J.Chen, H.Bi, M.Garzon, H.Rubin, D.F.Wood, *A PCR based protocol for in vitro selection of non-crosshybridizing oligonucleotides*, DNA Computing Proceedings of the 8th International Meeting on DNA Based Computers (M.Hagiya, A.Ohuchi editors), Springer LNCS 2568 (2003) 196-204.
4. R.Deaton, J.Chen, M.Garzon, J.Kim, D.Wood, H.Bi, D.Carpenter, Y.Wang, *Characterization of Non-Crosshybridizing DNA Oligonucleotides Manufactured in Vitro*, DNA computing Preliminary Proceedings of the 10th International Meeting on DNA Based Computers, (C.Ferretti, G.Mauri, C.Zandron editors) LNCS 3384 (2005) 50-61.
5. R. Deaton et al, *A DNA based implementation of an evolutionary search for good encodings for DNA computation*, Proc. IEEE Conference on Evolutionary Computation ICEC-97 (1997) 267-271.
6. D. Faulhammer, A. R. Cukras, R. J. Lipton, L. F.Landweber, *Molecular computation: RNA solutions to chess problems*, Proceedings of the National Academy of Sciences, USA, 97-4 (2000) 1385-1389.
7. C. Ferreti and G. Mauri, *Remarks on Relativisations and DNA Encodings* Aspects of Molecular Computing, N.Jonoska, G.Paun, G.Rozenberg editors, Springer LNCS 2950 (2004) 132-138.
8. M. Garzon, R. Deaton, D. Reanult, *Virtual test tubes: a new methodology for computing*, Proc. 7th. Int. Symposium on String Processing and Information retrieval, A Coruña, Spain. IEEE Computing Society Press (2000) 116-121.
9. T.Head, *Unique decipherability relative to a language*, Tamkang J. Math, 11(1980) 59-66.
10. T. Head, *Relativized code concepts and multi-tube DNA dictionaries*, Finite vs Infinite, C.S. Calude and G. Paun editors, (2000): 175-186.

11. S. Hussini, L. Kari, S. Konstantinidis, *Coding properties of DNA languages*, DNA Computing: Proceedings of the 7th International Meeting on DNA Based Computers (N. Jonoska, N.C. Seeman editors), LNCS 2340 (2002) 57-69.
12. N. Jonoska, D. Kephart, K. Mahalingam, *Generating DNA code words*, Congressus Numerantium 156 (2002) 99-110.
13. N. Jonoska and K. Mahalingam, *Languages od DNA based code words*, Proceedings of the 9th International Meeting on DNA Based Computers, (J.Chen, J.Reif editors), LNCS 2943 (2004) 61-73.
14. N.Jonoska, K.Mahalingam, J.Chen *Involution codes: with application to DNA coded languages*, Natural Computing, Vol 4-2 (2005) 141-162.
15. H.Jürgensen, S.S.Yu, *Solid Codes*, Journal of Information Processing Cybernatics, EIK 26 (1990) 10, 563-574.
16. H.Jürgensen, S.Konstantinidis, *Codes*, Handbook of Formal Languages, Vol 1, (G.Rozenberg and A.Salomaa editors), Springer Verlag (1997) 511-608.
17. H.Jürgensen, M.Katsura and S.Konstantinidis, *Maximal solid codes*, Journal of Automata, Languages and Combinatorics 6(1) (2001) 25-50.
18. L.Kari, S.Konstantinidis, E.Losseva, G.Wozniak, *Sticky-free and overhang-free DNA languages*, Acta Informatica 40 (2003) 119-157.
19. L.Kari and K.Mahalingam, *DNA Codes and their properties*, Submitted.
20. L.Kari, S.Konstantinidis and P.Sosik, *Bond-free languages: formalizations, maximality and construction methods*, Proceedings of the 10th International Meeting on DNA Computing, (C.Ferreti, G.Mauri, C.Zandron editors) LNCS 3384 (2005) 169-181.
21. L.Kari, S.Konstantinidis, P.Sosik, *Preventing undesirable bonds between DNA codewords* Proceedings of the 10th International Meeting on DNA Computing, LNCS 3384 (2005) 182-191.
22. K.Mahalingam, *Involution Codes: With Application to DNA Strand Design* Ph.d. Thesis, University of South Florida, Tampa, FL, 2004.
23. A. Marathe, A.E. Condon, R.M. Corn, *On combinatorial word design*, Preproceedings of the 5th International Meeting on DNA Based Computers, Boston (1999) 75-88.
24. G.Mauri, C.Ferretti, *Word design for molecular computing: a survey*. Proceedings of the 9th International Meeting on DNA Based Computers, (J.Chen, J.H.Reif editors), LNCS 2943 (2004) 37-47.
25. H.J, Shyr, S.S.Yu , *Solid codes and disjunctive domains*, Semigroup Forum 41(1990) 23-37.
26. H.J.Shyr, *Free Monoids and Languages*, Hon Min Book Company 2001.

Well-Founded Semantics for Boolean Grammars[*]

Vassilis Kountouriotis[1], Christos Nomikos[2], and Panos Rondogiannis[1]

[1] Department of Informatics & Telecommunications
University of Athens, Athens, Greece
{grad0771, prondo}@di.uoa.gr
[2] Department of Computer Science, University of Ioannina,
P.O. Box 1186, 45 110 Ioannina, Greece
cnomikos@cs.uoi.gr

Abstract. Boolean grammars [*A. Okhotin, Information and Computation 194 (2004) 19-48*] are a promising extension of context-free grammars that supports conjunction and negation. In this paper we give a novel semantics for boolean grammars which applies to *all* such grammars, independently of their syntax. The key idea of our proposal comes from the area of *negation in logic programming*, and in particular from the so-called *well-founded semantics* which is widely accepted in this area to be the "correct" approach to negation. We show that for every boolean grammar there exists a distinguished (three-valued) language which is a *model* of the grammar and at the same time the least fixed point of an operator associated with the grammar. Every boolean grammar can be transformed into an equivalent (under the new semantics) grammar in normal form. Based on this normal form, we propose an $\mathcal{O}(n^3)$ algorithm for parsing that applies to any such normalized boolean grammar. In summary, the main contribution of this paper is to provide a semantics which applies to *all* boolean grammars while at the same time retaining the complexity of parsing associated with this type of grammars.

1 Introduction

Boolean grammars constitute a new and promising formalism, proposed by A. Okhotin in [Okh04], which extends the class of conjunctive grammars introduced by the same author in [Okh01]. The basic idea behind this new formalism is to allow intersection and negation in the right-hand side of (context-free) rules. It is immediately obvious that the class of languages that can be produced by boolean grammars is a proper superset of the class of context-free languages.

Despite their syntactical simplicity, boolean grammars appear to be non-trivial from a semantic point of view. As we are going to see in the next section, the existing approaches for assigning meaning to boolean grammars suffer from certain shortcomings (one of which is that they do not give a meaning to all such grammars).

---

[*] The first author's work is being supported by a doctoral research grant by the General Secretariat for Research and Technology under the program $\Pi ENE\Delta$ (grant number $03E\Delta$ 330).

In this paper we propose a new semantics (the *well-founded* semantics) which applies to all boolean grammars. More specifically, we demonstrate that for every boolean grammar there exists a distinguished (*three-valued*, see below) language that can be taken as the meaning of this grammar; this language is the unique least fixed point of an appropriate operator associated with the grammar (and therefore it is easy to see that it satisfies all the rules of the grammar).

Our ideas originate from an important area of research in the theory of logic programming, that has been very active for more than two decades (references such as [AB94, PP90] provide nice surveys). In this area, there is nowadays an almost unanimous agreement that if one seeks a unique model of a logic program with negation, then one has to search for a three-valued one. In other words, classical two-valued logic is not sufficient in order to assign a proper meaning to logic programs with negation. Actually, it can be demonstrated that every logic program with negation has a distinguished three-valued model, which is usually termed the *well-founded model* [vGRS91].

We follow the same ideas here: we consider three-valued languages, namely languages in which the membership of strings may be characterized as true, false, or *unknown*. As we will see, this simple extension solves the semantic problems associated with negation in boolean grammars. Moreover, we demonstrate that under this new semantics, every boolean grammar has an equivalent grammar in normal form (similar to that of [Okh04]). Finally, we show that for every such normalized grammar, there is an $\mathcal{O}(n^3)$ parsing algorithm under our new semantics. Our results indicate that there may be other fruitful connections between formal language theory and the theory of logic programming.

2   Why an Alternative Semantics for Boolean grammars?

In [Okh04] A. Okhotin proposed the class of boolean grammars. Formally:

Definition 1 ([Okh04]). *A Boolean grammar is a quadruple* $G = (\Sigma, N, P, S)$, *where* Σ *and* N *are disjoint finite nonempty sets of terminal and nonterminal symbols respectively,* P *is a finite set of rules, each of the form*

$$A \rightarrow \alpha_1 \& \cdots \& \alpha_m \& \neg\beta_1 \& \cdots \& \neg\beta_n \qquad (m + n \geq 1, \alpha_i, \beta_i \in (\Sigma \cup N)^*),$$

and $S \in N$ *is the start symbol of the grammar. We will call the* α_i *'s positive literals and the* $\neg\beta_i$ *'s negative.*

To illustrate the use of Boolean grammars, consider a slightly modified example from [Okh04]:

Example 1. Let $\Sigma = \{a, b\}$. We define:

$$\begin{aligned}
S &\rightarrow \neg(AB) \ \& \ \neg(BA) \ \& \ \neg A \ \& \ \neg B \\
A &\rightarrow a \\
A &\rightarrow CAC \\
B &\rightarrow b \\
B &\rightarrow CBC \\
C &\rightarrow a \\
C &\rightarrow b
\end{aligned}$$

It can be shown that the above grammar defines the language $\{ww \mid w \in \{a, b\}^*\}$ (see [Okh04] for details). It is well-known that this language is not context-free.

Okhotin proposed two semantics intended to capture the meaning of boolean grammars. In this section we demonstrate some deficiencies of these two approaches, which led us to the definition of the well-founded semantics. Both semantics proposed in [Okh04] are defined using a system of equations, which is obtained from the given grammar.

In the first approach, the semantics is defined only in the case that the system of equations has a *unique solution*. This is a restrictive choice: actually most interesting grammars do not correspond to systems of equations having a unique solution. For example, even the simplest context-free grammars generating infinite languages, give systems of equations which have infinitely many solutions. For such grammars, it seems that the desired property is a form of *minimality* rather than uniqueness of the solution.

Apart from its limited applicability, the unique solution semantics also demonstrates a kind of instability. For example, let $\Sigma = \{0, 1\}$ and consider the boolean grammar consisting of the two rules $A \to \neg A \& \neg B$ and $B \to 0 \& 1$. The corresponding system of equations has no solution and therefore the unique solution semantics for this grammar is not defined. Suppose that we augment the above grammar with the rule $B \to B$. Seen from a constructive point of view, the new rule does not offer to the grammar any additional information. It is reasonable to expect that such a rule would not change the semantics of the grammar. However, the augmented grammar has unique solution semantics, namely $(A, B) = (\emptyset, \Sigma^*)$. On the other hand, suppose that we augment the initial grammar with the rule $A \to A$. Then, the unique solution semantics is also defined, but now the solution is $(A, B) = (\Sigma^*, \emptyset)$. Consequently by adding to an initially meaningless grammar two different information-free rules, we obtained two grammars defining complementary languages. To put it another way, three grammars that look equivalent, have completely different semantics.

Let's now turn to the second approach proposed in [Okh04], namely the naturally feasible solution semantics. Contrary to the unique solution semantics, the feasible solution semantics generalizes the semantics of context-free and conjunctive languages (see [Okh04][Theorem 3]). However, when negation appears, there are cases that this approach does not behave in an expected manner. Consider for example the boolean grammar with rules:

$$A \to \neg B, \quad B \to C \& \neg D, \quad C \to D, \quad D \to A$$

This grammar has the naturally feasible solution $(A, B, C, D) = (\Sigma^*, \emptyset, \Sigma^*, \Sigma^*)$. It is reasonable to expect that composing two rules would not affect the semantics of the grammar. For example in context-free grammars such a composition is a natural transformation rule that simply allows to perform two steps of the production in a single step. However, if we add $C \to A$ to the above set of rules, then the naturally feasible solution semantics of the resulting grammar is not defined. On the other hand, the technique we will define shortly, does not suffer from this shortcoming.

Furthermore, there exist grammars for which the naturally feasible solution semantics is undefined, although they may have a clear intuitive meaning. For example, let $\Sigma = \{a\}$ and consider the following set of eight rules:

$$A \to \neg B, \quad A \to D, \quad B \to \neg C, \quad B \to D,$$

$$C \to \neg A, \quad C \to D, \quad D \to aD, \quad D \to \epsilon$$

The semantics of this grammar should clearly be $(A, B, C, D) = (\Sigma^*, \Sigma^*, \Sigma^*, \Sigma^*)$, and actually this is what the well-founded semantics will produce. On the other hand the naturally feasible solution semantics is undefined.

The problem of giving semantics to recursive formalisms in the presence of negation has been extensively studied in the context of logic programming. Actually, the unique solution semantics can be paralleled with one of the early attempts to give semantics to logic programs with negation, namely what is now called *the Clark's completion semantics* (which actually presents similar shortcomings with the unique solution approach). On the other hand, the naturally feasible solution can be thought of as a first approximation to the procedure of constructing the intended minimal model of a logic program with negation (see also Theorem 3 that will follow). Since the most broadly accepted semantic approach for logic programs with negation is the well-founded semantics, we adopt this approach in this paper.

At this point we should also mention a recent work on the *stratified* semantics of Boolean grammars [Wro05], an idea that also originates from logic programming. However, the stratified semantics is less general than the well-founded one (since the former does not cover the whole class of Boolean grammars).

3   Interpretations and Models for Boolean Grammars

In this section we formally define the notion of *model* for boolean grammars. In context-free grammars, an interpretation is a function that assigns to each non-terminal symbol of the grammar a set of strings over the set of terminal symbols of the grammar. An interpretation of a context-free grammar is a model of the grammar if it satisfies all the rules of the grammar. The usual semantics of context-free grammars dictate that every such grammar has a minimum model, which is taken to be as its intended meaning.

When one considers boolean grammars, the situation becomes much more complicated. For example, a grammar with the unique rule $S \to \neg S$ appears to be meaningless. More generally, in many cases where negation is used in a circular way, the corresponding grammar looks problematic. However, these difficulties arise because we are trying to find *classical* models of boolean grammars, which are based on classical two-valued logic. If however we shift to three-valued models, every boolean grammar has a well-defined meaning. We need of course to redefine many notions, starting even from the notion of a language:

Definition 2. *Let Σ be a finite non-empty set of symbols. Then, a (three-valued) language over Σ is a function from Σ^* to the set $\{0, \frac{1}{2}, 1\}$.*

Intuitively, given a three-valued language L and a string w over the alphabet of L, there are three-cases: either $w \in L$ (ie., $L(w) = 1$), or $w \notin L$ (ie., $L(w) = 0$), or finally, the membership of w in L is unclear (ie., $L(w) = \frac{1}{2}$). Given this extended notion of language, it is now possible to interpret the grammar $S \to \neg S$: its meaning is the language which assigns to every string the value $\frac{1}{2}$.

The following definition, which generalizes the familiar notion of concatenation of languages, will be used in the following:

Definition 3. *Let Σ be a finite set of symbols and let $L_1, \ldots, L_n$ be (three-valued) languages over Σ. We define the three-valued concatenation of the languages $L_1, \ldots, L_n$ to be the language L such that:*

$$L(w) = \max_{\substack{(w_1,\ldots,w_n): \\ w=w_1\cdots w_n}} \left(\min_{1\leq i\leq n} L_i(w_i) \right)$$

The concatenation of $L_1, \ldots, L_n$ will be denoted by $L_1 \circ \cdots \circ L_n$.

We can now define the notion of *interpretation* of a given boolean grammar:

Definition 4. *An interpretation I of a boolean grammar $G = (\Sigma, N, P, S)$ is a function $I : N \to (\Sigma^* \to \{0, \frac{1}{2}, 1\})$.*

An interpretation I can be recursively extended to apply to expressions that appear as the right-hand sides of boolean grammar rules:

Definition 5. *Let $G = (\Sigma, N, P, S)$ be a boolean grammar and I be an interpretation of G. Then I can be extended to become a truth valuation $\hat{I}$ as follows:*

- *For the empty sequence ϵ and for all $w \in \Sigma^*$, it is $\hat{I}(\epsilon)(w) = 1$ if $w = \epsilon$ and 0 otherwise.*
- *Let $a \in \Sigma$ be a terminal symbol. Then, for every $w \in \Sigma^*$, $\hat{I}(a)(w) = 1$ if $w = a$ and 0 otherwise.*
- *Let $\alpha = \alpha_1 \cdots \alpha_n$, $n \geq 1$, be a sequence in $(\Sigma \cup N)^*$. Then, for every $w \in \Sigma^*$, it is $\hat{I}(\alpha)(w) = (\hat{I}(\alpha_1) \circ \cdots \circ \hat{I}(\alpha_n))(w)$.*
- *Let $\alpha \in (\Sigma \cup N)^*$. Then, for every $w \in \Sigma^*$, $\hat{I}(\neg \alpha)(w) = 1 - \hat{I}(\alpha)(w)$.*
- *Let $l_1, \ldots, l_n$ be literals. Then, for every string $w \in \Sigma^*$, $\hat{I}(l_1 \& \cdots \& l_n)(w) = \min\{\hat{I}(l_1)(w), \ldots, \hat{I}(l_n)(w)\}$.*

We are now in a position to define the notion of a model of a boolean grammar:

Definition 6. *Let $G = (\Sigma, N, P, S)$ be a boolean grammar and I an interpretation of G. Then, I is a model of G if for every rule $A \to l_1 \& \cdots \& l_n$ in P and for every $w \in \Sigma^*$, it is $\hat{I}(A)(w) \geq \hat{I}(l_1 \& \cdots \& l_n)(w)$.*

In the definition of the well-founded model, two orderings on interpretations play a crucial role (see [PP90]). Given two interpretations, the first ordering (usually called the *standard ordering*) compares their *degree of truth*:

Definition 7. *Let $G = (\Sigma, N, P, S)$ be a boolean grammar and I, J be two interpretations of G. Then, we say that $I \preceq J$ if for all $A \in N$ and for all $w \in \Sigma^*$, $I(A)(w) \leq J(A)(w)$.*

Among the interpretations of a given boolean grammar, there is one which is the least with respect to the $\preceq$ ordering, namely the interpretation $\perp$ which for all A and all w, $\perp(A)(w) = 0$.

The second ordering (usually called the *Fitting ordering*) compares the *degree of information* of two interpretations:

Definition 8. *Let $G = (\Sigma, N, P, S)$ be a boolean grammar and I, J be two interpretations of G. Then, we say that $I \preceq_F J$ if for all $A \in N$ and for all $w \in \Sigma^*$, if $I(A)(w) = 0$ then $J(A)(w) = 0$ and if $I(A)(w) = 1$ then $J(A)(w) = 1$.*

Among the interpretations of a given boolean grammar, there is one which is the least with respect to the $\preceq_F$ ordering, namely the interpretation $\perp_F$ which for all A and all w, $\perp_F (A)(w) = \frac{1}{2}$.

Given a set U of interpretations, we will write $lub_{\preceq}U$ (respectively $lub_{\preceq_F}U$) for the least upper bound of the members of U under the standard ordering (respectively, the Fitting ordering).

4   Well-Founded Semantics for Boolean Grammars

In this section we will define the well-founded semantics of boolean grammars. The basic idea behind the well-founded semantics is that the intended model of the grammar is constructed in stages, ie., there is a stratification process involved that is related to the levels of negation used by the grammar. For every nonterminal symbol, at each step of this process, the values of certain strings are computed and fixed (as either true or false); at each new level, the values of more and more strings become fixed (and this is a monotonic procedure in the sense that values of strings that have been fixed for a given nonterminal in a previous stage, cannot be be altered by the next stages). At the end of all the stages, certain strings for certain nonterminals may have not managed to get the status of either true or false (this will be due to circularities through negation in the grammar). Such strings are classified as unknown (ie., $\frac{1}{2}$).

Consider the boolean grammar G. Then, for any interpretation J of G we define the operator $\Theta_J : \mathcal{I} \to \mathcal{I}$ on the set $\mathcal{I}$ of all 3-valued interpretations of G. This operator is analogous to the one used in the logic programming domain (see for example [PP90]).

Definition 9. *Let $G = (\Sigma, N, P, S)$ be a boolean grammar, let $\mathcal{I}$ be the set of all three-valued interpretations of G and let $J \in \mathcal{I}$. The operator $\Theta_J : \mathcal{I} \to \mathcal{I}$ is defined as follows. For every $I \in \mathcal{I}$, for all $A \in N$ and for all $w \in \Sigma^*$:*

1. $\Theta_J(I)(A)(w) = 1$ *if there is a rule $A \to l_1 \& \cdots \& l_n$ in P such that, for all $i \leq n$, either $\hat{J}(l_i)(w) = 1$ or l_i is positive and $\hat{I}(l_i)(w) = 1$;*
2. $\Theta_J(I)(A)(w) = 0$ *if for every rule $A \to l_1 \& \cdots \& l_n$ in P, there is an $i \leq n$ such that either $\hat{J}(l_i)(w) = 0$ or l_i is positive and $\hat{I}(l_i)(w) = 0$;*
3. $\Theta_J(I)(A)(w) = \frac{1}{2}$, *otherwise.*

An important fact regarding the operator Θ_J is that it is monotonic with respect to the $\preceq$ ordering of interpretations:

Theorem 1. *Let G be a boolean grammar and let J be an interpretation of G. Then, the operator Θ_J is monotonic with respect to the $\preceq$ ordering of interpretations. Moreover, Θ_J has a unique least (with respect to $\preceq$) fixed point $\Theta_J^{\uparrow\omega}$ which is defined as follows:*

$$
\begin{aligned}
\Theta_J^{\uparrow 0} &= \bot \\
\Theta_J^{\uparrow n+1} &= \Theta_J(\Theta_J^{\uparrow n}) \\
\Theta_J^{\uparrow\omega} &= lub_{\preceq}\{\Theta_J^{\uparrow n} \mid n < \omega\}
\end{aligned}
$$

Proof. The proof essentially follows the same lines of thought as that of the logic programming case (see [Prz89]). □

We will denote by $\Omega(J)$ the least fixed point of Θ_J. Given a grammar G, we can use the Ω operator to construct a sequence of interpretations whose ω-limit M_G will prove to be a distinguished model of G:

$$
\begin{aligned}
M_0 &= \bot_F \\
M_{n+1} &= \Omega(M_n) \\
M_G &= lub_{\preceq_F}\{M_n \mid n < \omega\}
\end{aligned}
$$

Notice that here we have an essential difference with respect to the well-founded semantics of logic programming: there, the construction of the well-founded model may require a transfinite number of iterations which is greater than ω. In other words, the well-founded semantics of logic programs is not computable in the general case. However, in the case of Boolean grammars, the model is constructed in at most ω iterations:

Theorem 2. *Let G be a boolean grammar. Then, M_G is a model of G (which will be called the well-founded model of G). Moreover, M_G is the least (with respect to the $\preceq_F$ ordering) fixed point of the operator Ω.*

Proof. Technically, the proof is very similar to that of the logic programming case (see [Prz89]). □

Actually, it can be shown (following a similar reasoning as in [RW05]) that the model M_G is the *least* model of G according to a syntax-independent relation.

The construction of the well-founded model is illustrated by the following example:

Example 2. Let G be the grammar given in Example 1. Then, it is easy to see that $M_G = M_2$, ie., that in order to converge to the well-founded model of G we need exactly two iterations of Ω. More specifically, in $M_1 = \Omega(M_0)$ the denotations of the non-terminals A, B and C stabilize (notice that the definitions of these nonterminals are standard context-free rules). However, in order for the denotation of S to stabilize, an additional iteration of Ω is required. Notice that the language produced by this grammar is two-valued.

We can now state the relationship between the well-founded semantics and the naturally feasible semantics of boolean grammars:

Theorem 3. *Suppose that a boolean grammar G has a two-valued (ie., with values 0 and 1) well-founded semantics. Then the naturally feasible solution for this grammar either coincides with the well-founded semantics or is undefined.*

It is easy to see that if a boolean grammar has a naturally feasible solution semantics, then it is possible that this semantics differs from the well-founded one. For example, in the four-rule grammar of Section 2, the well-founded semantics assigns the $\perp_F$ interpretation to all the nonterminal symbols of the grammar. Notice that although the naturally feasible semantics for this grammar is defined, it appears to be counterintuitive.

5   Normal Form

In this section we demonstrate that every boolean grammar can be converted into an equivalent one that belongs to the following normal form:

Definition 10. *A Boolean grammar $G = (\Sigma, N, P, S)$ is said to be in binary normal form if P contains the rules $U \to \neg U$ and $T \to \neg\epsilon$, where U and T are two special symbols in $N - \{S\}$, and every other rule in P is of the form:*

$$A \to B_1C_1\& \cdots \&B_mC_m\&\neg(D_1E_1)\& \cdots \&\neg(D_nE_n)\&TT[\&U] \quad (m, n \geq 0)$$
$$A \to a[\&U]$$
$$S \to \epsilon[\&U] \quad \textit{(only if S does not appear in right-hand sides of rules)}$$

where $A, B_i, C_i, D_j, E_j \in N - \{U, T\}$, $a \in \Sigma$, and the brackets denote an optional part.

The basic theorem of this section states that for every boolean grammar G there exists a boolean grammar in binary normal form that defines the same language as G. More formally:

Theorem 4. *Let $G = (\Sigma, N, P, S)$ be a boolean grammar. Then there exists a grammar $G' = (\Sigma, N', P', S)$ in binary normal form such that $M_G(S) = M_{G'}(S)$.*

The proof of Theorem 4 is based on several transformations, justified by some lemmata given below. We give here an outline of how the binary normal form is constructed.

Consider a boolean grammar $G = (\Sigma, N, P, S)$. Without loss of generality we may assume that S does not appear in the right-hand side of any rule (otherwise we can replace S with S' in every rule, and add a rule $S \to S'$). Initially, we bring the grammar into a form, which we call *pre-normal form* (see Definition 11). This is performed using Lemmas 1,2 and 3. More specifically, Lemma 1 is used to eliminate terminal symbols from rules containing boolean connectives or concatenation; Lemma 2 separates boolean connectives from concatenation; and, Lemma 3 is used to eliminate "long" concatenations. Based on the pre-normal form, we then construct an ϵ-free version of the grammar (Definition 12). The ϵ-free version is then brought into binary-normal form (see Definition 10 above) using the technique described in Definition 15. Detailed proofs of lemmas comprising this procedure are lengthy, and are omitted in the current form of the paper.

Lemma 1. *Let $G = (\Sigma, N, P, S)$ be a boolean grammar, and let G' be the grammar $(\Sigma, N \cup \{A_a \mid a \in \Sigma\}, P' \cup \{A_a \to a \mid a \in \Sigma\}, S)$ where:*

- *$\{A_a \mid a \in \Sigma\} \cap N = \emptyset$.*
- *P' is obtained from P by replacing each occurrence of the terminal symbol a with A_a, in every rule that contains concatenation or boolean connectives.*

Then, for every $C \in N$, $M_G(C) = M_{G'}(C)$.

Lemma 2. *Let $G = (\Sigma, N, P, S)$ be a boolean grammar, and let $\beta \in N^k$, $k \geq 2$, be a sequence of non-terminal symbols. Let $G' = (\Sigma, N \cup \{B\}, P' \cup \{B \to \beta\}, S)$ where:*

- *$B \notin N$ is a new non-terminal symbol.*
- *For every rule $A \to \alpha_1 \& \cdots \& \alpha_m \& \neg \alpha_{m+1} \& \cdots \& \neg \alpha_n$ in P, P' contains the rule $A \to \alpha_1' \& \cdots \& \alpha_m' \& \neg \alpha_{m+1}' \& \cdots \& \neg \alpha_n'$, where $\alpha_i' = B$ if $\alpha_i = \beta$, otherwise $\alpha_i' = \alpha_i$.*

Then, for every $C \in N$, $M_G(C) = M_{G'}(C)$.

Lemma 3. *Let $G = (\Sigma, N, P, S)$ be a boolean grammar, let $A \to B_1 B_2 B_3 \ldots B_k$, $A, B_i \in N$, $k \geq 3$, be a rule of P and let $G' = (\Sigma, N \cup \{D\}, P', S)$ where:*

- *$D \notin N$ is a new non-terminal symbol.*
- *$P' = (P - \{A \to B_1 B_2 B_3 \ldots B_k\}) \cup \{A \to D B_3 \ldots B_k, D \to B_1 B_2\}$.*

Then, for every $C \in N$, $M_G(C) = M_{G'}(C)$.

Using the above lemmas it is straightforward to bring the initial grammar into the following form:

Definition 11. *A Boolean grammar $G = (\Sigma, N, P, S)$ is said to be in pre-normal form if every rule in P is of the form:*

$$
\begin{aligned}
&A \to B_1 \& \cdots \& B_m \& \neg C_1 \& \cdots \& \neg C_n &&(m + n \geq 1, B_i, C_j \in N \cup \{\epsilon\}) \\
&A \to BC &&(B, C \in N) \\
&A \to a &&(a \in \Sigma)
\end{aligned}
$$

Based on the pre-normal form of the grammar, we can now define its ϵ-free version:

Definition 12. *Let $G = (\Sigma, N, P, S)$ be a boolean grammar in pre-normal form. The ϵ-free version of G, denoted by G_ϵ, is the boolean grammar $(\Sigma, N \cup \{U\}, P', S)$ where P' is obtained as follows:*

- *P' contains a rule $U \to \neg U$, where $U \notin N$ is a special non-terminal symbol, which represents the set in which all strings have the value $\frac{1}{2}$.*
- *For every rule of the form $A \to B_1 \& \cdots \& B_m \& \neg C_1 \& \cdots \& \neg C_n$, $(m + n \geq 1, B_i, C_j \in N \cup \{\epsilon\})$ in P*
  - *If $B_i = \epsilon$ for some i, then the rule is ignored in the construction of P'.*
  - *Otherwise, if $C_i = \epsilon$ for some i, then the rule is included in P' as it is.*

- *Otherwise, P' contains the rule $A \rightarrow B_1\& \cdots \&B_m\&\neg C_1\& \cdots \&\neg C_n\&\neg\epsilon$*
- *For every rule of the form $A \rightarrow BC$ in P*
  - *P' contains the rule $A \rightarrow BC\&\neg\epsilon$*
  - *If $M_G(B)(\epsilon) = 1$ (respectively $M_G(C)(\epsilon) = 1$), then P' contains the rule $A \rightarrow C\&\neg\epsilon$ (respectively the rule $A \rightarrow B\&\neg\epsilon$).*
  - *If $M_G(B)(\epsilon) = \frac{1}{2}$ (respectively $M_G(C)(\epsilon) = \frac{1}{2}$), then P' contains the rule $A \rightarrow C\&U\&\neg\epsilon$ (respectively the rule $A \rightarrow B\&U\&\neg\epsilon$).*
- *For every $a \in \Sigma$ and $A \in N$, if $M_G(A)(a) = 1$ then P' contains the rule $A \rightarrow a$ and if $M_G(A)(a) = \frac{1}{2}$ then P' contains the rule $A \rightarrow a\&U$*

Lemma 4. *Let $G = (\Sigma, N, P, S)$ be a boolean grammar in pre-normal form, and let G_ϵ be its ϵ-free version. Then, for every $C \in N$ and for every $w \in \Sigma^*$, $w \neq \epsilon$ implies $M_G(C)(w) = M_{G_\epsilon}(C)(w)$.*

In order to obtain a grammar in binary normal form, we need to eliminate rules of the form $A \rightarrow B_1\& \cdots \&B_m\&\neg C_1\& \cdots \&\neg C_n\&\neg\epsilon$. Membership in $M_G(A)$ depends only on membership in each of $\hat{M}_G(BC)$, for all BC that appear in the right-hand sides of rules. We can express this dependency directly by a set of rules. In order to do this we treat each BC that appears in the right-hand side of a rule as a boolean variable (see also [Okh04]).

Definition 13. *Let X be a set of variables and let V, W be functions from X to $\{0, \frac{1}{2}, 1\}$. We denote by V_i the set $\{x \in X \mid V(x) = i\}$. We write $V \sqsubseteq W$ if $V_0 \subseteq W_0$ and $V_1 \subseteq W_1$*

Definition 14. *Let G be a grammar in pre-normal form and let $G_\epsilon = (\Sigma, N, P, S)$ be the ϵ-free version of G. Let $X = \{BC \mid A \rightarrow BC \in P\}$ and let V be a function from X to $\{0, \frac{1}{2}, 1\}$. Then, the extension of V to non-terminal symbols in N, denoted by $\hat{V}$, is defined as follows: $\hat{V}(A)$ is the value $M_{G_\epsilon}(A)(w)$ when $M_{G_\epsilon}(BC)(w) = V(BC)$, for all $BC \in X$ and for arbitrary w.*

Notice that $\hat{V}$ is well-defined and can be computed in finitely many steps from V.

Definition 15. *Let G be a grammar in pre-normal form, let $G_\epsilon = (\Sigma, N, P, S)$ be the ϵ-free version of G. Let $X = \{BC \mid A \rightarrow BC \in P\}$ and let $\mathcal{V}$ be the set of all functions from X to $\{0, \frac{1}{2}, 1\}$. The normal form $G_n = (\Sigma, N \cup \{T\}, P', S)$ of G is the grammar obtained from G_ϵ as follows:*

- *P' contains all the rules in P of the form $A \rightarrow a$ and $A \rightarrow a\&U$, where $a \in \Sigma$, the rule $U \rightarrow \neg U$ in P and the rule $T \rightarrow \neg\epsilon$, where $T \notin N$ is a special symbol which represents the set in which all non-empty strings have value 1.*
- *For every $A \in N$ let $\mathcal{T}_A = \{V \in \mathcal{V} \mid \hat{V}(A) = 1\}$. For every minimal (with respect to $\sqsubseteq$) element V of $\mathcal{T}_A$, P' contains the rule:*

$$A \rightarrow x_{i_1}\& \ldots \&x_{i_m}\&\neg y_{j_1}\& \ldots \&\neg y_{j_n}\&TT$$

where $\{x_{i_1}, \ldots, x_{i_m}\} = V_1$ and $\{y_{j_1}, \ldots, y_{j_n}\} = V_0$.

- *For every $A \in N$ let $\mathcal{U}_A = \{V \in \mathcal{V} \mid \hat{V}(A) = \frac{1}{2}\}$. For every maximal (with respect to $\sqsubseteq$) element V of $\mathcal{U}_A$, P' contains the rule:*

$$A \rightarrow x_{i_1} \& \neg x_{i_1} \& \ldots \& x_{i_m} \& \neg x_{i_m} \& TT \& U$$

where $\{x_{i_1}, \ldots, x_{i_m}\} = V_{\frac{1}{2}}$.

Notice that in the former case we consider only minimal elements, because if $V' \sqsubseteq V$ and $\hat{V}'(A) = 1$ then $\hat{V}(A) = 1$. Similarly in the latter case we consider only maximal elements, because if $V' \sqsubseteq V$ and $\hat{V}(A) = \frac{1}{2}$ then $\hat{V}'(A) = \frac{1}{2}$. The above properties follow from the monotonicity of the Ω operator, with respect to the $\preceq_F$ (Fitting) ordering.

Lemma 5. *Let $G = (\Sigma, N, P, S)$ be a boolean grammar in pre-normal form, and let G_n be its binary normal form. Then, for every $A \in N$ and for every $w \in \Sigma^*$, $w \neq \epsilon$ implies $M_G(A)(w) = M_{G_n}(A)(w)$.*

Given the above lemmas, a simple step remains in order to reach the statement of Theorem 4: if in the original grammar G it is $M_G(S)(\epsilon) \neq 0$, then an appropriate rule of the form $S \rightarrow \epsilon$ or $S \rightarrow \epsilon \& U$ is added to the grammar that has resulted after the processing implied by all the above lemmas. The resulting grammar is then in binary normal form and defines the same language as the initial one.

6   Parsing Under the Well-Founded Semantics

We next present an algorithm that computes the truth value of the membership of an input string $w \neq \epsilon$ in a language defined by a grammar G, which is assumed to be in binary normal form. The algorithm computes the value of $M_G(A)(u)$ for every non-terminal symbol A and every substring u of w in a bottom up manner. By convention $min_{i=1}^0 v_i = 1$.

Algorithm for parsing an input string $w = a_1 \cdots a_n$:

for $i := 1$ to n do begin
    for every $A \in N$ do
        if there exist a rule $A \rightarrow a_i$ then $M_G(A)(a_i) := 1$
        else if there exist a rule $A \rightarrow a_i \& U$ then $M_G(A)(a_i) := \frac{1}{2}$
        else $M_G(A)(a_i) := 0$
end

for $d := 2$ to n do
    for $i := 1$ to $n - d + 1$ do begin
        $j := i + d - 1$
        for every $B, C \in N$ such that BC appears in the right-hand side of a rule do
            $\hat{M}_G(BC)(a_i \ldots a_j) := \max_{k=i}^{j-1} \min\{M_G(B)(a_i \ldots a_k), M_G(C)(a_{k+1} \ldots a_j)\}$
        for every $A \in N$ do $M_G(A)(a_i \ldots a_j) := 0$
        for every rule $A \rightarrow B_1 C_1 \& \ldots \& B_m C_m \& \neg D_1 E_1 \& \ldots \& \neg D_r E_r \& TT \& U$ do begin
            $v := \min\{\frac{1}{2}, \min_{p=1}^m \hat{M}_G(B_p C_p)(a_i \ldots a_j), \min_{q=1}^r (1 - \hat{M}_G(D_q E_q)(a_i \ldots a_j))\}$

$\quad$ if $v > M_G(A)(a_i \ldots a_j)$ then $M_G(A)(a_i \ldots a_j) := v$
$\quad$ end
$\quad$ for every rule $A \rightarrow B_1 C_1 \& \ldots \& B_m C_m \& \neg D_1 E_1 \& \ldots \& \neg D_r E_r \& TT$ do begin
$\qquad$ $v := \min\{\min_{p=1}^{m} \hat{M}_G(B_p C_p)(a_i \ldots a_j), \min_{q=1}^{r}(1 - \hat{M}_G(D_q E_q)(a_i \ldots a_j))\}$
$\qquad$ if $v > M_G(A)(a_i \ldots a_j)$ then $M_G(A)(a_i \ldots a_j) := v$
$\quad$ end
$\quad$ end
return $M_G(S)(a_1 \cdots a_n)$

For a fixed grammar the above algorithm runs in time $\mathcal{O}(n^3)$: the value $M_G(A)(u)$ is computed for $\mathcal{O}(n^2)$ substrings u of w; each computation requires to break u in two parts in all possible ways, and there are $\mathcal{O}(n)$ appropriate breakpoints.

7 Conclusions

We have presented a novel semantics for boolean grammars, based on techniques that have been developed in the logic programming domain. Under this new semantics every boolean grammar has a distinguished language that satisfies its rules. Moreover, we have demonstrated that every boolean grammar can be transformed into an equivalent one in a binary normal form. For grammars in binary normal form, we have derived an $\mathcal{O}(n^3)$ parsing algorithm.

We believe that a further investigation of the connections between formal language theory and the theory of logic programming will prove to be very rewarding.

References

[AB94]   Apt, K., Bol, R.: Logic Programming and Negation: A Survey. Journal of Logic Programming **19,20** (1994) 9–71

[Okh01]   Okhotin, A.: Conjunctive Grammars. Journal of Automata, Languages and Combinatorics **6(4)** (2001) 519–535

[Okh04]   Okhotin, A.: Boolean Grammars. Information and Computation **194(1)** (2004) 19–48

[PP90]   Przymusinska, H., Przymusinski, T.: Semantic Issues in Deductive Databases and Logic Programs. In R. Banerji, editor, Formal Techniques in Artificial Intelligence: a Source-Book, North Holland (1990) 321–367

[Prz89]   Przymusinski, T., C,: Every Logic Program has a Natural Stratification and an Iterated Fixed Point Model. In Proceedings of the 8th Symposium on Principles of Database Systems ACM SIGACT-SIGMOD (1989) 11–21

[RW05]   Rondogiannis, P., Wadge, W., W.: Minimum Model Semantics for Logic Programs with Negation-as-Failure. ACM Transactions on Computational Logic **6(2)** (2005) 441–467

[vGRS91]   van Gelder, A, Ross, K., A., Schlipf, J., S.: The Well-Founded Semantics for General Logic Programs. Journal of the ACM **38(3)** (1991) 620–650

[Wro05]   Wrona M.: Stratified Boolean Grammars. MFCS (2005) 801–812

Hierarchies of Tree
Series Transformations Revisited

Andreas Maletti*

Technische Universität Dresden
Department of Computer Science
maletti@tcs.inf.tu-dresden.de

Abstract. Tree series transformations computed by polynomial top-down and bottom-up tree series transducers are considered. The hierarchy of tree series transformations obtained in [Fülöp, Gazdag, Vogler: *Hierarchies of Tree Series Transformations.* Theoret. Comput. Sci. 314(3), p. 387–429, 2004] for commutative izz-semirings (izz abbreviates idempotent, zero-sum and zero-divisor free) is generalized to arbitrary positive (*i. e.*, zero-sum and zero-divisor free) commutative semirings. The latter class of semirings includes prominent examples such as the natural numbers semiring and the least common multiple semiring, which are not members of the former class.

1 Introduction

Tree series transducers were introduced in [1, 2, 3] as a generalization of top-down and bottom-up tree transducers. With the advent of tree series [4, 5, 6, 7, 8], especially recognizable tree series [9, 10], in formal language theory also transducing devices capable of (finitely) representing transformations on tree series became interesting. For example, in [11] the power of (top-down) tree series transducers for natural language processing was recognized.

In the seminal paper [12] the hierarchy of top-down tree transformation classes was proved to be proper. This result lead to the hierarchy of top-down and bottom-up tree transformation classes (as, *e. g.*, displayed in [13]). This hierarchy was generalized to classes of top-down and bottom-up tree series transformations over izz-semirings (izz abbreviates idempotent, zero-divisor and zero-sum free) in [14]. Let us explain this generalization in some more detail.

By p–TOP$_\varepsilon(\mathcal{A})$ and p–BOT$_\varepsilon(\mathcal{A})$ we denote the classes of tree-to-tree-series transformations computable by polynomial top-down and bottom-up tree series transducers [2] over the semiring $\mathcal{A}$ [15, 16], respectively. Such a tree-to-tree-series transformation is a mapping $\tau \colon T_\Sigma \longrightarrow \mathcal{A}\langle\!\langle T_\Delta \rangle\!\rangle$ for some ranked alphabets Σ and Δ. Given ranked alphabets Σ, Δ, and Γ and $\tau_1 \colon T_\Sigma \longrightarrow \mathcal{A}\langle\!\langle T_\Delta \rangle\!\rangle$ and $\tau_2 \colon T_\Delta \longrightarrow \mathcal{A}\langle\!\langle T_\Gamma \rangle\!\rangle$, the composition of τ_1 with τ_2 is denoted by $\tau_1 \circ \tau_2$ and is a mapping $\tau \colon T_\Sigma \longrightarrow \mathcal{A}\langle\!\langle T_\Gamma \rangle\!\rangle$ (an output tree u produced by τ_1 is subjected to τ_2, and the result is multiplied by the weight of u in the series produced

* Financially supported by the German Research Foundation (DFG GK/334).

by τ_1). This composition is lifted to classes of transformations, and we write $\text{p--TOP}_\varepsilon^n(\mathcal{A})$ and $\text{p--BOT}_\varepsilon^n(\mathcal{A})$ for the n-fold composition of $\text{p--TOP}_\varepsilon(\mathcal{A})$ and $\text{p--BOT}_\varepsilon(\mathcal{A})$, respectively.

In [14] it is first proved that

$$\text{p--TOP}_\varepsilon^n(\mathcal{A}) \subseteq \text{p--BOT}_\varepsilon^{n+1}(\mathcal{A}) \quad \text{and} \quad \text{p--BOT}_\varepsilon^n(\mathcal{A}) \subseteq \text{p--TOP}_\varepsilon^{n+1}(\mathcal{A})$$

for every commutative semiring and $n \geqslant 1$ (see Theorems 5.1 and 5.7 in [14], respectively). Then in [14, Theorem 6.20] it is proved that

$$\text{p--TOP}_\varepsilon^n(\mathcal{A}) \nsubseteq \text{p--BOT}_\varepsilon^n(\mathcal{A}) \quad \text{and} \quad \text{p--BOT}_\varepsilon^n(\mathcal{A}) \nsubseteq \text{p--TOP}_\varepsilon^n(\mathcal{A})$$

for every izz-semiring and $n \geqslant 1$. Thus the hierarchy that is obtained in [14] is proved for commutative izz-semirings. We generalize the incomparability result to positive (i. e., zero-sum and zero-divisor free) semirings and thereby obtain the hierarchy for all positive and commutative semirings (see Figure 1 for the HASSE diagram).

Our approach used to prove the incomparability is (in essence) similar to the one presented in [14]. However, we carefully avoid the introduction of idempotency by a simpler proof method. We furthermore claim that our method of proof is more illustrative than the one of [14].

Apart from this introduction, the paper has 3 sections. Section 2 introduces the essential notation, Section 3 generalizes the mentioned incomparability result, and Section 4 presents the obtained hierarchy (see Figure 1).

2   Preliminaries

We use $\mathbb{N}$ to represent the nonnegative integers and $\mathbb{N}_+ = \mathbb{N} \setminus \{0\}$. In the sequel, let $k, n \in \mathbb{N}$ and $[k]$ be an abbreviation for $\{i \in \mathbb{N} \mid 1 \leqslant i \leqslant k\}$. A set Σ that is nonempty and finite is also called an *alphabet*, and the elements thereof are called *symbols*. As usual, Σ^* denotes the set of all finite sequences of symbols of Σ (also called Σ-words). Given $w \in \Sigma^*$, the *length of w* is denoted by $|w|$.

A *ranked alphabet* is an alphabet Σ with a mapping $\text{rk}_\Sigma \colon \Sigma \longrightarrow \mathbb{N}$, which associates to each symbol a *rank*. We use Σ_k to represent the set of symbols of Σ that have rank k. Moreover, we use the set $X = \{x_i \mid i \in \mathbb{N}_+\}$ of *(formal) variables* and $X_k = \{x_i \mid i \in [k]\}$. Given a ranked alphabet Σ and $V \subseteq X$, the set of Σ-*trees indexed by* V, denoted by $T_\Sigma(V)$, is inductively defined to be the smallest set T such that (i) $V \subseteq T$ and (ii) for every $k \in \mathbb{N}$, $\sigma \in \Sigma_k$, and $t_1, \ldots, t_k \in T$ also $\sigma(t_1, \ldots, t_k) \in T$. Since we generally assume that $\Sigma \cap X = \emptyset$, we write α instead of $\alpha()$ whenever $\alpha \in \Sigma_0$. Moreover, we also write T_Σ to denote $T_\Sigma(\emptyset)$.

Given $t_1, \ldots, t_n \in T_\Sigma(X)$, the expression $t[t_1, \ldots, t_n]$ denotes the result of substituting in t every x_i by t_i for every $i \in [n]$. Let $V \subseteq X$. We say that $t \in T_\Sigma(X)$ is *linear* and *nondeleting* in V, if every $x \in V$ occurs at most once and at least once in t, respectively.

A *semiring* is an algebraic structure $\mathcal{A} = (A, +, \cdot, 0, 1)$ consisting of a commutative monoid $(A, +, 0)$ and a monoid $(A, \cdot, 1)$ such that $\cdot$ distributes over $+$

and 0 is absorbing with respect to $\cdot$. The semiring is called commutative, if $\cdot$ is commutative. As usual we use $\sum_{i \in I} a_i$ for sums of families $(a_i)_{i \in I}$ of $a_i \in A$ where for only finitely many $i \in I$ we have $a_i \neq 0$. Let $\mathcal{A} = (A, +, \cdot, 0_{\mathcal{A}}, 1_{\mathcal{A}})$ and $\mathcal{B} = (B, \oplus, \odot, 0_{\mathcal{B}}, 1_{\mathcal{B}})$ be semirings and $h \colon A \longrightarrow B$. The mapping h is called *homomorphism from $\mathcal{A}$ to $\mathcal{B}$*, if

- $h(0_{\mathcal{A}}) = 0_{\mathcal{B}}$ and $h(1_{\mathcal{A}}) = 1_{\mathcal{B}}$, and
- $h(a + b) = h(a) \oplus h(b)$ and $h(a \cdot b) = h(a) \odot h(b)$ for every $a, b \in A$.

A semiring $\mathcal{A} = (A, +, \cdot, 0, 1)$ is called *idempotent*, if $1 + 1 = 1$. Moreover, we say that a semiring $\mathcal{A} = (A, +, \cdot, 0, 1)$ is *zero-sum free*, if $a + b = 0$ implies that $a = 0 = b$ for every $a, b \in A$. Moreover, $\mathcal{A}$ is *zero-divisor free*, if $a \cdot b = 0$ implies that $0 \in \{a, b\}$ for every $a, b \in A$. A zero-sum and zero-divisor free semiring is also called *positive*. The Boolean semiring $\mathbb{B} = (\{0, 1\}, \vee, \wedge, 0, 1)$ with the usual disjunction $\vee$ and conjunction $\wedge$ is an example of a positive semiring.

Let S be a set and $\mathcal{A} = (A, +, \cdot, 0, 1)$ be a semiring. A *(formal) power series* ψ is a mapping $\psi \colon S \longrightarrow A$. Given $s \in S$, we denote $\psi(s)$ also by (ψ, s) and write the series as $\sum_{s \in S} (\psi, s) \, s$. The *support* of ψ is $\mathrm{supp}(\psi) = \{s \in S \mid (\psi, s) \neq 0\}$. Power series with finite support are called *polynomials*. We denote the set of all power series by $\mathcal{A}\langle\langle S \rangle\rangle$ and the set of polynomials by $\mathcal{A}\langle S \rangle$. The polynomial with empty support is denoted by $\widetilde{0}$. Power series $\psi, \psi' \in \mathcal{A}\langle\langle S \rangle\rangle$ are added componentwise; i. e., $(\psi + \psi', s) = (\psi, s) + (\psi', s)$ for every $s \in S$, and we multiply ψ with a coefficient $a \in A$ componentwise; i. e., $(a \cdot \psi, s) = a \cdot (\psi, s)$ for every $s \in S$.

In this paper, we only consider power series in which the set S is a set of trees. Such power series are also called *tree series*. Let Δ be a ranked alphabet. A tree series $\psi \in \mathcal{A}\langle\langle T_\Delta(X) \rangle\rangle$ is said to be *linear* and *nondeleting* in $V \subseteq X$, if every $t \in \mathrm{supp}(\psi)$ is linear and nondeleting in V, respectively. Let $\psi \in \mathcal{A}\langle T_\Delta(X) \rangle$ and $\psi_1, \dots, \psi_n \in \mathcal{A}\langle T_\Delta(X) \rangle$. The *pure IO tree series substitution* (for short: pure substitution) (of $\psi_1, \dots, \psi_n$ into ψ) [17, 2], denoted by $\psi \xleftarrow{\ \varepsilon\ } (\psi_1, \dots, \psi_n)$, is defined by

$$\psi \xleftarrow{\ \varepsilon\ } (\psi_1, \dots, \psi_n) = \sum_{\substack{t \in T_\Delta(X), \\ t_1, \dots, t_n \in T_\Delta(X)}} (\psi, t) \cdot (\psi_1, t_1) \cdot \dots \cdot (\psi_n, t_n) \, t[t_1, \dots, t_n] \ .$$

Let Q be an alphabet. We write $Q(V)$ for $\{q(v) \mid q \in Q, v \in V\}$. We use the notions of linearity and nondeletion in V accordingly also for $w \in Q(X)^*$. Let $\mathcal{A} = (A, +, \cdot, 0, 1)$ be a semiring and Σ and Δ be ranked alphabets. A *tree representation* μ (over Q, Σ, Δ, and $\mathcal{A}$) [2] is a family $(\mu(\sigma))_{\sigma \in \Sigma}$ of matrices $\mu(\sigma) \in \mathcal{A}\langle\langle T_\Delta(X) \rangle\rangle^{Q \times Q(X_k)^*}$ where $k = \mathrm{rk}_\Sigma(\sigma)$ such that for every $q \in Q$ and $w \in Q(X_k)^*$ it holds that $\mu(\sigma)_{q,w} \in \mathcal{A}\langle\langle T_\Delta(X_n) \rangle\rangle$ with $n = |w|$, and $\mu(\sigma)_{q,w} \neq \widetilde{0}$ for only finitely many $(q, w) \in Q \times Q(X_k)^*$. A tree representation μ is said to be

- *polynomial*, if $\mu(\sigma)_{q,w}$ is polynomial for every $k \in \mathbb{N}$, $\sigma \in \Sigma_k$, $q \in Q$, and $w \in Q(X_k)^*$;
- *linear*, if $\mu(\sigma)_{q,w}$ is linear in $X_{|w|}$ and w is linear in X_k for every $k \in \mathbb{N}$, $\sigma \in \Sigma_k$, $q \in Q$, and $w \in Q(X_k)^*$ such that $\mu(\sigma)_{q,w} \neq \widetilde{0}$;

- *top-down* (respectively, *top-down with regular look-ahead*), if $\mu(\sigma)_{q,w}$ is linear and nondeleting (respectively, linear) in $X_{|w|}$ for every $k \in \mathbb{N}$, $\sigma \in \Sigma_k$, $q \in Q$, and $w \in Q(X_k)^*$; and
- *bottom-up*, if for every $k \in \mathbb{N}$, $\sigma \in \Sigma_k$, $q \in Q$, and $w \in Q(X_k)^*$ such that $\mu(\sigma)_{q,w} \neq \tilde{0}$ we have $w = q_1(x_1) \cdots q_k(x_k)$ for some $q_1, \ldots, q_k \in Q$.

A *tree series transducer* [2,6] (with designated states), in the sequel abbreviated to tst, is a sixtuple $M = (Q, \Sigma, \Delta, \mathcal{A}, F, \mu)$ consisting of

- an alphabet Q of *states*,
- ranked alphabets Σ and Δ, also called *input* and *output ranked alphabet*, respectively,
- a semiring $\mathcal{A} = (A, +, \cdot, 0, 1)$,
- a subset $F \subseteq Q$ of *designated states*, and
- a tree representation μ over Q, Σ, Δ, and $\mathcal{A}$.

Tst inherit the properties from their tree representation; *e. g.*, a tst with a polynomial bottom-up tree representation is called a polynomial bottom-up tst. Additionally, we abbreviate bottom-up tst to bu-tst and top-down tst to td-tst.

We introduce the semantics only for polynomial tst because we defined pure substitution only for polynomial tree series (in order to avoid a well-definedness issue related to infinite sums). Let $M = (Q, \Sigma, \Delta, \mathcal{A}, F, \mu)$ be a polynomial tst. Then M induces a mapping $\|M\|: T_\Sigma \longrightarrow \mathcal{A}\langle T_\Delta \rangle$ as follows. For every $k \in \mathbb{N}$, $\sigma \in \Sigma_k$, and $t_1, \ldots, t_k \in T_\Sigma$ we define the mapping $h_\mu: T_\Sigma \longrightarrow \mathcal{A}\langle T_\Delta \rangle^Q$ componentwise for every $q \in Q$ by

$$h_\mu(\sigma(t_1, \ldots, t_k))_q = \sum_{\substack{w \in Q(X_k)^*, \\ w = q_1(x_{i_1}) \cdots q_n(x_{i_n})}} \mu_k(\sigma)_{q,w} \xleftarrow{\ \varepsilon\ } (h_\mu(t_{i_1})_{q_1}, \ldots, h_\mu(t_{i_n})_{q_n}) \ .$$

For every $t \in T_\Sigma$ the *tree-to-tree-series (for short: ε-t-ts) transformation computed by M* is $\|M\|(t) = \sum_{q \in F} h_\mu(t)_q$.

By p–$\mathrm{TOP}_\varepsilon(\mathcal{A})$ and p–$\mathrm{BOT}_\varepsilon(\mathcal{A})$ we denote the class of ε-t-ts transformations computable by polynomial td-tst and bu-tst over the semiring $\mathcal{A}$, respectively. Likewise we use the prefix l for the linearity property and the stems $\mathrm{TOP}_\varepsilon^{\mathrm{R}}$ and GST_ε for td-tst with regular look-ahead and unrestricted tst, respectively.

We compose ε-t-ts transformations as follows. Let $\tau_1: T_\Sigma \longrightarrow \mathcal{A}\langle T_\Delta \rangle$ and $\tau_2: T_\Delta \longrightarrow \mathcal{A}\langle T_\Gamma \rangle$ then $(\tau_1 \circ \tau_2)(t) = \sum_{u \in T_\Delta}(\tau_1(t), u) \cdot \tau_2(u)$ for every $t \in T_\Sigma$. This composition is extended to classes of ε-t-ts transformations in the usual manner. By p–$\mathrm{TOP}_\varepsilon^n(\mathcal{A})$ and p–$\mathrm{BOT}_\varepsilon^n(\mathcal{A})$ with $n \in \mathbb{N}_+$ we denote the n-fold composition p–$\mathrm{TOP}_\varepsilon(\mathcal{A}) \circ \cdots \circ$ p–$\mathrm{TOP}_\varepsilon(\mathcal{A})$ and p–$\mathrm{BOT}_\varepsilon(\mathcal{A}) \circ \cdots \circ$ p–$\mathrm{BOT}_\varepsilon(\mathcal{A})$, respectively.

3   Incomparability Results

We show the incomparability of p–$\mathrm{TOP}_\varepsilon^n(\mathcal{A})$ and p–$\mathrm{BOT}_\varepsilon^n(\mathcal{A})$ for every $n \in \mathbb{N}_+$ and positive semiring $\mathcal{A}$. Together with the results of [14] this yields the HASSE diagram (see Figure 1) that displays the top-down, bottom-up, and alternating hierarchy of tree series transformations. We arrive at the same HASSE diagram as [14], but we can prove it for a distinctively larger class of semirings; namely positive commutative semirings instead of positive, idempotent, and commutative semirings as in [14].

First we show the main property that we exploit in the sequel. Roughly speaking, given a positive semiring $\mathcal{A}$ we present a specific homomorphism from $\mathcal{A}$ to the Boolean semiring $\mathbb{B}$. We later use this homomorphism to lift the incomparability of the top-down and bottom-up tree transformation classes to the level of ε-t-ts transformation classes.

Lemma 1. *Let* $\mathcal{A} = (A, +, \cdot, 0_{\mathcal{A}}, 1_{\mathcal{A}})$ *be a positive semiring. Let* $\chi \colon A \longrightarrow \{0, 1\}$ *be such that* $\chi(0_{\mathcal{A}}) = 0$ *and* $\chi(a) = 1$ *for every* $a \in A \setminus \{0_{\mathcal{A}}\}$. *Then* χ *is a homomorphism from* $\mathcal{A}$ *to* $\mathbb{B}$.

Let $\mathcal{A} = (A, +, \cdot, 0_{\mathcal{A}}, 1_{\mathcal{A}})$ and $\mathcal{B} = (B, \oplus, \odot, 0_{\mathcal{B}}, 1_{\mathcal{B}})$ be two semirings and $\tau \colon T_{\Sigma} \longrightarrow \mathcal{A}\langle\!\langle T_{\Delta} \rangle\!\rangle$ and $h \colon A \longrightarrow B$. The *image of* τ *under* h, denoted by $h(\tau)$, is defined by $(h(\tau)(t), u) = h((\tau(t), u))$ for every $t \in T_{\Sigma}$ and $u \in T_{\Delta}$. Clearly, $h(\tau) \colon T_{\Sigma} \longrightarrow \mathcal{B}\langle\!\langle T_{\Delta} \rangle\!\rangle$. If h is a homomorphism, then we also call $h(\tau)$ the *homomorphic image* of τ. This notion of (homomorphic) image is lifted to classes of ε-t-ts transformations in the usual manner.

Next we show that, given an ε-t-ts transformation τ computed by a polynomial td-tst or bu-tst M over the semiring $\mathcal{A}$ and a homomorphism h from $\mathcal{A}$ to $\mathcal{B}$, there exists a polynomial td-tst or bu-tst M' over the semiring $\mathcal{B}$ such that M' computes the homomorphic image of τ; i.e., h is applied to all coefficients in the range of the ε-t-ts transformation τ. This is also the main idea of the construction; we simply apply the homomorphism to all coefficients in the tree representation of M to obtain the tree representation of M'.

Moreover, we show that computable ε-t-ts transformations are also closed under inverse homomorphisms. For this we need the following definition. Let $h \colon A \longrightarrow B$ and $\tau' \colon T_{\Sigma} \longrightarrow \mathcal{B}\langle\!\langle T_{\Delta} \rangle\!\rangle$. By $h^{-1}(\tau')$ we denote the set

$$\{ \tau \in \mathcal{A}\langle\!\langle T_{\Delta} \rangle\!\rangle^{T_{\Sigma}} \mid h(\tau) = \tau' \} \ .$$

This is again lifted to classes as usual.

Lemma 2. *Let* $\mathcal{A}$ *and* $\mathcal{B}$ *be semirings and* h *be a homomorphism from* $\mathcal{A}$ *to* $\mathcal{B}$.

$$h(\text{p--TOP}_{\varepsilon}(\mathcal{A})) \subseteq \text{p--TOP}_{\varepsilon}(\mathcal{B}) \quad \text{and} \quad h(\text{p--BOT}_{\varepsilon}(\mathcal{A})) \subseteq \text{p--BOT}_{\varepsilon}(\mathcal{B})$$

If h *is surjective, then also*

$$h^{-1}(\text{p--TOP}_{\varepsilon}(\mathcal{B})) \subseteq \text{p--TOP}_{\varepsilon}(\mathcal{A}) \quad \text{and} \quad h^{-1}(\text{p--BOT}_{\varepsilon}(\mathcal{B})) \subseteq \text{p--BOT}_{\varepsilon}(\mathcal{A})$$

Proof. Let $\mathcal{C} = (C, +, \cdot, 0_{\mathcal{C}}, 1_{\mathcal{C}})$ and $\mathcal{D} = (D, \oplus, \odot, 0_{\mathcal{D}}, 1_{\mathcal{D}})$. Let $f \colon C \longrightarrow D$ and $M = (Q, \Sigma, \Delta, \mathcal{C}, F, \mu)$ be a tst. We construct the tst $f(M) = (Q, \Sigma, \Delta, \mathcal{D}, F, \mu')$ as follows. For every $k \in \mathbb{N}$, $\sigma \in \Sigma_k$, $q \in Q$, and $w \in Q(X_k)^*$

$$\mu'(\sigma)_{q,w} = \bigoplus_{u \in \text{supp}(\mu(\sigma)_{q,w})} f((\mu(\sigma)_{q,w}, u)) \, u \ .$$

Clearly, $f(M)$ is top-down and bottom-up whenever M is top-down and bottom-up, respectively.

Let us prove the former statement. Let $\tau \in \text{p–TOP}_\varepsilon(\mathcal{A})$ or $\tau \in \text{p–BOT}_\varepsilon(\mathcal{A})$. There exists a polynomial td-tst or bu-tst M such that $\|M\| = \tau$. We claim that $\|h(M)\| = h(\|M\|)$. The proof of this statement can be found below.

For the second statement, let $\tau \in \text{p–TOP}_\varepsilon(\mathcal{B})$ or $\tau \in \text{p–BOT}_\varepsilon(\mathcal{B})$. There exists a polynomial td-tst or bu-tst M such that $\|M\| = \tau$. Moreover, let $f\colon B \longrightarrow A$ be such that $h(f(b)) = b$ for every $b \in B$. Such an f exists, because h is surjective. The claim $\|f(M)\| \in h^{-1}(\|M\|)$ follows from $h(\|f(M)\|) = \|M\|$, whose proof can also be found below.

Now we prove the mentioned result. Let h be a homomorphism from $\mathcal{A}$ to $\mathcal{B}$ with $\mathcal{A} = (A, +, \cdot, 0_\mathcal{A}, 1_\mathcal{A})$ and $\mathcal{B} = (B, \oplus, \odot, 0_\mathcal{B}, 1_\mathcal{B})$. Let $M = (Q, \Sigma, \Delta, \mathcal{A}, F, \mu)$ be a tst. Then $\|h(M)\| = h(\|M\|)$. Let $h(M) = (Q, \Sigma, \Delta, \mathcal{B}, F, \mu')$. We first prove the auxiliary statement that $(h_{\mu'}(t)_q, u) = h((h_\mu(t)_q, u))$ for every $q \in Q$, $t \in T_\Sigma$, and $u \in T_\Delta$. This is proved inductively, so let $t = \sigma(t_1, \dots, t_k)$ for some $k \in \mathbb{N}$, $\sigma \in \Sigma_k$, and $t_1, \dots, t_k \in T_\Sigma$.

$$(h_{\mu'}(\sigma(t_1, \dots, t_k))_q, u)$$

$=$ (by definition of $h_{\mu'}$)

$$\left(\bigoplus_{\substack{w \in Q(X_k)^*, \\ w = q_1(x_{i_1}) \cdots q_n(x_{i_n})}} \mu'(\sigma)_{q,w} \underset{\varepsilon}{\longleftarrow} (h_{\mu'}(t_{i_1})_{q_1}, \dots, h_{\mu'}(t_{i_n})_{q_n}), u \right)$$

$=$ (by definition of $\underset{\varepsilon}{\longleftarrow}$)

$$\left(\bigoplus_{\substack{w \in Q(X_k)^*, \\ w = q_1(x_{i_1}) \cdots q_n(x_{i_n})}} \bigoplus_{\substack{u' \in T_\Delta(X_n), \\ u_1, \dots, u_n \in T_\Delta}} (\mu'(\sigma)_{q,w}, u') \odot \right.$$

$$\left. \odot (h_{\mu'}(t_{i_1})_{q_1}, u_1) \odot \cdots \odot (h_{\mu'}(t_{i_n})_{q_n}, u_n) \, u'[u_1, \dots, u_n], u \right)$$

$=$ (by definition of μ' and induction hypothesis)

$$\left(\bigoplus_{\substack{w \in Q(X_k)^*, \\ w = q_1(x_{i_1}) \cdots q_n(x_{i_n})}} \bigoplus_{\substack{u' \in T_\Delta(X_n), \\ u_1, \dots, u_n \in T_\Delta}} h((\mu(\sigma)_{q,w}, u')) \odot \right.$$

$$\left. \odot h((h_\mu(t_{i_1})_{q_1}, u_1)) \odot \cdots \odot h((h_\mu(t_{i_n})_{q_n}, u_n)) \, u'[u_1, \dots, u_n], u \right)$$

$=$ (by homomorphism property)

$$\bigoplus_{\substack{w \in Q(X_k)^*, \\ w = q_1(x_{i_1}) \cdots q_n(x_{i_n})}} \left(\bigoplus_{\substack{u' \in T_\Delta(X_n), \\ u_1, \dots, u_n \in T_\Delta}} h\Big((\mu(\sigma)_{q,w}, u') \cdot \right.$$

$$\left. \cdot (h_\mu(t_{i_1})_{q_1}, u_1) \cdot \ldots \cdot (h_\mu(t_{i_n})_{q_n}, u_n)\Big) \, u'[u_1, \dots, u_n], u \right)$$

$=$ (by homomorphism property and definition of $\underset{\varepsilon}{\longleftarrow}$)

$$\bigoplus_{\substack{w \in Q(X_k)^*, \\ w = q_1(x_{i_1}) \cdots q_n(x_{i_n})}} h\Big(\mu(\sigma)_{q,w} \underset{\varepsilon}{\longleftarrow} (h_\mu(t_{i_1})_{q_1}, \dots, h_\mu(t_{i_n})_{q_n}), u\Big)$$

$$= \quad \text{(by homomorphism property)}$$

$$h\Big(\sum_{\substack{w \in Q(X_k)^*, \\ w = q_1(x_{i_1}) \cdots q_n(x_{i_n})}} \mu(\sigma)_{q,w} \xleftarrow{\varepsilon} (h_\mu(t_{i_1})_{q_1}, \ldots, h_\mu(t_{i_n})_{q_n}), u \Big)$$

$$= \quad \text{(by definition of } h_\mu)$$

$$h((h_\mu(\sigma(t_1, \ldots, t_k))_q, u))$$

With this statement the proof is easy. We observe that for every $t \in T_\Sigma$ and $u \in T_\Delta$

$$(\|h(M)\|(t), u) = \Big(\bigoplus_{q \in F} h_{\mu'}(t)_q, u \Big) = \bigoplus_{q \in F} (h_{\mu'}(t)_q, u)$$

$$= \quad \text{(by the auxiliary statement)}$$

$$\bigoplus_{q \in F} h((h_\mu(t)_q, u)) = h\Big(\sum_{q \in F} (h_\mu(t)_q, u) \Big) = h\Big(\Big(\sum_{q \in F} h_\mu(t)_q, u \Big) \Big)$$

$$= h((\|M\|(t), u)) \ .$$

This lemma admits an important corollary, which will form the basis of our new lifting result. Roughly, the corollary states that every ε-t-ts transformation computed by a polynomial td-tst or bu-tst over $\mathbb{B}$ can also be computed as the homomorphic image (under χ) of the ε-t-ts transformation computed by a polynomial td-tst or bu-tst over the positive semiring $\mathcal{A}$. The statement also holds vice versa.

Corollary 1. *Let $\mathcal{A}$ be a positive semiring.*

$$\chi(\text{p--TOP}_\varepsilon(\mathcal{A})) = \text{p--TOP}_\varepsilon(\mathbb{B}) \quad \text{and} \quad \chi(\text{p--BOT}_\varepsilon(\mathcal{A})) = \text{p--BOT}_\varepsilon(\mathbb{B})$$

Proof. We have seen in Lemma 1 that χ is a homomorphism from $\mathcal{A}$ to $\mathbb{B}$. Consequently, the statement holds by Lemma 2 because χ is surjective.

Next we show that homomorphisms are compatible with the composition introduced for ε-t-ts transformations.

Lemma 3. *Let h be a homomorphism from the semiring $\mathcal{A}$ to the semiring $\mathcal{B}$. Moreover, let $\tau_1 : T_\Sigma \longrightarrow \mathcal{A}\langle T_\Delta \rangle$ and $\tau_2 : T_\Delta \longrightarrow \mathcal{A}\langle T_\Gamma \rangle$.*

$$h(\tau_1 \circ \tau_2) = h(\tau_1) \circ h(\tau_2)$$

Proof. Let $t \in T_\Sigma$ and $u' \in T_\Gamma$ be an input and output tree, respectively. Further, let $\mathcal{A} = (A, +, \cdot, 0_\mathcal{A}, 1_\mathcal{A})$ and $\mathcal{B} = (B, \oplus, \odot, 0_\mathcal{B}, 1_\mathcal{B})$.

$$h\big(((\tau_1 \circ \tau_2)(t), u')\big) = h\Big(\Big(\sum_{u \in T_\Delta} (\tau_1(t), u) \cdot \tau_2(u), u' \Big) \Big)$$

$$= \bigoplus_{u \in T_\Delta} h\big(((\tau_1(t), u) \cdot \tau_2(u), u')\big) = \bigoplus_{u \in T_\Delta} h((\tau_1(t), u)) \odot h((\tau_2(u), u'))$$

$$= \bigoplus_{u \in T_\Delta} (h(\tau_1)(t), u) \odot (h(\tau_2)(u), u') = ((h(\tau_1) \circ h(\tau_2))(t), u')$$

Now we ready to state our main theorem, which states the incomparability of $\text{p-TOP}_\varepsilon^n(\mathcal{A})$ and $\text{p-BOT}_\varepsilon^n(\mathcal{A})$ in all positive semirings.

Theorem 1. *Let $\mathcal{A}$ be a positive semiring and $n \in \mathbb{N}_+$.*

$$\text{p-TOP}_\varepsilon^n(\mathcal{A}) \nsubseteq \text{p-BOT}_\varepsilon^n(\mathcal{A}) \qquad \text{p-BOT}_\varepsilon^n(\mathcal{A}) \nsubseteq \text{p-TOP}_\varepsilon^n(\mathcal{A})$$

Proof. We prove the statement by contradiction. To this end, suppose that $\text{p-TOP}_\varepsilon^n(\mathcal{A}) \subseteq \text{p-BOT}_\varepsilon^n(\mathcal{A})$. Then

$$
\begin{aligned}
&\chi(\text{p-TOP}_\varepsilon^n(\mathcal{A})) \\
&= \chi(\text{p-TOP}_\varepsilon(\mathcal{A})) \circ \cdots \circ \chi(\text{p-TOP}_\varepsilon(\mathcal{A})) && \text{by Lemma 3} \\
&= \text{p-TOP}_\varepsilon(\mathbb{B}) \circ \cdots \circ \text{p-TOP}_\varepsilon(\mathbb{B}) && \text{by Corollary 1} \\
&= \text{p-TOP}_\varepsilon^n(\mathbb{B}) && \text{by definition}
\end{aligned}
$$

Analogously we obtain $\chi(\text{p-BOT}_\varepsilon^n(\mathcal{A})) = \text{p-BOT}_\varepsilon^n(\mathbb{B})$. It follows that we also have $\text{p-TOP}_\varepsilon^n(\mathbb{B}) \subseteq \text{p-BOT}_\varepsilon^n(\mathbb{B})$. This, however, contradicts the famous tree transducer hierarchy [18] due to [2, Corollaries 4.7 and 4.14]. The second statement is proved analogously.

4   Hierarchy Results

In this section we state the hierarchy result that can be obtained with the new incomparability result. First we recall the inclusion results of [14].

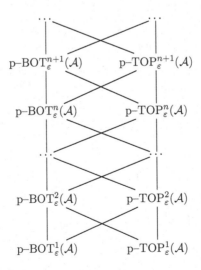

Fig. 1. HASSE diagram of the hierarchies

Proposition 1 (Theorems 5.1 and 5.7 of [14]). *Let $\mathcal{A}$ be commutative and* $n \in \mathbb{N}_+$.

$$\text{p--BOT}_\varepsilon^n(\mathcal{A}) \subseteq \text{p--TOP}_\varepsilon^{n+1}(\mathcal{A}) \qquad \text{p--TOP}_\varepsilon^n(\mathcal{A}) \subseteq \text{p--BOT}_\varepsilon^{n+1}(\mathcal{A})$$

With these inclusions and the incomparability results of Theorem 1 we obtain the following hierarchy result for positive and commutative semirings. Important semirings like

- the semiring of nonnegative integers $\mathbb{N} = (\mathbb{N}, +, \cdot, 0, 1)$,
- the least common multiple semiring $\text{Lcm} = (\mathbb{N}, \text{lcm}, \cdot, 0, 1)$, and
- the matrix semiring $\text{Mat}_n(\mathbb{N}_+) = (\mathbb{N}_+^{n \times n} \cup \{\underline{0}, \underline{1}\}, +, \cdot, \underline{0}, \underline{1})$ over $\mathbb{N}_+$ (where $\underline{0}$ is the $n \times n$ zero matrix and $\underline{1}$ is the $n \times n$ unit matrix)

are all positive, but not idempotent. However, the matrix semiring is not commutative.

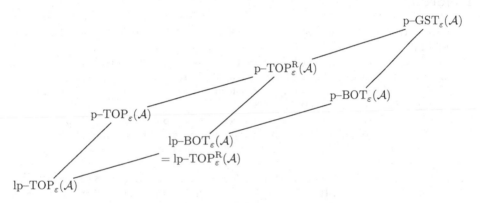

Fig. 2. Hasse diagram of general tst

Theorem 2. *Let $\mathcal{A}$ be a positive and commutative semiring. Figure 1 is the* Hasse *diagram for the depicted classes of transformations (ordered by inclusion).*

Proof. The inclusions are trivial or follow from Proposition 1. Incomparability is shown in Theorem 1.

Similarly, we can use the approach also for other incomparability results. For example, in [19] a diagram of inclusions is presented (for commutative semirings, *cf.* Section 6 of [20]), however the properness of the inclusions remained open. Using our approach we can now prove this diagram to be a Hasse diagram.

Theorem 3. *Let $\mathcal{A}$ be a positive and commutative semiring. Figure 2 is the* Hasse *diagram for the depicted classes of transformations (ordered by inclusion).*

Proof. Note that the inclusions are proved in [19]. It remains to prove strictness and incomparability.

First we note that the construction of Lemma 2 preserves all introduced properties (thus also linearity and top-down with regular look-ahead). Thus we obtain the following statements.

$$\chi(\text{p–TOP}_\varepsilon^R(\mathcal{A})) = \text{p–TOP}_\varepsilon^R(\mathbb{B}) \qquad \chi(\text{p–GST}_\varepsilon(\mathcal{A})) = \text{p–GST}_\varepsilon(\mathbb{B})$$
$$\chi(\text{lp–TOP}_\varepsilon^R(\mathcal{A})) = \text{lp–TOP}_\varepsilon^R(\mathbb{B}) \qquad \chi(\text{lp–GST}_\varepsilon(\mathcal{A})) = \text{lp–GST}_\varepsilon(\mathbb{B})$$
$$\chi(\text{lp–TOP}_\varepsilon(\mathcal{A})) = \text{lp–TOP}_\varepsilon(\mathbb{B}) \qquad \chi(\text{lp–BOT}_\varepsilon(\mathcal{A})) = \text{lp–BOT}_\varepsilon(\mathbb{B})$$

In Section 5 of [20] the diagram is proved to be HASSE diagram for the Boolean semiring and we lift the incomparability results of this diagram using the approach used in the proof of Theorem 2. This proves the correctness of the diagram presented in Figure 2.

References

1. Kuich, W.: Tree transducers and formal tree series. Acta Cybernet. **14** (1999) 135–149
2. Engelfriet, J., Fülöp, Z., Vogler, H.: Bottom-up and top-down tree series transformations. J. Autom. Lang. Combin. **7** (2002) 11–70
3. Fülöp, Z., Vogler, H.: Tree series transformations that respect copying. Theory Comput. Systems **36** (2003) 247–293
4. Kuich, W.: Formal power series over trees. In Bozapalidis, S., ed.: Proc. 3rd Int. Conf. Develop. in Lang. Theory, Aristotle University of Thessaloniki (1998) 61–101
5. Bozapalidis, S.: Equational elements in additive algebras. Theory Comput. Systems **32** (1999) 1–33
6. Ésik, Z., Kuich, W.: Formal tree series. J. Autom. Lang. Combin. **8** (2003) 219–285
7. Borchardt, B., Vogler, H.: Determinization of finite state weighted tree automata. J. Autom. Lang. Combin. **8** (2003) 417–463
8. Droste, M., Pech, C., Vogler, H.: A Kleene theorem for weighted tree automata. Theory Comput. Systems **38** (2005) 1–38
9. Berstel, J., Reutenauer, C.: Recognizable formal power series on trees. Theoret. Comput. Sci. **18** (1982) 115–148
10. Borchardt, B.: The Theory of Recognizable Tree Series. PhD thesis, Technische Universität Dresden (2005)
11. Graehl, J., Knight, K.: Training tree transducers. In: Human Lang. Tech. Conf. of the North American Chapter of the Assoc. for Computational Linguistics. (2004) 105–112
12. Engelfriet, J.: The copying power of one-state tree transducers. J. Comput. System Sci. **25** (1982) 418–435
13. Raoult, J.C.: A survey of tree transductions. In Nivat, M., Podelski, A., eds.: Tree Automata and Languages. Elsevier Science (1992)
14. Fülöp, Z., Gazdag, Z., Vogler, H.: Hierarchies of tree series transformations. Theoret. Comput. Sci. **314** (2004) 387–429
15. Hebisch, U., Weinert, H.J.: Semirings—Algebraic Theory and Applications in Computer Science. World Scientific (1998)
16. Golan, J.S.: Semirings and their Applications. Kluwer Academic, Dordrecht (1999)

17. Bozapalidis, S.: Context-free series on trees. Inform. and Comput. **169** (2001) 186–229
18. Engelfriet, J.: Three hierarchies of transducers. Math. Systems Theory **15** (1982) 95–125
19. Maletti, A.: The power of tree series transducers of type I and II. In de Felice, C., Restivo, A., eds.: Proc. 9th Int. Conf. Develop. in Lang. Theory. Volume 3572 of LNCS., Springer (2005) 338–349
20. Engelfriet, J.: Bottom-up and top-down tree transformations—a comparison. Math. Systems Theory **9** (1975) 198–231

Bag Context Tree Grammars*,**

Frank Drewes[1], Christine du Toit[2], Sigrid Ewert[3], Brink van der Merwe[2],
and Andries P.J. van der Walt[2]

[1] Department of Computing Science, Umeå University, S–901 87 Umeå, Sweden
`drewes@cs.umu.se`
[2] Department of Computer Science, University of Stellenbosch
7602 Stellenbosch, South Africa
`{cdutoit, abvdm, apjw}@cs.sun.ac.za`
[3] School of Computer Science, University of the Witwatersrand
2050 Wits, South Africa
`sigrid@cs.wits.ac.za`

Abstract. We introduce bag context, a device for regulated rewriting in tree grammars. Rather than being part of the developing tree, bag context (bc) evolves on its own during a derivation. We show that the class of bc tree languages is the closure of the class of random context tree languages under linear top-down tree transductions. Further, an interchange theorem for subtrees of dense trees in bc tree languages is established. This result implies that the class of bc tree languages is incomparable with the class of branching synchronization tree languages.

1 Introduction

In [DTE⁺05] we started an investigation into *random context* in tree grammars and tree transducers. In a random context (rc) tree grammar the derivations are regulated by context conditions that require the presence or absence of certain nonterminals in the current sentential form. If one is interested in the properties of the resulting language class one may, e.g., want to prove that it is closed under a certain operation. A standard approach would be to apply the operation to the right-hand sides of all rules. However, many natural operations on trees, such as extracting a path, delete parts of the trees they are applied to. In such a situation, the mentioned approach does not work, because it leads to a kind of grammar where the rewriting is regulated by context which has been deleted. Thus, the context is "out there" somewhere, rather than being part of the sentential form. These considerations lead naturally to what we call *bag context* in tree and in string grammars. The purpose of this paper is the introduction and investigation of the resulting tree grammars and languages.

---

* The work reported here has been part of the project *Random Context in the Generation and Transformation of Trees* funded by the National Research Foundation South Africa (NRF) and the Swedish International Development Cooperation Agency (Sida).
** This paper is a shortened version of [ETM⁺06].

O.H. Ibarra and Z. Dang (Eds.): DLT 2006, LNCS 4036, pp. 226–237, 2006.

A *bag context tree grammar* is a regular tree grammar in which the application of the rules is regulated by a vector of integers, the *bag*, which evolves with the tree under construction. The application of a particular rule is possible at a given stage if the bag at that stage is within the range defined, as part of the rule, by a *lower limit* and an *upper limit*. If a rule is applied, it affects not only the tree, but also the bag by means of a *bag adjustment* which is also part of the rule. (Bag context string grammars can, of course, be defined analogously by extending context-free grammars.)

Bag context tree grammars properly generalize random context tree grammars. One may say that their regulation mechanism is an extension of the one used in multiplicative valence grammars [DP89]. The latter correspond to the case where rules are always applicable (regardless of the bag contents), but the bag value must be the zero vector at the end of the derivation.

The layout of this paper is as follows. In the next section, we define bag context tree grammars and discuss an example as well as equivalent definitions. In Section 3, we compare random context and bag context tree grammars. In the last section, we show that bag context tree languages obey structural limitations. In particular, we prove an interchange theorem stating that sufficiently dense trees contain subtrees that can be interchanged without leaving the language.

2   Bag Context Tree Grammars

We denote the set of natural numbers (including 0) by $\mathbb{N}$ and the set of all integers by $\mathbb{Z}$. The sets $\mathbb{N} \cup \{\infty\}$ and $\mathbb{Z} \cup \{\pm\infty\}$ are denoted by $\mathbb{N}_\infty$ and $\mathbb{Z}_\infty$, resp. If $I = \{1, \ldots, k\}$ then elements of $\mathbb{Z}_\infty^I$ will be written as k-tuples. On $\mathbb{Z}_\infty^I$, arithmetic operations such as addition, subtraction, and scalar multiplication are defined componentwise. An element q of $\mathbb{Z}_\infty$ which occurs in the place of a vector denotes $(q, \ldots, q)$. For $\beta \in \mathbb{Z}^I$ and $A \in I$, $\beta(A)$ may also be denoted β_A.

A *signature* is a set Σ of ranked symbols $f^{(k)}$ (where $k \geq 0$ is the rank). The set T_Σ of all *trees over* Σ contains every a such that $a^{(0)} \in \Sigma$, and every $f[t_1, \ldots, t_k]$ such that $f^{(k)} \in \Sigma$ ($k \geq 1$) and $t_1, \ldots, t_k \in T_\Sigma$. For $t \in T_\Sigma$, $yield(t)$ denotes the string of leaves of t, read from left to right, and $height(t)$ and $nodes(t)$ denote its height and set of nodes, respectively. For a node v, $t(v)$ and t/v denote the symbol at node v and the subtree rooted at v, respectively. (The set of nodes of t equals $\{\epsilon\} \cup \{iv \mid 1 \leq i \leq k, \ v \in nodes(t_i)\}$, where k is the rank of the root symbol and $t_1, \ldots, t_k$ are the direct subtrees of t.)

Let $x_1^{(0)}, x_2^{(0)}, \ldots \notin \Sigma$ be pairwise distinct symbols, and let $t \in T_{\Sigma \cup \{x_1, \ldots, x_k\}}$ contain each of $x_1, \ldots, x_k$ exactly once. Given trees $t_1, \ldots, t_k$, we denote by $t[\![t_1, \ldots, t_n]\!]$ the tree obtained from t by replacing each x_i with t_i, for $i = 1, \ldots, k$. Thus, each t_i refers to a specific occurrence of this subtree in $t[\![t_1, \ldots, t_n]\!]$.

Definition 1. *A bag context (bc for short) tree grammar is a sextuple $G = (N, \Sigma, R, S, I, \beta_0)$ consisting of*

- *finite disjoint signatures N and Σ of nonterminals of rank 0 and terminals;*
- *a finite set R of rules of derivation;*

- *an initial nonterminal $S \in N$;*
- *a finite bag index set, I; and*
- *a vector $\beta_0 \in \mathbb{Z}^I$, the initial bag.*

A rule in R has the generic form $A \to t\,(\lambda, \mu; \alpha)$, where $A \in N$, $t \in T_{\Sigma \cup N}$, $\lambda, \mu \in \mathbb{Z}^I_\infty$ are the lower and upper limits respectively, and $\alpha \in \mathbb{Z}^I$ is the bag adjustment.

There is a derivation step from $(s, \beta) = (u[\![A]\!], \beta)$ to $(s', \beta') = (u[\![t]\!], \beta')$, written $(s, \beta) \Rightarrow_G (s', \beta')$ or just $(s, \beta) \Rightarrow (s', \beta')$, if there is a rule $A \to t\,(\lambda, \mu; \alpha)$ in R, with $\lambda \le \beta \le \mu$ and $\beta' = \beta + \alpha$. The relation $\Rightarrow^*$ is defined as usual, and the language generated by G is $L(G) = \{t \in T_\Sigma \mid (S, \beta_0) \Rightarrow^* (t, \beta), \beta \in \mathbb{Z}^I\}$. A rule of the form $A \to t\,(-\infty, \infty; 0)$ will usually be written simply as $A \to t$.

Let us briefly discuss an example. It exploits the bag context for generating a tree language which does not seem to belong to the class $\text{TBY}^+(\text{REGT})$, i.e., the closure of the regular tree languages under macro tree transductions (see [DE98] for a study of this class as well as for further references).

Let $\Sigma = \{f^{(2)}, a^{(0)}\}$. The aim is to generate the language L consisting of all trees t over Σ having the following property. Suppose the height of t is n and let, for $0 \le i \le n$, $\#_i(t)$ denote the number of nodes $v \in nodes(t)$ such that $|v| = i$ and $t(v) = a$. Intuitively, $\#_i(t)$ is the number of a's at level i in t. Now, L contains t if and only if, for all $i \in \{1, \ldots, n-1\}$, $\#_{i-1}(t) \le \#_i(t)$.

We generate an element of L levelwise, starting with the root which is the zeroth level, then the first level, and so on. Two bag positions are used to keep track of how many a's we still have to generate at the current level and how many have already been generated. The next two bag positions count nonterminals (to make sure that an entire level is generated), and the fifth holds some state information used to distinguish two stages in the generation of a level. Here is the grammar: $G = (\{A, B\}, \Sigma, R, A, \{1, \ldots, 5\}, (0, 0, 1, 0, 0))$, where

$$
\begin{aligned}
R = \{A &\to a & &((1, 0, 1, 0, 0), (\infty, \infty, \infty, \infty, 0); (-1, 1, -1, 0, 0)) \\
A &\to a & &((0, 0, 1, 0, 0), (0, \infty, \infty, \infty, 0); (0, 1, -1, 0, 0)) \\
A &\to f[B, B] & &((0, 0, 1, 0, 0), (\infty, \infty, \infty, \infty, 0); (0, 0, -1, 2, 0)) \\[4pt]
B &\to B & &((0, 0, 0, 0, 0), (0, \infty, 0, \infty, 0); (0, 0, 0, 0, 1)) \\[4pt]
B &\to A & &((0, 0, 0, 1, 1), (0, \infty, \infty, \infty, 1); (0, 0, 1, -1, 0)) \\
A &\to A & &((0, 1, 0, 0, 1), (\infty, \infty, \infty, \infty, 1); (1, -1, 0, 0, 0)) \\
A &\to A & &((0, 0, 0, 0, 1), (\infty, 0, \infty, 0, 1); (0, 0, 0, 0, -1))\}.
\end{aligned}
$$

At the first stage, which is implemented by the first three rules, enough A's must be replaced by a to decrease the first bag component to zero. The remaining A's must be replaced by $f[B, B]$ to decrease the third bag component to zero as well. Only then can the fourth rule be applied, thus turning to the next stage. Now, each B is turned into an A again, and the value of the second bag component is moved to the first. When this has been done, the last rule switches back to the first stage and the next level is generated. The generation of the third level may, for instance, take the following form:

$$A \qquad\qquad\qquad\qquad\qquad\qquad\qquad (0,0,1,0,0)$$
$$\Rightarrow^* f[f[f[A,A],a], f[a, f[A,A]]] \qquad (2,0,4,0,0)$$
$$\Rightarrow^4 f[f[f[a,a],a], f[a, f[f[B,B],a]]]\ (0,3,0,2,0)$$
$$\Rightarrow\ \ f[f[f[a,a],a], f[a, f[f[B,B],a]]]\ (0,3,0,2,1)$$
$$\Rightarrow^5 f[f[f[a,a],a], f[a, f[f[A,A],a]]]\ (3,0,2,0,1)$$
$$\Rightarrow\ \ f[f[f[a,a],a], f[a, f[f[A,A],a]]]\ (3,0,2,0,0)$$
$$\Rightarrow\ \ \cdots\ .$$

We now discuss some equivalent formulations of bc tree grammars. For this, we denote by BCTG the class of bag context tree grammars. The subset of BCTG consisting of bc tree grammars in which every rule $A \to t$ $(\lambda,\mu;\alpha)$ satisfies $\lambda,\mu,\lambda+\alpha \in \mathbb{N}_\infty^I$, and in which the initial bag is required to be an element of $\mathbb{N}^I$, is denoted by BCTG_+. (Clearly, this implies in particular that after any number of derivation steps the bag will be an element of $\mathbb{N}^I$.)

Given a bc tree grammar, we denote by $L^0(G)$ the set of all trees that can be generated by a derivation in which the final bag value is the zero vector.[1] Hence, $L^0(G) \subseteq L(G)$. Looking at all combinations, we get the following language classes:

$$\mathrm{BCTL}\ \ = \{L(G) \mid G \in \mathrm{BCTG}\}$$
$$\mathrm{BCTL}^0 = \{L^0(G) \mid G \in \mathrm{BCTG}\}$$
$$\mathrm{BCTL}_+ = \{L(G) \mid G \in \mathrm{BCTG}_+\}$$
$$\mathrm{BCTL}^0_+ = \{L^0(G) \mid G \in \mathrm{BCTG}_+\}.$$

We show in this section that $\mathrm{BCTL} = \mathrm{BCTL}^0 = \mathrm{BCTL}_+ = \mathrm{BCTL}^0_+$.

It often simplifies arguments if the assumption is made that a bc tree grammar is in the normal form described by the next theorem.

Lemma 2. *(Normal form) For every bc tree grammar $G = (N, \Sigma, R, S, I, \beta_0)$, there is a bc tree grammar $G' = (N', \Sigma, R', S, I, \beta_0)$ with $L(G) = L(G')$ and $L^0(G) = L^0(G')$, such that each rule in R' has one of the following formats:*
1. $A \to f[B_1, B_2, \ldots, B_r]\ (\lambda, \mu; \alpha)$
2. $A \to B\ (\lambda, \mu; \alpha)$
3. $A \to a$
where $A, B, B_1, B_2, \ldots, B_r \in N'$, $f^{(r)} \in \Sigma$ for some $r \geq 1$, and $a^{(0)} \in \Sigma$. Moreover, if $G \in \mathrm{BCTG}_+$, then $G' \in \mathrm{BCTG}_+$.

The proof is straightforward, using standard techniques. Using this normal form, it can be shown that the four variants of BCTL defined above are equal.

Theorem 3. $\mathrm{BCTL} = \mathrm{BCTL}^0 = \mathrm{BCTL}_+ = \mathrm{BCTL}^0_+$.

To prove the theorem, one may first establish that $\mathrm{BCTL} = \mathrm{BCTL}_+$ and $\mathrm{BCTL}^0 = \mathrm{BCTL}^0_+$ by showing how a bc tree grammar can be turned into an

---

[1] As mentioned in the introduction, this additional requirement is known from the definition of valence grammars [DP89].

equivalent one in BCTG_+. This can be done in a variety of ways. For example, replace each bag index i by indices i^+ and i^-, such that $\beta(i^+)$ and $\beta(i^-)$ are always nonnegative, and $\beta(i)$ in G corresponds to $\beta(i^+)-\beta(i^-)$ in G' at every point in a derivation. The reader should easily be able to work out the details. Now, to finish the proof of the theorem, it suffices to show that $\text{BCTL}^0 = \text{BCTL}$. Of the two inclusions, $\text{BCTL} \subseteq \text{BCTL}^0$ is the more interesting one. The reasoning is as follows.

Let $G \in \text{BCTG}$ be in normal form. We construct $G' \in \text{BCTG}$ by adding another bag position to the bag of G and making the following changes.

- The additional bag position is equal to 1 in the initial bag.
- Each rule $A \to t\,(\lambda,\mu;\alpha)$ in G is modified as follows to obtain a rule in G'. If t is not a leaf, then we add $A \to t(\lambda',\mu';\alpha')$ to the rules of G', where λ',μ' and α' are obtained from λ,μ and α, respectively, by setting the additional component equal to 1 in λ' and μ', and equal to 0 in α'. All other bag positions stay unchanged in λ',μ' and α'. If t is a leaf, then we add $A \to t\,(0,0;0)$ to the rules of G'.
- We add for each nonterminal B in G, the rule $B \to B\,(-\infty,+\infty;\alpha')$, where all positions in α' are equal to 0, except that the additional bag position has an entry of -1.
- We add for each nonterminal B in G the rule $B \to B\,(\lambda',\mu';\alpha')$, where the new bag position in α' is equal to 0 and all other positions in α', except for one, are equal to 0. The component in α' that is not equal to 0, is equal to 1 or -1. All bag positions in λ' are equal to $-\infty$, except for the new position, that has the value 0. All bag positions in μ' are equal to ∞, except for the new position, that has the value 0.

Consider a derivation of $s \in L(G)$. We may assume that this derivation is done in two stages. First we use rules of the form $A \to t\,(\lambda,\mu;\alpha)$, with t not a leaf, and then rules of the form $A \to a$, with $a^{(0)} \in \Sigma$. The first stage is used in the obvious manner to provide the first stage of a derivation for s in G'. Then we change the additional bag position to 0, then all other bag positions to 0, and finally we use the second stage of the derivation of s in G, in the obvious manner, to obtain the second stage of the derivation of s in G'. It should be clear that $L(G) = L(G') = L^0(G')$, as required.

3   Comparison of Bag Context and Random Context

In this section, we characterize the bc tree languages in terms of random context tree languages and top-down tree transducers.

A *random context tree grammar* (rc tree grammar) $G = (N, \Sigma, R, S)$ consists of finite signatures N and Σ of nonterminals and terminals, respectively, where the nonterminals have rank 0, an initial nonterminal $S \in N$, and a finite set R of rules. Each rule has the form $A \to t\,(P;F)$, where $A \in N$, $t \in T_{\Sigma \cup N}$, and $P, F \subseteq N$. The sets P and F are called the permitting context and the forbidding context, respectively. The rule is applicable to a tree $s[\![A]\!]$, thus giving rise to a derivation step $s[\![A]\!] \Rightarrow_G s[\![t]\!]$, if all nonterminals in P occur in s and none of those in F does. The language generated by G is then defined as usual: $L(G) = \{t \in T_\Sigma \mid S \Rightarrow_G^* t\}$.

Let us first define a notation useful for the technical constructions in this section. Suppose we are interested in a (bag or rc) tree grammar generating trees over Σ, using a set $N = \{A_1, \ldots, A_m\}$ of nonterminals (where the order of elements in N is arbitrary but fixed). For a tree $t \in T_{\Sigma \cup N}$, we denote by $\mathrm{cnt}(t)$ the m-tuple $(n_1, \ldots, n_m)$ such that n_i is equal to the number of occurrences of A_i in t. In particular, for $A \in N$, $\mathrm{cnt}(A)$ is the m-tuple in which the component corresponding to A is set to 1, whereas all others are set to 0. For a set $T \subseteq T_\Sigma$, we let $\mathrm{cnt}(T) = \sum_{t \in T} \mathrm{cnt}(t)$.

From an rc tree grammar $G = (N, \Sigma, R, S)$, we may construct a bc tree grammar with the bag index set N to record, at all times during a derivation, the number of occurrences of $A \in N$ in the A-entry of the bag. In this way, it is not difficult to simulate an rc tree grammar by a bc tree grammar, which yields the following result.

Lemma 4. *For every random context tree grammar $G = (N, \Sigma, R, S)$ there is an equivalent bag context tree grammar $G' = (N, \Sigma, R', S, N, \mathrm{cnt}(S))$.*

In the remainder of this section, we show that bc tree languages are precisely the images of the rc tree languages under linear top-down tree transductions. For this, let us first recall the definition of top-down tree transducers.

A *top-down tree transducer* (td transducer) $td = (\Sigma, \Sigma', Q, R, q_0)$ consists of finite signatures Σ, Σ' of input and output symbols, respectively, a finite signature Q of states of rank 1, an initial state $q_0 \in Q$, and a finite set of rules R. Each rule in R has the form $q\, f \to t[\![q_1[x_{i_1}], \ldots, q_l[x_{i_l}]]\!]$, where $q, q_1, \ldots, q_l \in Q$, $f^{(k)} \in \Sigma$, $i_1, \ldots, i_l \in \{1, \ldots, k\}$, and $t \in T_{\Sigma' \cup \{x_1, \ldots, x_l\}}$ for some $k, l \in \mathbb{N}$. The rules are interpreted as term rewrite rules in the usual way, where the left-hand side $q\, f$ is an abbreviation that stands for the tree $q[f[x_1, \ldots, x_k]]$. Thus, a computation step using the rule $q\, f \to t[\![q_1[x_{i_1}], \ldots, q_l[x_{i_l}]]\!]$ has the form

$$s[\![q[f[s_1, \ldots, s_k]]]\!] \Rightarrow_{td} s[\![t[\![q_1[s_{i_1}], \ldots, q_l[s_{i_l}]]\!]]\!].$$

For every tree $t \in T_\Sigma$, we let $td(t) = \{t' \in T_{\Sigma'} \mid q_0[t] \Rightarrow_{td}^* t'\}$. The transformation of a tree language $L \subseteq T_\Sigma$ is given by $td(L) = \bigcup_{t \in L} td(t)$.

Two special cases of td transducers will be of particular interest in the following. For this, call a rule $q\, f^{(k)} \to t[\![q_1[x_{i_1}], \ldots, q_l[x_{i_l}]]\!]$ *linear* if $x_{i_1}, \ldots, x_{i_l}$ are pairwise distinct and *nondeleting* if $\{x_{i_1}, \ldots, x_{i_l}\} = \{x_1, \ldots, x_k\}$. A td transducer is linear (nondeleting) if all of its rules are linear (nondeleting, respectively). Naturally, a td transduction is called linear if there is a linear td transducer computing it – and similarly for the nondeleting and linear nondeleting cases.

Below, we shall use the following notation for classes of tree languages and tree transductions: RCTL denotes the class of rc tree languages, i.e., the class of all languages generated by rc tree grammars. The classes of all td transductions, linear td transductions, nondeleting td transductions, and linear nondeleting td transductions are denoted by TD, lTD, nTD, and lnTD, resp. If $\mathcal{L}$ is a class of tree languages and $\mathcal{T}$ is a class of tree transductions, we let $\mathcal{T}(\mathcal{L}) = \{\tau(L) \mid \tau \in \mathcal{T} \text{ and } L \in \mathcal{L}\}$. If $\mathcal{T}$ is a singleton $\{\tau\}$, we may also denote $\mathcal{T}(\mathcal{L})$ by $\tau(\mathcal{L})$.

Next we show that the class of random context tree languages is closed under linear nondeleting td transductions. This will be used later on, in order to derive the main result of this section.

Theorem 5. lnTD(RCTL) = RCTL.

Proof. Of course, RCTL $\subseteq$ lnTD(RCTL) since lnTD contains the identity on T_Σ, for every finite signature Σ. For the other direction, let $G = (N, \Sigma, R, S)$ and $td = (\Sigma, \Sigma', Q, R_{td}, q_0)$ be an rc tree grammar and a linear nondeleting td transducer, respectively. We construct an rc tree grammar G' generating $td(L(G))$ by means of a standard technique. The nonterminals of G' are the pairs $(q, A) \in Q \times N$ and the rules are obtained by "running" td on the right-hand sides of the rules in R. For this, we extend td to input trees in $T_{\Sigma \cup N}$ by defining $q[A] \Rightarrow_{td} (q, A)$ for all $q \in Q$ and $A \in N$. Thus, the output trees of the extended td transducer are trees over $\Sigma' \cup Q \times N$. Now, $G' = (Q \times N, \Sigma', R', (q_0, S))$, where R' contains, for all $q \in Q$ and all rules $A \to t$ $(P; F)$ in R, all rules $(q, A) \to t'$ $(P'; F')$ such that

- $q[t] \Rightarrow_{td}^* t'$,
- $P = \{B \in N \mid (q', B) \in P' \text{ for some } q' \in Q\}$, and
- $F' = Q \times F$.

(Since the second item does not yield a unique set P', several copies of the rule, which are only distinguished by their permitting contexts, belong to R'.)

As td is linear and nondeleting, it preserves the occurrences of nonterminals. More precisely, if $t = s[A_1, \ldots, A_k]$, where $t \in T_{\Sigma \cup \{x_1, \ldots, x_k\}}$ and $A_1, \ldots, A_k \in N$, then $td(t)$ is of the form $s'[(q_1, A_1), \ldots, (q_k, A_k)]$, where $s' \in T_{\Sigma' \cup \{x_1, \ldots, x_k\}}$ and $q_1, \ldots, q_k \in Q$. Using this, and keeping in mind the way in which the permitting and forbidding contexts P' and F' were defined above, it should be obvious that $L(G') = L(G)$; the formal proof of this fact is thus omitted. $\square$

To investigate the relationship between the classes RCTL and BCTL, it is useful to introduce a generalization of rc tree grammars, called counting rc tree grammars. These grammars are defined in exactly the same way as rc tree grammars, but the permitting and forbidding contexts are multisets of nonterminals rather than ordinary sets. More precisely, every rule of a counting rc tree grammar is of the form $A \to t$ $(P; F)$, where P and F map the set N of nonterminals to $\mathbb{N}$ and $\mathbb{N} \cup \{\infty\}$, respectively. Such a rule is applicable to a tree $s[A]$ if each nonterminal $B \in N$ occurs at least $P(B)$ but at most $F(B)$ times in s. The remaining definitions carry over from the case of rc tree grammars; the class of all tree languages generated by counting rc tree grammars is denoted by cRCTL. Clearly, a rule $A \to t$ $(P; F)$ in an ordinary rc tree grammar can be identified with the rule $A \to t$ $(P'; F')$ where

$$P'(B) = \begin{cases} 1 \text{ if } B \in P \\ 0 \text{ otherwise} \end{cases} \qquad F'(B) = \begin{cases} 0 \text{ if } B \in F \\ \infty \text{ otherwise} \end{cases}$$

for all $B \in N$. Thus, RCTL $\subseteq$ cRCTL. To show that the converse holds as well, we may invent copies $A_1, \ldots, A_m$ of each nonterminal A, where m is the maximum finite multiplicity of elements in the sets P and F occurring in a counting rc tree grammar. In this way, an ordinary rc tree grammar can count up to m copies of A, and is thus able to express that a given sentential form contains at most or at least k occurrences of A ($k \in \{0, \ldots, m\}$). Thus, the following lemma is obtained.

Lemma 6. cRCTL = RCTL.

Now, we come to the main constructions of this section. Suppose we are given a signature Σ containing some binary symbol $\text{aux}^{(2)} \in \Sigma$. We are going to consider the td transduction 'clean$_{\text{aux}}$' which turns trees over Σ into trees over $\Sigma \setminus \{\text{aux}\}$, as follows: For every symbol $a^{(0)} \in \Sigma$, we let $\text{clean}_{\text{aux}}(a) = a$. Furthermore, for all trees $t = f[t_1, \ldots, t_k]$ with $k \geq 1$,

$$\text{clean}_{\text{aux}}(t) = \begin{cases} \text{clean}_{\text{aux}}(t_1) & \text{if } f = \text{aux} \\ f[\text{clean}_{\text{aux}}(t_1), \ldots, \text{clean}_{\text{aux}}(t_k)] & \text{otherwise.} \end{cases}$$

Thus, we simply remove all occurrences of aux, together with its second subtrees. Clearly, clean$_{\text{aux}}$ is a td transduction. By using additional bag positions to keep track of the number of nonterminals in deleted parts of the tree, one can show that BCTL is closed under clean, i.e., we have the following lemma.

Lemma 7. clean(BCTL) $\subseteq$ BCTL.

Continuing this chain, the application of clean to RCTL yields all of BCTL.

Lemma 8. BCTL $\subseteq$ clean(RCTL).

Proof. By Theorem 3 and Lemma 6, it suffices to prove the inclusion $\text{BCTL}^0_+ \subseteq$ clean(cRCTL). For this, let $G = (N, \Sigma, R, S, I, \beta_0) \in \text{BCTG}_+$. We may assume that $\beta_0 = (0, \ldots, 0)$. Our construction of a counting rc tree grammar G' with $\text{clean}_{\text{aux}}(L(G')) = L^0(G)$ (where $\text{aux}^{(2)}$ is any binary symbol not in Σ), is based on the following idea. The set of nonterminals is given by the disjoint union $N' = N \cup \bigcup_{i \in I} \{i, i', i_\downarrow, i'_\downarrow\}$. Intuitively, the nonterminals in N play the same role as in G, while the number of occurrences of $i \in I$ in a derived tree corresponds to the bag value $\beta(i)$. These copies of i will be stored in the parts of the tree that are deleted by clean$_{\text{aux}}$. The nonterminals of the form $i', i_\downarrow, i'_\downarrow$ are auxiliary ones used to decrease the number of occurrences of i in a controlled way if the bag value is adjusted downwards.

Following this intuition, the set R' of rules of G' is given by $R' = R'_1 \cup R'_2$, where R'_1 and R'_2 are constructed as follows:

- For each rule $A \to t$ $(\lambda, \mu; \alpha)$ in R, R'_1 contains the rule $A \to \text{aux}[t, t']$ $(P; F)$, where t', P, and F are given as follows. The tree t' is any tree over $\Sigma \cup \{\text{aux}\} \cup \bigcup_{i \in I} \{i, i_\downarrow\}$ such that the following holds for every $i \in I$. If $\alpha_i \geq 0$,

then t' contains α_i occurrences of i and no occurrence of $i_\downarrow$. If $\alpha_i \leq 0$, then t' contains $-\alpha_i$ occurrences of $i_\downarrow$ and no occurrence of i.[2]
The multisets P and F are given by

$$P(x) = \{\lambda_i i \mid i \in I\} \quad \text{and} \quad F(x) = \{\mu_i i \mid i \in I\} \cup \{\infty A \mid A \in N\}.$$

(Here, we define P and F using set notation extended by multiplicities.) Note that, since $F(i') = F(i_\downarrow) = F(i'_\downarrow) = 0$ for all $i \in I$, this rule is applicable to an occurrence of A only if the tree does not contain any of the symbols i', $i_\downarrow$, and $i'_\downarrow$. The rationale behind the additional subtree t' on the right-hand side is to adjust the number of occurrences of $i \in I$ according to α. If $\alpha_i \geq 0$, this can be done immediately by adding as many occurrences of i. However, if $\alpha_i < 0$, we have to delete some of the existing occurrences. Since this cannot be done in a single stroke, we add $-\alpha_i$ occurrences of $i_\downarrow$. The rules in R'_2 described next will make sure that i and $i_\downarrow$ cancel each other out.

- For every $i \in I$, we let R'_2 contain the rules

$$i \rightarrow i' \ (\emptyset; \{i', i'_\downarrow\})$$
$$i_\downarrow \rightarrow i'_\downarrow \ (\{i'\}; \{i'_\downarrow\})$$
$$i' \rightarrow a \ (\{i'_\downarrow\}; \emptyset)$$
$$i'_\downarrow \rightarrow a \ (\emptyset; \{i'\})$$

for some arbitrary symbol $a^{(0)} \in \Sigma$. These rules ensure that all occurrences of $i_\downarrow$ are turned into a, at the same time turning exactly as many occurrences of i into a. (Note that, since we started out with BCTL^0_+, the number of occurrences of i and $i_\downarrow$ must finally cancel each other out.)

Naturally, the initial nonterminal of G' is S. Once again, the correctness of the construction should be rather obvious, i.e., $\text{clean}_{\text{aux}}(L(G')) = L^0(G)$. $\square$

We can now prove the main result of this section.

Theorem 9. $\text{RCTL} \subsetneq \text{lTD}(\text{RCTL}) = \text{clean}(\text{RCTL}) = \text{BCTL}$.

Proof. Clearly, $\text{BCTL} \neq \text{RCTL}$ because the monadic tree languages in RCTL are regular (see also [DTE$^+$05]), whereas it is easy to generate, e.g., $\{a^n[b^n[\epsilon]] \mid n \in \mathbb{N}\}$ using a single bag position.

Now, consider the equality $\text{BCTL} = \text{clean}(\text{RCTL})$. Lemma 8 yields the inclusion $\text{BCTL} \subseteq \text{clean}(\text{RCTL})$. From Lemma 4, $\text{RCTL} \subseteq \text{BCTL}$ and thus, using Lemma 7, $\text{clean}(\text{RCTL}) \subseteq \text{clean}(\text{BCTL}) \subseteq \text{BCTL}$. Thus $\text{BCTL} = \text{clean}(\text{RCTL})$.

To complete the proof, note first that $\text{clean}(\text{RCTL}) \subseteq \text{lTD}(\text{RCTL})$ since $\text{clean} \in \text{lTD}$. Hence, it remains to show that $\text{lTD}(\text{RCTL}) \subseteq \text{BCTL}$. However, for $td \in \text{lTD}$ it is straightforward to construct a nondeleting variant $td' \in \text{lnTD}$ in such a way that $td = \text{clean}_{\text{aux}} \circ td'$, i.e., td is the composition of td' (first) and $\text{clean}_{\text{aux}}$ (second): td' simply preserves the subtrees that td deletes, storing them

---

[2] Clearly, we may assume that Σ contains at least one symbol of rank 0.

as the second subtrees of additional aux symbols. I.e., if td contains the rule $q\,f^{(2)} \to g[h[a], q'[x_1]]$, in td' this becomes $q\,f^{(2)} \to \text{aux}[g[h[a], q'[x_1]], q_{\text{id}}[x_2]]$. Here, q_{id} is a new state which computes the identity using linear nondeleting rules. Obviously, $td = \text{clean}_{\text{aux}} \circ td'$, as required.

Hence, $\text{lTD(RCTL)} \subseteq \text{clean(lnTD(RCTL))} \subseteq \text{clean(RCTL)} \subseteq \text{BCTL}$, using Theorem 5 for the second inclusion. $\qquad\square$

Owing to [Bak79, Corollary 4(2)] by Baker, we have $\text{lTD} \circ \text{clean} \subseteq \text{lTD}$. Using the previous theorem, we thus get $\text{lTD(BCTL)} = \text{lTD(clean(RCTL))} \subseteq \text{lTD(RCTL)} = \text{BCTL}$. Since the identity on T_Σ belongs to lTD, the converse inclusion holds as well, which yields the following corollary.

Corollary 10. $\text{lTD(BCTL)} = \text{BCTL}$.

4   The Interchange Theorem for bc Tree Languages

If an intermediate tree in a derivation in a bc tree grammar contains some nonterminal twice, then there is no means of distinguishing between the two. Consequently, in such a case two trees can be generated, each of which could be changed into the other by swapping appropriate subtrees. Clearly, such a situation must necessarily arise if the generated tree is sufficiently dense. Pursuing this line of thought, it is possible to find candidates for tree languages that do not seem to be bag context tree languages, such as the following.

For $n \geq 1$, let $t_{\langle n \rangle}$ be the complete binary tree with the following properties:

1. $t_{\langle n \rangle}$ is of height $2n$;
2. every internal node of $t_{\langle n \rangle}$ is labelled $f^{(2)}$;
3. $yield(t_{\langle n \rangle}) = w_1 w_2 \ldots w_{2^n}$, where $w_i = a^{i-1} b a^{2^n - i}$.

Let $L_{\text{bin}} = \{t_{\langle n \rangle} \mid n \geq 1\}$.

We shall prove that $L_{\text{bin}} \notin \text{BCTL}$. To begin with, the following lemma is easily proved by induction on the length of derivations.

Lemma 11. Let $(t[\![A_1, \ldots, A_m]\!], \beta) \overset{*}{\Rightarrow} (t[\![t_1, \ldots, t_m]\!], \beta')$ be a derivation in a bc tree grammar. If p is a permutation of $\{1, \ldots, m\}$ with $A_i = A_{p(i)}$ for all $i \in \{1, \ldots, m\}$, then $(t[\![A_1, \ldots, A_m]\!], \beta) \overset{*}{\Rightarrow} (t[\![t_{p(1)}, \ldots, t_{p(m)}]\!], \beta')$.

Now, consider a tree $t_{\langle n \rangle}$ in L_{bin}, for large enough n. Intuitively, the derivation must contain an intermediate tree $t[\![A_1, \ldots, A_m]\!]$ of height less than n, such that at least one nonterminal occurs twice among $A_1, \ldots, A_m$. Thus, the corresponding subtrees in $t_{\langle n \rangle}$ can be interchanged. Obviously, the resulting tree is not a member of L_{bin}.

In the following, we will develop general notions and results that formalize this intuition. We use the notion of a *run* to reflect the top-down manner in which bc tree grammars generate trees. To obtain a sufficiently general definition, we parametrize it by node properties. Here, we define a *node property* to be a predicate Ψ that determines, for every tree t, a nonempty prefix-closed subset Ψ_t of $nodes(t)$. For $v \in nodes(t)$, we say that v *satisfies* Ψ_t if $v \in \Psi_t$.

Definition 12 (run, density). *Let Ψ be a node property and t a tree. A Ψ-run on t is a sequence $N_0 \cdots N_m$ of subsets of nodes(t) such that (a) $N_0 = \{\lambda\}$, (b) for every $i \in \{1, \ldots, m\}$, there is some $v \in N_{i-1}$ with*

$$N_i = (N_{i-1} \setminus \{v\}) \cup \{vj \in nodes(t) \mid j \in \mathbb{N} \text{ and } vj \text{ satisfies } \Psi_t\},$$

and (c) $N_m = \emptyset$. The Ψ-density of t is given by

$$density_\Psi(t) = \min\{\max_{0 \le i \le m} |N_i| \mid N_0 \cdots N_m \text{ is a } \Psi\text{-run on } t\}.$$

Note that, if Ψ is the trivial property that holds for all nodes of a given tree, then a Ψ-run on t corresponds in a straightforward way to a node-by-node top-down construction of t. Choosing less general properties allows us to focus on particular types of nodes and to disregard all others. Intuitively, the density of t is the smallest number d such that t can be constructed by a run in which each node set contains at most d elements.

Consider a bc tree grammar in normal form, and let $(t_0, \beta_0) \Rightarrow \cdots \Rightarrow (t_m, \beta_m)$ be a derivation, where t_0 is a single nonterminal, and $t = t_m$ is a terminal tree. This derivation yields a Ψ-run on t, as follows: for $i \in \{1, \ldots, m\}$, let N_i be the set of all nonterminal nodes v of t_i that satisfy Ψ_t. The desired run is obtained from $N_0 \cdots N_m$ by discarding all N_i $(1 \le i \le m)$ for which $N_i = N_{i-1}$. By the definition of density, this means that the derivation must contain at least one tree t_i having $density_\Psi(t)$ pairwise distinct nodes v such that (a) $t_i(v)$ is a nonterminal and (b) v satisfies Ψ_t. Thus, if $density_\Psi(t)$ exceeds the number of nonterminals of the grammar, Lemma 11 implies that there exist two subtrees of t such that (a) the tree obtained by interchanging these two subtrees can also be generated from t_0, and (b) both nodes satisfy Ψ_t. Furthermore, none of these nodes is a prefix of the other – we say that they are *independent*. Clearly, the mentioned arguments are still valid if t occurs as a subtree within a larger tree generated. This proves the interchange theorem, which we state next. Here, we denote by $t_{u \leftrightarrow v}$ the tree obtained from t by interchanging the subtrees rooted at nodes u and v, respectively.

Theorem 13 (interchange theorem). *For every bc tree language L, there exists a constant k such that the following holds. Let Ψ be a node property, and assume that L contains a tree $s[\![t]\!]$, where $density_\Psi(t) > k$. Then there are independent nodes $u, v \in nodes(t)$ which satisfy Ψ_t, such that $s[\![t_{u \leftrightarrow v}]\!] \in L$.*

The interchange theorem proves that L_{bin} is not a bag context tree language: Let k be the constant of the theorem, and let a node $v \in nodes(t)$ satisfy Ψ if $|v| < height(t)/2$. Then consider $t = t_{\langle n \rangle}$ for some $n > k$.

The interchange theorem makes it possible to compare the class of bc tree languages with other classes of tree languages. One such class is the class of branching synchronization tree languages [DE04], which is identical to the closure of the class of regular tree languages under top-down tree transductions. Hence, there are bc tree languages which are not branching synchronization tree languages. (Recall that $\{a^n[b^n[\epsilon]] \mid n \in \mathbb{N}\} \in \text{BCTL}$.) To show that the converse holds as well, we use the interchange theorem. Let $\Sigma = \{f^{(2)}, g^{(1)}, h^{(0)}\}$, and let L_0 be the set of all trees over Σ such that every path from the root to a leaf contains exactly one g. Even a branching ET0L tree grammar (or, equivalently, a regular tree grammar followed by a top-down tree transducer) can generate the language $L_{\text{copy}} = \{f[t, t] \mid t \in L_0\}$. Assume that this language is a bc tree language, and let k be as in Theorem 13. For a tree $t \in T_\Sigma$, let Ψ_t be the node property that is satisfied by $v \in nodes(t)$ if $t(v') = f$ for all proper prefixes v' of v. Now, choose a tree $t \in L_0$ such that (a) the subtrees whose roots are labelled with g are pairwise distinct, and (b) $density_\Psi(t) > k$. Clearly, such a tree t exists. Choosing $s = f[t, x_1]$ in Theorem 13, it follows that $f[t, t_{u \leftrightarrow v}] \in L$, where u and v are independent nodes satisfying Ψ_t. However, by the choice of t and Ψ, the latter means that $t/u \neq t/v$, and thus $f[t, t_{u \leftrightarrow v}] \notin L$ – a contradiction. Thus, we have proved the following theorem.

Theorem 14. *There exist bc tree languages which are not branching synchronization tree languages (i.e., cannot be obtained by applying a finite sequence of top-down tree transductions to a regular tree language). Conversely, there are branching ET0L tree languages (i.e., images of regular tree languages under top-down tree transductions) which are not bc tree languages.*

Acknowledgment. We thank the referees for suggesting several improvements to the text.

References

[Bak79]   Brenda S. Baker. Composition of top-down and bottom-up tree transductions. *Information and Control*, 41:186–213, 1979.

[DE98]   Frank Drewes and Joost Engelfriet. Decidability of the finiteness of ranges of tree transductions. *Information and Computation*, 145:1–50, 1998.

[DE04]   Frank Drewes and Joost Engelfriet. Branching synchronization grammars with nested tables. *Journal of Computer and System Sciences*, 68:611–656, 2004.

[DP89]   Jürgen Dassow and Gheorghe Păun. *Regulated Rewriting in Formal Language Theory*. EATCS Monographs in Theoretical Computer Science. Springer, Berlin, 1989.

[DTE+05]   Frank Drewes, Christine du Toit, Sigrid Ewert, Johanna Högberg, Brink van der Merwe, and Andries P.J. van der Walt. Random context tree grammars and tree transducers. *South African Computer Journal*, 34:11–25, 2005.

[ETM+06]   Sigrid Ewert, Christine du Toit, Brink van der Merwe, Andries P.J. van der Walt, Frank Drewes, and Johanna Högberg. Bag tree grammars. Report UMINF 06.03, Umeå University, 2006.

Closure of Language Classes Under Bounded Duplication

Masami Ito[1], Peter Leupold[2,*], and Kayoko Shikishima-Tsuji[3]

[1] Department of Mathematics, Faculty of Science
Kyoto Sangyo University
Kyoto 603-8555, Japan
ito@ksuvx0.kyoto-su.ac.jp

[2] Research Group on Mathematical Linguistics
Rovira i Virgili University
Pça. Imperial Tàrraco 1, 43005 Tarragona, Catalunya, Spain
klauspeter.leupold@urv.cat

[3] Tenri University
Tenri 632-8510, Japan
tsuji@sta.tenri-u.ac.jp

Abstract. Duplication is an operation generating a language from a single word by iterated application of rewriting rules $u \to uu$ on factors. We extend this operation to entire languages and investigate, whether the classes of the Chomsky hierarchy are closed under duplication. Here we treat mainly bounded duplication, where the factors duplicated cannot exceed a given length.

While over two letters the regular languages are closed under bounded duplication, over three or more letters they are not, if the length bound is 4 or greater. For 2 they are closed under duplication, the case of 3 remains open. Finally, the class of context-free languages is closed under duplication over alphabets of any size.

1 Duplication

In a series of recent articles, languages generated from a single word by iteration of the duplication operation have been investigated. This operation was inspired by a behaviour observed in strands of DNA: certain factors of such sequences can be duplicated within their strand forming a so-called tandem repeat; from a formal language point of view, a word uvw is transformed into $uvvw$.

The first mechanism for generating languages derived from this behaviour were the so-called duplication grammars [9],[10]. Then a great deal of interest was paid to languages generated from a word by iterated application of the duplication operation as introduced by Dassow et al. [3]. The main focus in these investigations was on determining under which conditions those languages

---

* This work was done, while the author was funded by the Spanish Ministry of Culture, Education and Sport under the Programa Nacional de Formación de Profesorado Universitario.

O.H. Ibarra and Z. Dang (Eds.): DLT 2006, LNCS 4036, pp. 238–247, 2006.

are regular. In this context also the restriction of the duplicated factor's length to a maximum or one fixed length has been investigated [5],[6],[7],[8],[12].

The objective of this article will be the investigation of the duplication closure on entire languages rather than single words. The duplication closure of a language is the union of all the closures of the words contained in that language. Our main focus will be on the question, under which conditions this operation preserves regularity and context-freeness. It is rather obvious that all versions of duplication preserve context-sensitivity.

In Section 3 we establish that the class of regular languages is closed under 2-bounded duplication, but not under 4-bounded duplication; the case of 3 remains open. Further we show that over a two-letter alphabet 2-boundedness is equivalent to any longer bound and even to unbounded duplication. In combination with the preceding results this proves that over two letters regular languages are closed under general duplication and all bounded versions.

The class of context-free languages is the focus of Section 4. We establish its closure under bounded duplication, and further give a result that shows that this does not help us to answer the case of general duplication, because the n-bounded duplication languages of the word abc form an infinite hierarchy with the unbounded duplication language as its supremum.

2   Definitions

We now provide the formal definitions concerning the duplication operation. For this, we take for granted elementary concepts from the theory of formal languages as exposed, for example, by Harrison [4] or Salomaa [11]. A few notations we use are: $|w|$ for the length of the word w, $w[i]$ for the i-th letter of w, and $w[i \ldots j]$ for the factor of w starting at position i and ending at position j. A period of a word w is an integer k such that for all $i \leq |w| - k$ we have $w[i] = w[i + k]$. For an alphabet Σ, the set Σ^n consists of all the words of length n over this alphabet, further $\Sigma^{\leq n} := \bigcup_{i \leq n} \Sigma^i$.

An important notion that will be used is the relation $\sim_L$ over $\Sigma^* \times \Sigma^*$ for a language $L \subset \Sigma^*$, which is the *syntactic right-congruence* and is defined as follows:

$$u \sim_L v :\leftrightarrow \forall w \in \Sigma^*(uw \in L \leftrightarrow vw \in L).$$

This is obviously an equivalence relation. It is well-known from the Kleene-Myhill-Nerode Theorem that a language L is regular, if and only if the corresponding relation $\sim_L$ has a finite number of equivalence classes; this number is called the *index* of $\sim_L$.

Theorem 1 ([11]). *A language L is regular, if and only if $\sim_L$ has finite index.*

With this we come to the central notion of this article. We will use a rewriting relation to generate duplication languages. For details on string-rewriting systems we refer the reader to the book by Book and Otto [1], whose terminology we will follow here.

The relation we define will be denoted by $\heartsuit$; with the origin on the bottom expanding to two equal halves, this symbol seems quite appropriate for duplication. So, in detail, the duplication relation is defined as

$$u \heartsuit v :\Leftrightarrow \exists w[u = u_1 w u_2 \wedge v = u_1 w w u_2].$$

With $\heartsuit^*$ we denote the relation's reflexive and transitive closure. We generate languages with it in the following way.

Definition 2. The duplication language generated by a word w is

$$w^\heartsuit := \{u : w \heartsuit^* u\}.$$

Thus $w^\heartsuit$ is the language of all words that can be obtained from w by a finite number of duplications. Apart from general duplication, also two restricted variants have been investigated, namely *bounded* and *uniformly bounded duplication*. These are defined for some integer n as

$$u \heartsuit^{\leq n} v :\Leftrightarrow \exists w[u = u_1 w u_2 \wedge v = u_1 w w u_2 \wedge |w| \leq n]$$

and

$$u \heartsuit^{=n} v :\Leftrightarrow \exists w[u = u_1 w u_2 \wedge v = u_1 w w u_2 \wedge |w| = n]$$

respectively. So the n-bounded variant admits duplications of factors up to length n, the uniformly n-bounded one admits duplications of factors of length exactly n. The languages $w^{\heartsuit \leq n}$ and $w^{\heartsuit = n}$ are defined analogously to the unrestricted case. The latter variant we write also simply $w^{\heartsuit n}$.

We will now illustrate these definitions with two simple examples over the two-letter alphabet $\Sigma = \{a, b\}$. $(aba)^\heartsuit$ is the language $a\Sigma^* a$. $(aba)^{\heartsuit \leq 2} = a\Sigma^* a$. On the other hand $(aba)^{\heartsuit 2} = (ab)^* a$.

In the canonical way, the duplication operation is extended to sets of words, setting for such a set W its language generated by duplication as

$$W^\heartsuit := \bigcup_{w \in W} w^\heartsuit.$$

This is the form, in which we will now apply duplication to languages and then investigate, whether this preserves regularity and context-freeness.

3   Closure of Regular Languages

We start out with the closure of regular languages. Here the size of the alphabet will play an important role, and first we treat the three-letter case, where closure is not given in most of the cases. All results for this alphabet size also carry over to bigger alphabets.

It is known that the 4-bounded duplication closure of the word abc is not regular [8]. As one can see from the original proof, duplications longer than 4 do not affect the construction used, and therefore the result extends to longer bounds. Thus the class of regular languages is not closed under n-bounded duplication for $n \geq 4$, since singular sets are of course regular.

Proposition 3. *For $n \geq 4$ the class of regular languages is not closed under n-bounded duplication.*

On the other hand, it is trivial to see that 1-bounded duplication preserves regularity: the only possible change in the original word is that every letter a can be blown up to any word from a^+. We now take a look at the two cases inbetween, that is length-bounds of 2 and 3.

We now fix some notation, which will be convenient in the proof that follows. For a right-syntactic congruence $\sim_L$ we denote the set of all possible right contexts of a word u by $\sim_L (u) := \{w : uw \in L\}$. By $[u]_{\sim_L}$ we denote the congruence class of u; notice that for all $u_1, u_2 \in [u]_{\sim_L}$ we have $\sim_L (u_1) =\sim_L (u_2)$.

Proposition 4. *The class of regular languages is closed under 2-bounded duplication.*

Proof. Let L be a regular language, and $\sim_L$ the corresponding right-syntactic congruence. The right-syntactic congruence $\sim_{L^{\heartsuit \leq 2}}$ we will denote more simply by $\sim$. We will show that the number of congruence classes of $\sim$ is bounded by a function of the number of congruence classes of $\sim_L$.

First notice that always $(\sim_L (u))^{\heartsuit \leq 2} \subseteq \sim (u)$, i.e. if v is a possible right context of u in L, then all words in $v^{\heartsuit \leq 2}$ are possible right contexts of u in $L^{\heartsuit \leq 2}$. If the two sets are not equal, this can be caused only by some duplication transgressing the border between u and v. Duplications of length one cannot do this, thus the only possibility is one of length two affecting the last letter of u and the first letter of v.

If the two letters are the same, say a, then the result will be a^4, which could have been obtained also by duplicating twice the a in v, so the result is in $v^{\heartsuit \leq 2}$. If the two letters are distinct, say a and b, then the result of the duplication will be $abab$. If the following letter in v is an a, then we could have obtained the same by duplicating the prefix ba of v, so the result is in $v^{\heartsuit \leq 2}$.

Otherwise the result will be $ababc$ for some letter c different from a. The resulting right context is not in $v^{\heartsuit \leq 2}$, so in this case a new congruence class for u is created in $\sim$. More duplications on the right side will not lead to new classes, because now we have bab following the final a of u. The number of such constellations of two different letters at the border with a different one from the first one following is bounded by the total number of letters in the alphabet. Thus every congruence class of $\sim_L$ results in a finite number of congruence classes for $\sim$, except possibly for the one of words not being a prefix of a word in L.

Therefore it remains to show that the u, which are not prefixes of a word in L but are prefixes of a word in $L^{\heartsuit \leq 2}$, do not generate an infinite number of new congruence classes. So let $uv \in L^{\heartsuit \leq 2}$. If there exists $u'v'$ that $u \in u'^{\heartsuit \leq 2}$ and $v \in v'^{\heartsuit \leq 2}$, then we are done. Otherwise in the generation of uv from $u'v'$ there is a duplication transgressing the border between the two words.

Similarly as above, this is interesting only in the configuration $ca|b$, where $|$ denotes the border between u' and v' (or rather between the two intermediate words generated from them). The result of this duplication is $caba|b$. Let us call the word on the left u''. No further duplications transgressing the border can be necessary, since $(caba)^{\heartsuit \leq 2} b^{\heartsuit \leq 2} = (cabab)^{\heartsuit \leq 2}$. Thus for all words u here we have either $[u]_\sim = [u']_\sim$ or $[u]_\sim = [u'']_\sim$. Thus also here the increase of the index of $\sim$ compared to $\sim_L$ preserves finiteness, and thus the resulting language is regular by Theorem 1, if the original language was regular. $\square$

It appears possible to extend this proof technique to 3-bounded duplication under use of the fact that over two-letters the longest square-free word has length 3. While we leave this case open here, over an alphabet of only two letters things are not as complicated. To see this we first state a result that relates bounded and unbounded duplication. This will then allow us to state the closure of regular languages under these variants of duplication.

For the remainder of this section, $\rightarrow$ will denote the derivation relation of the string-rewriting system $R = \{a \rightarrow aa, b \rightarrow bb, ab \rightarrow abab, ba \rightarrow baba\}$, which generates the language $w^{\heartsuit \leq 2}$ for any word $w \in \{a, b\}$.

Lemma 5. *For every word $u \in \{a, b\}^*$ we have $ab \xrightarrow{*} abubab$, $ab \xrightarrow{*} abuaab$, and $ab \xrightarrow{*} abuab$.*

Proof. We prove this statement by induction on the length of u. For $|u| = 0$ the three derivations

$$ab \xrightarrow{ab \rightarrow abab} abab \xrightarrow{b \rightarrow bb} abbab = abubab$$
$$ab \xrightarrow{ab \rightarrow abab} abab \xrightarrow{a \rightarrow aa} abaab = abuaab$$
$$ab \xrightarrow{ab \rightarrow abab} abab \qquad\quad = abuab$$

show us that the lemma holds. So let us suppose it holds for all words, which are shorter than a number n. Any word u of length n has a factorization either as va or vb for a word v of length $n - 1$. For this word v the Lemma holds by our assumption. But then for $u = va$ the derivations

$$ab \xrightarrow{*} abvab \xrightarrow{ab \rightarrow abab} abvabab = abubab$$
$$ab \xrightarrow{*} abvaab \xrightarrow{a \rightarrow aa} abvaaab = abuaab$$
$$ab \xrightarrow{*} abvaab \qquad\quad = abuab$$

and for $u = vb$ the derivations

$$ab \xrightarrow{*} abvbab \xrightarrow{b \rightarrow bb} abvbbab = abubab$$
$$ab \xrightarrow{*} abvbab \xrightarrow{a \rightarrow aa} abvbaab = abuaab$$
$$ab \xrightarrow{*} abvbab \qquad\quad = abuab$$

show us that the lemma holds also for u and thus for all words. $\square$

Proposition 6. *Over an alphabet of two letters we have $w^{\heartsuit \leq n} = w^{\heartsuit \leq 2}$ and consequently $w^\heartsuit = w^{\heartsuit \leq 2}$ for all words w and for $n \geq 2$.*

Proof. From Lemma 5 we know that $ab \xrightarrow{*} abuab$ holds for every word u, and applying this to the initial factor ab in abu we obtain $abu \xrightarrow{*} abuabu$. Just interchanging the letters a and b everything still is valid, and thus we see that also $bau \xrightarrow{*} baubau$ holds.

Now we prove that $aau \xrightarrow{*} aauaau$. If $u \in a^*$, then the statement is obviously true. Otherwise there is at least one b in u, and therefore u can be factorized as $u = a^m bv$ for some word v and an integer $m \geq 0$. Now the derivation

$$aau = aa^m(ab)v \xrightarrow{\text{Lemma 5}} aa^m abvabv \xrightarrow{*} aaa^m bvaaa^m bv = aauaau$$

shows that the statement above holds. Interchanging the letters again provides us with the dual statement $bbu \xrightarrow{*} bbubbu$.

Because any word z longer than 1 has to start with either ab, ba, aa, or bb, this shows that we can always obtain by duplications of length at most 2 the word zz from z and thus $w^{\heartsuit \leq n} \subseteq w^{\heartsuit \leq 2}$. On the other hand, every duplication relation $\heartsuit^{\leq n}$ for $n \geq 2$ includes the relation $\heartsuit^{\leq 2}$ and so does $\heartsuit$. This suffices to prove that for all $n > 1$ we have $w^{\heartsuit \leq n} = w^{\heartsuit \leq 2}$, and $w^{\heartsuit} = w^{\heartsuit \leq 2}$ immediately follows from this, because in any derivation the length of duplications used is bounded. $\square$

Combining the results of this section we are now able to state the closure of regular languages under duplication.

Proposition 7. *The class of regular languages over two-letter alphabet is closed under n-bounded duplication and under general duplication.*

Proof. Proposition 4 states that regular languages are closed under 2-bounded duplication over any alphabet, and from Proposition 6 we see that in the two-letter case for any $n > 1$ the n-bounded and general duplication operations are equivalent to the 2-bounded one. $\square$

4   Closure of Context-Free Languages

When we talk about context-free languages, there is no difference between alphabets of size 2 and 3. It is already known that languages $w^{\heartsuit \leq n}$ are always context-free [8]. By further refining the push-down automaton used in that proof, we can establish the closure of context-free languages under bounded duplication.

Proposition 8. *The class of context-free languages is closed under bounded duplication.*

Proof. We will show this by constructing a Push-Down Automaton in a way rather analogous to the one used in earlier work for the bounded duplication closure of a single word [8]. There the PDA reduces the results of duplications uu to their origin u and matches the reduced string against the original word. Here, we also have to simulate a second PDA accepting the context-free input language. This can be done, because of the two components reducing duplications and accepting the original language, the latter one does not need to access the stack ever, while the first one is working. With this sketch of the proof idea we now proceed to the technical details.

We start out from a PDA M, which accepts the language L. Let the PDA be $M = [Q, \Sigma, \Gamma, \varphi, q_o, \perp]$, where Q is the set of states, Σ the tape alphabet, and Γ the stack alphabet. $\varphi : Q \times (\Sigma \cup \{\lambda\}) \times \Gamma \to Q \times \Gamma^*$ is the state transition function; i.e. we allow transitions without reading input and we always take the topmost symbol off the stack replacing it by an arbitrary number of stack symbols. q_0 is the start state, and $\perp$ marks the stack's bottom. The acceptance mode does not really need to be specified, since any common acceptance condition will carry over to the new PDA.

We now define the PDA A, which accepts $L^{\heartsuit \leq n}$. The state set is $S := Q \times (\underline{\Sigma} \cup \Sigma)^{\leq n} \times \Sigma^{\leq n}$, where $\underline{\Sigma} := \{\underline{a} : a \in \Sigma\}$ is a marked copy of the tape alphabet. States $s \in S$ we will denote in the way $s = q|_v^u$, where $q \in Q$, $u \in (\underline{\Sigma} \cup \Sigma)^{\leq n}$ is called the *match*, and $v \in \Sigma^{\leq n}$ the *memory*; then $q_0|_\lambda^\lambda$ is the start state of S. The stack alphabet is $\Gamma' := \Gamma \cup (\underline{\Sigma} \cup \Sigma)^{\leq n}$. The tape alphabet Σ and bottom-of-stack marker $\perp$ are as for M. What remains to be defined is the transition function δ. We first define the part

$$\delta(q|_\lambda^\lambda, x, \gamma) := (q'|_\lambda^\lambda, \alpha) \text{ where } \varphi(q, x, \gamma) = (q', \alpha) \tag{1}$$

for $x \in \Sigma \cup \{\lambda\}$, $\gamma \in \Gamma$, and $\alpha \in \Gamma^*$. We see that when guess and memory are empty, A works just as M; we will see that these are the only transitions changing the component from Q of A's states. Thus the simulation of M and the undoing of duplications, which uses match and memory leaving the component from Q unchanged, are done more or less independently. The next kind of transition makes a guess that the following letters on the input tape are the result of a duplication. Transitions

$$\delta(q|_v^u, x, \gamma) := (q|_v^w, u\gamma)$$

are defined for any words $u \in (\Sigma \cup \underline{\Sigma})^{\leq n}$ and $v, w \in \Sigma^{\leq n}$. Whatever is in the match is put on the stack to continue processing later. Note that the word u is put on the stack as a single symbol.

Next A checks whether the input continues with ww. This is done by matching the guess twice against the input, which is read, the first time underlining it in the guess, then undoing this. When both are matched, our PDA should continue as if there was one occurrence of w left on the input tape. However, both are already read. Thus we put w into the memory and read from there as if it was the input tape. Since in this construction the contents of the memory are thought to be situated in front of the input tape contents, nothing is ever read from the input tape, while the memory is not empty. For both situations all transitions are defined in parallel.

The variables used in the definition are quantified as follows: $q \in Q$, $x \in \Sigma$, $u, v, z \in \Sigma^*$, $\gamma \in \Gamma'$, $\beta \in \Gamma$, and $w \in \underline{\Sigma}^* \cdot \Sigma^* \cup \Sigma^* \cdot \underline{\Sigma}^*$ with $|w| \leq n$. Further, all catenations of words and letters are supposed to be no longer than n, and underlining a word from Σ^* shall signify the corresponding word over $\underline{\Sigma}$ obtained by underlining all the individual letters.

$$\delta(q|_{\lambda}^{zxu}, x, \gamma) := (q|_{\lambda}^{zxu}, \gamma) \quad \text{and} \quad \delta(q|_{xv}^{zxu}, \lambda, \gamma) := (q|_{v}^{zxu}, \gamma)$$

$$\delta(q|_{\lambda}^{xu}, x, \gamma) := (q|_{\lambda}^{xu}, \gamma) \quad \text{and} \quad \delta(q|_{xv}^{xu}, \lambda, \gamma) := (q|_{v}^{xu}, \gamma)$$

$$\delta(q|_{\lambda}^{zxu}, x, \gamma) := (q|_{\lambda}^{zxu}, \gamma) \quad \text{and} \quad \delta(q|_{xv}^{zxu}, \lambda, \gamma) := (q|_{v}^{zxu}, \gamma)$$

$$\delta(q|_{\lambda}^{zx}, x, w) := (q|_{zx}^{w}, \lambda) \quad \text{and} \quad \delta(q|_{xv}^{zx}, \lambda, w) := (q|_{zxv}^{w}, \lambda)$$

$$\delta(q|_{\lambda}^{zx}, x, \beta) := (q|_{zx}^{\lambda}, \beta) \quad \text{and} \quad \delta(q|_{xv}^{zx}, \lambda, \beta) := (q|_{zxv}^{\lambda}, \beta)$$

Finally, also the simulation of M must be possible, when the memory is not empty. Thus for $x \in \Sigma$ we define the analogue to transitions defined in 1 for reading from the tape:

$$\delta(q|_{xv}^{\lambda}, \lambda, \gamma) := (q'|_{v}^{\lambda}, \alpha) \text{ where } \varphi(q, x, \gamma) = (q', \alpha).$$

There are no other transitions than the ones defined above. We now prove that $L^{\heartsuit \leq n} \subseteq L(A)$. For this, one observation is essential, whose truth should be immediately comprehensible after what we have already said about the way that A works.

Lemma 9. *If from a state $q|_{\lambda}^{u}$ with vw next on the working tape and γ on the stack there exists an accepting computation for A, then from $q|_{v}^{u}$ with w next on the working tape and γ on the stack there also exists an accepting computation.*

With this we can prove $L^{\heartsuit \leq n} \subseteq L(A)$ by induction on the number of duplications used to reach a word $w \in L^{\heartsuit \leq n}$ from a word $u \in L$. While neither u nor the number need to be unique, they both must exist for all words in $L^{\heartsuit \leq n}$. So let u be a word such that $w \in u^{\heartsuit \leq n}$ via $k+1$ duplications. Then there exists a word u' reachable from u via k duplications such that $u' \heartsuit^{\leq n} w$.

Let us suppose that all words, which can be generated by k duplications from words in L, are accepted by A; then $u' \in L(A)$, and there exists an accepting computation of A for u', let us call it Ξ. Further let i, ℓ be integers such that the duplication of the factor of length ℓ starting at position i in u' results in w, i.e. $w = u'[1 \ldots i-1]u'[i \ldots i+\ell-1]^2 u'[i+\ell \ldots |u'|]$. Obviously A can on input w follow the computation Ξ on the prefix $u'[1 \ldots i-1]$. Let us call the configuration reached in the step before reading the next input letter ξ and let its state be s. Then in s the memory is empty, otherwise A would not read from the input tape.

Now instead of following Ξ further, we guess the duplication of $u'[i \ldots i+\ell-1]$ and reduce it in the manner described above. At the end of this process we will have reached a state equal to s except for the fact that its memory contains $u'[i \ldots i+\ell-1]$. On the tape we have left $u'[i+\ell \ldots |u'|]$. By Lemma 9 there is an accepting computation for this configuration if there is one for ξ. Since Ξ is such an accepting computation, also w is accepted by A.

Further, A can obviously simulate any computation of M and thus $L(M) \subseteq L(A)$, i.e. all words reachable by zero duplications are in $L(A)$. Thus also the basis for our induction is given and we have $L^{\heartsuit \leq n} \subseteq L(A)$.

We do not prove in detail that $L(A) \subseteq L^{\leq n}$. The two parts of A, the one deterministically reducing duplications and the one simulating the original PDA

M work practically independently, as the corresponding state sets are disjoint and separated by the match being filled or not. From these facts $L(A) \subseteq L^{\leq n}$ should be comprehensible rather easily. $\square$

Of course, the same construction works for any finite set of factors that can be duplicated, and we immediately obtain a corollary.

Corollary 10. *The class of context-free languages is closed under the operation of uniformly bounded duplication.*

For general duplication this proof technique does not apply, because over three letters there is no n such that $(abc)^\heartsuit = (abc)^{\heartsuit \leq n}$. In fact, n-bounded duplication grows more powerful with every increase of n. Here we will use the following two notions: a word w is *square-free*, if it does not contain any non-empty factor of the form $uu = u^2$; w is *circular square-free*, if the same holds true for w written along a circle, or equivalently if ww contains no square shorter than itself.

Proposition 11. *For two integers m and n with $17 < m < n$ the inclusion $(abc)^{\heartsuit \leq m} \subset (abc)^{\heartsuit \leq n}$ is proper.*

Proof. First we show that for every square-free word u over three letters starting with abc there exists a word v, such that $uv \in (abc)^{\heartsuit \leq k}$ for $k \geq 4$. This word is constructed from left to right in the following manner. The first three letters are abc and thus do not need to be constructed.

The fourth letter is created by going from the third letter left to the last occurrence of this desired letter. Since abc is a prefix of the word all three letters do have such an occurrence. Now the factor from this rightmost occurrence to the third letter is duplicated. In this way the fourth letter of the new word becomes the desired one. Then we move to the fifth letter, obtain it by duplicating the factor reaching back till its rightmost occurrence, and so on.

The last occurrence of any letter in the part of u already constructed can be at most four positions from the last, because there are only two more different letters and the longest square-free word over two letters has length three. Of course, if in some step more than one letter of u is produced, the process can advance to the next wrong one without further duplications.

We will illustrate this construction with a short example. From abc we construct $abcbacb$ as a prefix. Underlining signals the factor duplicated to obtain the following word, the horizontal bar signals the end of the prefix of $abcbacb$ constructed at the respective point. $a\underline{bc} \to ab\underline{cb}|c \to ab\underline{cb}a|bcbc \to abcbacb|abcbc$

We now establish some bounds for the number of additional symbols produced. Since abc is already there, $|u| - 3$ letters need to be constructed. In every step at most $2k - 1$ letters of u can be constructed, because u is square-free; thus at least one letter is added to v. At the same time at most $2k - 1$ letters are added to v, since no useless duplications are done. Thus we have $|u| - 3 \leq |v| \leq (|u| - 3)(2k - 1)$. Of course, every circular square-free word is square-free and can be constructed in this way, too. Starting from lengths of 18, such a word always exists [2].

Now we construct in this way a circular square-free word w of length n as a prefix of a word wv' in $(abc)^{\heartsuit \leq n}$. We can expand this prefix to w^i in $i-1$ steps for any given $i \geq 1$ by the rule $w \to ww$, so all $w^i v'$ are in $(abc)^{\heartsuit \leq n}$. Further, w^i contains no squares shorter than $2n$, because w is circular square-free. Thus for constructing the same prefix in $(abc)^{\heartsuit \leq m}$ also the bounds $|w^i| - 3 \leq |v| \leq (|w^i| - 3)(2m - 1)$ for the corresponding suffix v apply. For big enough i the shortest such v will be longer than v'. Thus such a $w^i v'$ cannot be in $(abc)^{\heartsuit \leq m}$, while it is in $(abc)^{\heartsuit \leq n}$. $\qquad\square$

5   Conclusions

Thus the problem, which has received most attention in investigations on duplication remains open: Is the general duplication closure of a word over three letters always context-free? Probably this is equivalent to asking whether context-free languages are closed under general duplication. Our investigations on the length-bounded case may have shed some more light on the nature of the problem, though.

Another problem is raised by Proposition 11: Are the inclusions $(abc)^{\heartsuit \leq m} \subset (abc)^{\heartsuit \leq n}$ for $m < n$ proper also for $n \leq 17$? Or do these inclusions hold only when a circular square-free word of the corresponding length exists?

References

1. R. BOOK and F. OTTO: *String-Rewriting Systems*. Springer, Berlin, 1988.
2. J.D. CURRIE: *There are Ternary Circular Square-free Words of Length n for n $\geq$ 18*. In: Electric Journal of Combinatorics, 9(1) N10, 2002.
3. J. DASSOW, V. MITRANA and GH. PĂUN: *On the Regularity of Duplication Closure*. Bull. EATCS 69, 1999, pp. 133–136.
4. M.A. HARRISON: *Introduction to Formal Language Theory*. Reading, Mass., 1978.
5. P. LEUPOLD: *n-Bounded Duplication Codes*. Proceedings of the ICALP-Workshop on Words, Avoidability, Complexity, Turku 2004. Technical Report 2004-07, Laboratoire de Recherche en Informatique d'Amiens, Amiens 2004.
6. P. LEUPOLD, C. MARTÍN VIDE and V. MITRANA: *Uniformly Bounded Duplication Languages*. In: Discrete Applied Mathematics Vol 146, Iss 3, 2005, pp. 301–310.
7. P. LEUPOLD and V. MITRANA: *Uniformly Bounded Duplication Codes*. Submitted.
8. P. LEUPOLD, V. MITRANA and J. SEMPERE: *Languages Arising from Gene Repeated Duplication*. In: Aspects of Molecular Computing. Essays in Honour Tom Head on his 70th Birthday. LNCS 2950, Springer Verlag, Berlin, 2004, pp. 297–308.
9. C. MARTÍN-VIDE and GH. PĂUN: *Duplication Grammars*. In: Acta Cybernetica 14, 1999, pp. 101–113.
10. V. MITRANA and G. ROZENBERG: *Some Properties of Duplication Grammars*. In: Acta Cybernetica 14, 1999, pp. 165–177.
11. A. SALOMAA : *Formal Languages*. Academic Press, Orlando, 1973.
12. M.-W. WANG: *On the Irregularity of the Duplication Closure*. Bull. EATCS 70, 2000, pp. 162–163.

The Boolean Closure of Growing
Context-Sensitive Languages

Tomasz Jurdziński

Institute of Computer Science, Wrocław University,
Przesmyckiego 20, PL-51-151 Wrocław, Poland
tju@ii.uni.wroc.pl

Abstract. The class of growing context-sensitive languages (GCSL) is a
naturally defined subclass of context-sensitive languages whose membership problem is solvable in polynomial time. GCSL and its deterministic counterpart called Church-Rosser Languages (CRL) complement the
Chomsky hierarchy in a natural way [9]. In this paper, the extension of
GCSL obtained by closures of this class under the boolean operations are
investigated. We show that there exists an infinite intersection hierarchy,
answering an open problem from [1]. Further, we compare the expressive
power of the boolean closures of GCSL, CRL, CFL and LOGCFL.

1 Introduction

Formalisms defining language classes located between context-free languages
(CFL) and context-sensitive languages (CSL) have been intensively studies for
many years. One of the motivations was to find families which possess an acceptable computational complexity, large expressibility, and natural characterizations by grammars and a machine model. Neither CSL nor CFL fulfil these
demands. For the first class the membership problem is PSPACE-complete what
makes it in its full generality too powerful. Context-free grammars are e.g. not
powerful enough to express all syntactical aspects of programming languages.

One of the most interesting proposals was presented by Dahlhaus and Warmuth [3], who considered grammars with strictly growing rules, i.e., such that the
righthand side of the production is longer than the lefthand side. They showed
a rather surprising result that each language generated by a growing grammar
can be recognized in deterministic polynomial time (and it is even included in
LOGCFL). Buntrock and Loryś showed that this class forms an abstract family
of languages. They also proved that the class GCSL can be characterized by less
restricted grammars.

Machine model characterizations of GCSL were investigated in the sequence of
papers [2, 12, 11]. Finally, the characterization by the so-called length-reducing
two-pushdown automata (lrTPDA) has been found. Recently, Holzer and Otto
considered generalizations of these automata [4].

The class of languages recognized by deterministic lrTPDAs is equal to the
class of Church-Rosser languages (CRL) [12], introduced by McNaughton et al.

O.H. Ibarra and Z. Dang (Eds.): DLT 2006, LNCS 4036, pp. 248–259, 2006.

[10] in terms of string-rewriting systems [2, 11]. So, CRL is a deterministic counterpart of GCSL. The class of Church-Rosser languages posses very useful properties, discussed in [11]. GCSL and CRL complement the Chomsky hierarchy in a natural way [9], as these classes fill the gap between CFL and CSL:

$$CFL \subsetneq GCSL \subsetneq CSL$$
$$\cup \qquad \cup \qquad \cup$$
$$DCFL \subsetneq CRL \subsetneq DCSL$$

where DCFL is the set of deterministic context-free languages and DCSL is the set of deterministic context-sensitive languages (equal to DLINSPACE(n)).

Though GCSL and CRL have many good properties, their weakness seems to be evident. For example, GCSL does not even contain the language $\{wcw \mid w \in \{a, b\}^*\}$, and CRL does not contain the language of palindromes [5]. In this paper, we study extensions of GCSL and CRL obtained by the closure of this classes under the boolean operations. Note that the recognition of each language from these closures is polynomial. So, a natural question arises how do they extend the appropriate classes. Such questions motivated the considerations of the boolean closures of many formal language classes [6, 14, 15, 16, 8]. An elegant generalization of the notion of the boolean closure of CFL was introduced by Okhotin [13].

First, we show that there exist infinite intersection hierarchies, $\Lambda_{\cap_i}(X) \subsetneq \Lambda_{\cap_{i+1}}(X)$ for each $i \geq 1$, $X \in \{CRL, GCSL\}$, where $\Lambda_{\cap_i}(X)$ is the class of languages obtained by intersections of i languages from the class X. This result solves an open problem from [1]. An analogous hierarchy for context-free languages is infinite as well [8]. Indeed, our result is related to this hierarchy, because the witness language for the inclusions $\Lambda_{\cap_i}(GCSL) \subsetneq \Lambda_{\cap_{i+1}}(GCSL)$ and $\Lambda_{\cap_i}(CRL) \subsetneq \Lambda_{\cap_{i+1}}(CRL)$ belongs to $\Lambda_{\cap_{i+1}}(DCFL) \setminus \Lambda_{\cap_i}(GCSL)$. (Similar languages were used in order to show the analogous hierarchy for context-free grammars.) Further, we investigate the expressive power of the closures of GCSL and CRL under the boolean operations. We compare these classes with the boolean closures of CFL and LOGCFL.

The paper is organized as follows. In Section 2 we introduce some basic notions and definitions. Sections 3 describes formal tools used in our proofs. Section 4 presents the intersection hierarchies. In Section 5 we compare the expressive power of the boolean closures of CRL, CFL, GCSL, and LOGCFL. Finally, in Section 6, we show that neither GCSL nor CRL is closed under the shuffle operation.

2 Preliminaries

Throughout the paper ε denotes the empty word, $\mathbb{N}$, $\mathbb{N}_+$ denote the set of non-negative and positive integers. For a word x, let $|x|$, $x[i]$ and $x[i, j]$ denote the length of x, the ith symbol of x and the factor $x[i] \ldots x[j]$ respectively, for $0 < i \leq j \leq |x|$. Further, let $[i, j] = \{l \in \mathbb{N} \mid i \leq l \leq j\}$, let x^R denote the reverse of the word x, that is, $x^R = x[n]x[n-1] \ldots x[2]x[1]$ for $|x| = n$.

Let $x = x_1 y_1 x_2 y_2 \ldots x_n y_n x_{n+1}$, $n > 0$, where $x_i, y_i \in \Sigma^*$ for the alphabet Σ ($i \in [1, n+1]$), let y_i in the above factorization of x denote the leftmost occurrence of y_i as a subword of x (so, the above factorization is possible only if the leftmost occurrence of y_{i+1} is located to the right of the leftmost occurrence of y_i). Then,

$$x - y_1 = x_1 \quad x_2 y_2 x_3 y_3 \ldots x_n y_n x_{n+1},$$
$$x - y_1 + z_1 = x_1 z_1 x_2 y_2 \ldots x_n y_n x_{n+1},$$
$$x - (y_1, \ldots, y_n) = x_1 x_2 \ldots x_{n+1},$$
$$x - (y_1, \ldots, y_n) + (z_1, \ldots, z_n) = x_1 z_1 x_2 z_2 \ldots x_n z_n x_{n+1}.$$

Let $\mathcal{L}$ be a family of languages and $op_1, \ldots, op_k$, $k \in \mathbb{N}$, be a finite number of operations defined on $\mathcal{L}$. Then $\Lambda_{op_1, \ldots, op_k}(\mathcal{L})$ denotes the least family of languages which contains $\mathcal{L}$ and is closed under $op_1, \ldots, op_k$. We consider the operations complementation ($\sim$), union ($\cup$), intersection ($\cap$) and shuffle ($\bowtie$). We write also Λ_{Bool} for $\Lambda_{\sim, \cap, \cup}$.

We will make use of the notion of Kolmogorov complexity $K(x)$ of words x over a binary alphabet [7].

Fact 1. *[7] 1. For each $n \in \mathbb{N}$, there exists a word $x \in \{0,1\}^n$ such that $K(x) > n - 1$.*
2. Let X be a set of words such that $|X| \geq m$. Then, there exists $x \in X$ such that $K(x) \geq \lceil \log m \rceil - 1$.
3. Assume that $K(x_1 x_2 x_3) > n - p$, where $n = |x_1 x_2 x_3|$. Then, $K(x_2 | x_1 x_3) > |x_2| - p - O(\log n)$, where $K(x|y)$ denotes Kolmogorov complexity of x when y is known.

Growing context-sensitive languages are basically defined by growing grammars and Church-Rosser languages are defined in terms of string-rewriting systems [3, 10]. We use characterizations of these classes by length-reducing two-pushdown automata.

A two-pushdown automaton (TPDA) $M = (Q, \Sigma, \Gamma, q_0, \perp, F, \delta)$ with a window of length $k = 2j$ is a nondeterministic automaton with two pushdown stores. It is defined by the set of states Q, the input alphabet Σ, the tape alphabet Γ ($\Sigma \subseteq \Gamma$), the initial state $q_0 \in Q$, the bottom marker of the pushdown stores $\perp \in \Gamma \backslash \Sigma$, the set of accepting states $F \subseteq Q$ and the transition relation δ : $Q \times \Gamma_{\perp, j} \times \Gamma_{j, \perp} \to \mathcal{P}(Q \times \Gamma^* \times \Gamma^*)$, where $\Gamma_{\perp, j} = \Gamma^j \cup \{\perp v : |v| \leq j - 1, v \in \Gamma^*\}$, $\Gamma_{j, \perp} = \Gamma^j \cup \{v \perp : |v| \leq j - 1, v \in \Gamma^*\}$, $\mathcal{P}(Q \times \Gamma^* \times \Gamma^*)$ denotes the set of finite subsets of $Q \times \Gamma^* \times \Gamma^*$. The automaton M is *deterministic* (DTPDA) if δ is a (partial) function from $Q \times \Gamma_{\perp, j} \times \Gamma_{j, \perp}$ into $Q \times \Gamma^* \times \Gamma^*$. A (D)TPDA is called *length-reducing* (lr(D)TPDA) if $(p, u', v') \in \delta(q, u, v)$ implies $|u'v'| < |uv|$, for all $q \in Q$, $u \in \Gamma_{\perp, j}$, and $v \in \Gamma_{j, \perp}$.

A *configuration* of a (D)TPDA M is described by the word $u q_i v^R$, where q_i is the current state, $u, v \in \Gamma^*$ are the contents of the first pushdown store and the second pushdown store, resp. Note that the bottom marker $\perp$ occurs on both ends of the word $u q_i v^R$. The transition $\delta(q, u, v) = (q', u', v')$ will be described as $u q v^R \to u' q' (v')^R$. We define a single step computation relation $\vdash_M$ on configurations in a natural way, i.e., $u z q x v \vdash_M u z' q' x' v$ if $z q x \to z' q' x'$.

For an input word $x \in \Sigma^*$, the corresponding *initial configuration* is $\perp q_0 x \perp$, i.e., the input word is given as the contents of the second pushdown store. The automaton M *finishes* its computation by empty pushdown stores. So, $L(M) = \{x \in \Sigma^* : \exists_{q \in F} \perp q_0 x \perp \vdash_M^* q\}$, where $L(M)$ is the language accepted by M. We also require that the special symbol $\perp$ occurs only on bottoms of the pushdowns and no other symbol can occur on the bottom.

Theorem 1 ([11, 12]). *A language is accepted by a (deterministic) lrTPDA if and only if it is a growing context-sensitive language (Church-Rosser language).*

The following table summarizes closure properties of GCSL and CRL [11]:

	$\cup$	$\cap$	$\sim$
GCSL	$+$	$-$	$-$
CRL	$-$	$-$	$+$

3   Lower Bounds Tools

We describe the notion of a computation graph and its properties, see [5]. (The similar notion of derivation graphs was considered in [3, 2].) Each computation of a lrTPDA $M = (Q, \Sigma, \Gamma, q_0, \perp, F, \delta)$ corresponds to a planar directed acyclic graph defined in the following way. Vertices are labeled with symbols, transitions and states, where $\omega(\pi)$ denotes the label of the vertex π, ω is a function from the set of the vertices of the graph to $\Gamma \cup Q \cup \delta$. Vertices labeled with symbols, states, and transitions are called symbol vertices, state vertices, and transition vertices, respectively.

A *computation graph* $G_{(j)} = (V_j, E_j)$ corresponding to the computation $C_0 \vdash \ldots \vdash C_j$ where C_0 denotes an initial configuration is defined inductively (see examples in [5]):

Case 1: $j = 0$. Let $C_0 = \perp q_0 x_1 x_2 \ldots x_n \perp$ be an initial configuration, where $x_i \in \Sigma$ for $i \in [1, n]$. Then $G_{(0)} = (V_0, E_0)$, where $E_0 = \emptyset$, $V_0 = \{\rho_i\}_{i=-2}^{n+2}$ such that $\omega(\rho_i) = x_i$ for $1 \leq i \leq n$, $\omega(\rho_i) = \perp$ for $i \in \{-2, -1, n+1, n+2\}$ and $\omega(\rho_0) = q_0$.

Case 2: $j > 0$. Assume that the computation $C_0 \vdash_M \ldots \vdash_M C_{j-1}$ corresponds to the graph $G_{(j-1)}$, and the transition $z \to z'$ is executed in $C_{j-1} \vdash C_j$ for $z, z' \in \Gamma^* Q \Gamma^*$, i.e., $C_{j-1} = y_1 z y_2 \vdash_M y_1 z' y_2 = C_j$. Let $|z| = p$ and $|z'| = p'$. The graph $G_{(j)}$ is constructed from $G_{(j-1)}$ by adding:

- the vertices $\pi'_1, \ldots, \pi'_{p'}$ which correspond to the word z', i.e., $\omega(\pi'_i) = z'[i]$ for $i \in [1, p']$;
- the vertex D_j which corresponds to the transition $z \to z'$;
- the edges $(\pi_1, D_j), \ldots, (\pi_p, D_j)$, where the vertices $\{\pi_i\}_{i=1}^{p}$ correspond to z;
- the edges $(D_j, \pi'_1), \ldots, (D_j, \pi'_{p'})$.

There is a natural left to right ordering among the sources of a computation graph, induced by the left to right ordering into the initial configuration. For two vertices π_1 and π_2, $\pi_1 \prec \pi_2$ denotes that π_1 precedes π_2 according to this ordering. There is also a left to right ordering among the in-neighbours and the out-neighbours of each transition vertex. This ordering induces a left to right ordering among the sinks of the computation graph.

Note that the *sources* (vertices with no incoming edges) of $G_{(j)}$ correspond to the initial configuration C_0, the sequence of their labels will be denoted as $src(G_{(j)})$. Similarly, the *sinks* (vertices with no outcoming edges) of $G_{(j)}$ correspond to the last configuration described by $G_{(j)}$ (i.e., C_j) and the sequence of their labels is denoted as $snk(G)$. (The vertices ρ_{-2} and ρ_{n+2} are the artificial vertices introduced for technical reasons only.)

We extend the single step transition relation $\vdash_M$ to computation graphs: $G \vdash_M G'$ if there exist the configurations C, C' and C_0 such that G corresponds to the particular computation $C_0 \vdash_M^* C$, and G' corresponds to the computation $C_0 \vdash^* C \vdash C'$.

We apply the term *path* exclusively to paths that start in a source vertex and finish in a sink. The relation $\prec$ among vertices of a graph induces a left-to-right partial ordering of paths. A path σ_1 is to the left of a path σ_2 iff none of the vertices of σ_1 is to the right of any vertex of σ_2. A path σ is the leftmost (rightmost) path in a set of paths S if it is to the left (right) of every path $\sigma' \in S$.

Let σ be a path in G with a sink π. We say that σ is *short* if there is no path with the sink π that is shorter than σ. A sink π is *i-successor* if one of the short paths with the sink in π starts in ρ_i, the source vertex associated to the ith symbol of the initial configuration ($i \in [-2, n + 2]$). Let us note here that a sink of a graph may be i-successor for many i's; on the other hand, it is possible that no sink is j-successor for some j.

Lemma 1. *[5] Let V_{sr} and V_{sn} be subsets of the set of sources of the computation graph G and the set of sinks of G, resp. Then, the set P of short paths with sources in V_{sr} and sinks in V_{sn} contains (if $P \neq \emptyset$) the path which is to the right/left of all other paths in P.*

Lemma 2. *[5] The length of each short path in a computation graph which describes the computation on an input word of length n is $O(\log n)$.*

Now, we introduce definitions needed to formulate cut and paste technique for computation graphs. The *description of the path* $\sigma = \pi_1, \pi_2 \ldots, \pi_{2l+1}$, denoted $desc(\sigma)$, consists of the sequence $(\omega(\pi_1), p_1), \ldots (\omega(\pi_{2l+1}), p_{2l+1})$ such that π_i is the p_i-th in-neighbour of π_{i+1} for odd $i < 2l$, and π_{i+1} is the p_{i+1}-st out-neighbour of π_i for even i, $p_{2l+1} = 0$. A *full description* of the path σ in the computation graph G, $descf(\sigma)$, is equal to $(desc(\sigma), p)$, where p is the position of the source of σ in the initial configuration.

We say that the descriptions $\gamma_1, \ldots, \gamma_{l-1}$ (for $l > 1$) *decompose* the computation graph G into the subgraphs $G_1, \ldots, G_l$ if G contains paths $\{\sigma_i\}_{i=1}^{l-1}$ such that: $desc(\sigma_i) = \gamma_i$ for $i \in [1, l-1]$, σ_i is located to the left of σ_{i+1} for $i \in [1, l-2]$, G_i is equal to the subgraph of G which contains all vertices and all edges located between σ_{i-1} and σ_i (where σ_0 and σ_l are the "artificial" empty paths located to the left/right of all other vertices of G). If the descriptions $\gamma_1, \ldots, \gamma_{l-1}$ decompose G into the subgraphs $G_1, \ldots, G_l$, we write $G = G_1 \ldots G_l$. Note that

$src(G) = src(G_1) \dots src(G_l)$ and $snk(G) = snk(G_1) \dots snk(G_l)$ for $G = G_1 \dots G_l$ (where the last symbol of $src(G_i)/snk(G_i)$ is identified with the first symbol of $src(G_{i+1})/snk(G_{i+1})$ for $i \in [1, l-1]$).

Lemma 3 (Cut and Paste Lemma). *[5] Assume that the descriptions γ_1, γ_2 decompose a graph G into G_1, G_2, G_3 and a graph H into H_1, H_2, H_3. Then, $J = G_1 H_2 G_3$ is the computation graph corresponding to the computation $src(G_1)$ $src(H_2)src(G_3) \vdash_M^* snk(G_1)snk(H_2)snk(G_3)$. Moreover, γ_1, γ_2 decompose J into G_1, H_2, G_3.*

Let G_i be a subgraph of $G = G_1 \dots G_l$ $(l > 1)$. Then, the surface of G_i, $\mathrm{srf}(G_i)$, is defined as the tuple $(\mathrm{desc}(\sigma_1), \mathrm{desc}(\sigma_2), \omega(snk(G_i)))$, where σ_1, σ_2 are paths on the borders of G_i.

Let G be a computation graph corresponding to the computation on the input word $x = x_1 x_2 x_3$, let σ_1, and σ_2 be the rightmost short path with the source in $x_1 x_2[1]$ and the leftmost short path with the source in $x_2[|x_2|]x_3$, resp. (such σ_1, σ_2 exist by Lemma 1). Then, if σ_1 is not to the left of σ_2, the *image* of x_2 (or $(|x_1| + 1, |x_1 x_2|)$-image) in G is *undefined*. Otherwise, let $G = G_1 G_2 G_3$, where σ_1, σ_2 are the paths on the borders of G_2. If $snk(G_2) < 3$, then the image of x_2 is undefined as well. Otherwise, the image of x_2 is equal to $\mathrm{srf}(G_2)$, and the length of the image is equal to $|snk(G_2)|$. Below, we enumerate some basic properties of images.

Proposition 1. *[5] Let G, G' be computation graphs corresponding to a computation on an input word of length n, $G \vdash G'$, and $0 \le l < r \le n+1$. Assume that the (l, r)-image is defined in G and it is equal to $(\sigma_1, \sigma_2, \tau)$. Then,*
(a) For every $1 < i < |\tau|$, if $\tau[i]$ is j-successor then $l < j < r$.
(b) If $|\tau| > 2k$ then the (l, r)-image is defined in G', and its length is in $[|\tau| - k, |\tau| + k]$.
(c) Assume that the (l', r')-image is defined in G for $r < l' < r' \le n+1$ and it is equal to $(\sigma_1', \sigma_2', \tau')$. Then, $|\tau \cap \tau'| \le 2$.
(d) If the (l, r)-image is undefined in G, it is undefined in G' as well.

Now, we present some new technical notions and results. Let $v = y_1 \, x_1 \, y_2 \, x_2 \, y_3 \in \Sigma^*$ be the input word of length n, let $0 < c < 1$, let $\mathcal{C}$ be a computation of a lrTPDA on v. We say that the pair (x_1, x_2) is *checked with level c* during the computation $\mathcal{C}$, if the images of x_1 and x_2 are not shorter then $c|x_1|$ and $c|x_2|$, as long as the image of y_2 is defined and longer that $2k$ (where k is equal to the size of the window).

Proposition 2. *Let $\mathcal{C}$ be a computation of a length-reducing two-pushdown automaton M on the input word $v = y_1 \, x_1 \, y_2 \, x_2 \, y_3 \in \Sigma^*$, such that the pair (x_1, x_2) is not checked in $\mathcal{C}$ with the level c' and $c'|x_1|, c'|x_2| > 2 + k$. Let G be the first (according to the computation relation $\vdash$) graph in the computation $\mathcal{C}$ such that the image of x_1 is shorter than $c'|x_1|$ or the image of x_2 is shorter than $c'|x_2|$. Then, the path on the right border of the subgraph of G defining the image of x_1 is to the left of the path on the left border of the (subgraph defining) the image of x_2.*

The proof of the above proposition is based on the observation that the image of y_2 is defined in G, so there exists a short path which starts in y_2 (by Proposition 1(a)). Then, the final statement follows from Lemma 1.

Proposition 3. *Let $\{M_i\}_{i=1}^r$ be lrTPDAs over the alphabet Σ. Let $w, w' \in L(M_i)$ for each $i \in [1, r]$, where $w = w_1w_2w_3w_4w_5$, $w' = w_1w_2'w_3w_4'w_5$, $|w_i| = |w_i'|$ for $i = 2, 4$. Assume that, for each $i \in [1, r]$, there exists an accepting computation of M_i on w (w', resp.) which contains the computation graph $G_i = G_{i,1}G_{i,2}G_{i,3}G_{i,4}G_{i,5}$ ($G_i' = G_{i,1}'G_{i,2}'G_{i,3}'G_{i,4}'G_{i,5}'$, resp.) such that:*

- *$\mathrm{descf}(\sigma_{i,j}) = \mathrm{descf}(\sigma_{i,j}')$ for $j \in [1, 4]$, where $\sigma_{i,j}$ ($\sigma_{i,j}'$, resp.) is the path on the border between $G_{i,j}$ and $G_{i,j+1}$ (between $G_{i,j}'$ and $G_{i,j+1}'$, resp.);*
- *the surface of $G_{i,l}$ ($G_{i,l}'$, resp.) forms the image of w_l (w_l', resp.) for $l \in \{2, 4\}$;*
- *$\mathrm{srf}(G_{i,2}) = \mathrm{srf}(G_{i,2}')$ or $\mathrm{srf}(G_{i,4}) = \mathrm{srf}(G_{i,4}')$.*

Then, each of $\{M_i\}_{i=1}^r$ accepts $w_1w_2'w_3w_4w_5$ and $w_1w_2w_3w_4'w_5$.

Proof. We show that $w_1w_2'w_3w_4w_5 \in L(M_i)$ for each $i \in [1, r]$. Note that the equality of the full descriptions of the paths $\sigma_{i,j}$ and $\sigma_{i,j}'$ for each $i \in [1, r]$, $j \in [1, 4]$ implies that $H_1H_2H_3H_4H_5$ is a computation graph, for each $H_j \in \{G_{i,j}, G_{i,j}'\}$ (see Cut and Paste Lemma). Consider two cases:

Case 1: $\mathrm{srf}(G_{i,2}) = \mathrm{srf}(G_{i,2}')$. Then, the input word corresponding to the computation graph $G'' = G_{i,1}G_{i,2}'G_{i,3}\,G_{i,4}G_{i,5}$ is equal to $w_1w_2'w_3w_4w_5$. Moreover, the (last) configuration corresponding to the graph G_i (i.e., $\mathrm{snk}(G_i)$) and the configuration corresponding to the graph G'' (i.e., $\mathrm{snk}(G'')$) are equal, by the assumption that $\mathrm{srf}(G_{i,2}) = \mathrm{srf}(G_{i,2}')$. As the configuration $\mathrm{snk}(G_i)$ occurs in the accepting computation of M_i (by the choice of the graphs $G_1, \ldots, G_r$), the automaton M_i can accept starting from the last configuration of G''. Thus, $w_1w_2'w_3w_4w_5 \in L(M_i)$.

Case 2: $\mathrm{srf}(G_{i,4}) = \mathrm{srf}(G_{i,4}')$. Then, the input word of the computation graph $G'' = G_{i,1}'G_{i,2}'G_{i,3}'\,G_{i,4}G_{i,5}'$ is equal to $w_1w_2'w_3w_4w_5$. Moreover, the (last) configurations corresponding to the graphs G_i' and G'' are equal, by the assumption that $\mathrm{srf}(G_{i,4}) = \mathrm{srf}(G_{i,4}')$. As $\mathrm{snk}(G_i')$ occurs in the accepting computation of M_i, the automaton M_i can accept starting from the last configuration of G''. Thus, $w_1w_2'w_3w_4w_5 \in L(M_i)$.   □

4   Intersection Hierarchies and Their Consequences

Let $L_j = \{x_1\#\ldots\#x_j\#x_1^R\#\ldots\#x_j^R \mid x_i \in \{0,1\}^* \text{ for } i \in [1,j]\}$. Below, we show that the family $\{L_j\}_{j \in \mathbb{N}}$ certifies the infinite intersection hierarchies.

Lemma 4. *$L_j \in \Lambda_{\cap_j}(\mathsf{DCFL}) \setminus \Lambda_{\cap_{j-1}}(\mathsf{GCSL})$ for each $j > 1$.*

Proof. First, observe that $L_j \in \Lambda_{\cap_j}(\mathsf{DCFL})$. Indeed, $L_j = \bigcap_{i=1}^j L(M_i')$, where M_i' is a deterministic pushdown automaton that checks if an input word has the form $x_1\#\ldots\#x_j\#\,y_1\#\ldots\#y_j$ such that $x_i = y_i^R$ and $x_l, y_l \in \{0,1\}^*$ for each $l \in [1, j]$.

For the sake of contradiction assume that there exist $j - 1$ (nondeterministic) length-reducing two-pushdown automata $M_1, \ldots, M_{j-1}$ such that $L_j = \bigcap_{i=1}^{j-1} L(M_i)$. Now, for m large enough, consider the input word $w = x_1 \# \ldots \# x_j \# y_1 \# y_2 \# \ldots \# y_j \in L_j$, where $|x_i| = |y_i| = m$, $x_i = y_i^R$ for each $1 \leq i \leq j$, and $K(x) > jm - 1$, where $x = x_1 \ldots x_j$. (In other words, the concatenation of $x_1, \ldots, x_j$ is an incompressible word.)

As $w \in L_j$, it is accepted by each M_i for $i \in [1, j-1]$. Let us choose an accepting computation of M_i on w, C_i, for each $i \in [1, j-1]$. Let p_i be such that the images of x_{p_i} and y_{p_i} are longer than $c'm$ for the longest time in the computation C_i, where $c' = 1/(16j \log g)$, $g = |\Gamma \cup Q|$. (If there is more than one such p_i, we choose one arbitrarily.)

Note that M_i can check (with level c') only the pair (x_{p_i}, y_{p_i}) in the computation C_i, among the pairs $\{(x_l, y_l)\}_{l \in [1,j]}$. In fact, for each other pair (x_l, y_l) $(l \neq p_i)$, x_{p_i} or y_{p_i} is located between x_l and y_l in the input word w. So, the image of the word separating x_l and y_l is longer than $2k$, until the first configuration in which the image of x_l or the image of y_l is shorter than $c'm$. Indeed, the word separating x_l and y_l contains x_{p_i} or y_{p_i}, so its image is longer than $c'm - k > 2k$ (for m large enough) until the image of x_l or the image of y_l becomes shorter than $c'm$ for the first time (by the choice of p_i and Proposition 1(b)).

As we have $j-1$ automata and j pairs of words, there exists $l \notin \{p_1, \ldots, p_{j-1}\}$. That is, the pair (x_l, y_l) is not checked with level c' in any of the computations C_i for $i \in [1, j-1]$.

Now, for each C_i, let us consider the first graph G_i (according to the computation relation $\vdash_M$) in the computation C_i such that the image of x_l or the image of y_l is shorter than $c'm$ in G_i. Note that, according to this choice, $G_i = G_{i,1} G_{i,2} G_{i,3} G_{i,4} G_{i,5}$, where $G_{i,2}$, $G_{i,4}$ are subgraphs defining the image of x_l and the image of y_l, respectively (by Proposition 2).

We will store a kind of "compact" description $D = (D_1, \ldots, D_{j-1})$ of the above decompositions of $\{G_i\}_{i=1}^{j-1}$. Assume that the image of x_l is shorter than $c'm$ in G_i. Then, D_i consists of the image of x_l (i.e., the surface of $G_{i,2}$) and the full descriptions of the paths on the borders between $G_{i,l}$ and $G_{i,l+1}$ for $l \in [1, 4]$. If only the image of y_l is shorter than $c'm$ in G_i, then the roles of x_l and y_l are replaced. The description D consists of $O(j)$ full descriptions of short paths, and $j - 1$ words, each shorter than $c'm$ and written over the alphabet $\Gamma \cup Q$. Thus, it requires $O(\log m) + jc'm \cdot \lceil \log g \rceil) \leq O(\log m) + \frac{2jm \log g}{16j \log g} < m/4$ bits for m large enough.

We say that the pair $(x_l', y_l') \in \Sigma^m \times \Sigma^m$ *agrees* with the description D if the following condition is satisfied for each automaton M_i, $i \in [1, j-1]$. There exists an accepting computation of M_i on $w' = w - (x_l, y_l) + (x_l', y_l')$, such that the graph $H_i = H_{i,1} H_{i,2} H_{i,3} H_{i,4} H_{i,5}$ occurs during this computation, and:

- the full description of the path $\sigma_{i,l}^H$ is equal to the full description of the path $\sigma_{i,l}^G$ for $l \in [1, 4]$ where $\sigma_{i,l}^X$ is the path on the right border of $X_{i,l}$.
- the surface of $H_{i,2}$ ($H_{i,4}$, resp.) defines the image of x_l' (y_l', respectively);
- if the image of x_l in G_i is shorter than $c'm$, then $\mathrm{srf}(G_{i,2}) = \mathrm{srf}(H_{i,2})$, otherwise $\mathrm{srf}(G_{i,4}) = \mathrm{srf}(H_{i,4})$.

As x_l is incompressible and the description D has $\leq m/4$ bits, there exists another word $x'_l \neq x_l$ such that $(x'_l, (x'_l)^R)$ agrees with the description D. Indeed, otherwise, we would be able to determine x_l, knowing all x_i's except x_l, and the above description D (by checking if z agrees with D for each $z \in \Sigma^m$). It implies that $K(x_l|(x - x_l)) \leq m/4$. On the other hand, $K(x_l|(x - x_l)) \geq m - O(\log m)$ for m large enough by Fact 1.3.

So, let $x'_l \neq x_l$ be such that $(x'_l, (x'_l)^R)$ agrees with D. Then, each of the automata $M_1, \ldots, M_{j-1}$ accepts $w' = w - x_l + x'_l$ by Proposition 3. However, $w' \notin L_j$. Contradiction. $\qquad \Box$

As a consequence of the above lemma, we obtain the following result.

Theorem 2. *For each $i \geq 0$,*

$$
\begin{aligned}
&(a) &&\Lambda_{\cap_i}(X) \subsetneq \Lambda_{\cap_{i+1}}(X) &&\text{for } X \in \{\mathsf{CRL}, \mathsf{GCSL}\} \\
&(b) &&\Lambda_{\cap_i}(X) \subsetneq \Lambda_{\cap_i}(\mathsf{GCSL}) &&\text{for } X \in \{\mathsf{CRL}, \mathsf{CFL}\} \\
&(c) &&\Lambda_{\cap_i}(\mathsf{CFL}) \text{ and } \Lambda_{\cap_i}(\mathsf{CRL}) &&\text{are incomparable} \\
&(d) &&\Lambda_{\cap_{i+1}}(\mathsf{DCFL}) \not\subset \Lambda_{\cap_i}(\mathsf{GCSL})
\end{aligned}
$$

Proof. (sketch) The items (a) and (d) follow directly from Lemma 4. The remaining inequalities follow from the fact that unary context-free languages are regular, while there exist non-regular unary languages in the classes GCSL and CRL [11], and the fact that CRL is closed under complement while CFL and GCSL are not closed under complement. (Observe that the complement of L_j is the context-free language.) $\qquad \Box$

5   The Boolean Closures

Theorem 3. *For $X \in \{\mathsf{CFL}, \mathsf{CRL}\}$, we have the following relationships: $\Lambda_{Bool}(X) \subsetneq \Lambda_{Bool}(\mathsf{GCSL})$; $\Lambda_{Bool}(\mathsf{CRL})$ and $\Lambda_{Bool}(\mathsf{CFL})$ are incomparable.*

Proof. Recall that all unary languages in $\Lambda_{Bool}(\mathsf{CFL})$ are regular, while $\mathsf{CRL} \subset \mathsf{GCSL}$ contains unary languages which are not regular [11]. This fact certifies $\Lambda_{Bool}(\mathsf{CRL}) \not\subset \Lambda_{Bool}(\mathsf{CFL})$ and $\Lambda_{Bool}(\mathsf{CFL}) \subsetneq \Lambda_{Bool}(\mathsf{GCSL})$.

Next, we show that the language $L_{copy} = \{w\#w \mid w \in \{0,1\}^*\}$ does not belong to $\Lambda_{Bool}(\mathsf{CRL})$. First, we prove that L_{copy} does not belong to $\Lambda_{\cap,\cup}(\mathsf{GCSL})$. Observe that $\Lambda_{\cap,\cup}(\mathsf{GCSL}) = \Lambda_{\cap}(\mathsf{GCSL})$ what follows from the distributive law (i.e., the fact that $\bigcap_i A_i \cup \bigcap_j B_j = \bigcap_{i,j}(A_i \cup B_j)$) and the fact that GCSL is closed under union. One can show that $L_{copy} \notin \Lambda_{\cap}(\mathsf{GCSL})$ by a simple modification of the proof of Lemma 4. Indeed, it follows from the observation that one can define L_{copy} as $\{w_1 \ldots w_l \# w_1 \ldots w_l \mid w_i \in \{0,1\}^* \text{ for } i \in [1, l]\}$, and consider the inputs in which $|w_i| = m$ for each $i \in [1, l]$ and m large enough. So, assuming that $L_{copy} = L(M_1) \cap \ldots \cap L(M_p)$ for the IrTPDAs $M_1, \ldots, M_p$, we can fool all automata by choosing an incompressible input $w_1 \ldots w_{p+1} \# w_1 \ldots w_{p+1}$ (fooling method as in the proof of Lemma 4).

As CRL is closed under complement, each language from $\Lambda_{Bool}(\mathsf{CRL})$ belongs to $\Lambda_{\cap,\cup}(\mathsf{CRL})$ which is contained in $\Lambda_{\cap,\cup}(\mathsf{GCSL}) = \Lambda_{\cap}(\mathsf{GCSL})$. So, $L_{copy} \notin$

$\Lambda_{\text{Bool}}(\text{CRL})$. On the other hand, L_{copy} belongs to coCFL, as one can construct a pushdown automaton which "guesses" the position on which the word preceding $\#$ and the word following $\#$ differ. As CFL $\subset$ GCSL, L_{copy} belongs to $\Lambda_{\text{Bool}}(\text{GCSL})$ as well. $\qquad\square$

Recall that GCSL is a strict subset of LOGCFL [3]. A natural question is whether an analogous inclusion remains strict for the classes $\Lambda_{\text{Bool}}(\text{GCSL})$ and $\Lambda_{\text{Bool}}(\text{LOGCFL})$. As the class LOGCFL is closed under union, intersection and complement, it remains to verify whether $\Lambda_{\text{Bool}}(\text{GCSL})$ is a strict subclass of LOGCFL. We give only a partial answer to this question.

Theorem 4. *The class $\Lambda_{\cup,\cap}(\text{GCSL}) \cup \Lambda_{\cup,\cap}(\text{coGCSL})$ is a strict subset of* LOGCFL.

Proof. The following language $L_\oplus = \{x\#y \mid |x| = |y| \text{ and } x \circ y = 1\}$ is the witness certifying the above relationship, where $x \circ y$ denotes the scalar product of x and y over Z_2, that is, $x \circ y = (\sum_{i=1}^n x[i]y[i]) \bmod 2$, where $x[i]y[i]$ denotes the product of $x[i]$ and $y[i]$. First, we show by contradiction that $L_\oplus \notin \Lambda_{\cup,\cap}(\text{GCSL})$. As we observed in the proof of Theorem 3, $\Lambda_{\cup,\cap}(\text{GCSL}) = \Lambda_\cap(\text{GCSL})$. So, each language L from $\Lambda_{\cup,\cap}(\text{GCSL})$ can be expressed as $L = \bigcap_{l=1}^r L_j$ where $L_j \in \text{GCSL}$ for $l \in [1,r]$, $r \in \mathbb{N}$.

As $L_\oplus \in \Lambda_\cap(\text{GCSL})$, $L_\oplus = \bigcap_{l=1}^r L(M_l)$ for the length-reducing two-pushdown automata $M_1, \ldots, M_r$. Let $w = x_1x_2 \ldots x_{r+1}\#y_1 \ldots y_{r+1}$ be the word in $L_\oplus$ such that $|x_i| = |y_i| = m$ for $i \in [1, r+1]$, m is large enough and $K(z) > 2(r+1)m-3$, where $z = x_1 \ldots x_{r+1}y_1 \ldots y_{r+1}$. (The number of words xy such that $x\#y \in L_\oplus$ and $|x| = |y| = (r+1)m$ is $2^{2(r+1)m-1} - 2^{(r+1)m-1} \geq 2^{2(r+1)m-2}$, so there exists a word with Kolmogorov complexity not smaller than $2(r+1)m-3$ among them, by Fact 1.) Now, following the arguments from the proof of Lemma 4, we choose the accepting computations $C_1, \ldots, C_r$ of $M_1, \ldots, M_r$ on w, the computation graphs $G_1, \ldots, G_r$ which appear in the computations $C_1, \ldots, C_r$ and some fixed value $s \in [1, r+1]$ such that there exists the factorization of G_i into $G_{i,1}G_{i,2} \ldots G_{i,5}$ for each $i \in [1, r]$, where $G_{i,2}$ ($G_{i,4}$, resp.) defines the image of x_s (y_s, resp.). Moreover, the image of x_s or the image of y_s is shorter than $c'm$ in G_i (where c' is the constant from Lemma 4). Then, we define the description $D = (D_1, \ldots, D_r)$ of the decompositions of $\{G_i\}_{i=1}^r$ of size $< m/4$ bits such that one can efficiently check whether the pair (x'_s, y_s) agrees with D and, if (x'_s, y_s) agrees with D, then $w - x_s + x'_s$ belongs to $L_\oplus$ (see Proposition 3 and the proof of Lemma 4).

Let X be the set of such elements x'_s that (x'_s, y_s) agrees with D. We show that the set X contains at least $2^{m/2}$ elements. Indeed, otherwise we could encode x_s with $m/4 + m/2$ bits by the description D and the number of x_s in the lexicographic order in the set X. This contradicts the fact that $K(z) \geq 2(r + 1)m - 3$ by Fact 1.3. Thus, each of the automata $M_1, \ldots, M_r$ accepts $w - x_s + x'_s$ for each $x'_s \in X$ by Proposition 3. So, as we assumed that $\bigcap_{l=1}^r L(M_l)$ is equal to $L_\oplus$, $x'_s \circ y_s = x_s \circ y_s$ for each $x'_s \in X$. Indeed, otherwise $w - x_s + x'_s \notin L_\oplus$. Let $c = x_s \circ y_s$. As $|X| \geq 2^{m/2}$, X contains at least $m/2$ linearly independent (over Z_2) elements $v_1, \ldots, v_{m/2}$. So, x_s is one of the solutions of the system of the linear equations $S = \{v_l \circ y^T = (c^{m/2})^T\}_{l=1}^{m/2}$. The linear independence of the set

$\{v_1, \ldots, v_{m/2}\}$ implies that the number of solutions of S is at most $2^{m-m/2} = 2^{m/2}$. These observations lead to the following algorithm which "compresses" x_s (knowing $w - x_s$):

- determine the set X, defined above;
- find the largest linearly independent subset $V = \{v_1, \ldots, v_p\}$ of X; if there are many linearly independent subsets of X of the largest size, choose the first one according to the lexicographic order (where the concatenations of elements of the subsets are compared);
- give the description of x_s as: the description D, and the number (in lexicographic order) of x_s in the set of solutions of the system $S = \{v_l \circ y^T = (c^{m/2})^T\}_{l=1}^{m/2}$.

According to the above discussion, we obtain the description of size at most $m/2 + m/4$. So, if this description determines x_s uniquely, we have $K(x_s|w-x_s) \leq 3m/4$, what contradicts Fact 1.3. On the other hand, having D and $w - x_s$, one can determine the set X, then the set V, and finally having the number of x_s in the set of solutions of the system S (in the lexicographic order), we determine x_s uniquely.

In order to show that $L_\oplus \notin \Lambda_{\cup,\cap}(\text{coGCSL})$, it is sufficient to prove that $\overline{L_\oplus} \notin \Lambda_{\cup,\cap}(\text{GCSL})$. Assume that it is not the case, i.e., $\overline{L_\oplus} \in \Lambda_{\cup,\cap}(\text{GCSL})$. Note that $\overline{L_\oplus} \cap \{\{0,1\}^n \# \{0,1\}^n \mid n \in \mathbb{N}\}$ is equal to $L_\ominus = \{x \# y \mid |x| = |x|, x \circ y = 0\}$. Following the above arguments, we can show that if a language from $\Lambda_{\cup,\cap}(\text{GCSL})$ contains $L_\ominus$, it contains a word from $\{\{0,1\}^n \# \{0,1\}^n \mid n \in \mathbb{N}\} \setminus L_\ominus$ for n large enough. Contradiction to $\overline{L_\oplus} \in \Lambda_{\cup,\cap}(\text{GCSL})$. $\qquad\square$

6 Closure Under the Shuffle Operation

Using a variant of the language L_2 from Section 4, we obtain the following result which solves an open problem from [11].

Theorem 5. *The classes* CRL *and* GCSL *are not closed under the shuffle operation.*

Proof. Let $\Sigma_i = \{a_{i,0}, a_{i,1}\}$ for each $i \in [1,4]$. Let $\varphi(a_{i,j}) = j$ for each $i \in [1,4]$, $j \in [0,1]$. Finally, let

$$L_2' = \{x_1 x_2 x_3 x_4 \mid x_i \in \Sigma_i^*, \varphi(x_1) = (\varphi(x_3))^R \wedge \varphi(x_2) = (\varphi(x_4))^R\},$$

and $L_{2,1}' = \{x_1 x_3 \mid \varphi(x_1) = (\varphi(x_3))^R, x_i \in \Sigma_i^*\}$, $L_{2,2}' = \{x_2 x_4 \mid \varphi(x_2) = (\varphi(x_4))^R, x_i \in \Sigma_i^*\}$. Observe that $L_2' = (L_{2,1}' \bowtie L_{2,2}') \cap \Sigma_1^* \Sigma_2^* \Sigma_3^* \Sigma_4^*$.

Proposition 4. $L_2' \notin$ GCSL

Proof. For the sake of contradiction assume that the length-reducing two-push-down automaton M recognizes L_2'. It is known that GCSL is equal to the set of languages recognized by so-called shrinking two-pushdown automata (sTPDA)

[2]. Shrinking TPDAs do not need to be length-reducing, but each step should reduce the weight, where the weight is a function from the alphabet into natural numbers (and the weight of a word is equal to the sum of the weights of its elements). On base of our assumption, we construct a shrinking two-pushdown automaton M' which recognizes L_2: first, the automaton rewrites the input $x_1 \# x_2 \# x_3 \# x_4$ ($x_i \in \{0,1\}^*$) into $\eta_1(x_1)\eta_2(x_2)\eta_3(x_3)\eta_4(x_4)$, where $\eta_i(j) = a_{i,j} \in \Sigma_i$ for $i \in [1,4]$, $j \in [0,1]$. Next, it simulates M. One can easily verify that M' is shrinking and it recognizes the language L_2. So, $L_2 \in$ GCSL, what contradicts Lemma 4. □

Assume that CRL or GCSL is closed under shuffle operation. As $L'_{2,1}, L'_{2,2} \in$ CRL, also $L_{new} = L'_{2,1} \bowtie L'_{2,2} \in$ CRL. So, $L'_2 \in$ CRL, because $L'_2 = L_{new} \cap \Sigma_1^* \Sigma_2^* \Sigma_3^* \Sigma_4^*$ and CRL is closed under intersection with regular languages. But this contradicts Proposition 4. □

References

1. G. Buntrock, K. Loryś, *On growing context-sensitive languages*, ICALP 1992, LNCS 623, 77–88.
2. G. Buntrock, F. Otto, *Growing Context-Sensitive Languages and Church-Rosser Languages*, Information and Compation 141(1), 1998, 1–36.
3. E. Dahlhaus, M.K. Warmuth, *Membership for growing context-sensitive grammars is polynomial*, Journal of Computer and System Sciences, 33(3), 1986, 456–472.
4. M. Holzer, F. Otto, *Shrinking Multi-pushdown Automata*, Proc. of Fundamentals of Computation Theory (FCT), 2005, LNCS 3623, 305–316.
5. T. Jurdzinski, K. Lorys, *Church-Rosser Languages vs. UCFL*, in Proc. of International Colloquium on Automata, Languages and Programming (ICALP), 2002, LNCS 2380, 147–158. (full version: www.ii.uni.wroc.pl/~tju/FullCRL.pdf).
6. M. Kutrib, A. Malcher, D. Wotschke, *The Boolean Closure of Linear Context-Free Languages*, DLT 2004, LNCS 3340, 2004, 284–295.
7. M. Li, P. Vitanyi, *An Introduction to Kolmogorov Complexity and its Applications*, Springer-Verlag 1993.
8. Leonard Y. Liu, Peter Weiner, *An infinite hierarchy of intersections of context-free languages*, Mathematical Systems Theory, 7 (1973), 185–192.
9. R. McNaughton, *An insertion into the Chomsky hierarchy?*, Jewels are Forever, Contributions on Theoretical Computer Science in Honour of A.Salomaa, Springer, 1999, 204–212.
10. R. McNaughton, P. Narendran, F. Otto, *Church-Rosser Thue systems and formal languages*, Journal of the Association Computing Machinery, 35 (1988), 324–344.
11. G. Niemann, *Church-Rosser Languages and Related Classes*, PhD Thesis, Univ. Kassel, 2002.
12. G.Niemann, F.Otto, *The Church-Rosser languages are the deterministic variants of the growing context-sensitive languages*, Inform. and Comp., 197(1-2), 2005, 1-21.
13. A. Okhotin, *Boolean grammars*, Information and Computation 194(1), 2004, 19-48.
14. D. Wotschke, *The Boolean Closures of the Deterministisc and Nondeterministic Context-free Languages*, GI Jahrestagung, LNCS 1, 1973, 113–121.
15. D. Wotschke, *A Characterization of Boolean Closures of Families of Languages*, Automatentheorie und Formale Sprachen, LNCS 2, 1973, 191–200.
16. D. Wotschke, *Nondeterminism and Boolean Operations in PDAs*, J. Comput. Syst. Sci. 16(3), 1978, 456–461.

Well Quasi Orders and the Shuffle Closure of Finite Sets*

Flavio D'Alessandro[1], Gwénaël Richomme[2], and Stefano Varricchio[3]

[1] Dipartimento di Matematica,
Università di Roma "La Sapienza"
Piazzale Aldo Moro 2, 00185 Roma, Italy
dalessan@mat.uniroma1.it
http://www.mat.uniroma1.it/people/dalessandro
[2] LaRIA, UPJV
33 Saint Leu, 80039 Amiens Cedex 01, France
gwenael.richomme@u-picardie.fr
http://www.laria.u-picardie.fr/~richomme/
[3] Dipartimento di Matematica,
Università di Roma "Tor Vergata",
via della Ricerca Scientifica, 00133 Roma, Italy
varricch@mat.uniroma2.it
http://www.mat.uniroma2.it/~varricch/

Abstract. Given a set I of words, the set $L_{\vdash_I}^\varepsilon$ of all words obtained by the shuffle of (copies of) words of I is naturally provided with a partial order: for u, v in $L_{\vdash_I}^\varepsilon$, $u \vdash_I^* v$ if and only if v is the shuffle of u and another word of $L_{\vdash_I}^\varepsilon$. In [3], the authors have stated the problem of the characterization of the finite sets I such that $\vdash_I^*$ is a well quasi-order on $L_{\vdash_I}^\varepsilon$. In this paper we give the answer in the case when I consists of a single word w.

1 Introduction

A *quasi-order* on a set S is called a *well quasi-order* (*wqo*) if every non-empty subset X of S has at least one minimal element in X but no more than a finite number of (non-equivalent) minimal elements. Well quasi-orders have been widely investigated in the past. We recall the celebrated Higman and Kruskal results [9, 14]. Higman gives a very general theorem on division orders in abstract algebras from which one derives that the *subsequence ordering* in free monoids is a wqo. Kruskal extends Higman's result, proving that certain embeddings on finite trees are well quasi-orders. Some remarkable extensions of the Kruskal theorem are given in [11, 16].

In the last years many papers have been devoted to the application of wqo's to formal language theory [1, 2, 4, 5, 6, 7, 10, 12, 13].

---

* This work was partially supported by MIUR project ''Linguaggi formali e automi: teoria e applicazioni''.

In [6], a remarkable class of grammars, called *unitary grammars*, has been introduced in order to study the relationships between the classes of context-free and regular languages. If I is a finite set of words then we can consider the set of productions

$$\{\epsilon \to u, \ u \in I\}$$

and the derivation relation $\Rightarrow_I^*$ of the semi-Thue system associated with I. The language generated by the unitary grammar associated with I is $L_I^\epsilon = \{w \in A^* \mid \epsilon \Rightarrow_I^* w\}$. Unavoidable sets of words are characterized in terms of the wqo property of the unitary grammars. Precisely it is proved that I is unavoidable if and only if the derivation relation $\Rightarrow_I^*$ is a wqo.

In [8], Haussler investigated the relation $\vdash_I^*$ defined as the transitive and reflexive closure of $\vdash_I$ where, for every pair v, w of words, $v \vdash_I w$ if

$$v = v_1 v_2 \cdots v_{n+1},$$

$$w = v_1 a_1 v_2 a_2 \cdots v_n a_n v_{n+1},$$

where the a_i's are letters, and $a_1 a_2 \cdots a_n \in I$. In particular, a characterization of the wqo property of $\vdash_I^*$ in terms of subsequence unavoidable sets of words was given in [8]. Let $L_{\vdash_I}^\epsilon$ be the set of all words derived from the empty word by applying $\vdash_I^*$.

A remarkable result proved in [2] states that for any finite set I the derivation relation $\vdash_I^*$ is a wqo on the language L_I^ϵ. It is also proved that, in general, $\Rightarrow_I^*$ is not a wqo on L_I^ϵ and $\vdash_I^*$ is not a wqo on $L_{\vdash_I}^\epsilon$. In [3] the authors characterize the finite sets I such that $\Rightarrow_I^*$ is a wqo on L_I^ϵ. Moreover, they have left the following problem open: *characterize the finite sets I such that $\vdash_I^*$ is a wqo on $L_{\vdash_I}^\epsilon$.* In this paper we give the answer in the case when I consists of a single word w.

In this context, it is worth noticing that in [3] the authors prove that $\vdash_{\{w\}}^*$ is not a wqo on $L_{\vdash_{\{w\}}}^\epsilon$ if $w = abc$. A simple argument allows one to extend the result above in the case that $w = a^i b^j c^h$, $i, j, h \geq 1$. By using Lemma 2, this implies that if a word w contains three distinct letters at least, then $\vdash_{\{w\}}^*$ is not a wqo on $L_{\vdash_{\{w\}}}^\epsilon$. Therefore, in order to prove our main result, we can focus our attention to the case where w is a word on the binary alphabet $\{a, b\}$. Let E be the exchange morphism ($E(a) = b$, $E(b) = a$), and let $\tilde{w}$ be the mirror image of w.

Definition 1. *A word w is called* bad *if one of the words w, $\tilde{w}$, $E(w)$ and $E(\tilde{w})$ has a factor of one of the two following forms*

$$a^k b^h \quad \text{with } k, h \geq 2 \tag{1}$$

$$a^k b a^l b^m \quad \text{with } k > l \geq 1, m \geq 1 \tag{2}$$

A word w is called good *if it is not bad.*

Although it is immediate that a word w is bad if and only if one of the words w, $\tilde{w}$, $E(w)$ and $E(\tilde{w})$ contains a factor $a^2 b^2$ or $a^{k+1} b a^k b$, with $k \geq 1$, it will be useful to consider the definition as above. Moreover, we observe that, by Lemma 4 a word is good if and only if it is a factor of $(ba^n)^\omega$ or $(ab^n)^\omega$ for some $n \geq 0$.

Theorem 1. *Let w be a word over the alphabet $\{a, b\}$. The derivation relation $\vdash^*_{\{w\}}$ is a wqo on $L^\epsilon_{\vdash_{\{w\}}}$ if and only if w is good.*

In the rest of the paper, we present the main steps of the proof of this theorem. For length reason, most of the technical parts are not included.

We assume the reader to be familiar with the basic theory of combinatorics on words as well as with the theory of well quasi-orders (*see also* [5, 15]).

2   Bad Words

In order to prove the "only if" part of Theorem 1, we first recall the following theorem which gives a useful characterization of the concept of well quasi-order.

Theorem 2. *Let S be a set quasi-ordered by $\leq$. The following conditions are equivalent:*

   i. $\leq$ is a well quasi-order;
  ii. if $s_1, s_2, \ldots, s_n, \ldots$ is an infinite sequence of elements of S, then there exist integers i, j such that $i < j$ and $s_i \leq s_j$.

Let $\sigma = (s_i)_{i \geq 1}$ be an infinite sequence of elements of a set S. Then σ is called *good* if it satisfies condition ii of Theorem 2 and it is called *bad* otherwise, that is, for all integers i, j such that $i < j$, $s_i \not\leq s_j$. It is worth noting that, by condition ii above, a useful technique to prove that $\leq$ is a wqo on S is to prove that no bad sequence exists in S. Conversely, in order to prove that $\leq$ is not a wqo on S, it is enough to show the existence of a bad sequence of S.

Using this technique, the next two propositions show that for any bad word w of one of the two forms considered in Definition 1, $\vdash^*_{\{w\}}$ is not a wqo on $L^\epsilon_{\vdash_{\{w\}}}$.

Proposition 1. *Let $w = a^h b^k$ with $h, k \geq 2$, and consider the sequence $(S_n)_{n \geq 1}$ of words of A^* defined as: for every $n \geq 1$,*

$$S_n = a^h (a^{2h} b^{2k})(aba^{h-1} b^{k-1})^n (a^{2h} b^{2k}) b^k.$$

$(S_n)_{n \geq 1}$ is a bad sequence of $L^\epsilon_{\vdash_{\{w\}}}$ with respect to $\vdash^*_{\{w\}}$. In particular $\vdash^*_{\{w\}}$ is not a wqo on $L^\epsilon_{\vdash_{\{w\}}}$.

Proposition 2. *Let $w = a^k ba^\ell b^m$ with $k > \ell \geq 1$, $m \geq 1$, and consider the sequence $(S_n)_{n \geq 1}$ of words of A^* defined as: for every $n \geq 1$,*

$$S_n = a^k ba^\ell a^k ba^\ell b^m (a^k b^{m+1} a^\ell)^n a^k ba^\ell b^m b^m.$$

$(S_n)_{n \geq 1}$ is a bad sequence of $L^\epsilon_{\vdash_{\{w\}}}$ with respect to $\vdash^*_{\{w\}}$. In particular $\vdash^*_{\{w\}}$ is not a wqo on $L^\epsilon_{\vdash_{\{w\}}}$.

Let us say few words about the proofs of the two previous propositions. First the reader can easily verify that in each case, for every $n \geq 1$, $S_n \in L^\epsilon_{\vdash_{\{w\}}}$.

The techniques used to proved that the two sequences above are bad, are based upon some combinatorial properties of the words of the language $L^\epsilon_{\vdash_{\{w\}}}$. More precisely, for Proposition 1, this can be done by using the following characterization of the words of $L^\epsilon_{\vdash_{\{w\}}}$ given in [3]:

Proposition 3. *Let* $u \in \{a, b\}^*$ *and* $h, k \geq 1$. *Set*

$$q_a^u = \lfloor |u|_a / h \rfloor, \quad q_b^u = \lfloor |u|_b / k \rfloor, \quad and$$

$$r_a^u = |u|_a \bmod h, \quad r_b^u = |u|_b \bmod k.$$

Let $w = a^h b^k$, $h, k \geq 2$. *Then* $u \in L^\epsilon_{\vdash_{\{w\}}}$ *if and only if* $q_a^u = q_b^u$, *and for every prefix* p *of* u, *either* $q_a^p > q_b^p$ *or* $q_a^p = q_b^p$ *and* $r_b^u = 0$.

In the case of Proposition 2, the proof is essentialy based on the following lemma:

Lemma 1. *Let* $u \in L^\epsilon_{\vdash_{\{a^k b a^\ell b^m\}}}$, *with* $k > \ell \geq 1$, $m \geq 1$. *For every non empty prefix* p *of* u, *we have*

$$\frac{|p|_a}{|p|_b} \geq \frac{k + \ell}{m + 1}.$$

Propositions 1 and 2 allow to prove that if w is of the forms (1) or (2) of Definition 1, then $\vdash^*_{\{w\}}$ is not a wqo on $L^\epsilon_{\vdash_{\{w\}}}$. This does not suffice to prove the "only if" part of our main theorem. In order to complete the proof, the following lemma (and its symmetric version, say Lemma 3) provides a key result: indeed it shows that the property "$\vdash^*_{\{w\}}$ is not a wqo on $L^\epsilon_{\vdash_{\{w\}}}$" is preserved by the factor order.

Lemma 2. *Let* b *be a letter of an alphabet* A *and let* u *be a word over* A *not ending with* b. *Assume* $\vdash^*_{\{u\}}$ *is not a wqo on* $L^\epsilon_{\vdash_{\{u\}}}$. *Then, for every* $k \geq 1$, $\vdash^*_{\{ub^k\}}$ *is not a wqo on* $L^\epsilon_{\vdash_{\{ub^k\}}}$.

Proof. Let $(w_n)_{n \geq 0}$ be a bad sequence of $L^\epsilon_{\vdash_{\{u\}}}$ with respect to $\vdash^*_{\{u\}}$ and, for every $n \geq 1$, let us denote ℓ_n the positive integer such that

$$\epsilon \vdash^{\ell_n}_{\{u\}} w_n. \tag{3}$$

Since $(w_n)_{n \geq 0}$ is a bad sequence, by using a standard argument, we may choose the sequence $(w_n)_{n \geq 0}$ so that $(\ell_n)_{n \geq 0}$ is a strictly increasing sequence of positive integers. Let k be a positive integer and define the sequence of words $(w_n(b^k)^{\ell_n})_{n \geq 0}$. It is easily checked that, for every $n \geq 1$,

$$\epsilon \vdash^{\ell_n}_{\{ub^k\}} w_n(b^k)^{\ell_n},$$

so that all the words of the sequence defined above belong to the language $L^\epsilon_{\vdash_{\{ub^k\}}}$. Now we prove that this sequence is bad with respect to $\vdash^*_{\{ub^k\}}$. By contradiction, suppose the claim false. Thus there exist positive integers n, m such that

$$w_n(b^k)^{\ell_n} \vdash^*_{\{ub^k\}} w_{n+m}(b^k)^{\ell_{n+m}}. \tag{4}$$

Since, for every $n \geq 1$,

$$|w_n(b^k)^{\ell_n}| = \ell_n k + |w_n| = \ell_n(k + |u|),$$

we have that the length L of the derivation (4) is

$$L = \ell_{n+m} - \ell_n. \tag{5}$$

Now it is useful to do the following remarks. First observe that, since u does not end with the letter b, for every $n \geq 1$, $(b^k)^{\ell_n}$ is the longest power of b which is a suffix of $w_n(b^k)^{\ell_n}$. Second: at each step

$$v \vdash_{\{ub^k\}} v',$$

of the derivation (4), the exponent of the longest power of b which is a suffix of the word v' increases of k at most (with respect to v). Moreover this upper bound can be obtained by performing the insertion of ub^k in the word v only if its suffix b^k is inserted after the last letter of v which is different from b. By the previous remark and by (5), all the insertions of the derivation (4) must be done in this way. This implies that the derivation (4) defines in an obvious way a new one with respect to the relation $\vdash^*_{\{u\}}$ such that

$$w_n \vdash^*_{\{u\}} w_{n+\ell}.$$

The latter condition contradicts the fact that the sequence of words $(w_n)_{n \geq 0}$ is bad. □

By using a symmetric argument, we can prove the following.

Lemma 3. *Let b be a letter of an alphabet A and let u be a word over A not beginning with b. Assume $\vdash^*_{\{u\}}$ is not a wqo on $L^\epsilon_{\vdash_{\{u\}}}$. Then, for every $k \geq 1$, $\vdash^*_{\{b^k u\}}$ is not a wqo on $L^\epsilon_{\vdash_{\{b^k u\}}}$.*

The "only if part" of Theorem 1 is a consequence of the following:

Theorem 3. *If w is a bad word then $\vdash^*_{\{w\}}$ is not a wqo on the language $L^\epsilon_{\vdash_{\{w\}}}$.*

Proof. If w has a factor of the form $a^k b^h$ with $h, k \geq 2$, or $a^k b a^\ell b^m$, with $k > \ell \geq 1$, $m \geq 1$, then the claim is a straightforward consequence of Lemma 2, Lemma 3, Proposition 1, and Proposition 2.

In the general case, that is whenever $\tilde{w}$ or $E(w)$ or $E(\tilde{w})$ has a factor of the previous two forms, the proof is similar since the property of wqo is preserved under taking exchange morphism and mirror image of the word w. □

3   Good Words

We now present the proof of the "if" part of Theorem 1. This is divided into three steps. In the first we analyze the form of good words and the languages of words derivable from a good word. Their characterizations are given in Section 3.1. In the second step we prove that $\vdash^*_I$ is a wqo on $L^\epsilon_{\vdash_I}$ for some particular sets of words I of cardinality 2. This is stated by Theorem 5 whose proof is based on some intermediary results presented in Section 3.2. The third step, Section 3.3, is the proof of the "if" part of Theorem 1 as a direct consequence of Theorem 5.

3.1   Form of Good Words

Lemma 4. *A word w is good if and only if $w = \epsilon$ or there exist some integers n, e, i, f such that $w = a^i(ba^n)^e ba^f$ or $w = b^i(ab^n)^e ab^f$, $e \geq 0$, $0 \leq i, f \leq n$, and if $e = 0$ then $n = \max(i, f)$.*

Proof. Clearly if w is a bad word, then w cannot be decomposed as in the lemma.

Assume now that w is a good word. This means that w has no factor of the form $aabb$, $bbaa$, $a^{n+1}ba^n b$, $ba^n ba^{n+1}$, $b^{n+1}ab^n a$, $ab^n ab^{n+1}$ for an integer $n \geq 1$. If $|w|_a = 0$, then $w = \epsilon$ or $w = a^i(ba^n)^e ba^f$ with $i = n = f = 0$. If $|w|_a = 1$, $w = a^p ba^q$ with $\max(p, q) = 1$, that is $w = a^i(ba^n)^e ba^f$ with $i = p$, $f = q$, $n = \max(p, q)$, $e = 0$. Similarly if $|w|_b \leq 1$, w is a good word.

Assume from now on that $|w|_a \geq 2$ and $|w|_b \geq 2$. If both aa and bb are not factors of w, then w is a factor of $(ab)^\omega$ and so $w = a^i(ba^n)^e ba^f$ with $n = 1$.

Let us prove that aa and bb cannot be simultaneously factors of w. Assume the contrary. We have $w = w_1 aa w_2 bb w_3$ (or $w = w_1 bb w_2 aa w_3$ which leads to the same conclusion) for some words w_1, w_2, w_3. Without loss of generality we can assume that aa is not a factor of aw_2 and bb is not a factor of $w_2 b$. This implies that $w_2 = (ba)^m$ for an integer $m \geq 0$. This is not possible since $aabab$ and $aabb$ are not factors of w.

Assume from now on that bb is not a factor of w (the case where aa is not a factor is similar). This implies that $w = a^{i_0}ba^{i_1}ba^{i_2}b\ldots ba^{i_p}ba^{i_{p+1}}$ for some integers $i_0, i_1, \ldots, i_{p+1}$ such that $i_j \neq 0$ for each $j \in \{1, \ldots, p\}$. Let j be an integer such that $1 \leq j < j + 1 \leq p$. Since $a^{i_{j+1}+1}ba^{i_{j+1}}b$ and $ba^{i_j}ba^{i_j+1}$ are not factors of w, we have $i_j = i_{j+1}$. Thus set $n = i_1 = \cdots = i_p$ and write $w = a^{i_0}(ba^n)^p ba^{i_{p+1}}$. Since $a^{n+1}ba^n b$ and $ba^n ba^{n+1}$ are not factors of w, we have $i_0 \leq n$, $i_{p+1} \leq n$. This ends the proof. □

For X a set of words and n an integer, let $X^{\leq n} = \bigcup_{i=0}^{n} X^i$. Lemma 4 can be reformulated: the set of good words w is the set

$$\{\epsilon\} \cup \bigcup_{n \geq 0} a^{\leq n}(ba^n)^* ba^{\leq n} \cup \bigcup_{n \geq 0} b^{\leq n}(ab^n)^* ab^{\leq n}.$$

The following proposition gives a characterization of the languages of the words derivable from a good word. The proof is rather technical and it will be omitted.

Proposition 4. *Let w be a word over $\{a, b\}$ and n_w, e_w, i_w, f_w be integers such that $|w|_a \geq 1$, $|w|_b \geq 1$, $w = a^{i_w}(ba^{n_w})^{e_w} ba^{f_w}$, $0 \leq i_w, f_w \leq n_w$, $e_w \geq 0$ and if $e_w = 0$ then $n_w = \max(i_w, f_w)$.*

Let u be a word : $u \in L^\epsilon_{\vdash \{w\}}$ if and only if

1. $\dfrac{|u|_a}{|w|_a} = \dfrac{|u|_b}{|w|_b}$;
2. *for all words p, s, if $u = ps$ then*

   2.1) $|p|_a \geq i_w |p|_b + \max(0, |p|_b - \frac{|u|_b}{|w|_b})(n_w - i_w)$;

   2.2) $|s|_a \geq f_w |s|_b + \max(0, |s|_b - \frac{|u|_b}{|w|_b})(n_w - f_w)$.

3.2 Some Useful wqo's

In this section, we present some useful wqo's. First we recall the following result.

Proposition 5. [3] *For any integer $n \geq 0$, if $w \in \{a^n b, ab^n, ba^n, b^n a\}$, then $\vdash^*_{\{w\}}$ is a wqo on $L^\epsilon_{\vdash_{\{w\}}} = L^\epsilon_{\{w\}}$.*

This result allows us to state:

Lemma 5. *Let $n \geq 0$ be an integer. Let I be one of the following sets: $\{a^n b, b\}$, $\{a^n b, a\}$, $\{b^n a, a\}$, $\{b^n a, b\}$, $\{ba^n, b\}$, $\{ba^n, a\}$, $\{ab^n, a\}$, $\{ab^n, b\}$. Then*

$$L^\epsilon_{\vdash_I} = L^\epsilon_I.$$

Proof. Assume $I = \{a^n b, a\}$. It is immediate that $L^\epsilon_I \subseteq L^\epsilon_{\vdash_I}$. Let w be a word in $L^\epsilon_{\vdash_I}$. There exists a word w_1 such that $\epsilon \vdash^*_{\{a^n b\}} w_1 \vdash^*_{\{a\}} w$. By Proposition 5, $w_1 \in L^\epsilon_{\{a^n b\}}$, and so $w \in L^\epsilon_I$.

The proof for the other values of I is similar.   □

Lemma 6. *Let $n \geq 1$ be an integer. The three following assertions are equivalent for a word w:*

1. *$w \in L^\epsilon_{\vdash_{\{a^n b, a^n\}}}$;*
2. *$|w|_a = 0 \mod n$, and, for any prefix p of w, $|p|_a \geq n|p|_b$;*
3. *$w \in L^\epsilon_{\{a^n b, a^n\}}$.*

In particular, $L^\epsilon_{\vdash_{\{a^n b, a^n\}}} = L^\epsilon_{\{a^n b, a^n\}}$.

Proof. $3 \Rightarrow 1$ is immediate.

For any word w in $L^\epsilon_{\vdash_{\{a^n b, a^n\}}}$, obviously $|w|_a = 0 \mod n$. Moreover there exists a word u such that $\epsilon \vdash^*_{\{a^n b\}} u \vdash^*_{\{a^n\}} w$. Thus $1 \Rightarrow 2$ is a direct consequence of Proposition 4 (taking $w = a^n b$, $n_w = n = i_w$, and $e_w = f_w = 0$, Condition 2.1 of Proposition 4 says that for any prefix of a word in $L^\epsilon_{\vdash_{\{ab\}}}$, $|p|_a \geq i_w|p|_b = n|p|_b$).

We now prove $2 \Rightarrow 3$ by induction on $|w|_b$. Since $|w|_a = 0 \mod n$, the result is immediate if $|w|_b = 0$. Assume $|w|_b > 1$. Assertion 2 on w implies the existence of an integer $k \geq 0$ and a word w' such that $w = a^k a^n b w'$. Let p be a prefix of $a^k w'$. If $|p| \leq k$, then $|p|_b = 0 \leq n|p|_a$. If $|p| > k$, $p = a^k p'$ for a prefix p' of w'. Assertion 2 on w implies that $|a^k a^n b p'|_a \geq n|a^k a^n b p'|_b$ that is $|a^k p'|_a \geq n|a^k p'|_b$. Thus $a^k w'$ satisfies Assertion 2 and so by the inductive hypothesis, $a^n w' \in L^\epsilon_{\{a^n b, a^n\}}$. It follows that $w \in L^\epsilon_{\{a^n b, a^n\}}$.   □

Similarly to Lemma 6, one can state that $L^\epsilon_{\vdash_{\{ba^n, a^n\}}} = L^\epsilon_{\{ba^n, a^n\}}$ (this needs to exchange prefixes by suffixes), and, exchanging the roles of a and b, $L^\epsilon_{\vdash_{\{b^n a, b^n\}}} = L^\epsilon_{\{b^n a, b^n\}}$, $L^\epsilon_{\vdash_{\{ab^n, b^n\}}} = L^\epsilon_{\{ab^n, b^n\}}$.

Let us recall that:

Theorem 4. [1, 2] *For any finite set I, $\vdash^*_I$ is a wqo on L^ϵ_I.*

Hence from this theorem and the previous lemma, we deduce:

Proposition 6. *Let $n \geq 0$ be an integer. Suppose that I is one of the following sets: $\{a^n b, b\}$, $\{a^n b, a\}$, $\{b^n a, a\}$, $\{b^n a, b\}$, $\{ba^n, b\}$, $\{ba^n, a\}$, $\{ab^n, a\}$, $\{ab^n, b\}$, $\{a^n b, a^n\}$, $\{ba^n, a^n\}$, $\{b^n a, b^n\}$, $\{ab^n, b^n\}$. Then the derivation relation $\vdash_I^*$ is a wqo on $L_{\vdash_I}^\epsilon$.*

We end this section by stating the following two important results used in the proof of Theorem 5.

Proposition 7. *Let n, m be two integers such that $n, m \geq 1$ and let w be a word in $a^{\leq n}(ba^n)^* b \cup \{\epsilon\}$ such that $wa^n ba^m$ is a good word. If $\vdash_{\{wa^n, wa^n b\}}^*$ is a wqo on $L_{\vdash_{\{wa^n, wa^n b\}}}^\epsilon$ then $\vdash_{\{wa^n b, wa^n ba^m\}}^*$ is a wqo on $L_{\vdash_{\{wa^n b, wa^n ba^m\}}}^\epsilon$.*

Observe that the hypothesis "$wa^n ba^m$ is a good word" means only $1 \leq m \leq n$ when $w \neq \epsilon$.

Proposition 8. *Let $n \geq 1$ be an integer and let w be a word in $a^{\leq n}(ba^n)^*$. If $\vdash_{\{wb, wba^n\}}^*$ is a wqo on $L_{\vdash_{\{wb, wba^n\}}}^\epsilon$ then $\vdash_{\{wba^n, wba^n b\}}^*$ is a wqo on $L_{\vdash_{\{wba^n, wba^n b\}}}^\epsilon$.*

3.3 Proof of the "If" Part of Theorem 1

From the results of the previous section we can deduce:

Theorem 5. *For any integers $n, m \geq 1$, and for any word w in $a^{\leq n}(ba^n)^* b \cup \{\epsilon\}$ such that $wa^n ba^m$ is a good word, one has:*

1. *$\vdash_{\{wa^n, wa^n b\}}^*$ is a wqo on $L_{\vdash_{\{wa^n, wa^n b\}}}^\epsilon$;*
2. *$\vdash_{\{wa^n b, wa^n ba^m\}}^*$ is a wqo on $L_{\vdash_{\{wa^n b, wa^n ba^m\}}}^\epsilon$.*

Proof. We act by induction on $|w|_b$.

When $|w|_b = 0$, $w = \epsilon$ and we know by Proposition 6 that $\vdash_{\{a^n, a^n b\}}^*$ is a wqo on $L_{\vdash_{\{a^n, a^n b\}}}^\epsilon$. By Proposition 7, we deduce that $\vdash_{\{a^n b, a^n ba^m\}}^*$ is a wqo on $L_{\vdash_{\{a^n b, a^n ba^m\}}}^\epsilon$.

Assume now $|w|_b \geq 1$. Then $w = a^h b$ with $0 \leq h \leq n$ or $w = w'a^n b$ with $w' \in a^{\leq n}(ba^n)^* b$. If $w = b$, then by Proposition 6, $\vdash_{\{b, ba^n\}}^*$ is a wqo on $L_{\vdash_{\{b, ba^n\}}}^\epsilon$. In the other cases, by inductive hypothesis, $\vdash_{\{w, wa^n\}}^*$ is a wqo on $L_{\vdash_{\{w, wa^n\}}}^\epsilon$. Hence by Proposition 7, $\vdash_{\{wa^n, wa^n b\}}^*$ is a wqo on $L_{\vdash_{\{wa^n, wa^n b\}}}^\epsilon$, and by Proposition 8, we deduce that $\vdash_{\{wa^n b, wa^n ba^m\}}^*$ is a wqo on $L_{\vdash_{\{wa^n, wa^n ba^m\}}}^\epsilon$. $\square$

Corollary 1. *Let $n \geq 1$ be an integer. For any word w in $a^{\leq n}(ba^n)^* ba^{\leq n}$, $\vdash_{\{w\}}^*$ is a wqo on $L_{\vdash_{\{w\}}}^\epsilon$.*

Proof. The result is immediate if $|w|_b = 0$. Assume from now on $|w|_b > 0$.

First we consider the case where w ends with b. Two cases are possible: $w = a^m b$ with $1 \leq m \leq n$ or $w = w'ba^n b$ with w' in $a^{\leq n}(ba^n)^*$. If $w = a^m b$, the result is stated in Proposition 5.

Assume $w = w'ba^n b$. By Theorem 5, we know that $\vdash_{\{w'ba^n, w'ba^n b\}}^*$ is a wqo on $L_{\vdash_{\{w'ba^n, w'ba^n b\}}}^\epsilon$. Let $(u_k)_{k \geq 0}$ be a sequence of words in $L_{\vdash_{\{w'ba^n b\}}}^\epsilon$. Since

$L^{\epsilon}_{\vdash_{\{w'ba^nb\}}} \subseteq L^{\epsilon}_{\vdash_{\{w'ba^n,w'ba^nb\}}}$, $u_k \in L^{\epsilon}_{\vdash_{\{w'ba^n,w'ba^nb\}}}$ and so we can replace the sequence $(u_k)_{k \geq 0}$ by a subsequence such that $u_k \vdash^*_{\{w'ba^n,w'ba^nb\}} u_{k+1}$ for each $k \geq 0$. For any k this means that there exists a word v_k in $L^{\epsilon}_{\vdash_{\{w'ba^n,w'ba^nb\}}}$ such that u_{k+1} is in the shuffle of u_k and v_k. The word v_k is the shuffle of α_k occurrences of $w'ba^n$ and β_k occurrences of $w'ba^nb$, and the words u_k and u_{k+1} are the shuffle of γ_k and γ_{k+1} occurrences of $w'ba^nb$ respectively. From $|v_k|_a = |u_{k+1}|_a - |u_k|_a$ and $|v_k|_b = |u_{k+1}|_b - |u_k|_b$, we deduce respectively $\alpha_k + \beta_k = \gamma_{k+1} - \gamma_k$ and $(\gamma_{k+1} - \gamma_k)|w'ba^nb|_b = (\alpha_k + \beta_k)|w'ba^nb|_b - \alpha_k$ which imply $\alpha_k = 0$, that is, $v_k \in L^{\epsilon}_{\vdash_{\{w'ba^nb\}}}$. Hence $u_k \vdash^*_{\{w'ba^nb\}} u_{k+1}$, so that $\vdash^*_{\{w'ba^nb\}}$ is a wqo on $L^{\epsilon}_{\vdash_{\{w'ba^nb\}}}$.

Now we consider the case where w ends with a so that $w = w'ba^m$ with $w' \in a^{\leq n}(ba^n)^*b \cup \{\epsilon\}$ and $n \geq m \geq 1$. By Theorem 5, $\vdash^*_{\{w'b,w'ba^m\}}$ is a wqo on $L^{\epsilon}_{\vdash_{\{w'b,w'ba^m\}}}$. The proof ends as the previous case. □

We are now able to prove the "if" part of Theorem 1.

Proof of the "if" part of Theorem 1. Assume w is a word such that w, $\tilde{w}$, $E(w)$ and $E(\tilde{w})$ have no factor of the two possible forms 1 and 2 of Definition 1. By Lemma 4, we know that

$$w \in \{\epsilon\} \cup \bigcup_{n \geq 0} a^{\leq n}(ba^n)^*ba^{\leq n} \cup \bigcup_{n \geq 0} b^{\leq n}(ab^n)^*ab^{\leq n}.$$

The result is trivial if $w = \epsilon$ and stated by Corollary 1 if $w \in a^{\leq n}(ba^n)^*ba^{\leq n}$. The case $w \in b^{\leq n}(ab^n)^*ab^{\leq n}$ is treated as the previous case by exchanging the roles of a and b. □

References

1. F. D'Alessandro and S. Varricchio. On Well Quasi-orders On Languages. volume 2710 of *Lecture Notes in Computer Science*, pages 230–241. Springer-Verlag, Berlin, 2003.

2. F. D'Alessandro and S. Varricchio. Well quasi-orders and context-free grammars. *Theoretical Computer Science*, 327(3):255–268, 2004.

3. F. D'Alessandro and S. Varricchio. Well quasi-orders, unavoidable sets, and derivation systems. *RAIRO Theoretical Informatics and Applications*, to appear.

4. A. de Luca and S. Varricchio. Well quasi-orders and regular languages. *Acta Informatica*, 31:539–557, 1994.

5. A. de Luca and S. Varricchio. *Finiteness and regularity in semigroups and formal languages*. EATCS Monographs on Theoretical Computer Science. Springer, Berlin, 1999.

6. A. Ehrenfeucht, D. Haussler, and G. Rozenberg. On regularity of context-free languages. *Theoretical Computer Science*, 27:311–332, 1983.

7. T. Harju and L. Ilie. On quasi orders of words and the confluence property. *Theoretical Computer Science*, 200:205–224, 1998.

8. D. Haussler. Another generalization of Higman's well quasi-order result on Σ^*. *Discrete Mathematics*, 57:237–243, 1985.

9. G. H. Higman. Ordering by divisibility in abstract algebras. *Proc. London Math. Soc.*, 3:326–336, 1952.

10. L. Ilie and A. Salomaa. On well quasi orders of free monoids. *Theoretical Computer Science*, 204:131–152, 1998.

11. B. Intrigila and S. Varricchio. On the generalization of Higman and Kruskal's theorems to regular languages and rational trees. *Acta Informatica*, 36:817–835, 2000.

12. M. Ito, L. Kari, and G. Thierrin. Shuffle and scattered deletion closure of languages. *Theoretical Computer Science*, 245(1):115–133, 2000.

13. M. Jantzen. Extending regular expressions with iterated shuffle. *Theoretical Computer Science*, 38:223–247, 1985.

14. J. Kruskal. The theory of well-quasi-ordering: a frequently discovered concept. *J. Combin. Theory, Ser. A*, 13:297–305, 1972.

15. Lothaire. *Combinatorics on words*, volume 17 of *Series Encyclopedia of Mathematics and its Applications*. Addison-Wesley, Reading, Mass., 1983.

16. L. Puel. Using unavoidable sets of trees to generalize Kruskal's theorem. *J. Symbolic Comput.*, 8(4):335–382, 1989.

The Growth Ratio of Synchronous Rational Relations Is Unique

Olivier Carton

LIAFA, Université Paris 7 & CNRS,
2, pl. Jussieu – 75251 Paris Cedex 05
Olivier.Carton@liafa.jussieu.fr
http://www.liafa.jussieu.fr/~carton

Abstract. We introduce α-synchronous relations for a rational number α. We show that if a rational relation is both α- and α'-synchronous for two different numbers α and α', then it is recognizable. We give a synchronization algorithm for α-synchronous transducers. We also prove the closure under boolean operations and composition of α-synchronous relations.

1 Introduction

We introduce α-synchronous relations for a rational number α. They extend the classical notion of synchronous relations, which are rational relations realized by letter to letter transducers. In these usual synchronous transducers, the ratio between the output length and the input length is always 1 whereas we allow it to be any fixed rational number α. These relations have already been mentioned by Sakarovitch [1, p. 660].

The main result about these α-synchronous relations is a Cobham-like theorem. We show that if a relation is both α- and α'-synchronous for two distinct rational number α and α', then it is recognizable. This question was raised by Sakarovitch [1, p. 660]. Recall that Cobham's result states that if the base k representation of a set of integers is regular for two multiplicatively independent bases k, it is ultimately periodic [2]. If k and k' are multiplicatively independent, conversion from base k to base k' is not rational. Otherwise, this conversion is a $(\log k' / \log k)$-synchronous relation.

We also study the synchronization of transducers. We show that if a relation is realized by a transducer in which the ratio between the output length and the input length is α for any cycle, it can be also realized by another transducer in which the ratio between the output length and the input length is α for any transition. This algorithm can be viewed as a normalization of α-synchronous transducers.

The question of the synchronization of transducers goes back to the paper of Elgot and Mezei [3] about rational relations realized by finite automata, and to the result of Eilenberg and Schützenberger [4], which states that a length preserving rational relation of $A^* \times B^*$ is a rational subset of $(A \times B)^*$, or,

O.H. Ibarra and Z. Dang (Eds.): DLT 2006, LNCS 4036, pp. 270–279, 2006.

equivalently, is realized by a synchronous automaton. The proof of Eilenberg is effective but is done on regular expressions and not directly on automata. In [5], Frougny and Sakarovitch give an algorithm for synchronization of relations with bounded length difference, the relations being between finite words or between one-sided infinite words. This constitutes another proof of Eilenberg and Schützenberger's result. Their algorithm operates directly on the transducer that realizes the relation. Our synchronization algorithm is an extension to α-synchronous transducer of the algorithm given in [5].

One main ingredient of the synchronization algorithm is the state splitting transformation. The notion of state splitting, which appeared early in information theory, has been introduced to symbolic dynamics by Williams. Since then, it has been widely used, for example to solve certain coding problems [6, 7].

Finally, we show that the class of α-synchronous relations is a boolean algebra as in the case of classical synchronous relations. This result is a crucial property in the theory of automatic structures [8].

The paper is organized as follows. Sect. 2 introduces the basic definitions of rational relations and transducers. The α-synchronous transducers and relations are defined in Sect. 3. The synchronization algorithm is described in Sect. 4. Some closure properties of these relations are proved in section 5 and the main theorem is proved in Sect. 6.

2 Preliminaries

When a rational number α is written $\alpha = p/q$, we always assume that the integers p and q are relatively prime.

In what follows, A and B denote finite alphabets. The free monoid A^* is the set of finite *words* or sequences of letters from A. The *empty word* is denoted by ε. The *length* of a word $u \in A^*$ is denoted by $|u|$. In this paper, we study relations, that is, subsets of the product monoid $A^* \times B^*$. For a relation $R \subseteq A^* \times B^*$, we denote by R^{-1} the relation $\{(v, u) \mid (u, v) \in R\}$.

A *transducer* (also known as a *two-tape automaton*) is a non-deterministic automaton whose transitions are labeled by pairs of words. A *transducer* over the monoid $A^* \times B^*$ is composed of a finite set Q of *states*, a finite set $E \subset Q \times A^* \times B^* \times Q$ of *transitions* and two sets $I, F \subseteq Q$ of *initial* and *final* states. A transition $\tau = (s, u, v, t)$ from s to t is denoted by $s \xrightarrow{u|v} t$.

A *path* in a transducer $\mathcal{T}$ is a sequence

$$s_0 \xrightarrow{u_0|v_0} s_1 \xrightarrow{u_1|v_1} \cdots \xrightarrow{u_n|v_n} s_n$$

of consecutive transitions. The *label* of this path is the pair (u, v) where its *input label* u is the word $u_1 u_2 \cdots u_n$ and its *output label* v is the word $v_1 v_2 \cdots v_n$. Such a path is denoted $s_0 \xrightarrow{u|v} s_n$. This path is *accepting* if s_0 is initial and s_n is final. The set *accepted* by the transducer is the set of labels of its accepting paths, which is a relation $R \subseteq A^* \times B^*$. We say that the relation R is *realized* by the transducer $\mathcal{T}$.

A subset of $A^* \times B^*$ is *rational* if it can be obtained from some finite subsets using union, concatenation and star iteration. It is a consequence of Kleene's theorem that a subset of $A^* \times B^*$ is a rational relation if and only if it is the relation realized by a transducer.

We now recall the definition of a class of very simple rational relations. A relation $R \subseteq A^* \times B^*$ is *recognizable* if there are two families $K_1, \ldots, K_n$ and $L_1, \ldots, L_n$ of rational subsets of A^* and B^* such that $R = \bigcup_{i=1}^n K_i \times L_i$.

3 α-Synchronous Relations

In this section, we first define the notion of α-synchronous transducer and α-synchronous relation. Through the section, a positive rational number $\alpha = p/q$ is fixed.

A transducer $\mathcal{T}$ is α-*synchronous* if for each transition $s \xrightarrow{u|v} t$, the lengths of the input and output labels satisfy $|u| = q$ and $|v| = p$. If follows immediately that α is the ratio between the output length and input length of any path. For any path $s \xrightarrow{u|v} t$ in $\mathcal{T}$, the equality $|v|/|u| = \alpha$ holds.

Let $\#$ be a padding symbol that does not belong to the alphabets A and B. The *padding* of a pair (u, v) of words over A and B is the pair $(u\#^m, v\#^n)$ where m and n are the least integers such that $|v\#^n|/|u\#^m| = \alpha$. The integers m and n are actually given by $m = qr - |u|$ and $n = pr - |v|$ where the integer r is defined by $r = \max(\lceil |u|/q \rceil, \lceil |v|/p \rceil)$. The padding of (u, v) is denoted $(u, v)^\#$ without any reference to α although it depends on α. For a relation $R \subseteq A^* \times B^*$, we denote by $R^\#$ the following relation

$$R^\# = \{(u, v)^\# \mid (u, v) \in R\}.$$

A relation $R \subseteq A^* \times B^*$ is α-*synchronous* if the relation $R^\#$ can be realized by an α-synchronous transducer over the alphabets $A \cup \{\#\}$ and $B \cup \{\#\}$.

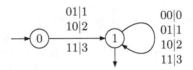

Fig. 1. Conversion from base 2 to base 4

Example 1. The transducer of Fig. 1 performs conversion of the integers from base 2 to base 4. Indeed, its accepts the pairs (u, v) where u is the base-2 expansion of some integer n (padded with a leading 0 to make it of even length) and v is the base-4 expansion of n. This transducer is $1/2$-synchronous.

4   Synchronization

A transducer has the α-*cycle property* if for any cycle $s \xrightarrow{u|v} s$, the equality $|v|/|u| = \alpha$ holds. Of course, an α-synchronous transducer satisfies the α-cycle property. The property of being α-synchronous is a local property whereas the α-cycle property is a global one. The following proposition states that these two properties are in fact equivalent.

Proposition 1. *A relation realized by a transducer with the α-cycle property is α-synchronous.*

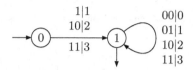

Fig. 2. Conversion from base 2 to base 4 (cont.)

Example 2. In the transducer of Fig. 1, a leading 0 has been added in the transition $0 \xrightarrow{01|1} 1$ to make the transducer 1/2-synchronous. If this 0 is removed, this yields the transducer of Fig. 2, which is not 1/2-synchronous but has the 1/2-cycle property.

The proof of the proposition is separated into several lemmas.

Lemma 1. *Let $\alpha = p/q$ be a rational number. For any transducer T with the α-cycle property, there is a constant K such that the inequality $\big|q|v| - p|u|\big| \le K$ holds for any path $s \xrightarrow{u|v} t$ in T.*

We say that a transducer has the α-*balance property* if for any two paths $s \xrightarrow{u|v} t$ and $s \xrightarrow{u'|v'} t$ with the same starting and ending states, the equality $q|v| - p|u| = q|v'| - p|u'|$ holds. Note that the α-balance property is stronger than the α-cycle property since we always assume an empty path from s to s labeled by the pair $(\varepsilon, \varepsilon)$. The following lemma states that they are in fact equivalent.

Lemma 2. *Let $\alpha = p/q$ be a rational number. For any transducer T with the α-cycle property, there is a transducer T' with the α-balance property and a single initial state that realizes the same relation as T.*

Example 3. The transducer of Fig. 2 has the 1/2-cycle property but it does not have the 1/2-balance property. If the construction given in the proof of the lemma is applied to this transducer, one gets the transducer of Fig. 3, which has the 1/2-balance property.

Let $\alpha = p/q$ be a rational number and let T be a transducer with the α-balance property and a single initial state i. Define the function b from Q to $\mathbb{N}$ by $b(s) = q|v| - p|u|$ where $i \xrightarrow{u|v} s$ is a path from i to s. The α-balance property ensures that the value $q|v| - p|u|$ does not depend on the path and the function b is well-defined. The value $b(s)$ is called the *balance* of the state s. The function b has the following properties. The balance of the initial state is 0 and for any path $s \xrightarrow{u|v} t$, $q|v| - p|u|$ is equal to $b(t) - b(s)$.

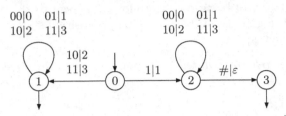

Fig. 3. Conversion from base 2 to base 4 (cont.)

If a transducer $\mathcal{T}$ has the property that the balance of any state is 0, the label of any transition is a pair (u, v) such that $q|v| - p|u|$ equals 0. Since p and q are relatively prime, there is an integer k such that $|u| = kq$ and $|v| = kp$. If k equals 0, the label of the transition is $(\varepsilon, \varepsilon)$. These transitions can be removed using standard techniques to remove ε-transitions. If $k \geq 2$, the words u and v can be factorized $u = u_1 \cdots u_k$ where $|u_i| = q$ and $v = v_1 \cdots v_k$ where $|v_i| = p$. The transition $s \xrightarrow{u|v} t$ can be replaced by the k transitions $s_{i-1} \xrightarrow{u_i|v_i} s_i$ where $s_0 = s$, $s_k = t$ and $s_1, \ldots, s_{k-1}$ are newly introduced states. Applying this transformation to each transition yields an α-synchronous transducer that realizes the same relation.

Once a transducer with the α-balance property has been obtained, it must be ensured that all final states have a 0-balance. Some final states may have a balance different from 0 since a pair $(u, v) \in R$ may not satisfy $|v|/|u| = \alpha$. Since any pair $(u, v)^\# \in R^\#$ satisfies this equality, the transducer must be slightly transformed. A new state f, which becomes the unique final state is added. Furthermore, for any former final s state with balance $qa - bp$, a new transition $s \xrightarrow{\#^m|\#^n} f$ is also added where $m = qr - b$, $n = pr - a$ and $r = \max(\lceil b/q \rceil, \lceil a/p \rceil)$.

Let $\alpha = p/q$ be a rational number. Since p and q are relatively prime, any integer n is equal to $qa - pb$ for some integers a and b. Furthermore, both integers a and b can be chosen positive since n is also equal to $q(a + kp) - p(b + kq)$ for any integer k. For any integer n, let us define $||n||$ as follows.

$$||n|| = \min\{a + b \mid n = qa - pb \text{ with } a, b \geq 0\}.$$

We call $||n||$ the *weight* of n. It is true that $||n|| = 0$ if and only if $n = 0$ but the weight of $-n$ is not necessarily equal to the weight of n. Note that if $n = qa - pb$ and $||n|| = a + b$, then $a < p$ or $b < q$ holds. Otherwise $n = qa' - pb'$ where $a' = a - p$ and $b' = b - q$ and $a' + b' < a + b$.

Example 4. If $\alpha = 2/3$, the weight of some small integers is the following.

n	$\cdots$	-5	-4	-3	-2	-1	0	1	2	3	4	5	6	$\cdots$
$||n||$		5	2	4	1	3	0	2	4	1	3	5	2	

For a transducer $\mathcal{T}$ with the α-balance property, we denote by $||\mathcal{T}||$ the integer $\max_{s \in Q} ||b(s)||$. Note that $||\mathcal{T}||$ equals 0 if and only the balance of any state is 0.

Lemma 3. *Let $\mathcal{A}$ be transducer with the α-balance property such that the balance of any final state is 0. If $||\mathcal{T}|| > 0$, there is a transducer $\mathcal{T}'$ realizing the same relation and such that $||\mathcal{T}'|| < ||\mathcal{T}||$.*

The proof of this lemma is based on two transformations of transducers called state splitting and letter shifting. We first define the operation of out-state splitting in a transducer $\mathcal{T}$. Let s be a state of $\mathcal{T}$ and let $O = (O_1, \ldots, O_n)$ be a partition of all transitions leaving s. The operation of *out-state splitting* relative to the partition O transforms $\mathcal{T}$ into the transducer $\mathcal{T}'$ where $Q' = (Q \setminus \{s\}) \cup \{s_1, \ldots, s_n\}$ is obtained from Q by splitting state s into n new states $s_1, \ldots, s_n$.

- all transitions of $\mathcal{T}$ that are not incident to s are left unchanged.
- each s_i has a copy of the transitions entering s.
- the transitions leaving s are distributed among $s_1, \ldots, s_n$ according to the partition O.

Note that transitions from s to s are considered as both leaving and entering s. Note also that the balance of the new states $s_1, \ldots, s_n$ is the same as the balance of s. The operation of in-state splitting is obtained by reversing the roles played by transitions entering and leaving s.

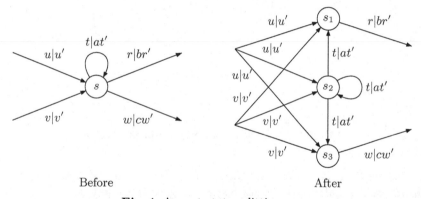

Before   After

Fig. 4. An out-state splitting

An out-state splitting is shown in Fig. 4. There are three transitions τ_1, τ_2 and τ_3, leaving the state s. The state s is split according to the partition $O = (\{\tau_1\}, \{\tau_2\}, \{\tau_3\})$ into three states s_1, s_2 and s_3.

We now describe the letter shifting operation. There are actually four variants whether input or output labels are considered and whether letters are shifted forwards or backwards. We describe the backwards shifting of the output letters. Let s be state that is neither initial nor final and such that all outgoing transitions have a non-empty output label and that all these output labels start with the same letter a. If these conditions are not fulfilled, the operation cannot be performed. This letter a is removed from the output labels of all outgoing transitions and it is added as the last letter of the output labels of all ingoing

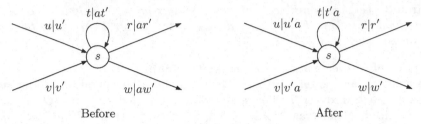

Before   After

Fig. 5. An output backwards shifting

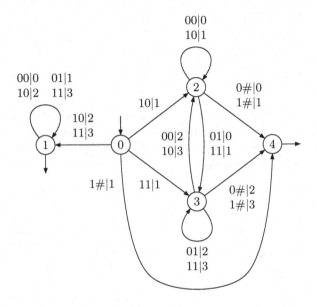

Fig. 6. Conversion from base 2 to base 4 (cont.)

transitions. Note that transitions from s to s are considered as both leaving and entering s. The balance of the state s is changed by the letter shifting. If its balance is r before the operation, it becomes $r+q$ after an output backwards shifting and $r - p$ after an input backwards shifting. An output backwards shifting is shown in Fig. 5.

Example 5. If the proof of the lemma is applied to the transducer of Fig. 3, one gets the transducer of Fig. 6. State 2 has been split into states 2, 3 and 4 and there has been one backwards input shifting.

5   Closure Properties

The class of α-synchronous relations is closed under several operations. Here, we prove the closure properties that are needed in the proof of the main theorem. We consider only the boolean operations and composition.

If R and R' are two α-synchronous relations (for the same α), it is clear that the relation $R \cup R'$ is also α-synchronous. For a relation R, we denote by $\check{R}$ the complement relation $\{(u, v) \mid (u, v) \notin R\}$.

Proposition 2. *If the relation R is α-synchronous, the relation $\check{R}$ is also α-synchronous.*

Proof. Let $\mathcal{T}$ be an α-synchronous transducer realizing the relation $R^{\#}$. Since each transition of $\mathcal{T}$ is labeled by a pair (u, v) of words such that $|u| = q$ and $|v| = p$, the transducer $\mathcal{T}$ can be viewed as an automaton over the alphabet $C = (A \cup \{\#\})^q \times (B \cup \{\#\})^p$. This automaton accepts a rational set of words L over C. Let $\mathcal{T}'$ be an automaton over C accepting the complement of L. This automaton can be modified to accept only words of the form $A^* \#^* \times B^* \#^*$. This modified automaton over C is actually an α-synchronous transducer realizing $\check{R}^{\#}$. □

For two relations $R \subseteq A^* \times B^*$ and $R' \subseteq B^* \times C^*$, we denote by RR' the relation obtained by composition of R and R', that is, the relation $\{(u, w) \mid \exists v \ (u, v) \in R \text{ and } (v, w) \in R'\}$.

Proposition 3. *If the relations $R \subseteq A^* \times B^*$ and $R' \subseteq B^* \times C^*$ are respectively α- and α'-synchronous, the relation RR' is $\alpha\alpha'$-synchronous.*

Proof. Let $\mathcal{T}$ be an α-synchronous transducer realizing $R^{\#}$ and $\mathcal{T}'$ be an α'-synchronous transducer realizing $R'^{\#}$. The following operation is performed on $\mathcal{T}$ and $\mathcal{T}'$ to get $\mathcal{S}$ and $\mathcal{S}'$. Each transition $s \xrightarrow{u|v} t$ where $u = u_1 \cdots u_q$ and $v = v_1 \cdots v_p$ is replaced by the following path of length $q + p$

$$s_0 \xrightarrow{u_1|\varepsilon} s_1 \xrightarrow{u_2|\varepsilon} \cdots \xrightarrow{u_q|\varepsilon} s_q \xrightarrow{\varepsilon|v_1} s_{q+1} \xrightarrow{\varepsilon|v_2} \cdots \xrightarrow{\varepsilon|v_p} s_{q+p}$$

where $s_0 = s$, $s_{q+p} = t$ and $s_1, \ldots, s_{q+p-1}$ are the newly introduced states. The transducers $\mathcal{S}$ and $\mathcal{S}'$ still realize the relations $R^{\#}$ and $R'^{\#}$. They are no longer α- and α'-synchronous but they have the α- and α'-cycle properties. A transducer $\mathcal{R}$ realizing $(RR')^{\#}$ is defined as follows. Its set of states is $Q \times Q'$ where Q and Q' are the sets of states of $\mathcal{S}$ and $\mathcal{S}'$. Its initial and final states are $I \times I'$ and $F \times F'$ where I, F, I' and F' are the initial and final set of states of $\mathcal{S}$ and $\mathcal{S}'$. If E and E' are the sets of transitions of $\mathcal{S}$ and $\mathcal{S}'$, the transitions of $\mathcal{R}$ are defined as follows.

$$\left\{(s, s') \xrightarrow{u|w} (t, t') \mid \exists v \in B \cup \{\#, \varepsilon\} \ s \xrightarrow{u|v} t \in E \text{ and } s' \xrightarrow{v|w} t' \in E'\right\}$$

It is straightforward to verify that $\mathcal{R}$ realizes the relation $(RR')^{\#}$ and that it has the $\alpha\alpha'$-cycle property. By Proposition 1, the relation RR' is $\alpha\alpha'$-synchronous. □

6 Main Theorem

It is clear that a recognizable relation is α-synchronous for any $\alpha > 0$. The following theorem gives a converse.

Theorem 1. *If a rational relation $R \subseteq A^* \times B^*$ is α- and α'-synchronous for $\alpha \neq \alpha'$, then R is recognizable.*

The proof of the theorem is based on the following lemma.

Lemma 4. *Let $\sim$ be an equivalence relation on A^* that is α-synchronous for $\alpha \neq 1$. The relation $\sim$ is of finite index (it has finitely many classes).*

Proof. We claim that there is a constant K such that if $w \in A^*$ satisfies $|w| > K$, there is $w' \in A^*$ such that $w \sim w'$ and $|w'| < |w|$. This proves that the relation $\sim$ is of finite index since the number of classes is bounded by the number of words of length smaller than K.

Since the relation $\sim^{-1}$ is equal to the relation $\sim$, we may assume that $\alpha = p/q > 1$. Let $\mathcal{T}$ be an α-synchronous transducer realizing the relation $\sim$. For each state s of $\mathcal{T}$, we define $c(s)$ by

$$c(s) = \min\{|u| \mid s \xrightarrow{u|v} f \text{ is a path where } v \in \#^* \text{ and } f \text{ is final}\}.$$

By convention, we set $c(s) = \infty$ if no such word u exists. Let C and K be defined by $C = \max_{c(s) \neq \infty} c(s)$ and $K = \alpha(p + C)/(1 - \alpha)$.

Let w be a word such that $w > K$. Since $w \sim w$ holds, there is an accepting path labeled by $(w, w)^\#$ in $\mathcal{T}$. This path can be decomposed

$$i \xrightarrow{u|v} s \xrightarrow{u'|v'} f$$

where $(uu', vv') = (w, w)^\#$, $v = w\#^n$ with $n < p$ and $v' \in \#^*$. Since $|v| \leq |w| + p$ and $|v|/|u| = \alpha$. One has $|u| < p + |w|/\alpha$. By definition of C, there is a path $s \xrightarrow{u''|v''} f'$ such that $|u''| \leq C$, $v'' \in \#^*$ and f' is a final state. Then there is an accepting path labeled by (uu'', uv'') in $\mathcal{T}$. Furthermore, the word uu'' satisfies $|uv''| \leq |u| + C < p + C + |w|/\alpha \leq |w|$. There is a word w' such that $(uu'', vv'') = (w', w)^\#$ and $|w'| < |w|$. We have found a word w' such that $w \sim w'$ and $|w'| < |w|$. $\square$

We come now to the proof of the main theorem.

Proof. Let R be a relation that is α- and α'-synchronous for $\alpha \neq \alpha'$. For a word u in A^*, we denote by $R(u)$ the set $\{v \mid (u, v) \in R\}$. We define the equivalence relation $\sim$ on A^* as follows. For any word u and u', the relation $u \sim u'$ holds if and only if $R(u) = R(u')$. We claim that the relation $\sim$ is α/α'-synchronous.

By Proposition 2, it suffices to prove that the relation $\not\sim$ is α/α'-synchronous. By definition of $\sim$, one has the following equivalence

$$u \not\sim u' \iff \exists v \begin{cases} (u, v) \in R \wedge (u', v) \notin R \\ \vee \\ (u, v) \notin R \wedge (u', v) \in R \end{cases},$$

which shows that the relation $\not\sim$ is equal to $R\check{R}^{-1} \cup \check{R}R^{-1}$ where $\check{R} = \{(u, v) \mid (u, v) \notin R\}$. Since R is α-synchronous, the relation $\check{R}$ is also α-synchronous by

Proposition 2. Since R is α'-synchronous, both relations $\check{R}^{-1}$ and R^{-1} are also $1/\alpha'$-synchronous by Proposition 2. By Proposition 3, both relations $R\check{R}^{-1}$ and $\check{R}R^{-1}$ are α/α'-synchronous.

By the previous lemma, the relation $\sim$ is of finite index. This proves that R is recognizable.   □

References

1. Sakarovitch, J.: Éléments de théorie des automates. Vuibert (2003)
2. Cobham, A.: On the base-dependance of sets of numbers recognizable by finite automata. Math. Systems Theor. **3** (1969) 186–192
3. Elgot, C.C., Mezei, J.E.: On relations defined by generalized finite automata. IBM Journal Res. and Dev. **9** (1965) 47–68
4. Eilenberg, S.: Automata, Languages and Machines. Volume A. Academic Press, New York (1972)
5. Frougny, C., Sakarovitch, J.: Synchronized relations of finite words. Theoret. Comput. Sci. **108** (1993) 45–82
6. Marcus, B.: Factors and extensions of full shifts. Monats. Math **88** (1979) 239–247
7. Adler, R.L., Coppersmith, D., Hassner, M.: Algorithms for sliding block codes. IEEE Trans. Inform. Theory **IT-29** (1983) 5–22
8. Khoussainov, B., Rubin, S.: Automatic structures: Overview and future directions. Journal of Automata, Languages and Combinatorics **8**(2) (2003) 287–301

On Critical Exponents in Fixed Points of Non-erasing Morphisms

Dalia Krieger

School of Computer Science
University of Waterloo
Waterloo, ON N2L 3G1, Canada
d2kriege@cs.uwaterloo.ca

Abstract. Let Σ be an alphabet of size t, let $f : \Sigma^* \to \Sigma^*$ be a non-erasing morphism, let $\mathbf{w}$ be an infinite fixed point of f, and let $E(\mathbf{w})$ be the critical exponent of $\mathbf{w}$. We prove that if $E(\mathbf{w})$ is finite, then for a uniform f it is rational, and for a non-uniform f it lies in the field extension $\mathbb{Q}[\lambda_1, \dots, \lambda_\ell]$, where $\lambda_1, \dots, \lambda_\ell$ are the eigenvalues of the incidence matrix of f. In particular, $E(\mathbf{w})$ is algebraic of degree at most t. Under certain conditions, our proof implies an algorithm for computing $E(\mathbf{w})$.

MSC: 68R15
Keywords: Critical exponent; Circular D0L languages.

1 Introduction

A non-empty finite word z over a finite alphabet Σ is a *fractional power* if it has the form $z = x \cdots xy$, where x is a non-empty word and y is a prefix of x. If $r = |z|/|x|$, we say that z is a power with *exponent r*, or an *r-power*. Let α be a positive real number. A right-infinite word $\mathbf{w}$ over Σ is said to be α-*power-free* if no subword of it is an r-power for any rational $r \geq \alpha$. Otherwise, $\mathbf{w}$ *contains an α-power*. The *critical exponent* of $\mathbf{w}$, denoted by $E(\mathbf{w})$, is the supremum of the set of exponents $r \in \mathbb{Q}_{\geq 1}$, such that $\mathbf{w}$ contains an r-power. If $\mathbf{w}$ is an arbitrary infinite word, $E(\mathbf{w})$ can be any real number greater than 1 [13]. In this work, we are interested in the critical exponents of a more restricted set of words, namely, words generated by iterating a morphism, also known as *pure morphic sequences* or *D0L-words*.

Examples of infinite words for which the critical exponent has been computed include the *Thue-Morse word* $\mathbf{t}$, proved by Thue in 1912 to have $E(\mathbf{t}) = 2$ [18,3], and the *Fibonacci word* $\mathbf{f}$, proved by Mignosi and Pirillo in 1992 to have $E(\mathbf{f}) = 2 + \varphi$, where $\varphi = (1 + \sqrt{5})/2$ is the golden mean [15]. Both words are fixed points of morphisms defined over $\Sigma = \{0, 1\}$: $\mathbf{t}$ is the fixed point beginning with 0 of the *Thue-Morse morphism*, defined by $\mu(0) = 01$, $\mu(1) = 10$; $\mathbf{f}$ is the unique fixed point of the *Fibonacci morphism*, defined by $f(0) = 01$, $f(1) = 0$. The Fibonacci word gives an example of an irrational critical exponent; however, $E(\mathbf{f})$ is algebraic of degree $|\Sigma|$. As we shall see, this is the general case.

O.H. Ibarra and Z. Dang (Eds.): DLT 2006, LNCS 4036, pp. 280–291, 2006.

In a general setting, critical exponents have been studied mainly in relation to Sturmian words; see [4, 5, 7, 8, 10, 14, 20]. Let **s** be a Sturmian word. The main results, proved a few times independently in the papers mentioned above, are a criterion for $E(\mathbf{s})$ to be bounded, and a formula for computing $E(\mathbf{s})$ when it is bounded. Both results make use of the continued fraction expansion of the slope of **s**. For morphic sequences, most of the research has focused on deciding whether a given word has bounded critical exponent; see [6, 9, 11, 16, 17].

Our goal is to characterize and compute critical exponents of pure morphic sequences. In our previous paper [12], we completely characterized $E(\mathbf{w})$ for fixed points of binary k-uniform morphisms. In this paper we extend our results to fixed points of non-erasing morphisms over a finite alphabet. Let $\Sigma = \Sigma_t = \{0, 1, \ldots, t-1\}$, let $f : \Sigma^* \to \Sigma^*$ be a non-erasing morphism, and let **w** be an infinite fixed point of f. We show that if $E(\mathbf{w}) < \infty$, then it is rational for a uniform f, and algebraic of degree at most t for a non-uniform f. More specifically, $E(\mathbf{w}) \in \mathbb{Q}[\lambda_1, \ldots, \lambda_\ell]$, where $\lambda_1, \ldots, \lambda_\ell$ are the distinct eigenvalues of the incidence matrix of f. Under certain conditions, our proof implies an algorithm for computing $E(\mathbf{w})$. Based on our results, we give a short proof of the theorem of Mignosi and Pirillo mentioned above: the critical exponent of the Fibonacci word is $2 + \varphi$, where φ is the golden mean.

2 Basic Definitions and Notation

We use $\mathbb{Z}_{\geq r}$ (and similarly $\mathbb{Q}_{\geq r}, \mathbb{R}_{\geq r}$) to denote the integers (similarly rational or real numbers) greater than or equal to r. If S is a set of numbers, we denote by $M_{n \times m}(S)$ the set of $n \times m$ matrices with entries in S, and by $M_n(S)$ the set of square $n \times n$ matrices with entries in S. Let $A \in M_n(\mathbb{Z})$, and let $\lambda_1, \ldots, \lambda_\ell$ be the distinct eigenvalues of A. We denote by $\mathbb{Q}[A]$ the field extension $\mathbb{Q}[\lambda_1, \ldots, \lambda_\ell]$.

Let $\Sigma = \Sigma_t = \{0, \ldots, t-1\}$ be a finite alphabet. We use the notation Σ^*, Σ^+ and Σ^ω to denote the sets of finite words, non-empty finite words, and right-infinite words over Σ, respectively. We use ϵ to denote the empty word. A morphism $f : \Sigma_t^* \to \Sigma_s^*$ is called *non-erasing* if $f(a) \neq \epsilon$ for all $a \in \Sigma_t$. Infinite words are usually denoted by bold letters. For a finite word $w \in \Sigma^*$, $|w|$ is the length of w, and $|w|_a$ is the number of occurrences in w of the letter $a \in \Sigma$. For both finite and infinite words, w_i is the letter at position i, starting from zero; e.g., $w = w_0 w_1 \cdots w_n$, $\mathbf{w} = w_0 w_1 w_2 \cdots$, where $w_i \in \Sigma$. The set of subwords of a word $w \in \Sigma^* \cup \Sigma^\omega$ is denoted by $S(w)$.

Let $z = a_0 \cdots a_{n-1} \in \Sigma^+$. A positive integer $q \leq |z|$ is a *period* of z if $a_{i+q} = a_i$ for $i = 0, 1, \cdots, n-1-q$. An infinite word $\mathbf{z} = a_0 a_1 \cdots \in \Sigma^\omega$ has a period $q \in \mathbb{Z}_{\geq 1}$ if $a_{i+q} = a_i$ for all $i \geq 0$; in this case, $\mathbf{z}$ is *periodic*, and we write $\mathbf{z} = x^\omega$, where $x = a_0 \cdots a_{q-1}$. We say that $\mathbf{z}$ is *ultimately periodic* if it has a periodic suffix.

A *fractional power* is a word of the form $z = x^n y$, where $n \in \mathbb{Z}_{\geq 1}$, $x \in \Sigma^+$, and y is a proper prefix of x. Equivalently, z has a $|x|$-period and $|y| = |z|$ mod $|x|$. If $|z| = p$ and $|x| = q$, we say that z is a p/q-power, or $z = x^{p/q}$. Since q stands for both the fraction's denominator and the period, we use non-reduced fractions to denote fractional powers: for example, 10101 is a $\frac{5}{2}$-power (as well as a $\frac{5}{4}$-power), while 1010101010 is a $\frac{10}{4}$-power (as well as a $\frac{10}{2}$-power). The word x is referred to as the *power block*.

Let α be a real number. We say that a word w (finite or not) is α-*power-free* if no subword of it is an r-power for any rational $r \geq \alpha$; otherwise, w *contains an α-power*. The *critical exponent* of an infinite word $\mathbf{w}$ is defined by

$$E(\mathbf{w}) = \sup\{r \in \mathbb{Q}_{\geq 1} : \mathbf{w} \text{ contains an } r\text{-power}\} . \tag{1}$$

By this definition, $\mathbf{w}$ contains α-powers for all $1 \leq \alpha < E(\mathbf{w})$, but no α-powers for $\alpha > E(\mathbf{w})$; it may or may not contain $E(\mathbf{w})$-powers.

Let f be a morphism defined over Σ. If $f(a) = ax$ for some $a \in \Sigma$ and $x \in \Sigma^{+}$, and furthermore $f^{n}(x) \neq \epsilon$ for all $n \geq 0$ (f is *prolongable* on a), then $f^{n}(a)$ is a proper prefix of $f^{n+1}(a)$ for all $n \geq 0$, and by applying f successively we get an infinite fixed point of f, $f^{\omega}(a) = \lim_{n \to \infty} f^{n}(a) = axf(x)f^{2}(x)f^{3}(x) \cdots$. Such fixed points are called *pure morphic sequences* or *D0L words*. In this work we consider powers in fixed points of non-erasing morphisms, and so we assume that f is prolongable on 0.

3   Preliminary Results

3.1   The Incidence Matrix Associated with a Morphism

Definition 1. Let $u \in \Sigma_{t}^{*}$. The *Parikh vector* of u, denoted by $[u]$, is a vector $[u] \in M_{t \times 1}(\mathbb{Z}_{\geq 0})$, defined by $[u] = (|u|_{0}, |u|_{1}, \ldots, |u|_{t-1})^{T}$.

Definition 2. Let $f : \Sigma_{k}^{*} \to \Sigma_{n}^{*}$ be a morphism. The *incidence matrix* associated with f, denoted by $F(f)$, is a matrix $F(f) \in M_{n \times k}(\mathbb{Z}_{\geq 0})$, defined by

$$F(f) = (F_{i,j})_{0 \leq i < n,\, 0 \leq j < k} ; \quad F_{i,j} = |f(j)|_{i} . \tag{2}$$

In other words, column j of $F(f)$ is the Parikh vector of $f(j)$.

The next proposition follows directly from the two definitions above:

Proposition 3. Let $f : \Sigma_{t}^{*} \to \Sigma_{s}^{*}$ be a morphism, and let $F = F(f)$ be the incidence matrix of f. Then for all $u \in \Sigma_{t}^{*}$, we have $[f(u)] = F[u]$. If $t = s$, then for all $n \in \mathbb{N}$, we have $F(f^{n}) = F^{n}$.

The proof for the following theorem can be found in [1, Theorem 8.3.11].

Theorem 4 (Perron-Frobenius). *Let $A \in M_{n}(\mathbb{R}_{\geq 0})$, and let r be the spectral radius of A, i.e., $r = \max\{|\lambda| : \lambda \text{ is an eigenvalue of } A\}$. Then*

1. *r is an eigenvalue of A, with a real non-negative corresponding eigenvector;*
2. *there exists a positive integer n such that any eigenvalue λ of A with $|\lambda| = r$ satisfies $\lambda^{n} = r^{n}$.*

Definition 5. The number r described in the above theorem is called the *Perron-Frobenius eigenvalue of A*. We denote it by $r(A)$. If $f : \Sigma_t^* \to \Sigma_t^*$ is a morphism, we denote by $r(f)$ the Perron-Frobenius eigenvalue of $F(f)$.

The next two propositions follow easily from Proposition 3 and Theorem 4.

Proposition 6. *Let $f : \Sigma_t^* \to \Sigma_t^*$ be a morphism prolongable on 0, and let $\mathbf{w} = f^\omega(0)$. Then there exists a morphism $f' : \Sigma_t^* \to \Sigma_t^*$ such that $\mathbf{w} = f'^\omega(0)$ and $F(f')$ has no eigenvalue λ satisfying $|\lambda| = |r(f')|$ and $\lambda \neq r(f')$.*

Proposition 7. *Let $A \in M_n(\mathbb{Z}_{\geq 0})$. Then either $r(A) = 0$ or $r(A) \geq 1$.*

In the next theorem, the notation "U/V", where U, V are column vectors, stands for $\sum_{i=1}^{N} u_i / \sum_{i=1}^{N} v_i$, where u_i, v_i are the components of U, V, respectively.

Theorem 8. *Let $A \in M_N(\mathbb{Z}_{\geq 0})$ be a matrix with no zero columns, and let $r(A) = r, \lambda_1, \ldots, \lambda_\ell$ be its distinct eigenvalues. Assume $r \geq 1$, and $|\lambda_i| < r$ for $i = 1, 2, \ldots, \ell$. Let $U, V, W \in M_{N \times 1}(\mathbb{Z}_{\geq 0})$ be arbitrary vectors with $W \neq 0$, and let*

$$L = \lim_{m \to \infty} \frac{A^m U + (\sum_{i=0}^{m-1} A^i)V}{A^m W}. \tag{3}$$

Then either $L = \infty$, or L exists and is finite. If the second case holds, then $L \in \mathbb{Q}[A]$. In particular, it is algebraic of degree at most N.

The proof of this theorem relies on the Jordan decomposition of A, in particular the asymptotic behavior of powers of Jordan blocks. Details are omitted.

3.2 Circular D0L Languages

A *D0L system* is a triple $G = (\Sigma, f, w)$, where Σ is a finite alphabet, f is a morphism defined over Σ, and $w \in \Sigma^+$ is a word known as the system's *axiom*. If f is non-erasing then G is called a *PD0L system*; in this paper, when referring to a D0L system, we always mean a PD0L system. The system's *language* is the set $L(G) = \{f^n(w) : n \geq 0\}$; thus an infinite fixed point of f represents a D0L language for which f is prolongable on the axiom. A D0L language is α-power-free if all of its elements are α-power-free.

The definition of circular D0L languages we present here is the one used by Mignosi and Séébold in [16], where they prove that every non-repetitive D0L language is circular; their result is the main tool we use in this paper. For injective morphisms, the definition coincides with Mossé's recognizable substitutions [17] and Cassaigne's circular D0L-systems [6].

Roughly speaking, a D0L language $L(G)$ is circular if every sufficiently long word $v \in S(L(G))$ can be decomposed unambiguously into images under f, except perhaps a prefix and a suffix of bounded length. The bound on the length of these prefix and suffix is called the *synchronization delay*. More formally, we have the following definitions:

Definition 9. Let $G = (\Sigma, f, w)$ be a D0L system. We say that a word v admits an *interpretation* by G if there exists a word $v' = a_0 a_1 \cdots a_n a_{n+1} \in S(L(G))$, $a_i \in \Sigma$, such that $v = y_0 f(a_1) \cdots f(a_n) x_{n+1}$, where y_0 is a suffix of $f(a_0)$ and x_{n+1} is a prefix of $f(a_{n+1})$. The word v' is called an *ancestor* of v.

Definition 10. Let $v = y_0 f(a_1) \cdots f(a_n) x_{n+1} = s_0 f(b_1) \cdots f(b_m) r_{m+1}$, where

- $a_i, b_j \in \Sigma$ for $0 \le i \le n+1$ and $0 \le j \le m+1$;
- y_0, s_0 are suffixes of $f(a_0), f(b_0)$, respectively;
- x_{n+1}, r_{m+1} are prefixes of $f(a_{n+1}), f(b_{m+1})$, respectively.

We say that $L(G)$ is *circular* with *synchronization delay* D if whenever we have $|y_0 f(a_1) \cdots f(a_{i-1})| > D$ and $|f(a_{i+1}) \cdots f(a_n) x_{n+1}| > D$ for some $1 \le i \le n$, then $y_0 f(a_1) \cdots f(a_{i-1}) = s_0 f(b_1) \cdots f(b_{j-1})$ for some $0 \le j \le m$, and $a_i = b_j$ (see Fig. 1).

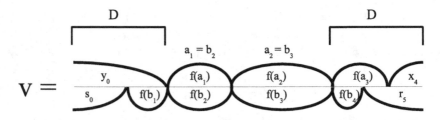

Fig. 1. Synchronization of two interpretations

Theorem 11 (Mignosi and Séébold [16]). *If a D0L language is K-power-free for some number K, then it is circular.*

As mentioned above, when f is injective, Definition 10 coincides with the definition of circular D0L-systems in [6]. We now present this alternative definition, since it will be more convenient to use when computing $E(\mathbf{w})$ for injective morphisms.

Definition 12. Let $h : \Sigma^* \to \Sigma^*$ be a morphism injective over Σ^*, and let $w \in \Sigma^*$. We say that (w_1, w_2) is a *synchronization point* of w (for h), if $w = w_1 w_2$, and for all $v_1, v_2, u \in \Sigma^*$,

$$v_1 w v_2 = h(u) \Rightarrow u = u_1 u_2, \text{ and } v_1 w_1 = h(u_1), \ w_2 v_2 = h(u_2).$$

We denote a synchronization point by $w = w_1 | w_2$.

Definition 13. Let $G = (\Sigma, h, w)$ be a D0L system. We say that G is circular with synchronization delay D if h is injective on $S(L(G))$, and every word $u \in S(L(G))$ with $|u| \ge D$ has at least one synchronization point.

Note that if f is injective over $S(L(G))$ and $u \in S(L(G))$ has at least two synchronization points, then any two distinct interpretations of u must synchronize.

4   Algebraicity of $E(\mathbf{w})$ for Non-erasing Morphisms

In this section we prove our main result, which is the following theorem:

Theorem 14. *Let $f : \Sigma_t^* \to \Sigma_t^*$ be a non-erasing morphism, prolongable on 0, and let $\mathbf{w} = f^\omega(0)$. Let F be the incidence matrix associated with f, and let $\mathbb{Q}[F] = \mathbb{Q}[r, \lambda_1, \cdots, \lambda_\ell]$, where $r, \lambda_1, \cdots, \lambda_\ell$ are the eigenvalues of F. Suppose $E(\mathbf{w}) < \infty$. Then $E(\mathbf{w}) \in \mathbb{Q}[F]$. In particular, $E(\mathbf{w})$ is algebraic of degree at most t.*

To prove Theorem 14 we need the following definitions:

Definition 15. An *occurrence* of a subword within a word $\mathbf{w} \in \Sigma^\omega$ is a triple (z, i, j), where $z \in S(\mathbf{w})$, $0 \le i \le j$, and $w_i \cdots w_j = z$. In other words, z occurs in $\mathbf{w}$ at positions $i, \cdots, j$. For convenience, we usually omit the indices, and refer to an occurrence (z, i, j) as $z = w_i \cdots w_j$. The set of all occurrences of subwords within $\mathbf{w}$ is denoted by $OC(\mathbf{w})$. We say that an occurrence (z, i, j) *contains* an occurrence (z', i', j'), and denote it by $z \rhd z'$, if $i \le i'$ and $j \ge j'$.

Definition 16. Let $z = w_i \cdots w_j \in OC(\mathbf{w})$ be a p/q-power. We say that (z, q) is *left stretchable* (resp. *right stretchable*) if the q-period of z can be stretched left (resp. right), i.e., if $w_{i-1} \cdots w_j$ (resp. $w_i \cdots w_{j+1}$) is a $(p+1)/q$-power. If (z, q) can be stretched left by $c > 0$ letters and no more, then the *left stretch* of (z, q) is defined by $\sigma(z, q) = w_{i-c} \cdots w_{i-1}$; otherwise, if (z, q) is not left stretchable, then $\sigma(z, q) = \epsilon$. Similarly, the *right stretch* of (z, q) is given by $\rho(z, q) = w_{j+1} \cdots w_{j+d}$ if (z, q) can be stretched right by exactly $0 < d < \infty$ letters, by $\rho(z, q) = \epsilon$ if (z, q) is not right stretchable, and by $\rho(z, q) = (w_m)_{m>j}$ if (z, q) can be stretched right infinitely (i.e., $(w_m)_{m \ge i}$ is periodic with period q). The *stretch vector* of (z, q), denoted by $\Lambda(z, q)$, is the Parikh vector of the left and right stretch combined:

$$\Lambda(z, q) = [\sigma(z, q)\rho(z, q)].$$

If $\rho(z, q) \in \Sigma^\omega$, then $\Lambda(z, q)$ is not defined.

Outline of Proof of Theorem 14: Since $E(\mathbf{w})$ is an upper bound, it is enough to consider unstretchable powers when computing it. The idea of the proof is as follows:

1. Take an unstretchable power $z \in OC(\mathbf{w})$, apply f to it, and stretch the result to an unstretchable power. Repeat the process to get an infinite sequence of unstretchable powers in $OC(\mathbf{w})$. For reasons that will be clear shortly, we refer to such sequences as "π-sequences".
2. Show that when considered as a sequence of rational numbers, each of the π-sequences has its lim sup in $\mathbb{Q}[F]$.
3. Show that every sufficiently long unstretchable power in $OC(\mathbf{w})$ belongs to one of finitely many π-sequences.

Clearly, the three steps above suffice to prove Theorem 14: if $E(\mathbf{w})$ is attained by some power $z \in S(\mathbf{w})$, then it is rational; otherwise, there exists a sequence of unstretchable powers $\mathcal{A} = \{z_i\}_{i \geq 0} \subset OC(\mathbf{w})$, such that $E(\mathbf{w}) = \lim_{i \to \infty}(z_i)$. Since every sufficiently long element of $\mathcal{A}$ belongs to one of finitely many π-sequences, there must be an infinite subsequence of $\mathcal{A}$ which belongs to one π-sequence, hence its limit must belong to $\mathbb{Q}[F]$.

Sketch of Proof of Theorem 14: Let $z = x^{p/q} \in OC(\mathbf{w})$ be an unstretchable p/q-power. Let $P = [z]$ and $Q = [x]$. In order to keep track of the components of P and Q, we introduce the notation "z is a P/Q-power", where

$$\frac{P}{Q} := \frac{\sum_{0 \leq i < t} |z|_i}{\sum_{0 \leq i < t} |x|_i} = \frac{p}{q}.$$

Recall that by Proposition 3, $[f(z)] = FP$ and $[f(x)] = FQ$; thus under this notation, $f(z)$ is an FP/FQ-power. This power can be stretched by $\sigma(f(z), FQ)$ on the left and $\rho(f(z), FQ)$ on the right; the result (provided that ρ is finite) is an unstretchable $(FP + \Lambda(f(z), FQ))/FQ$-power. Let us define the following mapping:

$$\pi : OC(\mathbf{w}) \times \mathbb{Q} \to OC(\mathbf{w}) \times \mathbb{Q},$$

$$\pi\left(z, \frac{P}{Q}\right) = \left(\sigma f(z)\rho, \frac{FP + \Lambda}{FQ}\right). \tag{4}$$

Here $\sigma = \sigma(f(z), FQ)$, $\rho = \rho(f(z), FQ)$, and $\Lambda = \Lambda(f(z), FQ)$. In what follows, we use $\pi(z)$ and $\pi(P/Q)$ to denote the first and second component, respectively.

Iterating π on an initial unstretchable P/Q-power z, we get a sequence of unstretchable powers, $\{\pi^m(z, P/Q)\}_{m \geq 0}$. We refer to such a sequence as a "π-sequence". Let $\{\Lambda_m\}_{m \geq 0}$ be the sequence of stretch vectors generated by successive applications of π. Then

$$\pi^m\left(\frac{P}{Q}\right) = \frac{F^m P + \sum_{i=0}^{m-1} F^{m-1-i}\Lambda_i}{F^m Q}. \tag{5}$$

Lemma 17. *Suppose that* $E(\mathbf{w}) < \infty$. *Let* $z \in OC(\mathbf{w})$ *be an unstretchable P/Q- power, let* $\{\pi^m(z, P/Q)\}_{m \geq 0}$ *be the π-sequence generated by z, and let* $\mathcal{S} = \{\Lambda_m\}_{m \geq 0}$ *be the associated stretch sequence. Then $\mathcal{S}$ is ultimately periodic.*

Proof (sketch). $E(\mathbf{w}) < \infty$ implies that $\mathbf{w}$ is circular (Theorem 11) and not *pushy* [9]. We use this two properties, together with the fact that f is non-erasing, to show the following property: there exists a constant C, such that for every unstretchable P/Q-power $z \in OC(\mathbf{w})$, applying f to the C characters to the left and right of z results in occurrences that contain both the left and right stretch of $(f(z), FQ)$, and the C characters to the left and right of $\pi(z, P/Q)$. We call these C characters the *left* and *right context* of $(z, P/Q)$. Since there are only finitely many words of length C, and the context of $(z, P/Q)$ determines the context of $\pi(z, P/Q)$, the sequence of contexts must be ultimately periodic. This, in turn, implies that the stretch sequence is ultimately periodic. ☐

Corollary 18. *Suppose that* $E(\mathbf{w}) < \infty$. *Let* $z = w_i \cdots w_j \in OC(\mathbf{w})$ *be an unstretchable P/Q- power. Then* $\limsup_{m \to \infty} \pi^m(P/Q) \in \mathbb{Q}[F]$.

Proof (sketch). Let h be the period of the stretch sequence. Then $\{\pi^m(P/Q)\}_{m \geq 0}$ can be partitioned into h subsequences, $\{\pi^{mh+j}(z, P/Q)\}_{m \geq 0}$, where $0 \leq j < h$. Each of these subsequences can be reduced to an expression of the form of (3). We then apply Theorem 8. $\qquad\square$

Lemma 19. *Suppose $E(\mathbf{w}) < \infty$. Let $e = \lceil E(\mathbf{w}) \rceil$, let D be the synchronization delay, and let $M = \max\{D, \{|f(a)| : a \in \Sigma\}\}$. Let $K = e(2D + M + 1)$. Then*

1. *every unstretchable power $z = x^{p/q} \in OC(\mathbf{w})$ satisfying $|z| \geq K$ is an image under the π mapping;*
2. *for $|z| < K$, $OC(\mathbf{w})$ contains only finitely many different sequences of the form $\{\pi^m(z, P/Q)\}_{m \geq 0}$.*

Proof (sketch). Use the circularity of $\mathbf{w}$ to show that if $q > 2D + M$, then all power blocks must have the same inverse image under f; use Lemma 17 to show that for $|z| < K$, there are finitely many relevant contexts. $\qquad\square$

Lemma 19 completes the proof of Theorem 14 as was outlined in the beginning of this section.

4.1 The Uniform Case

When f is a k-*uniform* morphism (i.e., $|f(a)| = k$ for all $a \in \Sigma$), the π-sequences have a simpler form: if z is a p/q-power, then $f(z)$ is a kp/kq-power, and the vector notation we used in the general case is unnecessary. Let $\sigma = \sigma(f(z), kq)$, $\rho = \rho(f(z), kq)$, and let $\lambda = \lambda(f(z), kq) = |\sigma| + |\rho|$ be the stretch size. The π mapping now has the form $\pi(z, p/q) = (\sigma f(z)\rho, (kp + \lambda)/kq)$, and

$$\pi^m\left(\frac{p}{q}\right) = \frac{k^m p + \sum_{i=0}^{m-1} k^{m-1-i}\lambda_i}{k^m q} .$$

As in the general case, when applying π successively we get an ultimately periodic sequence of stretch sizes. The π sequence can be thus partitioned into a finite number of subsequences, each of which converges to a rational limit (in fact, all subsequences converge to the same limit). We have thus proved the following theorem, which generalizes our result in [12] regarding binary alphabets:

Theorem 20. *Let f be a uniform morphism over $\Sigma = \Sigma_t$, prolongable on 0, and let $\mathbf{w} = f^\omega(0)$. Then either $E(\mathbf{w}) = \infty$, or $E(\mathbf{w}) \in \mathbb{Q}$.*

5 Computing Critical Exponents

When trying to apply Theorem 14 for actually computing $E(\mathbf{w})$, the main problem is that the π-subsequences, while converging to a computable limit, don't necessarily *increase* towards the limit; in other words, $\lim_{m \to \infty} \pi^{mh+j}(z, P/Q)$ and $\sup\{\pi^{mh+j}(z, P/Q)\}_{m \geq 0}$ are not necessarily the same, and it is not clear how to compute the second value. However, if for a given morphism f we can show that the π-subsequences are increasing, or if we know how to compute their suprema, then Theorem 14 suggests an algorithm for computing $E(\mathbf{w})$:

Input: An integer t denoting the alphabet size, and a morphism f prolongable on 0.

Algorithm:

1. Check whether or not $E(\mathbf{w}) < \infty$. If $E(\mathbf{w}) = \infty$, return 0.
2. Compute the number $k = 2D + M$, where D is the synchronization delay, $M = \max\{D, m\}$, and $m = \{|f(a)| : a \in \Sigma\}$. Alternately, if f is injective, k is an integer such that every subword of $\mathbf{w}$ with length at least k has two or more synchronization points (Definition 12).
3. Compute the set of powers

$$S_k(\mathbf{w}) := \{z = x^{p/q} \in OC(\mathbf{w}) : z \text{ is unstretchable and } q < k\}.$$

4. For every unstretchable power $z \in S_k(\mathbf{w})$:
   (a) compute the period h of the stretch sequence generated by iterating π on z;
   (b) compute the supremum of each of the subsequences $\{\pi^{mh+j}(z, P/Q)\}$, $m \geq 0$, $j = 0, 1, \cdots, h-1$.

Output: The maximum of the values computed in 4(b).

Notes:

1. It is decidable whether or not $E(\mathbf{w}) < \infty$; see [9, 16, 11].
2. If $E(\mathbf{w}) < \infty$, the synchronization delay D is effectively computable [16].
3. Given a fixed integer n, it is decidable whether $\mathbf{w}$ is n-power-free [16]. Therefore, given that $E(\mathbf{w}) < \infty$, the number $e := \lceil E(\mathbf{w}) \rceil$ is computable. The set $S_k(\mathbf{w})$ can thus be computed by computing the set $\{z \in S(\mathbf{w}) : |z| < ek + 2C\}$. The $2C$ factor is added to cover all possible contexts (see Lemma 17).

The complexity of this algorithm is not clear. In particular, we don't have a bound on D, nor on the prefix of $\mathbf{w}$ which contains $S_k(\mathbf{w})$ (however, if f is a uniform binary morphism, we know by [12] that it is enough to consider $f^4(0)$). Also, the complexity of computing $S_k(\mathbf{w})$ is not clear. In some cases, however, computing $E(\mathbf{w})$ becomes very simple. In particular, if it is easy to show that $\mathbf{w}$ is circular with delay D, and the set $S_k(\mathbf{w})$ is easy to compute and is shown to be finite, step 1 of the algorithm becomes unnecessary. In the next section we give a specific example.

5.1   Computing the Critical Exponent of the Fibonacci Word

Let f be the Fibonacci morphism, $f(0) = 01$, $f(1) = 0$, and let $\mathbf{f} = f^\omega(0)$. In [15], Mignosi and Pirillo showed that $E(\mathbf{f}) = 2 + \varphi$, where $\varphi = (1 + \sqrt{5})/2$ is the golden mean. We give an alternative proof.

Let $z = w_i \cdots w_j \in OC(\mathbf{f})$ be an unstretchable P/Q-power, $z = x^{p/q}$. First, observe that if $q \geq 3$, then x has at least 2 synchronization points. If $q = 2$, then the only possible power block is $x = 01$, since it is easy to see that 11 and $(00)^2$ are not subwords of $\mathbf{f}$, and a power of the form $(10)^r$ will be left stretchable. The word 01 has two synchronization points, $01 = \epsilon|01|\epsilon$. Since f is injective, this means that in order to compute $E(\mathbf{f})$ it is enough to start from powers with $q = 1$. The only such power in $\mathbf{f}$ is the square 00. We conclude that $k = 2$ and $S_k(\mathbf{f}) = \{0^2\}$. Note that we do not need to check separately that $E(\mathbf{f}) < \infty$.

Next, let us compute the stretch sequence of $\pi^m(z)$. Assume w.l.o.g that $w_{i-1} = w_{j+1} = 1$, and $w_{i+q-1} = w_{j-q+1} = 0$. Since $f(0)$ and $f(1)$ don't have a common suffix, $f(z)$ cannot be stretched left, and $\sigma(f(z), FQ) = \epsilon$. To the right, we can always stretch by the letter 0, which is the common prefix of $f(0)$ and $f(1)$; however, we cannot stretch by more, since we must have $w_{j+2} = 0$, or else we would get $11 \in \mathbf{f}$. Thus $f(w_{j-q+1}) = 01$, $f(w_{j+1}w_{j+2}) = 001$, and $\rho(f(z), FQ) = 0$. We get that the stretch vector is always $\begin{bmatrix} 1 \\ 0 \end{bmatrix}$, and the π mapping is given by

$$\pi^m(P/Q) = \frac{F^m P + \left(\sum_{i=0}^{m-1} F^i\right)\begin{bmatrix} 1 \\ 0 \end{bmatrix}}{F^m Q}.$$

The incidence matrix of the Fibonacci morphism is given by $F = \begin{pmatrix} 1 & 1 \\ 1 & 0 \end{pmatrix}$. To compute $\lim_{m \to \infty} \pi^m(P/Q)$ we can use the Jordan decomposition of F; however, because of the special properties of the Fibonacci sequence, we can also compute it directly. Let $\{f_n\}_{n \geq 0}$ be the Fibonacci sequence, defined by $f_0 = 0$, $f_1 = 1$, and $f_n = f_{n-1} + f_{n-2}$ for all $n \geq 2$. It is an easy induction to show that for all $m \geq 1$,

$$F^m = \begin{pmatrix} f_{m+1} & f_m \\ f_m & f_{m-1} \end{pmatrix}.$$

Using the identity $\sum_{i=1}^{n} f_i = f_{n+2} - 1$ (see e.g. [19]), we get that

$$\sum_{i=0}^{m-1} F^i = \begin{pmatrix} \sum_{i=1}^{m} f_i & \sum_{i=1}^{m-1} f_i \\ \sum_{i=1}^{m-1} f_i & \sum_{i=1}^{m-2} f_i + 1 \end{pmatrix} = \begin{pmatrix} f_{m+2} - 1 & f_{m+1} - 1 \\ f_{m+1} - 1 & f_m \end{pmatrix}.$$

In the case of $z = 0^2$, we have $P = \begin{bmatrix} 2 \\ 0 \end{bmatrix}$ and $Q = \begin{bmatrix} 1 \\ 0 \end{bmatrix}$, and so

$$\pi^m(P/Q) = \frac{2(f_{m+1} + f_m) + f_{m+1} + f_{m+2} - 2}{f_{m+1} + f_m} = \frac{3f_{m+2} + f_{m+1} - 2}{f_{m+2}} =$$

$$3 + \frac{f_{m+1} - 2}{f_{m+2}} \xrightarrow{m \to \infty} 2 + \varphi.$$

Also, using the identity $f_{n-1}f_{n+1} - f_n^2 = (-1)^n$ [19], we get that

$$\pi^m(P/Q) - \pi^{m-1}(P/Q) = \frac{f_{m+1}^2 - f_m f_{m+2} + 2f_m}{f_{m+2}f_{m+1}} = \frac{(-1)^m + 2f_m}{f_{m+2}f_{m+1}} > 0,$$

i.e., $\{\pi^m(\begin{bmatrix} 2 \\ 0 \end{bmatrix}/\begin{bmatrix} 1 \\ 0 \end{bmatrix})\}_{m \geq 0}$ is an increasing sequence, and

$$E(\mathbf{f}) = \lim_{m \to \infty} \pi^m \left(0^2, \frac{\begin{bmatrix} 2 \\ 0 \end{bmatrix}}{\begin{bmatrix} 1 \\ 0 \end{bmatrix}}\right) = 2 + \varphi.$$

6 Some Open Problems

1. We have shown that given an infinite fixed point $\mathbf{w}$ of a non-erasing morphism defined over an alphabet of size t, $E(\mathbf{w})$ is either infinite or algebraic of degree at most t. It yet remains to prove the result for erasing morphisms.

Another generalization which seems plausible is *morphic sequences*, that is, images of pure morphic sequences under *codings*. A coding is a 1-uniform morphism $\tau : \Sigma_t \to \Sigma_s$, where typically $s < t$. If $\mathbf{w} = f^\omega(0)$ and $\mathbf{v} = \tau(\mathbf{w})$, then obviously $E(\mathbf{v}) \geq E(\mathbf{w})$. Computer tests suggest that when the inequality is strict then $E(\mathbf{v})$ is attained, i.e., $E(\mathbf{v}) \in \mathbb{Q}$. Proving Theorem 14 for morphic sequences will cover the erasing case as well, since every word generated by iterating a morphism is the image under a coding of a word generated by iterating a non-erasing morphism [1, Theorem 7.5.1].

2. Given an algebraic number α of degree d, can we construct a morphism $f : \Sigma_t \to \Sigma_t$ for some $t \geq d$ such that $E(f^\omega(0)) = \alpha$?

Acknowledgement

The author would like to thank Jeffrey Shallit, for his comments and suggestions; Anna Frid, for first mentioning circularity; and Kalle Saari, for his help with incidence matrices.

References

1. J-P. Allouche and J. Shallit. *Automatic Sequences: Theory, Applications, Generalizations.* Cambridge University Press (2003).
2. T. M. Apostol. *Introduction to Analytic Number Theory.* Springer-Verlag New York Inc. (1976).
3. J. Berstel. Axel Thue's Papers on Repetitions in Words: A Translation. Publications du Laboratoire de Combinatoire et d'Informatique Mathématique **20**, Université du Québec à Montréal (1995).
4. J. Berstel. On the Index of Sturmian Words. In *Jewels are Forever*, Springer, Berlin (1999), 287–294.
5. A. Carpi and A. de Luca. Special factors, periodicity, and an application to Sturmian words. *Acta Informatica* **36** (2000), 983–1006.
6. J. Cassaigne. An algorithm to test if a given circular HD0L-language avoids a pattern. In *IFIP World Computer Congress'94*, Vol **1**, Elsevier (North-Holland) (1994), 459–464.
7. W.-T. Cao and Z.-Y. Wen. Some properties of the factors of Sturmian sequences. *Theoret. Comput. Sci.* **304** (2003), 365–385.
8. D. Damanik and D. Lenz. The Index of Sturmian Sequences. *European J. Combin.* **23** (2002), 23–29
9. A. Ehrenfeucht and G. Rozenberg. Repetition of subwords in D0L languages. *Information and Control* **59** (1983), 13–35
10. J. Justin and G. Pirillo. Fractional powers in Sturmian words. *Theoret. Comput. Sci.* **255** (2001), 363–376.

11. Y. Kobayashi and F. Otto. Repetitiveness of languages generated by morphisms. *Theoret. Comput. Sci.* **240** (2000), 337–378.
12. D. Krieger. On critical exponents in fixed points of binary k-uniform morphisms. *STACS'06, LNCS* **3884** (2006), 104–114.
13. D. Krieger and J. Shallit. Every real number greater than 1 is a critical exponent. Preprint.
14. F. Mignosi. Infinite words with linear subword complexity. *Theoret. Comput. Sci.* **65** (1989), 221–242.
15. F. Mignosi and G. Pirillo. Repetitions in the Fibonacci infinite word. *RAIRO Inform. Théor.* **26** (1992), 199–204.
16. F. Mignosi and P. Séébold. If a D0L Language is k-Power Free then it is Circular. *ICALP'93, LNCS* **700** (1993), 507–518.
17. B. Mossé. Puissances de mots et reconnaissabilité des points fixes d'une substitution. *Theoret. Comput. Sci.* **99** (1992), 327–334.
18. A. Thue, Über die gegenseitige Lage gleicher Teile gewisser Zeichenreihen. *Norske vid. Selsk. Skr. Mat. Nat. Kl.* **1** (1912), 1–67.
19. S. Vajda. *Fibonacci & Lucas Numbers, and the Golden Section: Theory and Applications.* Ellis Horwood Limited (1989).
20. D. Vandeth. Sturmian Words and Words with Critical Exponent. *Theoret. Comput. Sci.* **242** (2000), 283–300.

P Systems with Proteins on Membranes and Membrane Division

Andrei Păun and Bianca Popa

Department of Computer Science/IfM, Louisiana Tech University,
P.O. Box 10348, Ruston, LA 71272
{apaun, bdp010}@latech.edu

Abstract. In this paper we present a method for solving the **NP**-complete SAT problem using the type of P systems that is defined in [9]. The SAT problem is solved in $O(nm)$ time, where n is the number of boolean variables and m is the number of clauses for a instance written in conjunctive normal form. Thus we can say that the solution for each given instance is obtained in linear time. We succeeded in solving SAT by a uniform construction of a deterministic P system which uses rules involving objects in regions, proteins on membranes, and membrane division. We also investigate the computational power of the systems with proteins on membranes and show that the universality can be reached even in the case of systems that do not even use the membrane division and have only one membrane.

1 Introduction

In the recent years we have witnessed an explosion in the data accumulated and describing the biological processes that happen in cells. Many new research projects in the direction of studying and understanding the wonderful and extremely complex system that we call *a cell* have been started recently. They form the core of the new area of research named *Systems Biology*. A complete understanding of the *inner-workings* of the cells could yield immense benefits in all the areas related to disease prevention, cure, health care etc. Even a few years ago the research in the Human Genome Project has benefitted from the successful manipulation of the replication machinery of procaryotic cells. We believe that our understanding of the replication mechanisms in (at least some of) the eucaryotic cells is imminent. In such a case, one can envision that soon we will be able to re-wire the replication pathway in such a cell, leading to interesting possibilities for the computing field. We propose in this paper a way to solve NP complete problems by using such a cell replication machinery. Since the cell replication is usually linked to a protein receptor on the plasma membrane, we are modeling this process with a sequence of steps: a rule simulating the binding of the signalling molecule to its corresponding receptor will be simulated, and then, the bound receptor is viewed as a new protein that starts the division process for the cell. More details will be given in section 2 about all the rules that can be used in the proposed systems.

O.H. Ibarra and Z. Dang (Eds.): DLT 2006, LNCS 4036, pp. 292–303, 2006.

This work is a continuation of the investigations aiming to bridge membrane computing (where in a compartmental cell-like structure the chemicals to evolve are placed in the compartments defined by the membranes) and brane calculi recently introduced in [4] (where one considers again a compartmental cell-like structure with the chemicals/proteins placed on the membranes themselves).

Using the membrane division rules we are able to solve hard problems (NP-complete) such as SAT in polynomial time. Several such results have been obtained recently (see for example [1], [6]), all using the trade-off between space and time made possible by the membrane division rules. Our approach is novel as it refers to the systems in which the parallelism is restricted by the number of proteins embedded in membranes. Even in this case we were able to obtain fast solutions for SAT. Once the biology research gives way to manipulation of cell division, we believe that such an approach could be both feasible and energy efficient thus being the best approach in solving computationally hard problems.

Satisfiability (SAT) is the problem of deciding whether a boolean formula in propositional logic has an assignment that evaluates to true. SAT occurs as a problem and as a tool in applications, and it is considered a fundamental problem in theory, since many problems can be reduced to it. Traditional methods treat SAT as a discrete decision problem.

2   Definition of the New Type of P Systems

In what follows we assume that the reader is familiar with membrane computing basic notions and terminology, e.g., from [10] and [12], as well as with basic elements of computability, so that we only mention here a few notations we use (for a more detailed discussion we refer the interested reader to [11]).

In the P systems which we consider below, we use two types of objects, *proteins* and usual *objects*; the former are placed **on** the membranes, the latter are placed **in** the regions delimited by membranes. The fact that a protein p is on a membrane (with label) i is written in the form $[_i p|\,]_i$. Both the regions of a membrane structure and the membranes can contain multisets of objects and of proteins, respectively.

We consider the types of rules introduced in [9]. In all of these rules, a, b, c, d are objects, p is a protein, and i is a label ("res" stands for "restricted"):

Type	Rule	Effect
1res	$[_i p|a]_i \to [_i p|b]_i$ $a[_i p|\,]_i \to b[_i p|\,]_i$	modify an object, but not move
2res	$[_i p|a]_i \to a[_i p|\,]_i$ $a[_i p|\,]_i \to [_i p|a]_i$	move an object, but not modify
3res	$[_i p|a]_i \to b[_i p|\,]_i$ $a[_i p|\,]_i \to [_i p|b]_i$	modify and move one object
4res	$a[_i p|b]_i \to b[_i p|a]_i$	interchange two objects
5res	$a[_i p|b]_i \to c[_i p|d]_i$	interchange and modify two objects

In all cases above, the protein is not changed, it plays the role of a catalyst, just assisting the evolution of objects. A generalization is to allow rules of the forms below (now, "cp" means "change protein"):

Type	Rule	Effect (besides changing also the protein)
1cp	$[_i p \vert a]_i \rightarrow [_i p' \vert b]_i$	
	$a[_i p \vert \ ]_i \rightarrow b[_i p' \vert \ ]_i$	modify an object, but not move
2cp	$[_i p \vert a]_i \rightarrow a[_i p' \vert \ ]_i$	
	$a[_i p \vert \ ]_i \rightarrow [_i p' \vert a]_i$	move an object, but not modify
3cp	$[_i p \vert a]_i \rightarrow b[_i p' \vert \ ]_i$	
	$a[_i p \vert \ ]_i \rightarrow [_i p' \vert b]_i$	modify and move one object
4cp	$a[_i p \vert b]_i \rightarrow b[_i p' \vert a]_i$	interchange two objects
5cp	$a[_i p \vert b]_i \rightarrow c[_i p' \vert d]_i$	interchange and modify two objects

where p, p' are two proteins (possibly equal, and then we have rules of type *res*).

An intermediate case can be that of changing proteins, but in a restricted manner, by allowing at most two states for each protein, $p, \bar{p}$, and the rules either as in the first table (without changing the protein), or changing from p to $\bar{p}$ and back (like in the case of bistable catalysts). Rules with such flip-flop proteins are denoted by $nff, n = 1, 2, 3, 4, 5$ (note that in this case we allow both rules which do not change the protein and rules which switch from p to $\bar{p}$ and back).

Both in the case of rules of type ff and of type cp we can ask that the proteins are always moved in their complementary state (from p into $\bar{p}$ and vice versa). Such rules are said to be of *pure* ff or cp type, and we indicate the use of pure ff or cp rules by writing ffp and cpp, respectively.

To *divide a membrane*, we use the following type of rule, where p, p', p'' are proteins (possible equal): $[_i p \vert \ ]_i \rightarrow [_i p' \vert \ ]_i [_i p'' \vert \ ]_i$

The membrane i is assumed not to have any polarization and it can be non-elementary. The rule doesn't change the membrane label i and instead of one membrane, at next step, will have two membranes with the same label i and the same contents, objects and/or other membranes (although the rule specifies only the proteins involved).

A *P system with proteins on membranes and membrane division* is a system of the form $\Pi = (O, P, \mu, w_1/z_1, \ldots, w_m/z_m, E, R_1, \ldots, R_m, i_o)$, where m is the degree of the system (the number of membranes), O is the set of objects, P is the set of proteins (with $O \cap P = \emptyset$), μ is the membrane structure, $w_1, \ldots, w_m$ are the (strings representing the) multisets of objects present in the m regions of the membrane structure μ, $z_1, \ldots, z_m$ are the multisets of proteins present on the m membranes of μ, $E \subseteq O$ is the set of objects present in the environment (in an arbitrarily large number of copies each), $R_1, \ldots, R_m$ are finite sets of rules associated with the m membranes of μ, and i_o is the label of the output membrane.

The rules are used in the non-deterministic maximally parallel way: in each step, a maximal multiset of rules is used, that is, no rule can be applied to the objects and the proteins which remain unused by the chosen multiset. At each step we have the condition that each object and each protein can be involved in the application of at most one rule, but the membranes are not considered as involved in the rule applications, hence the same membrane can appear in any number of rules at the same time. By halting computation we understand a sequence of configurations that ends with a halting configuration (there is no rule that can be applied considering the objects and proteins present at that moment in the system). With a halting computation we associate a result, in the form of the multiplicity of objects present in region i_o at the moment when the system halts. We denote by $N(\Pi)$ the set of numbers computed in this way by a given system Π. We denote, in the usual way, by $NOP_m(pro_r;list\text{-}of\text{-}types\text{-}of\text{-}rules)$ the family of sets of numbers $N(\Pi)$ generated by systems Π with at most m membranes, using rules as specified in the list-of-types-of-rules, and with at most r proteins present on a membrane. When parameters m or r are not bounded, we use $*$ as a subscript.

3   Preliminary Observations

We assume that all SAT instances are in *conjunctive normal form*, i.e., the conjunction of clauses, where each clause is a disjunction of variables or of their negation. We may write an instance γ, with n variables, in conjunctive normal form using m clauses, as follows: $\gamma = c_1 \wedge c_2 \wedge \ldots \wedge c_m$, $c_i = y_{i,1} \vee y_{i,2} \vee \ldots \vee y_{i,k_i}$, where $y_{i,j} \in \{x_l, \neg x_l \mid 1 \leq l \leq n\}, 1 \leq j \leq k_i, 1 \leq i \leq m$.

Example 1. For $n = 3$ variables, we may have the instance $\gamma = c_1 \wedge c_2$ with $m = 2$ clauses, where $c_1 = y_{1,1} \vee y_{1,2} \vee y_{1,3}$, $c_2 = y_{2,1}$ and $y_{1,1} = x_1$, $y_{1,2} = x_2$, $y_{1,3} = \neg x_3$, $y_{2,1} = \neg x_2$. If $(x_1, x_2, x_3) = (0,0,0)$, we have $c_1 = 1$ and $c_2 = 1$, thus $\gamma = 1$.

We now proceed to solving SAT using P systems with proteins on membranes, we will need to encode the instance to be solved by the system using multisets of objects (since in the P system one cannot have a order imposed on the objects such that they become strings). A solution to the encoding issue is given in the following and will be used in the construction given in the next section:

To encode an instance γ, we use the following notations.

$code(\gamma) = code(c_1)code(c_2)\ldots code(c_m)$, $code(c_i) = \alpha_{i1}\alpha_{i2}\ldots\alpha_{in}$, with

$$\alpha_{ij} = \begin{cases} b_{i,j}, & \text{if } x_j \text{ appears in } c_i; \\ b'_{i,j}, & \text{if } \neg x_j \text{ appears in } c_i; \\ d_{i,j}, & \text{if } x_j \text{ and } \neg x_j \text{ do not appear in } c_i, \text{ for } 1 \leq i \leq m, 1 \leq j \leq n. \end{cases}$$

Example 2. For instance, if we have γ as in example 1, we obtain the following when using the encoding idea given above: $code(c_1) = b_{1,1}b_{1,2}b'_{1,3}$ $code(c_2) = d_{2,1}b'_{2,2}d_{2,3}$ $code(\gamma) = b_{1,1}b_{1,2}b'_{1,3}d_{2,1}b'_{2,2}d_{2,3}$.

We can now pass to the construction for solving SAT using membrane division, before doing so, we state some basic observations.

A clause is satisfied if at least one of the positive variables contained in the clause is assigned the value *true* or a negated variable is assigned the value *false*.

If a clause is not satisfied by one variable (i.e. a positive variable with the assignment *false* or a negated variable assigned the value *true*), then we will move to the next variable in order and check that one whether it satisfies the clause. If we reach the nth variable and it still does not satisfy the clause, then the particular truth assignment does not satisfy the whole instance γ. On the other hand, as soon as we satisfy a clause i by the variable j, we move immediately to the clause $i+1$ and variable 1 to continue this process. When reaching and satisfying the last clause (the clause m), then we know that the instance γ is satisfied by the current truth assignment.

We give the construction and explain in detail the idea in the next section.

4   Solving SAT

We start with an instance γ with n variables and m clauses, encoded as above into $code(\gamma)$. We construct the P system with protein on membranes and membrane division $\Pi = (O, P, \mu, w_1/z_1, \ldots, w_5/z_5, E, R_1, \ldots, R_5)$, where

$$O = \{d, e, f, g, g', \mathbf{yes}, \mathbf{no}\} \cup \{a_i, f_i, t_i \mid 1 \le i \le n\}$$
$$\cup \{b_{i,j}, b'_{i,j}, d_{i,j} \mid 1 \le i \le m, 1 \le j \le n\},$$
$$P = \{p, p', p'_n, r_0\} \cup \{q_i \mid 1 \le i \le n+2\} \cup \{p_i \mid -(3n+2) \le i \le n\}$$
$$\cup \{p_i^t, p_i^f \mid 1 \le i \le n\}$$
$$\cup \{c_i \mid 0 \le i \le 2nm + 5n + 8\} \cup \{r_{i,j}, r'_{i,j}, r''_{i,j} \mid 1 \le i \le m+1, 1 \le j \le n+s1\},$$
$$E = \emptyset, \quad \mu = [_1[_2[_3 \;]_3]_2[_4 \;]_4[_5 \;]_5]_1,$$
$$w_1 = dg, \; w_2 = a_1 a_2 \ldots a_n d, \; w_3 = code(\gamma)f, \; w_4 = et_1 t_2 \ldots t_n f_1 f_2 \ldots f_n, \; w_5 = g,$$
$$z_1 = p, \quad z_2 = p_{-(3n+2)}, \quad z_3 = r_0, \quad z_4 = q_1, \quad z_5 = c_0,$$
$$R_1 = \{[_1 p | \mathbf{yes}]_1 \to \mathbf{yes}[_1 p' | \;]_1, \; [_1 p | g']_1 \to \mathbf{no}[_1 p | \;]_1\},$$
$$R_2 = \{d[_2 p_i | d]_2 \to d[_2 p_{i+1} | d]_2 \mid -(3n+2) \le i \le -1\}$$
$$\cup \{[_2 p_i | \;]_2 \to [_2 p_{i+1}^t | \;]_2[_2 p_{i+1}^f | \;]_2 \mid 0 \le i \le n-1\}$$
$$\cup \{t_i[_2 p_i^t | a_i]_2 \to a_i[_2 p_i | t_i]_2, \; f_i[_2 p_i^f | a_i]_2 \to a_i[_2 p_i | f_i]_2 \mid 1 \le i \le n\}$$
$$\cup \{e[_2 p_n | \;]_2 \to [_2 p_n | e]_2, \; [_2 p_n | \mathbf{yes}]_2 \to \mathbf{yes}[_2 p'_n | \;]_2\},$$
$$R_3 = \{e[_3 r_0 | \;]_3 \to [_3 r_{1,1} | e]_3, \; [_3 r_{m+1,1} | f]_3 \to \mathbf{yes}[_3 r_{m+1,1} | \;]_3\}$$
$$\cup \{[_3 r_{i,j} | d_{i,j}]_3 \to d_{i,j}[_3 r'_{i,j+1} | \;]_3, \; d_{i,j}[_3 r'_{i,j+1} | \;]_3 \to [_3 r_{i,j+1} | d_{i,j}]_3,$$
$$t_j[_3 r_{i,j} | b_{i,j}]_3 \to b_{i,j}[_3 r'_{i+1,j} | t_j]_3, \; [_3 r'_{i+1,j} | t_j]_3 \to t_j[_3 r_{i+1,1} | \;]_3,$$
$$f_j[_3 r_{i,j} | b'_{i,j}]_3 \to b'_{i,j}[_3 r'_{i+1,j} | f_j]_3, \; [_3 r'_{i+1,j} | f_j]_3 \to f_j[_3 r_{i+1,1} | \;]_3,$$
$$f_j[_3 r_{i,j} | b_{i,j}]_3 \to b_{i,j}[_3 r''_{i,j} | f_j]_3, \; [_3 r''_{i,j} | f_j]_3 \to f_j[_3 r_{i,j+1} | \;]_3,$$
$$t_j[_3 r_{i,j} | b'_{i,j}]_3 \to b'_{i,j}[_3 r''_{i,j} | t_j]_3, \; [_3 r''_{i,j} | t_j]_3 \to t_j[_3 r_{i,j+1} | \;]_3,$$
$$[_3 r_{i,j} | t_j]_3 \to t_j[_3 r_{i,j+1} | \;]_3, \; [_3 r_{i,j} | f_j]_3 \to f_j[_3 r_{i,j+1} | \;]_3$$
$$\mid \text{ for } 1 \le i \le m, \; 1 \le j \le n\},$$
$$R_4 = \{[_4 q_i | \;]_4 \to [_4 q_{i+1} | \;]_4[_4 q_{i+1} | \;]_4 \mid 1 \le i \le n+1\} \cup \{[_4 q_{n+2} | e]_4 \to e[_4 q_{n+2} | \;]_4\}$$
$$\cup \{[_4 q_{n+2} | t_i]_4 \to t_i[_4 q_{n+2} | \;]_4, \; [_4 q_{n+2} | f_i]_4 \to f_i[_4 q_{n+2} | \;]_4 \mid 1 \le i \le n\},$$

$$R_5 = \{g[_5c_i|g]_5 \rightarrow g[_5c_{i+1}|g]_5 \mid 0 \le i \le 2nm + 5n + 5\}$$
$$\cup \{g[_5c_{2nm+5n+7}|g]_5 \rightarrow g'[_5c_{2nm+5n+8}|g']_5\}.$$

The rules that are used by the system Π are of one of the forms $3res$, $2cp$, $5cp$, or membrane division. Note that on each membrane in the system we have only one protein. Initially the environment is empty and will be used to receive the output, the answer **yes** or **no** (no other objects are sent out in the environment during the computation).

We start with the *preliminary phase*, by generating 2^{n+1} copies of t_i and f_i, $1 \le i \le n$, in region 4. In the first $n+1$ steps we apply the following membrane division rules. $[_4q_i|]_4 \rightarrow [_4q_{i+1}|]_4[_4q_{i+1}|]_4, 1 \le i \le n+1$.

In the initial configuration we have protein q_1 on membrane 4 and after applying the membrane division rules, in the first $n+1$ steps, we get protein q_{n+2} on all 2^{n+1} membranes labeled 4 . Now, we can send out, to membrane 1, all objects from the elementary membranes 4, in $2n+1$ steps, by applying the following $3res$ rules. $[_4q_{n+2}|t_i]_4 \rightarrow t_i[_4q_{n+2}|]_4$, $[_4q_{n+2}|f_i]_4 \rightarrow f_i[_4q_{n+2}|]_4$, and $[_4q_{n+2}|e]_4 \rightarrow e[_4q_{n+2}|]_4$, $1 \le i \le n$.

In parallel with these rules, in the first $3n+2$ steps, we apply the following $5cp$ rule in membrane 2. $d[_2p_i|d]_2 \rightarrow d[_2p_{i+1}|d]_2, -(3n+2) \le i \le -1$.

When protein p_0 is present on membrane 2, we start the *generating truth-assignments phase*. The following sequence of rules is applied, and after $2n$ steps, we get 2^n membranes labeled 2, all having a similar contents: the initial membrane 3 and the multiset of objects $x_1x_2 \ldots x_n d$, where $x_i \in \{t_i, f_i\}$ for $1 \le i \le n$:

$$[_2p_i|]_2 \rightarrow [_2p_{i+1}^t|]_2[_2p_{i+1}^f|]_2, 0 \le i \le n-1,$$
$$t_i[_2p_i^t|a_i]_2 \rightarrow a_i[_2p_i|t_i]_2, \ f_i[_2p_i^f|a_i]_2 \rightarrow a_i[_2p_i|f_i]_2, 1 \le i \le n.$$

So we are now $5n+2$ steps from the start of the simulation. We can we now check the clauses, starting with the first one. The computation will take place in region 3, where we have the input, $code(\gamma)$. At this moment we have in the membranes labeled 2 all the possible truth-assignments for the n boolean variables appearing in γ. On the membrane 3 we currently have the protein r_0. We start checking each clause by changing the protein (which will be some variant of r) on membrane 3. We change the protein $r_{i,j}$ according with the ith clause and the jth variable that we check. In order to have clause c_i satisfied, we need at least one variable $y_{i,j}$ present in c_i to be true; for γ to be satisfied, we need all clauses to be true.

When we finish generating truth-assignments in region 2, we have protein p_n on membrane 2, and in 2 steps we start the *checking phase* moving the object e from region 1 to region 2 and then into region 3: $e[_2p_n|]_2 \rightarrow [_2p_n|e]_2$, and then $e[_3r_0|]_3 \rightarrow [_3r_{1,1}|e]_3$.

Now we start a sequence of pairs of steps, an even step followed by an odd one and so on. At each moment, there is one protein on membrane 3 that gets primed (or double primed) in the even steps and then lose the prime in the odd steps.

If x_j and $\neg x_j$ are not present in c_i, in the even steps we apply the rules $[_3 r_{i,j} | d_{i,j}]_3 \rightarrow d_{i,j} [_3 r'_{i,j+1} |]_3$, $1 \leq i \leq m$, $1 \leq j \leq n$. and in the odd steps the following rules are used and we move to the next variable to check. $d_{i,j} [_3 r'_{i,j+1} |]_3 \rightarrow [_3 r_{i,j+1} | d_{i,j}]_3$, $1 \leq i \leq m$, $1 \leq j \leq n$.

If x_j is present and it is true, then clause c_i is satisfied and we move to the next clause. In the even steps the following rules are used. $t_j [_3 r_{i,j} | b_{i,j}]_3 \rightarrow b_{i,j} [_3 r'_{i+1,j} | t_j]_3$, $1 \leq i \leq m$, $1 \leq j \leq n$. In the odd steps, t_j is sent back to region 2 and we move to check the next clause by applying the following rules. $[_3 r'_{i+1,j} | t_j]_3 \rightarrow t_j [_3 r_{i+1,1} |]_3$, $1 \leq i \leq m$, $1 \leq j \leq n$.

If $\neg x_j$ is present and x_j is false, then clause c_i is satisfied and, in the even steps, the following rules are applied. $f_j [_3 r_{i,j} | b'_{i,j}]_3 \rightarrow b'_{i,j} [_3 r'_{i+1,j} | f_j]_3$, $1 \leq i \leq m$, $1 \leq j \leq n$. We move to the next clause and f_j is sent back to membrane 2 by using the following rules in the odd steps. $[_3 r'_{i+1,j} | f_j]_3 \rightarrow f_j [_3 r_{i+1,1} |]_3$, $1 \leq i \leq m$, $1 \leq j \leq n$.

For the cases when the current variable j does not make the clause true we use the following rules at the even steps (the move to the next variable happens at the next step): $f_j [_3 r_{i,j} | b_{i,j}]_3 \rightarrow b_{i,j} [_3 r''_{i,j} | f_j]_3$, $1 \leq i \leq m$, $1 \leq j \leq n$, or $t_j [_3 r_{i,j} | b'_{i,j}]_3 \rightarrow b'_{i,j} [_3 r''_{i,j} | t_j]_3$, $1 \leq i \leq m$, $1 \leq j \leq n$.

At the next step the protein $r''_{i,j}$ will be changed into $r_{i,j+1}$ so that the checking can continue with the next variable: $[_3 r''_{i,j} | f_j]_3 \rightarrow f_j [_3 r_{i,j+1} |]_3$, $1 \leq i \leq m$, $1 \leq j \leq n$, or $[_3 r''_{i,j} | t_j]_3 \rightarrow t_j [_3 r_{i,j+1} |]_3$, $1 \leq i \leq m$, $1 \leq j \leq n$.

If the protein $r_{i,n+1}$ is reached on membrane 3, then the clause c_i is not satisfied and there is no further move (γ is false). If the protein $r_{m+1,1}$ is reached, then all clauses $c_1, c_2, \ldots, c_m$ are satisfied and we stop (γ is true). The checking phase takes $2nm$ steps.

At the end of the computation, in the *answering phase*, we have to send out the answer, **yes** or **no**. In three steps the object **yes** is sent out in the environment, and the total number of steps needed to get the **yes** answer is $2nm + 5n + 7$.

First we apply $[_3 r_{m+1,1} | f]_3 \rightarrow$ **yes** $[_3 r_{m+1,1} |]_3$, then $[_2 p_n | \text{yes}]_2 \rightarrow$ **yes** $[_2 p'_n |]_2$ and finally $[_1 p | \text{yes}]_1 \rightarrow$ **yes** $[_1 p' |]_1$.

In parallel, in the membrane 5 (which is used as a counter), the following rules are applied. $g [_5 c_i | g]_5 \rightarrow g [_5 c_{i+1} | g]_5$, $0 \leq i \leq 2nm + 5n + 5$

Simultaneously with sending out the object **yes** from region 1, the following rule is applied. $g [_5 c_{2nm+5n+7} | g]_5 \rightarrow g' [_5 c_{2nm+5n+8} | g']_5$

If the object **yes** is not sent out at the time step $2nm + 5n + 7$, (thus we still have the protein p, not p', on membrane 1), then, in the step $2nm + 5n + 8$, we apply the rule $[_1 p | g']_1 \rightarrow$ **no** $[_1 p |]_1$ and the computation is completed.

It is now clear that the solution to the satisfiability problem of the instance γ is given by the system in linear time, observation that completes the proof. $\square$

It is also interesting to investigate the computational power of such devices, the following section is a step in this direction. A more detailed discussion about the Turing equivalence of such P systems using proteins on membranes is reported in [9].

5   Universality Results

We investigate in this section the classes of P systems with rules considering the proteins on membranes (as above) which are computationally complete, able to characterize NRE, and we begin by considering systems in which only one type of rules is used.

Theorem 1. $NOP_1(pro_2; 2cpp) = NRE$.

Proof. The construction simulates a register machine (see for example [7] for their definition and properties) $M = (m, B, l_0, l_h, R)$ without direct loops (we can assume without losing generality that the register machine to be simulated does not have any instructions that have a jump the themselves, if that is not the case, one can modify the program of the register machine by adding new instructions to remove all the "direct loops") in the ADD instructions and we construct the system $\Pi = (O, P, [_1\ ]_1, \lambda/l_0 p, E, R_1, 1)$ with the following components

$$O = \{a_r \mid 1 \le r \le m\} \cup \{c_l \mid l \in B\} \cup \{c, d\},$$
$$P = \{l, l', l'' \mid l \in B\} \cup \{p, p', p''\} \cup \{p_l \mid l \in B\},$$
$$E = \{a_r \mid 1 \le r \le m\} \cup \{c_l \mid l \in B\} \cup \{c, d\}, \text{ and the following rules.}$$

1. For an ADD instruction $l_1 : (\text{ADD}(r), l_2, l_3) \in R$, we consider the rules
   $a_r[_1 l_1| \to [_1 l_2| a_r$, and $a_r[_1 l_1| \to [_1 l_3| a_r$.

   When protein l_1 is in membrane 1, one of these two rules is applied non-deterministically. This leads to a copy of object a_r being brought in region 1 and protein l_1 being changed into l_2 or l_3.
2. For a SUB instruction $l_1 : (\text{SUB}(r), l_2, l_3) \in R$ we consider the following rules (we also specify the proteins present on the membrane):

Step	Proteins	Rules
1	l_1 and p	$c_{l_1}[_1 l_1| \to [_1 l_1'| c_{l_1}$
2	l_1' and p	$[_1 p| c_{l_1} \to c_{l_1}[_1 p_{l_1}|$ and $d[_1 l_1'| \to [_1 p'| d$
3	p' and p_{l_1}	$[_1 p_{l_1}| a_r \to a_r[_1 l_2''|$, if a_r exists, and $c[_1 p'| \to [_1 p''| c$
4	$(l_2''$ or $p_{l_1})$ and p''	$[_1 l_2''| c \to c[_1 l_2|$ or $[_1 p_{l_1}| c \to c[_1 l_3|$, and $[_1 p''| d \to d[_1 p|$

When protein l_1 is present on the membrane, we apply the rule $c_{l_1}[_1 l_1| \to [_1 l_1'| c_{l_1}$, which changes the protein l_1 into l_1', and moves one copy of c_{l_1} inside. In the second step, we apply both rules at the same time. By applying the first rule, object c_{l_1} is sent out and protein p is changed into p_{l_1}, while by applying the second rule, object d is brought inside and protein l_1' is changed into p'.

At step 3, we send inside the object c and we change protein p' into p''. If we have at least one copy of object a_r inside the region, we can also apply the rule $[_1 p_{l_1} | a_r \rightarrow a_r [_1 l_2'' |$, which sends out object a_r and changes the protein p_{l_1} into l_2''.

At the last step, we send out object d while changing the protein p'' into its original form, p. If at step 3 we have sent out a copy of object a_r, then we can apply rule $[_1 l_2'' | c \rightarrow c [_1 l_2 |$, which sends out object c and changes protein l_2'' into l_2. If at step 3 we have not applied rule $[_1 p_{l_1} | a_r \rightarrow a_r [_1 l_2'' |$, then we still have protein p_{l_1} on the membrane, and we apply rule $[_1 p_{l_1} | c \rightarrow c [_1 l_3 |$, which sends out object c and changes protein p_{l_1} into l_3.

After applying all these rules we change the protein l_1 into l_2 or l_3 depending whether we can send out an object a_r or not, and this way we simulate the SUB instruction.

3. When the halt label l_h is present on the membrane, no further instruction can be simulated, and the number of copies of a_1 in membrane 1 is equal to the value of register 1 of M.   □

A similar result can be obtained for rules of type $3ff$ even in the pure form (but without a bound on the number of proteins).

Theorem 2. $NOP_1(pro_*; 3ffp) = NRE$.

Proof. We consider a register machine $M = (m, B, l_0, l_h, R)$. For each label $l \in B$ we consider a "clone" g; in the proof, we indicate the fact that l, g form such a pair label-clone by using the same subscripts for l and g. The set of all clones of labels from B is denoted by C. Let $w(B)$ be the multiset which contains each $l \in B$ exactly once. We construct the system $\Pi = (O, P, [_1]_1, g_0/w(B)p, E, R_1, 1)$ with the following components

$$O = \{g, g', g'', g''', g^{iv} \mid g \in C\} \cup \{a_r \mid 1 \le r \le m\},$$
$$P = \{l, l' \mid l \in B\} \cup \{p, p'\}, \quad E = \{a_r \mid 1 \le r \le m\},$$

and the following rules. (We start with one copy of each $l \in B$ present on the membrane, together with the protein p, and with the clone of the initial label l_0, that is g_0, in the region.)

1. For an ADD instruction $l_1 : (\text{ADD}(r), l_2, l_3) \in R$, we consider the rules (when specifying the proteins, we mention only those of interest for the use of the rules in that step):

Step	Proteins	Obj. inside	Rules
1	l_1 and $(p$ or $p')$	g_1	$[_1 l_1 | g_1 \rightarrow g_1' [_1 l_1' |$
2	l_1' and $(p$ or $p')$		$a_r [_1 l_1' | \rightarrow [_1 l_1 | a_r$ and $g_1' [_1 p | \rightarrow [_1 p' | g_2$, or $g_1' [_1 p' | \rightarrow [_1 p | g_2$, $g_1' [_1 p | \rightarrow [_1 p' | g_3$, $g_1' [_1 p' | \rightarrow [_1 p | g_3$

We start with g_1 inside and all labels from B on the membrane, and we end with one of the symbols g_2, or g_3 inside plus one extra copy of a_r and again with all labels on the membrane.

2. For a SUB instruction $l_1 : (\text{SUB}(r), l_2, l_3) \in R$, we consider the rules from the next table:

Step	Proteins	Obj. inside	Rules
1	l_1 and (p or p')	g_1	$[_1 l_1\|g_1 \to g_1'[_1 l_1'\|$
2	l_1' and (p or p')		$[_1 l_1'\|a_r \to a_r[_1 l_1\|$, if a_r exists, and
			$g_1'[_1 p\| \to [_1 p'\|g_1''$
			$g_1'[_1 p'\| \to [_1 p\|g_1''$
3	(l_1 or l_1') and (p or p')	g_1''	$[_1 l_1\|g_1'' \to g_2'''[_1 l_1'\|$, or
			$[_1 l_1'\|g_1'' \to g_3^{iv}[_1 l_1\|$
4	(l_1 or l_1') and (p or p')		$g_2'''[_1 l_1'\| \to [_1 l_1\|g_2$ or
			$g_3^{iv}[_1 p\| \to [_1 p'\|g_3$
			$g_3^{iv}[_1 p'\| \to [_1 p\|g_3$

We start with object g_1 inside and at step 1 we send it out modified into g_1'. The rule $[_1 l_1\|g_1 \to g_1'[_1 l_1'\|$ also changes the protein l_1 into l_1'.

At step 2, object g_1' is moved inside and changed into g_1'', and the protein p is changed between its non-primed version and its primed version. If there is at least one copy of a_r inside, we can also apply the rule $[_1 l_1'\|a_r \to a_r[_1 l_1\|$, which sends out a_r and changes protein l_1' into l_1.

We now have two possibilities: one when, at previous step, we have sent out a_r, and one when we have not. In the first case, at step 3 we have protein l_1 on the membrane, we change it into l_1', and we also send out object g_1'', modified into g_2'''. At step 4, we change back protein l_1' into l_1, and we bring in object g_2''', modified into g_2. In the second case, at step 3 we have protein l_1' on the membrane, we change it into l_1, and we also send out object g_1'', modified into g_3^{iv}. At step 4, we bring in g_3^{iv}, modified into g_3, but we do not bring it through the l_1 protein as we need it unchanged, so we bring it using the protein p/p'.

3. We also consider the following final rules: $[_1 p\|g_h \to g_h[_1 p'\|$, $[_1 p'\|g_h \to g_h[_1 p\|$, which remove the clone of the halt label leaving in the system only objects from the output register.

When the computation in M stops, that is, the clone of l_h is introduced in the system, the final rule is used and the computation in Π also stops. The number of copies of a_1 in membrane 1 is equal to the value of register 1 of M.   □

Several other results have been also discussed in this direction in [9], we mention here the results without providing the proofs as the constructions are similar with the constructions from theorems 1 and 2.

We pass now to the case when rules of two types are used to reach universality.

The rules of type 2*res* correspond to uniport rules, while rules of type 4*res* correspond to minimal antiport rules (for basic definitions of symport, uniport, or antiport rules see [8]). Is is important however to note that in our case the number of proteins never changes, hence at a given step the number of rules which can be used is bounded by the number of proteins (hence the parallelism is restricted in this way). Fortunately enough, in the proof of the universality of P systems with minimal symport and antiport rules from [2], $NOP_3(sym_1, anti_1) = NRE$, the parallelism is also bounded. Consequently, we have the following result:

Theorem 3. $NOP_3(pro_*; \alpha\beta, \gamma\delta) = NRE$, where we have that $\alpha \in \{2,3\}$, and $\beta, \delta \in \{res, ff, cp\}, \gamma \in \{4,5\}$.

As expected, when we use rules where also proteins can be used, controlling the operations of passing objects through membranes in the symport/antiport manner, improvements of this result can be obtained (while the proof is much simplified).

Theorem 4. $NOP_1(pro_2; \alpha\beta, \gamma) = NRE$ for all $\alpha \in \{2,3\}, \beta \in \{res, ff, cp\}$, $\gamma \in \{4cpp, 4cp, 5cpp, 5cp\}$.

The problem of obtaining a similar result remains open for α, β as in the previous theorem and $\gamma \in \{4res, 5res, 4ff, 5ff\}$.

The next result shows that universality can be obtained even when using only $1cpp$ rules (objects are changed but are not transported at the same time when the protein is changed) and rules of type $2res$ (uniport rules) for transporting objects between environment and the membrane.

Theorem 5. $NOP_1(pro_2; 2res, 1cpp) = NRE$.

The last result obtained in this area is dealing with only uniports that are only flip-flopping the protein $2ffp$ and catalytic rules of type $1res$ that do not move objects between regions.

Theorem 6. $NOP_1(pro_*; 1res, 2ffp) = NRE$.

In this way, many pairs of types of rules lead to characterizations of NRE, but the problem remains open (even for the case of several membranes being used) for the following pairs of types of rules $(1res, 3res)$, $(1ff, 2res)$, $(1ff, 3res)$, $(2ff, 3res)$, as well as for pairs involving rules of types $4\beta, 5\beta, \beta \in \{res, cp, ff\}$.

6   Final Remarks

We have introduced and investigated a class of P systems where the multisets of objects from the compartments of the membrane structure evolve under the control of multisets of proteins placed on the membranes. We showed that the membrane division is an important feature that could hold the power to solving computationally hard problems in polynomial time using the cell's replication mechanisms. Several universality results have been also obtained, even for systems with the minimal number of membranes, one; in many cases, also the number of proteins present at any moment on the membrane is rather small (this can be considered as a descriptive complexity measure of the systems). For the universality results, we did not even use the membrane division rules. It remains as an open problem to bound the number of proteins also in Theorems 2 and 6.

Another question is whether rules of pure forms are strictly weaker than rules of the general form of types cp and ff.

Besides the open problems mentioned above, several other research topics remain to be considered, such as other types of rules, other strategies to use them (sequentially, in the minimally parallel way, etc.), other ways of using the systems (in the accepting mode, then looking for proofs based on deterministic systems, for solving decision problems, and so on), or of defining the result of a computation (e.g., taking as the result the length of the computation, in the sense of [5]). We will consider such topics in our forthcoming work.

Acknowledgements

A. Păun gratefully acknowledges the support in part by LA BoR RSC grant LEQSF (2004-07)-RD-A-23 and NSF Grants IMR-0414903 and CCF-0523572.

References

1. A. Alhazov: Solving SAT by symport/antiport P systems with memebrane division. In *Cellular Computing. Complexity Aspects* (M.A. Gutierrez-Naranjo, Gh. Păun, M.J. Perez-Jimenez, eds.), Fenix Editora, Sevilla, 2005, 1–6.
2. A. Alhazov, M. Margenstern, V. Rogozhin, Y. Rogozhin, S. Verlan: Communicative P systems with minimal cooperation. In *Membrane Computing. 5th International Workshop, WMC2004, Milan, Italy, June 2004. Revised Selected and Invited Papers* (G. Mauri, Gh. Păun, M.J. Pérez-Jiménez, G. Rozenberg, A. Salomaa, eds.), LNCS 3365, Springer-Verlag, Berlin, 2005, 161–177.
3. N. Busi: On the computational power of the mate/bud/drip brane calculus: interleaving vs, maximal parallelism. In *Pre-Proceedings of Sixth Workshop on Membrane Computing, WMC6*, Vienna, July 2005, 235–252.
4. L. Cardelli: Brane calculi – interactions of biological membranes. In *Computational Methods in Systems Biology. International Conference CMSB 2004, Paris, France, May 2004, Revised Selected Papers. Lecture Notes in Computer Science*, 3082, Springer-Verlag, Berlin, 2005, 257–280.
5. O.H. Ibarra, A. Păun: Counting time in computing with cells. In *Proceedings of DNA Based Computing, DNA11*, London, Ontario, 2005, 25–36.
6. A. Leporati, C. Zandron: A family of P systems which solve 3-SAT. In *Cellular Computing. Complexity Aspects* (M.A. Gutierrez-Naranjo, Gh. Păun, M.J. Perez-Jimenez, eds.), Fenix Editora, Sevilla, 2005, 247–256.
7. M.L. Minsky: *Computation: Finite and Infinite Machines.* Prentice Hall, Englewood Cliffs, New Jersey, 1967.
8. A. Păun, Gh. Păun: The power of communication: P systems with symport/antiport. *New Generation Computing*, 20, 3 (2002), 295–306.
9. A. Păun, B. Popa: P systems with proteins on membranes. *Fundamenta Informaticae* accepted 2006.
10. Gh. Păun: *Membrane Computing – An Introduction.* Springer-Verlag, Berlin, 2002.
11. G. Rozenberg, A. Salomaa, eds.: *Handbook of Formal Languages.* Springer-Verlag, Berlin, 1987.
12. The P Systems Website: `http://psystems.disco.unimib.it`.

Computing by Only Observing

Matteo Cavaliere[1], Pierluigi Frisco[2], and Hendrik Jan Hoogeboom[3]

[1] Microsoft Research - University of Trento,
Centre for Computational and Systems Biology, Trento, Italy
matteo.cavaliere@msr-unitn.unitn.it
[2] School of Mathematical and Computer Sciences,
Heriot-Watt University, Edinburgh, EH14 4AS, U.K.
P.Frisco@macs.hw.ac.uk
[3] Leiden Institute for Advanced Computer Science,
Universiteit Leiden, Leiden, The Netherlands
hoogeboom@liacs.nl

Abstract. The paradigm of *evolution/observation* is based on the idea
that a computing device can be obtained by combining a basic system
and an observer that transforms the evolution of the basic system into
a readable output. In this framework we investigate what can be computed by changing the observer but not the basic observed system. We
consider grammars as basic systems combined with finite state automata
as observers, watching either the sequence of sentential forms or the productions used by the grammar. It is possible to obtain computational
completeness only varying the observer, without modifying the basic system, which is a fixed context-free grammar.

1 Introduction

In [4] a new way to look at computation named *evolution/observation* has been
introduced. This approach stresses the role of an observer in computation. The
basic idea being that a computing device can be constructed using two systems:
the *basic system* and the *observer*. Following a set of rules the observer translates
the behaviour of the underlying basic system into a readable output.

The evolution/observation framework was originally [4] applied to a membrane system. There, the evolution of a membrane system was observed by using
a multiset finite automaton. Following the same idea, new bio-inspired models
of computation have been obtained in [1] (sticker systems) and in [2] (splicing
systems). The generalisation of the framework to formal language theory has
been proposed in [3, 5, 6, 12], where derivations of grammars are observed by
finite state automata. Generally speaking, in all the mentioned works, already
rather simple basic systems observed by simple observers were proved to be
computationally universal.

However, in most of the results one needs to design both a specific basic
system and a specific observer to produce a specific result. This strategy causes
problems if one plans to implement the framework, for instance, in laboratory by
using bio-systems as basic systems. The main problem comes from the fact that

O.H. Ibarra and Z. Dang (Eds.): DLT 2006, LNCS 4036, pp. 304–314, 2006.
© Springer-Verlag Berlin Heidelberg 2006

most bio-systems cannot be (easily) programmed or modified. These bio-systems evolve following their fixed rules and it is possible, however, that their evolution can be monitored by using different observers.

Therefore it is natural to ask how much one can compute by fixing a basic system and choosing different observers. This paper tries to answer this question in the framework of formal language theory, extending, in this way, the work done in [3, 5, 6, 12].

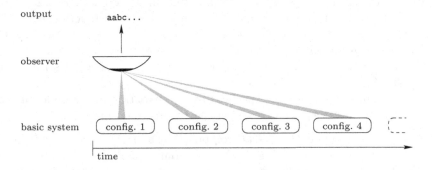

Fig. 1. Sketch of the evolution/observation framework

2   Basic Definitions

A generative grammar is a quadruple $G = (N, T, S, P)$, where N is the alphabet of non-terminals, T is the terminal alphabet disjoint from N, $S \in N$ is the start symbol, and P is the set of productions. By REG, CF and RE we denote the classes of languages generated by regular, context-free, and type-0 grammars, respectively.

Given the set of productions P in a grammar G, we can assign a unique label to each production. We write $x \Rightarrow_\pi y$ if x can be rewritten into y using production with label π. The Szilard language [8, 13] of a grammar G contains only words over the alphabet of labels of productions in G, each word being the sequence of the labels of the productions used to derive the words $w \in L(G)$ from S. Formally, it is defined as $Sz(G) = \{\pi_1 \ldots \pi_h \mid S = w_0 \Rightarrow_{\pi_1} w_1 \Rightarrow_{\pi_2} w_2 \ldots \Rightarrow_{\pi_h} w_h$ and $w_h \in T^*\}$.

A finite state transducer *(fst)* is a system $T = (Z, V_1, V_2, z_0, F, \delta)$, where Z is a finite set of states, $z_0 \in Z$ is the initial state, $F \subseteq Z$ the set of final states, and V_1, V_2 are the input and output alphabet, respectively. The transition relation δ is a finite subset of $Z \times (V_1 \cup \{\lambda\}) \times V_2^* \times Z$, λ denotes the empty string; $(z, a, w, z') \in \delta$ meaning that T in state $z \in Z$, reading $a \in V_1 \cup \{\lambda\}$, goes to state z' and outputs $w \in V_2^*$. The step relation is defined by by $(ax, z, y) \longmapsto (x, z', yw)$ for $(z, a, w, z') \in \delta$ and $x \in V_1^*$, $y \in V_2^*$. This extends to a relation in $V_1^* \times V_2^*$ by setting $T(x) = \{y \in V_2^* \mid (x, z_0, \lambda) \longmapsto^* (\lambda, z, y), z \in F\}$.

3 Observing Sentential Forms

Following [6], an observer is a finite automaton whose set of states is labelled with the symbols of an output alphabet Σ, with λ, or with the symbol $\bot \notin \Sigma$. Any computation of the automaton produces as output the label of the state it halts in. We consider only deterministic and complete automata. We do not specify final states but $\bot$ is present in rejected computations.

Formally, an observer is a tuple $O = (Z, V, \Sigma, z_0, \delta, \sigma)$ with state set Z, input alphabet V, initial state $z_0 \in Z$, and a complete transition function $\delta : Z \times V \to Z$; further there is the output alphabet Σ and a labelling function $\sigma : Z \mapsto \Sigma \cup \{\lambda, \bot\}$. Given a string $w \in V^*$ and an observer O we indicate with $\Theta_O(w) \in \Sigma \cup \{\lambda, \bot\}$ the output of O, that is the label $\sigma(z)$ of the state $z = \delta(z_0, w)$ in which O with w as input halts. For a sequence $w_1, \ldots, w_n$ of $n \geq 1$ strings over V^* we write $\Theta_O(w_1, \ldots, w_n)$ for the string $\Theta_O(w_1) \cdots \Theta_O(w_n)$.

In what follows, observers are specified by the partition into regular subsets of V^* they induce (rather then by giving their transition function).

A *G/O system* is a pair $\Omega = (G, O)$ composed by a grammar $G = (N, T, S, P)$ and an observer $O = (Z, V, \Sigma, z_0, \delta, \sigma)$; the input alphabet of the observer is $V = N \cup T$. The output alphabet Σ of the observer is also the output alphabet of the G/O system.

In [6] three models of G/O systems (always writing, initial, and free) have been defined and investigated. In this paper we only consider free G/O systems: the symbol λ can be associated (by σ) to any state of the observer. Formally, a *free G/O system* defines a language in the following manner:

$$L_{\bot,f}(\Omega) = \{\Theta_O(w_0, w_1, \ldots, w_n) \in \Sigma^* \mid S = w_0 \Rightarrow_G w_1 \Rightarrow_G \ldots \Rightarrow_G w_n, w_n \in T^*\}$$

Thus $L_{\bot,f}(\Omega)$ contains all and only the words over Σ that the observer produces during the possible derivations of the underlying grammar ending in a terminal word. Derivations during which O produces the symbol $\bot$ are then filtered out.

With $\mathcal{L}^G_{\bot,f}(\text{FA})$ we denote the families of languages defined by G/O systems working in the free mode, using the grammar G and a finite state observer. Similarly $\mathcal{L}^G_f(\text{FA})$ denotes the families where only observers are considered that do not use $\bot$ as label: intuitively they cannot 'reject' their input.

Given a class of grammars $\mathcal{G}$ we are interested in those grammars in $\mathcal{G}$ that are equivalent, under suitable observation, to any other grammar in $\mathcal{G}$: we call these grammars *universal modulo observation*. Thus a single grammar is universal (modulo observation) for a family of languages if it can generate every language in the family when observed in an appropriate way.

Definition 1. *A grammar G is universal modulo observation (m.o.) for a family of languages $\mathcal{L}$ if $\mathcal{L} = \mathcal{L}^G_{\bot,f}(\text{FA})$.*

We show by means of an example that a G/O system can generate different classes of languages if the observer is changed while the grammar remains fixed.

Let us consider the following context-free grammar:

$$G = (\{S, A, B, C\}, \{t, p\}, S, \{S \rightarrow pS, S \rightarrow p,$$
$$S \rightarrow A, A \rightarrow AB, A \rightarrow C, B \rightarrow C, C \rightarrow t\}).$$

If G is coupled with the observer O' such that $\Theta_{O'}(w) = a$ if $w \in \{S, A, B, C, t, p\}^+$, then $\Omega = (G, O')$ defines the language $L_{\perp,f}(\Omega) = \{a^i \mid i \geq 2\}$, a regular language. In fact, the derivation $S \rightarrow pS \stackrel{n-2}{\Rightarrow} p^{n-1}S \rightarrow p^n$ produces (when observed) the string a^{n+1}.

Keeping the same grammar G we change the observer into O'' such that:

$$\Theta_{O''}(w) = \begin{cases} \lambda & \text{if } w = S, \\ a & \text{if } w \in AB^*, \\ b & \text{if } w \in C^+B^*, \\ c & \text{if } w \in t^+C^*, \\ \perp & \text{else} \end{cases}$$

In this case $\Omega = (G, O'')$ generates the language $L_{\perp,f}(\Omega) = \{a^n b^n c^n \mid n > 0\}$, a context-sensitive language. In fact, the derivations that result in the output of words $a^n b^n c^n$ are the ones of the form

$$S \rightarrow A \stackrel{n-1}{\Rightarrow} AB^{n-1} \Rightarrow CB^{n-1} \stackrel{n-1}{\Rightarrow} C^n \Rightarrow tC^{n-1} \stackrel{n-1}{\Rightarrow} t^n.$$

The observer outputs the labels according to the described mapping and in this way it rules out all other derivations (including the ones that derive sentential forms containing the terminal p).

This example suffices to underline that part of the computation can be done by choosing the right observer, keeping unchanged the underlying basic system. In the following we try to find out how crucial this 'part' can be.

4   Observing Context-Free Grammars

G/O systems cannot *directly* restrict the sequence of productions that is applied, but they can 'test' each of the separate sentential forms that are generated by the grammar. As the observer can only 'see' sentential forms, (in general) it cannot distinguish the productions applied by the grammar. This can be achieved by changing the single production $\pi : A \rightarrow \alpha$ into the pair $A \rightarrow \pi$, $\pi \rightarrow \alpha$, where π is used as a nonterminal symbol to reflect the use of the production of the same name.

If the observer outputs π when observed, $\perp$ when the string contains more than one production name, and λ otherwise, we observe the *Szilard language* of the original grammar.

Extending this idea we can put regular control on the derivations, and observe the original Szilard language under a finite state transduction. Note that the result follows the 'only observing' paradigm already faithfully. We introduce a single context-free grammar that works for all finite state transductions, which are taken care of by a suitable observer for each fst.

Lemma 2. *For each context-free grammar G there exists a context-free grammar G' such that, for each fst T there exists an observer O_T with $L_{\perp,f}(G', O_T) = T(Sz(G))$.*

Proof. We assume a normal form of the transducer, and allow only transitions of the form (p, a, b, q) where both a and b are either a single letter or the empty string. Without loss of generality we may assume that the states p, q are nonnegative integers. We add a new nonterminal N to the sentential forms, and the 'superfluous' transition $0 : N \to N$. We replace transitions of the form (p, λ, b, q) in the transducer by transitions $(p, 0, b, q)$ to avoid the case of λ-transitions.

As already discussed, we should make productions in G 'visible' to the observer by splitting a production $\pi : A \to \alpha$ into the pair $A \to \pi$, $\pi \to \alpha$, where π is introduced into the grammar G' as a 'name' for the production. In fact we need more 'intermediate states' of π in order to synchronise with a new component we add to the grammar.

As the observer only inspects single sentential forms, it cannot verify by itself what is the state of the fst T in reading the sequence of productions in $Sz(G)$. Instead the context-free grammar G' generates a representation of the consecutive states, and the observer verifies that these match the possible state-transitions of T.

We sketch the construction of the grammar G' and of the observer O_T.

First, introduce new symbols I and X to the alphabet of the context-free grammar G'. We encode the state p of the fst by appending $X I^p X$ to the sentential form produced by G'. The observed derivation of G' should go through the following phases, when simulating a transition (p, π, b, q) of the fst T, where $\pi : A \to \alpha$ is a production of the grammar G:

$\dots A \dots$	$\dots X I^p X$	the fst is in state p;
$\dots A \dots$	$\dots X I^p X I^q X$	guess next state q with $X \to XX$, $X \to IX$;
$\dots \pi \dots$	$\dots X I^p X I^q X$	now the observer checks transition, outputs b;
$\dots \pi \dots$	$\dots X I^q X$	delete old state p with $X \to \lambda$, $I \to \lambda$;
$\dots \alpha \dots$	$\dots X I^q X$	back to normal.

The observer O_T can guarantee that the derivations of the grammar G' go through the successive phases (in fact, only a regular checking is needed).

For more details the reader can refer to [6, Theorem 3] where it is shown how the rewriting of two nonterminals can be synchronised by using a regular observer.   □

In the construction of the above proof, the observer checks whether each production is used in a sentential form belonging to a certain regular language (which depends on the production used). This is a regulating feature similar to that of conditional grammars, see e.g., [7]. In fact, it is easy to see that this feature can be incorporated into the proof, in the simulation of a transition of the fst. Thus, the lemma is equally valid for context-free *conditional* grammars. Szilard languages of these grammars where introduced in [7], but their properties where left open.

We have shown that regular control of a given context-free grammar is possible by adapting that grammar, but the context-free grammar we obtain in that way does not depend on the control itself.

Theorem 3. *There exists a context-free grammar that is universal m.o. for* RE.

Proof. The G/O system will simulate a *one-way two-counter automaton* [14] by observing a fixed context-free grammar. The grammar will essentially act as the storage of the two-counter automaton, while the transitions of the two-counter automaton are simulated by a form of regular control.

To model the contents (p, q) of two counters the sentential form of the constructed grammar equals $A^p S B^q$, adding one to the first counter can be performed by the production $+_A : S \rightarrow AS$, subtracting one by $-_A : A \rightarrow \lambda$. Similar productions apply to B for the second counter.

Now each transition of the two-counter automaton that increases or decreases one of the counters corresponds to a transition of a fst checking the derivation process of the grammar.

This shows that as consequence of Lemma 2, a G/O system can simulate a *partially blind counter automaton* [11], a finite state automaton extended with a finite number of counters holding a non-negative integer that can be incremented and decremented, but not tested for zero. For the full power of the recursively enumerable language the zero-test has to be simulated. This is rather straightforward. To test the first counter A for zero, the observer forces the two consecutive productions $S \rightarrow S_A$, $S_A \rightarrow S$, and additionally, when the sentential form contains S_A checks whether it does *not* contain any A's. Obviously that is a regular test, so the observer can implement it.   □

The universal m.o. grammar constructed here does not have terminal symbols (e.g., to store information on the past of the derivation).

This result is intuitively very close to the AFA and AFL theory, see [9], where a family of languages is called a *full principal trio* if there exists a single language L such that the family equals the languages that can be obtained as finite state transductions of that language. Most famous in that area is the result that the context-free languages are the full principal trio generated by D_2 the Dyck language on two pairs of brackets, i.e., the context-free language generated by $S \rightarrow a S \bar{a} S$, $S \rightarrow b S \bar{b} S$, $S \rightarrow \lambda$. Intuitively that is due to the fact that D_2 models the behaviour of a push-down stack. Here, the instructions of increasing and decreasing the counters are taken care of by a basic grammar with sentential forms in $A^* S B^*$. The power of regular control is added by the construction of Lemma 2, with an extra provision for the zero test.

In the previous proof the ability of the observer to emit $\perp$ in order to reject some computations seems to be crucial. Actually this feature is not necessary if one wants to obtain non-recursive languages by changing the observer of a fixed context-free grammar.

Corollary 4. *There is a context-free grammar G such that $\mathcal{L}_f^G(\text{FA})$ contains non-recursive languages.*

Proof. Any observer O (with output alphabet Σ) using $\bot$ can be changed into an observer O' that does not use it, by replacing $\bot$ by a symbol $1 \notin \Sigma$. The new observer cannot reject its input, and has output alphabet $\Sigma \cup \{1\}$. The results now comes from Theorem 3 and from the fact that recursive languages are closed under intersection with regular languages: $L_{\bot,f}(G,O) = L_{\bot,f}(G,O') \cap \Sigma^*$.   □

5   Restrictions on Context-Free Grammars

We consider two restrictions on context-free grammars: bounding the number of nonterminals and considering leftmost derivations only. In these cases we observe REG and CF, respectively.

Note that the context-free grammar used as a universal grammar for the type-0 grammars has no bound on the number of nonterminals in its sentential forms. The next result shows that indeed this is a necessary property of context-free grammars that are observationally complete for type-0 grammars. In fact, when a bound is imposed, our observations constitute regular languages only. Recall that a context-free grammar is *nonterminal bounded* if there exists a constant k such that all sentential forms generated by the grammar have at most k nonterminals. Incidentally, a context-free grammar has a regular Szilard language iff it is nonterminal bounded [8].

The following result solves a conjecture from [12].

Theorem 5. *For every G/O system $\Omega = (G,O)$ with G nonterminal bounded context-free, $L_{\bot,f}(\Omega)$ is regular.*

Proof. Given $\Omega = (G,O)$ with G nonterminal bounded we will directly construct a finite state automaton accepting $L_{\bot,f}(\Omega)$. Let k be a constant such that every sentential form of G has at most k nonterminals.

The states of the observer (finite state automaton) are of the form $\nu = [\nu_0, A_1, \nu_1, \ldots, A_\ell, \nu_\ell]$, $\ell \leq k$ such that $A_i \in N$ for $1 \leq i \leq \ell$, and ν_i is a mapping from Z into Z, for $0 \leq i \leq \ell$. The interpretation of such a state is that it keeps, for a sentential form $\alpha = \alpha_0 A_1 \alpha_1 A_2 \ldots A_\ell \alpha_\ell$, the state of the new automaton records the sequence of nonterminals, and for each of the segments of terminals α_i the mapping $z \mapsto \delta(z, \alpha_i)$ maps each state z of the observer to the state the observer after reading α_i when starting in z.

Two observations can be made. First, ν as given above, representing sentential form α, the state reached by the observer after reading α can be deduced. The state reached after reading α_0 equals $z'_0 = \nu_0(z_0)$, then A_1 via O reaches $z_1 = \delta(z'_0, A_1)$, and so on. Consequently, ν fixes the observation of O on α.

Second, if a production of the kind $B \to \beta$ is applied to any of the nonterminals $A_i = B$ of a sentential form α, obtaining α', then the state ν' representing α' can be obtained from state ν representing α, by computing the mappings $z \mapsto \delta(z, \gamma)$ for any segment of terminals in β, and composing the mappings of the outer segments with ν_{i-1} and ν_i. In case, $\beta = \lambda$, we compose ν_{i-1} and ν_i.

In this way the G/O system is able to simulate the nonterminal bounded grammar by a finite state automaton. The transitions of the grammar are sketched above, we step from ν to ν' simulating a production, while reading from the tape the observation determined by ν'. The initial state equals $[\iota, S, \iota]$ with ι the identity on Z, representing the initial S, and final states are of the form $[\nu_0]$ representing terminal strings.   □

A *leftmost* generative grammar is a quadruple $G = (N, T, S, P)$, as before, except that G has only leftmost derivations: we assume that $P \subseteq N^+ \times (N \cup T)^*$, and $\alpha \to \beta \in P$ implies $w\alpha x \Rightarrow w\beta x$ for $w \in T^*$ and $x \in (N \cup T)^*$.

A pushdown automaton corresponds to leftmost derivations in a type 0 grammar with productions of the form $pA \to aq\alpha$ for the instruction from state p to q, reading a, popping A and pushing α back to the stack. The following result implies that pushdown automata are less computationally powerful than context-free grammars when observed.

Lemma 6. *For every G/O system $\Omega = (G, O)$ with G a leftmost type-0 grammar, $L_{\perp,f}(\Omega)$ is context-free.*

Proof. Any leftmost type 0 grammar can be simulated by a pushdown automaton. In case of the sentential form $wA\alpha$, with $w \in T^*$, $A \in N$, and $\alpha \in N(N \cup T)^*$, the automaton has read w from the tape and the 'tail' of symbols $A\alpha$ on the stack. The result now is based on the fact that, for a pushdown automaton, regular information on the stack can be stored on the topmost symbol [10], and can be kept up-to-date while popping and pushing.

Let M be the pushdown automaton simulating a leftmost grammar G. From it we construct a pushdown automaton M_O that accepts the output of the observer O on the sentential forms of G (rather than the string generated by G).

For a sentential form $wA\alpha$ as above M_O 'knows' w that has been derived, so the state of the observer O on reading w can be deduced from it. Moreover, the behaviour of O on $A\alpha$ is 'regular', and it can be stored with the topmost symbol of the stack A, known to M_O. Consequently, M_O can deduce the output of observer O on $wA\alpha$, and read that symbol (instead of its usual symbol).   □

All 'leftmost observations' are context-free, and there is a single leftmost context-free grammar that captures them all. This result is similar to the fact that every context-free language is a finite state transduction of the Dyck set D_2, which is used to obtain the universality result for CF.

Theorem 7. *There is a leftmost context-free grammar universal m.o. for CF.*

Proof. Consider a (leftmost) context-free grammar in Greibach normal form for the Dyck language D_2 interleaved with $\{c, d\}^*$. For a fst on D_2 we create an observer that simulates the transducer and produces its output. Due to the Greibach normal form the last symbol from $\{a, \bar{a}, b, \bar{b}\}$ is the last one generated, so the observer 'knows' which step of the fst is observing. The additional symbols c, d make it possible to produce more than one symbol output on a single letter of the Dyck language, like the fst may do. If not enough symbols are derived for the observer, it aborts the computation by outputting $\perp$. Finally, the observer is deterministic, whereas the finite state transducer it simulates is not. Instead we put the nondeterminism in the observed grammar: the occurrences of c and d are (also) used as coin tosses to decide which path to take in the transducer.   □

In a similar way, any regular language can be obtained by observing derivations for the language $\{a, b\}^*$.

Corollary 8. *There exists a rightlinear grammar that is universal m.o. for* REG.

The last results show universality m.o. for CF and REG. Due to Theorem 5 linear grammars always yield regular observations. Hence a linear grammar cannot be universal m.o. for the linear languages. A context-free grammar on the other hand, is likely too powerful, as the example in Section 3 shows.

6   Observing Productions

In this section we briefly (and somewhat informally) consider the case in which the observer does not 'see' the sequence of sentential forms obtained by the basic system, instead it can only 'see' the sequence of productions implemented by the grammar. Such an observer is less powerful than the ones considered until now, as it cannot distinguish the specific position to which the production was applied.

Let us consider, for instance, the G/O system $\Omega = (G, O'')$ defined in Section 3 and the following two derivations of G:

$$1 : S \Rightarrow A \Rightarrow AB \Rightarrow ABB \Rightarrow CBB \Rightarrow CCB \Rightarrow CCC \Rightarrow tCC \Rightarrow ttC \Rightarrow ttt$$

$$2 : S \Rightarrow A \Rightarrow AB \Rightarrow ABB \Rightarrow CBB \Rightarrow CBC \Rightarrow CCC \Rightarrow CCt \Rightarrow tCt \Rightarrow ttt$$

Both derivations are obtained with the application of the same sequence of productions but in Ω derivation 1 renders a result while derivation 2 does not (the derivation is rejected on the word CBC).

To avoid new definitions, let us consider here as observations of the sequences of productions of grammar G any finite state transduction of the Szilard language $Sz(G)$, cf. Lemma 2.

The observation of the productions of context-free grammars is indeed less powerful than observing the sentential forms generated by them. For the Szilard language only the number of nonterminals in the sentential form is important, not their ordering. This implies that the Szilard language of a context-free grammar can be recognised by a partially blind multi-counter automaton, as was observed in the proof of Theorem 3. Recall that these language are all semilinear and contained in the context-sensitive languages.

For each context-free grammar G it is possible to bound the number of counters needed to recognise any of its observations by the number of nonterminals of the grammar. Thus we have the following result, which (together with the previous paragraph) should be contrasted to Theorem 3.

Proposition 9. *For each k, there exists a context-free grammar that is universal modulo observations of productions for the family of languages accepted by partially blind k-counter automata.*

In order to obtain all languages in RE as observations of productions, we can use context-sensitive grammars. The proof of the following result is based on the fact that every recursively enumerable language can be obtained from a context-sensitive language by erasing all occurrences of a fixed symbol, used as 'padding' in the strings.

Proposition 10. *There exists a context-sensitive grammar that is universal m.o. of productions for* RE.

7   Final Remarks

The approach proposed in this paper can be applied to get a more general theory about computing by observation. If we have a fixed system, that evolves according to certain rules, how much differs the (observed) evolution of the system, when watched by 'different' observers? In other words, what is the importance of the observer? Theorem 3 proves that the observer can be crucial: it is possible to 'observe' every recursively enumerable language from a context-free grammar.

It would be then extremely interesting to find grammars that can characterise (by changing the observer) families with certain specified properties. These grammars should have the property that their observed evolution, independently from the observer, always respects certain properties. For instance,is there a grammar universal m.o. for linear languages?

Other directions of investigations are possible: for instance, what classes can be obtained without making use of the symbol ⊥ (largely used in most of the proofs presented here), or with G/O systems using weaker observers (considering interesting restrictions on finite state automata).

Acknowledgements. Parts of this research was performed while M. Cavaliere, PhD student at University of Seville, was visiting the Leiden Institute for Advanced Computer Science during the Spring of 2005.

The authors thank the referees of DLT2006 for their constructive comments.

References

1. A. Alhazov and M. Cavaliere. Computing by observing bio-systems: The case of sticker systems. *10th International Workshop on DNA Computing, Lecture Notes in Computer Science* 3384, pages 1–13. Springer-Verlag, 2004.
2. M. Cavaliere, N. Jonoska, and P. Leupold. Recognizing DNA splicing. *Preproceedings 11th International Workshop on DNA Computing, 2005.* University of Western Ontario, 2005.
3. M. Cavaliere and P. Leupold. Observation of string-rewriting systems. Fundamenta Informaticae. to appear.
4. M. Cavaliere and P. Leupold. Evolution and observation - A new way to look at membrane systems. *Workshop on Membrane Computing, Lecture Notes in Computer Science* 2933, pages 70–87. Springer-Verlag, 2003.

5. M. Cavaliere and P. Leupold. Evolution and observation - A non-standard way to accept formal languages. *Machines, Computations, and Universality, Lecture Notes in Computer Science* 3354, pages 152–162. Springer-Verlag, 2004.
6. M. Cavaliere and P. Leupold. Evolution and observer: A non-standard way to generate formal languages. *Theoretical Computer Science*, 321 (2004) 233–248.
7. J. Dassow and G. Păun. *Regulated rewriting in formal language theory*, volume 18 of *EATCS monographs in Theoretical computer science*. Springer-Verlag, 1989.
8. A.C. Fleck, *An analysis of Grammars by their derivation sets, Information and Control*, 24 (1974) 389–398.
9. S. Ginsburg. *Algebraic and automata-theoretic properties of formal languages, Foundamental studies in computer science*, volume 2. North-Holland, 1975.
10. S. A. Greibach. A Note on Pushdown Store Automata and Regular Systems. *Proceedings of the American Mathematical Society*, 18 (1967) 263-268.
11. S.A. Greibach, Remarks on Blind and Partially Blind One-Way Multicounter Machines, *Theoretical Computer Science* 7 (1978) 311–324.
12. P. Leupold. Non-universal G/O systems. submitted.
13. E. Mäkinen. *On contect-free derivations*. PhD thesis, Acta Universitatis Tamperensis, 1985.
14. M. L. Minsky. *Computation: Finite and Infinite Machines*. Automatic computation. Prentice-Hall, 1967.

A Decision Procedure for Reflexive Regular Splicing Languages

Paola Bonizzoni and Giancarlo Mauri

Dipartimento di Informatica Sistemistica e Comunicazione,
Università degli Studi di Milano – Bicocca,
Via Bicocca degli Arcimboldi 8, 20126 Milano – Italy
{bonizzoni, mauri}@disco.unimib.it

Abstract. A structural characterization of reflexive splicing languages has been recently given in [1] and [5] showing surprising connections between long standing notions in formal language theory, the syntactic monoid and Schützenberger constant and the splicing operation.

In this paper, we provide a procedure to decide whether a regular language is a reflexive splicing language, based on the above mentioned characterization that is given in terms of a finite set of constants for the language. The procedure relies on a basic result showing how to determine, given a regular language L, a finite set of representatives for constant classes of the syntactic monoid of L. This finite set provides the splice sites of splicing rules generating language L. Indeed, we recall that in [1] it is shown that a regular splicing language is reflexive iff splice sites of the rules are constants for the language.

1 Introduction

Splicing systems theory is a formal framework to model and investigate the fundamental biochemical process involved in molecular cut and paste phenomenon occurring in nature and known as *splicing operation*. Starting from the original formal definition of a finite splicing system introduced by T. Head in [7] and later reformulated by G. Paun in [10], the investigation of the splicing operation has been carried out by using tools from formal language theory thus establishing a link between biomolecular sciences and language theory [11]. Since a splicing system is a formal device to generate languages, called *splicing languages*, a lot of research has been devoted to characterization of classes of formal languages in terms of the splicing operation, even showing that recursively enumerable languages can be generated by a special type of splicing systems [11]. In spite of the vast literature on the topic, the real computational power of finite splicing systems is still partially unknown as the characterization of languages generated by these systems is an open problem. The original concept of *finite splicing system* is close to the real biological process: the splicing operation is meant to act by a finite set of rules (modelling enzymes) on a finite set of initial strings (modelling DNA sequences). Under this formal model, a splicing system is a generative

O.H. Ibarra and Z. Dang (Eds.): DLT 2006, LNCS 4036, pp. 315–326, 2006.

mechanism of languages which have been proved to be *regular languages* (see [6] and [13]).

On the other hand, recently progress have been made towards the solution of the problem of determining the computational power of finite splicing systems in [1] and in [9] by giving a characterization of the class of *reflexive regular splicing languages*.

Formally, a splicing system is a triple $S = (A, I, R)$, where A is a finite alphabet, $I \subseteq A^*$ is the finite initial set of words and R is the finite set of rules. A *rule* consists of an ordered pair of factored words, denoted as $((u_1, u_2)\$(u_3, u_4))$, where $u_1 u_2$, $u_3 u_4$ are called splicing *sites*. The set R specifies a binary relation between factored sites that can be *reflexive* or *symmetric* [3]. More precisely, a set R of rules is *reflexive* iff $((u_1, u_2)\$(u_3, u_4)) \in R$ implies that $((u_1, u_2)\$(u_1, u_2)) \in R$ and R is *symmetric* iff $((u_1, u_2)\$(u_3, u_4)) \in R$ implies that $((u_3, u_4)\$(u_1, u_2)) \in R$. Given $x, y \in A^+$, then rule $r = ((u_1, u_2)\$(u_3, u_4))$ applies to x, y if the splice site $u_1 u_2$ is a factor of x and the splice site $u_3 u_4$ is a factor of y, that is $x = x_1 u_1 u_2 x_2$ and $y = y_1 u_3 u_4 y_2$. Then the application of a splicing rule r to x, y produces the word $w = x_1 u_1 u_4 y_2$ which is said to be generated by splicing of x, y by r.

The splicing language generated by a system S, denoted $L(S)$ is then defined by first giving the following notion of *closure of a language L by R, $cl(L, R)$*, which is the set of all words that result from the application of a splicing rule $r \in R$ to a pair of words in L.

Then the *splicing language* generated by iterated splicing is $L(S) = \sigma^*(I)$, where $\sigma^0(I) = I$ and for $i \geq 0$, $\sigma^{i+1}(I) = \sigma^i(I) \cup cl(\sigma^i(I), R)$ while $\sigma^*(I) = \cup_{i \geq 0} \sigma^i(I)$.

In the case R is *reflexive or symmetric*, the splicing language $L(S)$ is said to be *reflexive* or *symmetric*, respectively.

In [9] an example of regular splicing language that is neither reflexive nor symmetric is provided, and it has been proved that we can decide whether a regular language is a reflexive splicing language. A quite different characterization of reflexive symmetric splicing languages is given in [1] and it has been extended to the general class of reflexive regular languages in [5]. Surprisingly, this characterization has been given by using the concept of constant introduced by Schützenberger [14]. Indeed, the class of reflexive splicing language is shown to be equivalent to a class of regular languages, the PA-con-split languages. Each language L in this class is constructed from a finite set of constants for L, as L is expressed by a finite union of *constant languages* and *split languages*, where a split language is a language obtained by a single iteration of a splicing operation over constant languages. Such constants will be called *generating constants* for the splicing language.

In this paper, we provide a decision procedure for reflexive splicing languages that is based on the definition of such languages as PA-con-split languages. Thus, such procedure differs from the one proposed in [9] which is based on a different characterization of reflexive splicing languages. We achieve this result by investigating the set of constants of a regular language in terms of the well known concept of syntactic congruence. More precisely, we are able to exhibit a notion of equivalence relation among words that leads to a refinement of constant classes of the syntactic monoid into classes whose smallest representatives directly provide the finite set of generating constants for the splicing language. Indeed, our decision procedure is based on the idea of finding the finite set of rules used to build the splice systems generating the reflexive language. We believe that the results achieved in the paper to compute the finite set of constants generating a reflexive splicing language can help to give a deeper insight into the question of characterizing all regular splicing languages.

The paper is organized as follows. In section 2 preliminaries are given: the notion of reflexive regular splicing language and properties of reflexive rules are introduced. Then section 3 states the basic Theorem leading to the decision procedure detailed in the section.

2 Preliminaries

Let A be a finite alphabet. We denote the empty word over A by 1. In the paper, when dealing with a finite state automaton $\mathcal{A} = (A, Q, \delta, q_0, F)$ recognizing a regular language L, where δ is the transition function, q_0 is the initial state, F the set of final states, we assume that $\mathcal{A}$ is *deterministic, trim*, that is each state is accessible and coaccessible, and is the minimal automaton for L (see [12] for basic notions). A *path* π in the automaton $\mathcal{A}$ is a finite sequence $\pi = (q, a_1, q_1)(q_1, a_2, q_2) \ldots (q_{n-1}, a_n, q_n)$ where $\delta(q, a_1) = q_1$ and $\delta(q_i, a_{i+1}) = q_{i+1}$ for each $i = 1, \ldots, n - 1$. An abbreviated notation for a path is $\pi = (q, a_1 a_2 \cdots a_n, q_n)$ and $x = a_1 a_2 \ldots a_n$ is called the *label* of π. A path $\pi = (q, x, q_n)$ with $x \in A^+$, is a *closed path* iff $q = q_n$. Moreover, we say that $q, q_1, \ldots, q_n$ are the *states crossed* by the path $(q, a_1 \cdots a_n, q_n)$ and, for each $i \in \{1, \ldots, n - 1\}$, q_i is an *internal state crossed* by the same path.

Given $w \in A^+$, then Q_w denotes the set $\{q \in Q : \delta(q, w) = q', q' \in Q\}$.

Definition 1

(q-label) Let $q \in Q$. The word $c \in A^+$ is a *q-label in $\mathcal{A}$* (or simply a *q-label, if $\mathcal{A}$ is understood*) if c is the label of a closed path $\pi = (q, c, q)$ in $\mathcal{A}$.

(Q'-label) Let $Q' \subseteq Q$. The word $c \in A^+$ is a label w.r.t. the set Q' of states, or simply *Q'-label*, if c is a *q'-label* for each state $q' \in Q'$.

(g-label) Let $y \in A^+$ be a word such that $y = xcz$. The word $c \in A^+$ is a general label, in short *g-label* w.r.t. x and z, if c is a *Q'-label* for $Q' = \{q' : \delta(q, x) = q', q \in \cup_{0 \leq i \leq |Q|} Q_{xc^i z}, \}$.

The investigation of a regular language $L \subseteq A^*$ has been thoroughly developed by using the algebraic theory of finite monoids via the so-called *syntactic monoid* associated with L [12]. This is the quotient monoid $\mathcal{M}(L)$ with respect to the *syntactic congruence* $\equiv_L$, defined as follows: two words w, w' are equivalent with respect to the syntactic congruence if they have the same set of contexts, i.e.,

$$w \equiv_L w' \Leftrightarrow [\forall x, y \in A^*, xwy \in L \Leftrightarrow xw'y \in L] \Leftrightarrow C(L, w) = C(L, w').$$

Here, $C(L, w) = \{(x, y) \in A^* \times A^* \mid xwy \in L\}$ denotes the set of *contexts* $C(L, w)$ of $w \in A^*$ for $L \subseteq A^*$ and $[w] = \{w' \in A^* \mid w' \equiv_L w\}$ the class of w modulo $\equiv_L$. Analogously, $C_{\mathcal{L}}(L, w) = \{x \in A^* \mid xwy \in L\}$ and $C_{\mathcal{R}}(L, w) = \{y \in A^* \mid xwy \in L\}$ denote the left and right contexts of $w \in A^*$ for $L \subseteq A^*$.

A known result in theory of formal languages is that L is a regular language iff the index (i.e., the number of congruence classes) of the syntactic congruence is finite and so $\mathcal{M}(L)$ is a finite monoid.

Finally, we recall that a word $w \in A^*$ is a *constant* for a regular language L if $A^* w A^* \cap L \neq \emptyset$ and $C(L, w) = C_{\mathcal{L}}(L, w) \times C_{\mathcal{R}}(L, w)$ [14]. Given m a constant for language L, then $[m]$ is called constant class.

2.1   Reflexive Splicing Languages and Constant Languages

In this section, we recall the characterization of reflexive splicing languages (also called Paun reflexive splicing or *PA-reflexive* languages in [1]) stated in [1] in terms of constant languages that have been introduced by T. Head in [8].

Definition 2 (constant languages). *Given a regular language L and a constant m for L with $m \in A^+$, the constant language associated with m and L is the set $L(L, m) = L \cap A^* m A^*$. Then L is called a constant language if $L = L(L, m)$, for some constant m.*

A constant language $L(L, m)$ associated with a constant m and L is also simply denoted by $L(m)$, whenever L is known from the context. Given a constant language $L(m)$ associated with a constant m and L, by a well known result on constants, there exists a unique state $q_m \in Q$ such that, for each $q \in Q_m$, $\delta(q, m) = q_m$ (see [1]). Then we have $L(m) = C_{\mathcal{L}}(m, L) m C_{\mathcal{R}}(m, L)$, i.e., $L(m) = L_1 m L_2$, where $L_1 = C_{\mathcal{L}}(m, L) = L(m)(mL_2)^{-1}$ and $L_2 = C_{\mathcal{R}}(m, L) = (L_1 m)^{-1} L(m)$ are regular languages.

As a consequence of properties of constant languages, given two constant languages $L(m), L(m')$ (associated with m and m' and a regular language L), we can define a regular language obtained "by splicing" languages $L(m), L(m')$: this is formalized in the notion of *PA-split language* given in [1].

Given a constant m, we pose $F(m) = \{(\alpha_1, \alpha_2) \mid \alpha_1 \alpha_2 = m\}$.

Definition 3 (PA-split language). *Let L be a regular language and let m and m' be two constants for L. Given $\alpha = (\alpha_1, \alpha_2) \in F(m)$ and $\beta = (\beta_1, \beta_2) \in F(m')$, the PA-split language generated by (α, β), with respect to L, is the language:*

$$L_{(\alpha, \beta)} = C_{\mathcal{L}}(L, m) \, \alpha_1 \beta_2 \, C_{\mathcal{R}}(L, m').$$

Then $s = (\alpha, \beta)$ is called split-rule *generating language L, or we say that L is generated by rule r.*

Remark 1. Observe that constant languages are special PA-split languages. Indeed, when we choose $m = m'$ and $\alpha = \beta$ in Definition 3, we obtain $L_{(\alpha, \beta)} = L(m)$.

Generalizing the notion of PA-split language, we can define a regular language obtained by "splicing" a finite set of constant languages using a finite set of different split-rules.

Definition 4 (PA-con-split language). *Let L be a regular language and let M be a finite set of constants for L. Let Y be a finite subset of L such that, for each $m \in M$, m is not a factor of a word in Y. Let $J \subseteq \{(\alpha, \beta) \mid \alpha \in F(m), \beta \in F(m'), m, m' \in M\}$. L is a PA-con-split language (associated with Y, M, J) if and only if*

$$L = Y \cup \bigcup_{m \in M} L(m) \cup \bigcup_{(\alpha,\beta) \in J} L_{(\alpha,\beta)}.$$

Then J is the set of split-rules generating language L, denoted as $\mathcal{R}(L)$.

Theorem 1 proved in [1] for symmetric reflexive splicing languages and generalized in [5] to reflexive splicing languages states that the class of PA-con-split languages is equivalent to the class of reflexive splicing languages.

Theorem 1. *A regular language $L \subseteq A^*$ is a reflexive splicing language if and only if L is a PA-con-split language.*

Thus Theorem 1 defines a reflexive regular splicing language L in terms of a finite union of constant languages for L and a finite union of languages obtained by one iteration of split-rules applied to constant languages.

2.2 Properties of Split-Rules

In this subsection, we state some basic properties of split-rules, i.e. rules generating PA-split languages, that will be used in proving the main Theorem 2 of the next section.

Actually, the notion of split-rule (see definitions 3 and 4) derives by extending the splicing operation on words to the case of constant languages and thus some properties of splicing rules naturally generalize to split-rules. A basic property of a splicing rule is preserving the closure of a language L under the splicing operation. More precisely, a language L *is closed under rule r* iff $cl(L, r) \subseteq L$.

Similarly, we say that a regular language L *is closed under a split-rule s* iff each PA-split language generated from two constants languages by rule s is contained in L.

Lemma 1. *Let $w_i = zc^i x$, $w_j = zc^j x$, with $0 \le j < i$ and c a g-label w.r.t. z and x. Then $w_j \in [w_i]$.*

Proof. In the following we show that for each state $q \in Q_{w_i} \cup Q_{w_j}$ it holds that $\delta(q, w_i) = \delta(q, w_j)$, which implies that $C(L, w_i) = C(L, w_j)$, that is $w_j \in [w_i]$, as required. Let $q \in Q_{w_i} \cup Q_{w_j}$ and $q' = \delta(q, z)$. Now, since c is a g-label w.r.t. z and x, it holds that c is a Q'-label, for $Q' = \{q' : \delta(q, z) = q', q \in Q_{w_i} \cup Q_{w_j}\}$. Consequently, it holds that $\delta(q', c^j) = q' = \delta(q', c^i)$, thus proving what required.

Lemma 2. *Let* $s = ((\alpha, \beta), (\gamma, \delta))$ *be a split-rule of a* PA-*con-split language* L, *where* $\alpha\beta = xc^ly$ *(or* $\gamma\delta = xc^ly$*),* c *g-label w.r.t.* x *and* y. *Let* $\beta = c_2c^iy$, $i \geq 1$, $c_1c_2 = c$, $c_2 \in A^*$, *(* $\gamma = xc^ic_1$, *respect.). Then* $s' = ((\alpha, \beta'), (\gamma, \delta))$ *(or* $s' = ((\alpha, \beta), (\gamma', \delta)))$, *where* $\beta' = c_2c^{i-1}y$ *(* $\gamma' = xc^{i-1}c_1$*) is a split-rule that generates the same language of* s.

Proof. By Lemma 1, it holds that $\alpha\beta' \in [\alpha\beta]$ ($\gamma'\delta \in [\gamma\delta]$) and thus s' is a split-rule since splice sites of s' are constants. Moreover, $C_{\mathcal{L}}(L, \alpha\beta') = C_{\mathcal{L}}(L, \alpha\beta)$ ($C_{\mathcal{R}}(L, \gamma'\delta) = C_{\mathcal{R}}(L, \gamma\delta)$, respectively). Thus it is immediate to show that L is closed under rule s' that generates the same language of s.

The proof of Lemma 3 follows from the fact that given a constant m for a language, then the word zmx, for $z, x \in A^+$ is also a constant for the language.

Lemma 3. *Let* L *be a regular language closed under the split-rule* $s = ((\alpha, \beta), (\gamma, \delta))$. *Then* L *is also closed under the split-rule* $s^* = ((z\alpha, \beta x), (u\gamma, \delta y))$, *for* $z, u, x, y \in A^*$.

3   A Decision Algorithm for Reflexive Splicing Languages

In this section we describe an effective procedure to decide whether a regular language is a reflexive splicing language. This procedure is based on the definition of reflexive splicing languages in terms of constant languages obtained by means of the equivalence of such languages with the class of PA-con-split languages. Indeed, we provide a decision procedure for the class of PA-con-split languages: it relies on an algorithm to compute the set M of constants that are used to define a PA-con-split-language L. Such constants are called *generating constants of language* L.

The approach we use to compute generating constants is based on the following idea. We introduce an equivalence relation among words of a regular language that is a refinement of the syntactic congruence. This relation allows us to compute a finite set of representatives of the constant congruence classes of the syntactic monoid: this set defines the generating constants of a PA-con-split language, i.e. of a reflexive splicing language.

We start the section by providing some preliminary notions used to compute generating constants of a reflexive regular splicing language.

Definition 5. *Let* L *be a regular language. A word* $y \in A^+$ *is called* special *in* $\mathcal{A}$ *(or simply special, if* $\mathcal{A}$ *is understood) iff for some* $i > 0$, $y = y_1c^iy_2$ *where* c *is a g-label w.r.t.* y_1 *and* y_2.

The following Lemmas are used to show that an infinite congruence class of a regular language contains an infinite number of special words and a finite number of non special ones.

Lemma 4. *Let* L *be a regular language. Let* $y \in A^+$ *such that* $y = xc^tz$, *with* $t > |Q|^{|Q|+1}$ *and* $z \neq cw$, $w \in A^*$. *Then* $y = x'(c_1)^jz$, *where* $c_1 = c^k$ *is a g-label w.r.t.* x' *and* 1 *(and w.r.t.* x' *and* z*), where* $x' = xx_1$, $k \leq |Q|^{|Q|}$ *and* $|x_1| < |Q| + k$.

Similarly, as for Lemma 4, we can show that the following stronger property holds:

Lemma 5. *Let L be a regular language. Let $y \in A^+$ such that $y = xc^t z$, with $t > |Q|^{|Q|+1}$ and $z \neq cw$, $w \in A^*$. Then $y = x'(c_1)^j z$, where $c_1 = c^k$ is a g-label w.r.t. x' and 1 and w.r.t. 1 and y, where $x' = xx_1$, $k \leq |Q|^{|Q|}$ and $|x_1| < |Q|^{|Q|+1}$.*

Lemma 6. *Let L be a regular language and $[w]$ a congruence class of L. Let $[w]$ be an infinite class, then there exists a word $c_1 \in A^+$ such that $y = x_1 c_1{}^i x_2 \in [w]$, $i > 0$ and c_1 is a g-label w.r.t. x_1 and x_2, that is y is a special word of $[w]$.*

Proof. Since $[w]$ infinite class, then there exists a word $y = x_1' c^t z$, with t arbitrarily large integer. By Lemma 4, it follows that $y = x_1 c_1{}^i x_2$, where c_1 a g-label w.r.t. x_1 and x_2, thus proving what required.

Lemma 7. *Let $[w]$ be a congruence class of L. Then the set of non special elements in $[w]$ is finite.*

Moreover, we can introduce a stronger notion of equivalence, label-equivalence; Lemma 8 states that an important property holds for label-equivalence classes of special words in $[w]$.

Definition 6 (label-equivalent). *Let L be a regular language and $\mathcal{A}$ the automaton for L. Let $y, x \in A^*$ be words in L. Then y, x are label-equivalent iff the following conditions hold:*

1. *$y = x$ and y, x are not special in $\mathcal{A}$, or*
2. *for each word c that is g-label of x w.r.t x_1, x_2 (or of y w.r.t. y_1, y_2), then $x = x_1 c^j x_2$ and $y = y_1 c^i y_2$, where y_1, x_1 are label-equivalent, y_2, x_2 are label-equivalent and c is a g-label w.r.t. y_1 and y_2 (or w.r.t. x_1, x_2, respectively).*

It is easy to verify that label-equivalence is an equivalence relation on a regular language L and that the quotient of L under label-equivalence is a refinement of the quotient monoid $\mathcal{M}(L)$ with respect to the *syntactic congruence* (the proof based on Lemma 6 is omitted).

Lemma 8. *The number of label-equivalence classes of special words in $[w] \in \mathcal{M}(L)$ is finite.*

We use representatives of the equivalence classes of this refinement to determine constants generating a PA-con-split language reflexive, as detailed below.

Definition 7 (representative of a finite class). *Let w be a word in A^+ and $[w]$ is the finite congruence class of w. Then every word $y \in [w]$ is a representative of $[w]$.*

Definition 8 (representative of an infinite class). *Let w be a word in A^+ and $[w]$ the infinite congruence class of w. A word $y \in [w]$ is a representative of $[w]$ iff*

1. y is not special in $[w]$ or
2. y is a shortest word in a label-equivalence class in $[w]$.

A consequence of the above Lemmas 8 and 7 is the following result.

Lemma 9. *The set $R[w]$ of representative elements in the class $[w]$ is finite.*

Lemma 10. *Let L be a PA-con-split language. Then L is generated by a set $\mathcal{R}(L)$ of split-rules such that for each split-rule $s = ((\alpha, \beta), (\gamma, \delta)) \in \mathcal{R}(L)$, given the paste site $\alpha\delta$, then if $\alpha\delta = xcy$, with c a g-label w.r.t. x and 1 and c a g-label w.r.t. 1 and y, then $y = 1$ or $x = 1$.*

Proof. (sketch of the proof) Let $\mathcal{R}(L)$ be a set of split-rules generating language L that minimizes the number of rules that are not of the required form. Let $s = ((\alpha, \beta), (\gamma, \delta))$ be a split-rule in $\mathcal{R}(L)$ that is not of the form required by the Lemma. Assume that $\alpha\delta = xcy$, c is a g-label w.r.t. x and 1 and c is a g-label w.r.t. 1 and y, but $y \neq 1$ and $x \neq 1$. Observe that c is also a g-label w.r.t. x and y. Clearly, being s a split-rule of L, given $Z_1 = C_{\mathcal{L}}(L, \alpha\beta)$, $Z_2 = C_{\mathcal{R}}(L, \gamma\delta)$, it holds that language $L_s = Z_1 xcy Z_2 \subseteq L$, as L_s is a PA-split language of L. Assume that $L_1 = Z_1' xcy Z_2'$ is a regular language contained in L_s such that no words in L_1 is contained in any other PA-split language of L generated by means of rules in $\mathcal{R}(L)$ distinct from rule s. Clearly, L_1 is not empty, otherwise we contradict the minimality of set $\mathcal{R}(L)$ of rules. Moreover, no word in L_1 is contained in a PA-split language generated by a split rule s' of the required form such that L is closed under rule s', otherwise again we contradict the assumption on set $\mathcal{R}(L)$ of rules for language L. Observe that $L_1^* = Z_1' xc^* y Z_2' \subseteq L$, being c a g-label w.r.t. x and y.

Since L is a PA-con-split language, that is by definition 4 L is a finite union of PA-split languages (indeed, by Remark 1, even constant languages are PA-split languages), clearly there exists a PA-split language L_2 such that $L_2 \cap L_1^* = W$, where we can assume that $W = \cup_{l \in I} Y xc^l y X$, for I an infinite set, $Y \subseteq Z_1'$ and $X \subseteq Z_2'$. Then, there exists a split-rule $s_2 \in \mathcal{R}(L)$ generating L_2, that is $s_2 = ((\alpha', \beta'), (\gamma', \delta'))$ with paste site $\alpha'\delta'$. Since $W \subseteq L_2$, it holds that $\alpha'\delta'$ is a substring or factor of all words in W.

Thus, the following cases must be considered: (1) $\alpha'\delta'$ is a factor of some words in Y, or (2) $\alpha'\delta'$ is a factor of some words in X or (3) $\alpha'\delta'$ is a factor of words in Yxc^l, but not in Y, and finally (4) $\alpha'\delta'$ is a factor of words in $c^l y X$, but not in X.

Case 1. Assume that $\alpha'\delta'$ is a factor of some words in Y, that is $Y_1 \alpha'\delta' Y_2 \subseteq Y$. Since s_2 generates language L_2 including W, it follows that $Y_2 xc^l y X \subseteq C_{\mathcal{R}}(L, \gamma'\delta')$, for some $l > 0$. But, being c a g-label w.r.t. x and y, clearly it follows that $Y_2 xc^* y X \subseteq C_{\mathcal{R}}(L, \gamma'\delta')$. By the above fact, it follows that language $W' = Y_1 \alpha'\delta' Y_2 xcy X$ is included in L_2, where $W' \cap L_1 \neq \emptyset$, that is W' is generated by the split rule s_2, thus contradicting the fact that L_1 is only included in L_s.

Case 2. Assume that $\alpha'\delta'$ is a factor of some word in X, that is $X = X_1 \alpha'\delta' X_2$. This case is similar to case 1.

Case 3. Assume that $\alpha'\delta'$ is a factor of words in Yxc^l, but not in Y. By using Lemma 3, we can assume that $\alpha'\delta' = x'c^t$, where $x'c^t$ is a factor of Yxc^l, $x' = x''x$, where eventually $x'' = 1$.

Let us first show that $yX \subseteq C_{\mathcal{R}}(L, \gamma'\delta')$. This fact follows from the observation that $Yxc^lyX \subseteq L_2$ for $l \in I$, I an infinite set of indices and thus $c^*yX \subseteq C_{\mathcal{R}}(L, \gamma'\delta')$, as c is a g-label w.r.t. 1 and y and hence $yX \subseteq C_{\mathcal{R}}(L, \gamma'\delta')$.

Assume first that $\alpha' = x'c^ic'$, with $i > 0$. Then, we can build the rule $s^* = ((x'c, c^{i-1}c'\beta'), (\gamma'\delta', 1))$. Clearly, rule s^* is a split-rule since splice sites of such rules are constants, being $\alpha'\beta'$ and $\gamma'\delta'$ constants. Let us now show that L is closed under rule s^*. Indeed, assume $A = C_{\mathcal{L}}(L, \alpha'\beta')$ and $B = C_{\mathcal{R}}(L, \gamma'\delta')$. Since c is a g-label w.r.t. x and 1 and by construction $x' = x''x$, it is immediate to verify that c is also a g-label for x' and 1. Since $Ax'c^tB = L_2 \subseteq L$, as c is a g-label w.r.t. x' and 1, it holds that $Ax'c^*B \subseteq L$. It follows that the language $Ax'cB$ generated by s^* is included in L. Observe that since $yX \subseteq B$ and $Y_1 \subseteq A$, language $Y_1x'cyX \subseteq L(s^*)$, where $Y_1x'cyX \subseteq YxcyX \subseteq L_1$ and $L(s^*)$ generated by the split-rule s^* which is of the required form. Consequently, we obtain a contradiction with the previous assumptions on L_1 (indeed we assumed that L_1 is not generated by split-rules of the required form and for which L is closed).

Now, assume that α' is a prefix of $x'c'$. Along the same lines of the above proof, we obtain a contradiction.

Case 4. Assume that $\alpha'\delta'$ is a factor of words in c^lyX, but not in X. This case is proved similarly as case 3 (the proof is omitted because of space constraints).

Finally, since all cases 1-4 lead to a contradiction, it must be that $\alpha\delta$ is of the required form.

The set of constants generating a PA-con-split language is defined by the notion of set $R_Q[m]$ of Q-representatives of a constant class $[m]$:

$$R_Q[m] = \{xc^ly : xcy \in R[m], c \text{ is a } g\text{-label w.r.t. } x \text{ and } y \text{ where}, l \leq 2|Q|^{|Q|+1}\}.$$

Notice that $R[m] \subseteq R_Q[m]$ and $R_Q[m]$ is finite.

Lemma 11. *Let L be a PA-con-split language. Then there exists a set $\mathcal{R}(L)$ of split-rules generating L such for each split-rule $s = ((\alpha, \beta), (\gamma, \delta)) \in \mathcal{R}(L))$ it holds that $\alpha\beta \in R_Q[\alpha\beta]$ and $\gamma\delta \in R_Q[\gamma\delta]$.*

Proof. Assume that M is a set of constants generating language L such that the sum of lengths of words in M is minimum. Let $s = ((\alpha, \beta), (\gamma, \delta))$ be a split-rule of language L. Then, given $m = \alpha\beta$ and $m' = \gamma\delta$, it holds that $m, m' \in M$.

Clearly, by definition 7, if $[m]$ is finite then $m \in R[m]$ and thus the Lemma holds. Hence, let $[m]$ be an infinite class. Assume to the contrary that $\alpha\beta \notin R_Q[m]$. By Lemma 6, it must be that $\alpha\beta$ is special i.e. there exists a g-label c_1 w.r.t. x and y such that $xc_1y \in R[m]$, but $m = xc_1^ly$, where $l > 2|Q|^{|Q|+1}$ as $m \notin R_Q[m]$. Assume first that $\alpha = x'$, where x' is a prefix of x. Given rule $s' = ((\alpha, \beta'), (\gamma, \delta))$, where $\beta' = x''c_1^{l-1}y$, if $\beta = x''c_1^ly$, then it holds that language L is closed under the split-rule s' that generates the same language as s. Indeed, $\alpha\beta' \in [m]$, by Lemma 1.

Now, assume that $\alpha = x(c_1)^i c'$, with $c'c'' = c_1$, $c'c'' \in A^*$. Then we must consider two cases:

(1) $\beta = c''(c_1)^j y$, with $j \geq 1$, or
(2) $\beta = c''(c_1)^j y$, with $j < 1$.

Assume first that condition (1) holds. In this case, since c_1 is a g-label w.r.t. x and y, Lemma 2 applies and thus language L is closed under the split-rule rule $s' = ((\alpha, \beta'), (\gamma, \delta))$, where $\beta' = c''(c_1)^{j-1} y$ and s' generates the same PA-split language of s. But $|\alpha\beta| > |\alpha\beta'|$, thus we contradict the minimality of set M.

Assume now that condition (2) holds. In this case, since $\beta = c''y$, with $c'c'' = c_1$, $c'' \in A^*$, it holds that $\alpha = x(c_1)^{l'} c'$, where $l' \geq 2|Q|^{|Q|+1}$.

Then, by applying Lemma 5 to $\alpha\delta$, then $\alpha\delta = x''((c_1)^k)^i z$, where $(c_1)^k$ g-label w.r.t. x'' and 1 and w.r.t. 1 and z, $x'' = xx'$ and $|x'| < k + |Q|$, $k < |Q|^{|Q|}$.

Now, $\alpha\delta$ verifies the hypothesis of Lemma 10. Thus, $\alpha\delta = (c_1)^k z$ or $\alpha\delta = x''(c_1)^k$. Thus it follows that $\beta = c''(c_1)^j y$, with $j \geq 1$, $c'', c' \in A^*, y \in A^*$, a contradiction with condition (2).

Since both cases (1) and (2) lead to a contradiction, it must be that $m \in R_Q[m]$, as required.

Now, let $m' = \gamma\delta$ and assume to the contrary that $m' \notin R_Q[m]$. Assume to the contrary that $\gamma\delta \notin R_Q[m']$. By Lemma 6, it must be that $\gamma\delta$ is special i.e. there exists a g-label c w.r.t. x and y such that $xcy \in R[m']$, but $m' = xc^l y$, where $l > 2|Q|^{|Q|+1}$ as $m' \notin R_Q[m]$.

Assume that δ is a suffix y'' of y, with $y = y'y''$. Given rule $s' = ((\alpha, \beta), (\gamma', \delta))$, where $\gamma' = xc^{l-1}y'$, if $\gamma = xc^l y'$, then it holds that language L is closed under the split-rule s' that generates the same language as s. Indeed, $\gamma'\delta \in [m]$, by Lemma 1.

Now, assume that $\delta = c''(c)^i y$, with $c'c'' = c$ and $c', c'' \in A^*$. Thus, we must consider two cases:

(1) $\gamma = x(c)^j c'$, with $j \geq 1$, or
(2) $\gamma = x(c)^j c'$, with $j < 1$.

Assume first that condition (1) holds. In this case, since c is a g-label w.r.t. x and y, Lemma 2 applies and thus language L is closed under the split-rule rule $s' = ((\alpha, \beta), (\gamma', \delta))$, where $\gamma' = x(c)^{j-1}c''$ and s' generates the same PA-split language of s. But $|\gamma\delta| > |\gamma\delta'|$, thus we contradict the minimality of set M.

Assume now that condition (2) holds. In this case, since $\gamma = xc'$, with $c'c'' = c$, $c'' \in A^*$, it holds that $\delta = c''(c)^{l'} y$, where $l' \geq 2|Q|^{|Q|+1}$.

Then, by applying Lemma 5 to $\alpha\delta$, then $\alpha\delta = x''((c)^k)^i y$, where $(c)^k$ g-label w.r.t. x'' and 1 and w.r.t. 1 and y, $x'' = xx'$ and $|x'| < k + |Q|$, $k < |Q|^{|Q|}$. Now, $\alpha\delta$ verifies the hypothesis of Lemma 10. Thus, $\alpha\delta = (c)^k y$. It follows that $\gamma = x(c)^j c'$, with $j \geq 1$, $c' \in A^*, x \in A^*$, a contradiction with condition (2).

Since both cases (1) and (2) lead to a contradiction, it must be that $m' \in R_Q[m']$, as required.

The main Theorem of the section follows.

Theorem 2. *Let L be a reflexive regular splicing language. Then language L is generated by a splicing system with constants M that are Q-representatives of each class $[m], m \in M$.*

Proof. Assume that M is a set of constants generating language L such that the sum of the lengths of constants in M is minimum. Since L is a reflexive language, by Theorem 1, and remark 1 then L is PA-con-split-language, that is L is the union of languages Y and L', where Y is finite, L' is the finite union of of PA-split languages. Thus by applying Lemma 11 the Theorem follows.

3.1   The Algorithm

In this subsection, we describe the main steps of the decision algorithm for reflexive splicing languages based on Theorem 2 proved in the previous section.

The decision algorithm is based on a fundamental decision result proved in [1]:

we can decide whether a language L is closed under a set R of rules, that is $cl(L, R) \subseteq L$.

Given a regular language L, the procedure to test whether L is a reflexive splicing language consists of the following basic steps:

1. Given the syntactic monoid, then let $\mathcal{C}_M = \{[m] : [m] \in \mathcal{M}(L), m$ is a constant for $L\}$ and let $M = \cup_{[m] \in \mathcal{C}_M} R_Q[m]$ be the finite set of all Q-representatives of constant classes in $\mathcal{C}_M$ as defined in Definitions 7 and 8.
2. Let $\mathcal{R}(M) = \{((\alpha, \beta), (\gamma, \delta)) \in F(m) \times F(m'), m, m' \in M\}$ be the set of candidate split-rules built from the set M of constants.
   Then, find the subset J of $\mathcal{R}(M)$ consisting of the split-rules under which the language L is closed (using the result stated in [1]).
3. Finally, verify whether L is a *PA-con-split language* (associated with Y, M, J), that is

$$L = Y \cup \bigcup_{m \in \mathcal{M}} L(m) \cup \bigcup_{(\alpha, \beta) \in J} L_{(\alpha, \beta)},$$

where Y is a finite set of words such that each word $m \in \mathcal{M}$ is not a factor of Y.

Theorem 2 can be applied to prove the correctness of the above steps to verify that a regular language L is a *PA*-con-split language, i.e. a reflexive splicing language by Theorem 1.

Acknowledgements. The authors are grateful to C. De Felice and R. Zizza for their helpful suggestions on the problem faced in the paper.

References

1. P. Bonizzoni, C. De Felice, R. Zizza, The structure of reflexive regular splicing languages via Schützenberger constants, *Theoretical Computer Science* 334 (2005) 71 − 98.
2. P. Bonizzoni, C. De Felice, G. Mauri, R. Zizza (2005), Linear splicing and syntactic monoid, *Discrete Applied Mathematics*, to appear.

3. P. Bonizzoni, C. De Felice, G. Mauri, R. Zizza, Regular Languages Generated by Reflexive Finite Linear Splicing Systems, Proc. DLT 2003, Lecture Notes in Computer Science, Vol. 2710 (Springer, Berlin, 2003) 134 − 145.
4. P. Bonizzoni, C. Ferretti, G. Mauri, R. Zizza (2001), Separating some splicing models, *Information Processing Letters* 76 (6), pp. 255 − 259.
5. P. Bonizzoni, G. Mauri, Regular splicing languages and subclasses, *Theoretical Computer Science* 340 (2005), pp. 349 − 363.
6. K. Culik, T. Harju, Splicing semigroups of dominoes and DNA, Discrete Applied Math. 31 (1991) 261 − 277.
7. T. Head, Formal Language Theory and DNA: an analysis of the generative capacity of specific recombinant behaviours, Bull. Math. Biol. 49 (1987) 737 − 759.
8. T. Head, Splicing languages generated with one sided context, in: Paun, Gh. ed., Computing with Bio-molecules. Theory and Experiments (Springer-Verlag, Singapore, 1998).
9. E. Goode, D. Pixton, Recognizing splicing languages: Syntactic Monoids and Simultaneous Pumping, submitted, 2004
   (available from http://www.math.binghamton.edu/dennis/Papers/index.html).
10. G. Paun, On the splicing operation, Discrete Applied Math. 70 (1996) 57 − 79.
11. G. Paun, G. Rozenberg, A. Salomaa, DNA computing, New Computing Paradigms, (Springer-Verlag, Berlin, 1998).
12. D. Perrin (1990), Finite Automata, *in* "Handbook of Theoretical Computer Science" (J. Van Leeuwen, Ed.), Vol. B, pp. 1 − 57, Elsevier.
13. D. Pixton (1996), Regularity of splicing languages, *Discrete Applied Math.* **69**, pp. 101 − 124.
14. M. P. Schützenberger (1975), Sur certaines opérations de fermeture dans les langages rationnels, *Symposia Mathematica* **15**, pp. 245 − 253.

Contextual Hypergraph Grammars – A New Approach to the Generation of Hypergraph Languages[*]

Adrian-Horia Dediu[1,4], Renate Klempien-Hinrichs[2],
Hans-Jörg Kreowski[3], and Benedek Nagy[1,5]

[1] Research Group on Mathematical Linguistics, Rovira i Virgili University
Pl. Imperial Tàrraco 1, 43005 Tarragona, Spain
adrian.dediu@urv.net
[2] University of Bremen, Department of Production Engineering
P.O.Box 33 04 40, 28334 Bremen, Germany
kh@biba.uni-bremen.de
[3] University of Bremen, Department of Computer Science
P.O.Box 33 04 40, 28334 Bremen, Germany
kreo@informatik.uni-bremen.de
[4] Faculty of Engineering in Foreign Languages
University "Politehnica" of Bucharest, Romania
[5] Faculty of Informatics, University of Debrecen, Hungary
nbenedek@inf.unideb.hu

Abstract. In this paper, we introduce contextual hypergraph grammars, which generalize the total contextual string grammars. We study the position of the class of languages generated by contextual hypergraph grammars in comparison with graph languages generated by hyperedge replacement grammars and double-pushout hypergraph grammars. Moreover, several examples show the potential of the new class of grammars.

Keywords: Contextual hypergraph grammars, contextual grammars, hypergraphs, hyperedge replacement grammars, double-pushout hypergraph grammars.

1 Introduction

Solomon Marcus introduced the *external* variant of contextual grammars in 1969. Initially designed to generate languages without nonterminal rewriting, only by

[*] This work was possible for the first author thanks to the grant 2002CAJAL-BURV4, provided by the University Rovira i Virgili. The third author's research was partially supported by the EC Research Training Network SegraVis (Syntactic and Semantic Integration of Visual Modeling Techniques) and by the Collaborative Research Centre 637 (Autonomous Cooperating Logistic Processes: A Paradigm Shift and Its Limitations) funded by the German Research Foundation (DFG). The fourth author acknowledges the financial support provided through the European Community's Human Potential Programme under contract HPRN-CT- 2002-00275, SegraVis and by the grants from the Hungarian National Foundation for Scientific Research (OTKA F043090 and T049409).

O.H. Ibarra and Z. Dang (Eds.): DLT 2006, LNCS 4036, pp. 327–338, 2006.

adjoining contexts using a selection procedure, contextual grammars were discovered to have some limitations. Yet, as the term contextual seems to be very appropriate to model linguistic aspects, soon a large variety of such grammars appeared. We mention here only several types, like the *internal* case, *total contextual, grammars with choices* respectively, etc. For more details, the reader is referred to [1]. In the basic case of a contextual grammar, a context is a pair (u, v) of strings that are to be inserted into axioms or derived strings. More explicitly, a string x directly derives a string y using the context (u, v) if $y = x_1 u x_2 v x_3$ for some decomposition $x = x_1 x_2 x_3$.

Despite the large variety of contextual grammars, it is difficult to put together strings and structures, a very important intrinsic quality of natural languages. There were several proposals to introduce bracketed contextual grammars [2, 3, 4] in order to enhance the words in the generated languages with a tree structure, or to add a dependency relation to contexts, axioms and to generated words. However, we believe that hypergraphs provide a more general structure that could enhance a textual string representation, and therefore we propose a new approach to the generation of hypergraph languages.

In this paper (Section 3), we introduce the concept of contextual hypergraph grammars as a generalization of contextual grammars by considering hypergraphs instead of strings as underlying data structures. The insertion operation is taken over by a merging operation. Two hypergraphs, each with a sequence of external vertices of the same length, are merged by identifying corresponding external vertices whereas all other items are kept disjoint. We can see the external vertices as gluing points, each external node "waiting for" another external node from another hypergraph in order to become an internal node.

A contextual hypergraph grammar is given by finite sets of axioms and contexts both being hypergraphs. Starting with an axiom, derivations are composed of iterated merging with contexts. While the contexts are equipped with external vertices by definition, axioms and intermediately derived hypergraphs do not have external vertices of their own. Therefore, before they can be merged with some context, some preparation is necessary that equips them with external vertices in a suitable way. For this purpose, a contextual hypergraph grammar provides an operator Θ that depends on the contexts and associates each hypergraph without external vertices with a set of hypergraphs each with a proper sequence of external vertices. The idea is that $\Theta_C(H)$ for some context C and some hypergraph H yields variants of H that can be merged with C in particular. Such a merging defines a derivation step if H is an axiom or some already derived hypergraph. In this way, the Θ-operator plays on the level of hypergraphs the role of decomposition on the level of strings.

As shown in Section 4, hyperedge replacement grammars in the sense of [5, 6] can be simulated as contextual hypergraph grammars using a suitable kind of variants which are constructed by the removal of single hyperedges. By means of a more sophisticated kind of variants that are constructed by the removal of homomorphic images of hypergraphs up to a certain set of vertices, one can also translate arbitrary hypergraph grammars in the double-pushout approach (see, e.g., [7]) into contextual hypergraph grammars. This proves in particular that all recursively enumerable sets of hypergraphs can be generated by contextual hypergraph grammars. In this sense, our new approach is computationally complete.

2 Preliminaries

In this section, we recall all the notions and notations of hypergraphs as needed in this paper.

We denote by $\mathbb{N}$ the set of natural numbers. For $n \in \mathbb{N}$, $[n]$ represents the finite set $\{1, \ldots, n\}$, with $[0] = \emptyset$.

For a given finite set A, $|A|$ denotes the number of elements in A. We use the notation $\mathcal{P}(A)$ for the powerset of A, i.e. the set of all subsets of A. A function $w : [n] \to A$ is called a *string* over A, also denoted by $w = a_1 \ldots a_n$ with $w(i) = a_i$ for $i \in [n]$. The set A is also called an *alphabet* and strings over A are *words*. For a given string $w = a_1 \ldots a_n$, $|w| = n$ is the length of the string and $w(i) = a_i$ denotes the i-th symbol of the string. We denote by λ the empty string with $|\lambda| = 0$. The set of all strings over A is denoted by A^*.

For a given function $f : A \to B$, we may define the canonical extension to strings as the function $f^* : A^* \to B^*$ defined as $f^*(\lambda) = \lambda$, and $f^*(aw) = f(a)f^*(w)$ for $a \in A$ and $w \in A^*$. By convention if $A = \emptyset$, then $\emptyset^* = \{\lambda\}$ and the canonical extension to strings is also defined.

A relation $R \subseteq A \times A$ is called an *equivalence relation* if R is reflexive, symmetric and transitive. For $x \in A$, the set $\langle x \rangle = \{z \in A \mid xRz\}$ of all elements related to x by R is called the *equivalence class* of x. The set of all equivalence classes $A/R = \{\langle x \rangle \mid x \in A\}$ is the *quotient set* of A by R.

Let Σ be an alphabet. A *hypergraph* over Σ is a system $H = (V, E, att, lab, ext)$ where V is a set of *vertices*, E is a set of *hyperedges*, $att : E \to V^*$, called the *attachment function*, is a mapping that assigns a sequence of vertices to every hyperedge, $lab : E \to \Sigma$ is a mapping that assigns a *label* to every hyperedge, and $ext \in V^*$ is a sequence of *external vertices* of H. For notational convenience, the components of a hypergraph H will often be written with index H, i.e. $H = (V_H, E_H, att_H, lab_H, ext_H)$.

The length $|ext_H|$ is called the *type* of H, denoted also by $type(H)$. The class of all hypergraphs of type $n \in \mathbb{N}$ over Σ is denoted by $\mathcal{H}_{\Sigma,n}$. The class of *type-0* hypergraphs $\mathcal{H}_{\Sigma,0}$ is also denoted by $\mathcal{H}_\Sigma$.

Analogously, for $e \in E_H$, the length $|att_H(e)|$ is called the type of e. A hypergraph with all hyperedges of type 2 is an ordinary directed graph. To denote this special case, we use $\mathcal{G}_\Sigma$ and $\mathcal{G}_{\Sigma,n}$ instead of $\mathcal{H}_\Sigma$ and $\mathcal{H}_{\Sigma,n}$ respectively. If the alphabet is not important we use simply the notations $\mathcal{H}$ or $\mathcal{G}$. Sometimes, to represent unlabelled hypergraphs we use the alphabet $\Sigma = \{*\}$.

A hypergraph $H \in \mathcal{H}_{\Sigma,n}$ is a *subhypergraph* of a hypergraph $\overline{H} \in \mathcal{H}_{\Sigma,n}$, denoted by $H \subseteq \overline{H}$, if $V_H \subseteq V_{\overline{H}}$, $E_H \subseteq E_{\overline{H}}$, $att_H(e) = att_{\overline{H}}(e)$ and $lab_H(e) = lab_{\overline{H}}(e)$ for all $e \in E_H$, and $ext_H = ext_{\overline{H}}$.

Given two hypergraphs $H, H' \in \mathcal{H}_{\Sigma,n}$, a *hypergraph morphism* $f : H \to H'$ is a pair $f = (f_V, f_E)$ of functions $f_V : V_H \to V_{H'}$ and $f_E : E_H \to E_{H'}$ such that we have $lab_H(e) = lab_{H'}(f_E(e))$ and $f_V^*(att_H(e)) = att_{H'}(f_E(e))$ for all $e \in E_H$, and $f_V^*(ext_H) = ext_{H'}$. The morphism is *injective* if the functions f_V, f_E are injective.

The following two operations on hypergraphs, namely disjoint union and merge, are similar to the hypergraph operations studied by Courcelle [8].

For two hypergraphs $H, \overline{H}$ we define the *disjoint union* denoted as $H + \overline{H}$ that yields a hypergraph $(V_H \uplus V_{\overline{H}}, E_H \uplus E_{\overline{H}}, att, lab, ext)$, where $\uplus$ denotes the disjoint union of sets,

$$att(e) = \begin{cases} att_H(e) & \text{if } e \in E_H, \\ att_{\overline{H}}(e) & \text{otherwise,} \end{cases} \quad lab(e) = \begin{cases} lab_H(e) & \text{if } e \in E_H, \\ lab_{\overline{H}}(e) & \text{otherwise,} \end{cases} \quad \text{and}$$

$$ext(i) = \begin{cases} ext_H(i) & \text{if } i \leq |ext_H|, \\ ext_{\overline{H}}(i - |ext_H|) & \text{otherwise} \end{cases} \quad \text{for } i \in [|ext_H| + |ext_{\overline{H}}|].$$

The disjoint union of hypergraphs is associative. It is commutative only if one component is of type 0, since the external sequence is the concatenation of the external sequences of the component hypergraphs.

For two hypergraphs $H, \overline{H} \in \mathcal{H}_{\Sigma,n}$ we denote by EXT the equivalence relation on $(V_H \uplus V_{\overline{H}}) \times (V_H \uplus V_{\overline{H}})$ induced by $ext_H(i) = ext_{\overline{H}}(i)$ for all $i \in [n]$. Then the *merge* operation $H \circ \overline{H}$ of H and $\overline{H}$ is defined by

$$H \circ \overline{H} = ((V_H \uplus V_{\overline{H}})/EXT, E_H \uplus E_{\overline{H}}, att, lab, \lambda) \in \mathcal{H}_{\Sigma}$$

where $att(e) = \begin{cases} att_H(e) & \text{if } e \in E_H, \\ att_{\overline{H}}(e) & \text{otherwise,} \end{cases}$ and $lab(e) = \begin{cases} lab_H(e) & \text{if } e \in E_H, \\ lab_{\overline{H}}(e) & \text{otherwise.} \end{cases}$

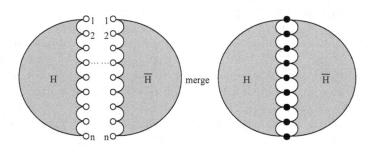

Fig. 1. The merging of H and $\overline{H}$

Figure 1 illustrates the merge operation. It should be noted that the merging is commutative, but not associative.

3   Contextual Hypergraph Grammars

In this section we introduce the new concept of contextual hypergraph grammars and their generated hypergraph languages. Besides a label alphabet and a subalphabet of terminal labels, a contextual hypergraph grammar consists of two finite sets of hypergraphs: a set of axioms and a set of contexts, as well as an operator that specifies how each context can be used to derive hypergraphs from hypergraphs. Moreover, various examples are given to illustrate the derivation mechanism of contextual hypergraph grammars.

Definition 1 (Contextual Hypergraph Grammar). *A contextual hypergraph grammar is a system $CHG = (\Sigma, T, \mathcal{A}, \mathcal{C}, \Theta)$ where Σ is a finite set of labels, $T \subseteq \Sigma$ is a set of terminal labels, $\mathcal{A} \subset \mathcal{H}_\Sigma$ is a finite set of axioms, $\mathcal{C}$ is a finite set of hypergraphs of various types called hypergraph contexts, and $\Theta = (\Theta_C)_{C \in \mathcal{C}}$ is a family of mappings where every $\Theta_C : \mathcal{H}_\Sigma \to \mathcal{P}(\mathcal{H}_{\Sigma, type(C)})$ is a selection function for every hypergraph context $C \in \mathcal{C}$. The elements of $\Theta_C(H)$ are called variants of the hypergraph H.*

A derivation relation is defined on $\mathcal{H}_\Sigma \times \mathcal{H}_\Sigma$ and the hypergraph G directly derives the hypergraph H, denoted by $G \Rightarrow H$, if there are $C \in \mathcal{C}$ and $G' \in \Theta_C(G)$ such that $H = C \circ G'$, i.e. H is a merging of a variant of G with some hypergraph context. We denote by $\Rightarrow^*$ the reflexive and transitive closure of the derivation relation $\Rightarrow$. The language generated by a contextual hypergraph grammar $CHG = (\Sigma, T, \mathcal{A}, \mathcal{C}, \Theta)$ consists of all terminal hypergraphs derived from some axiom, i.e. $L(CHG) = \{H \in \mathcal{H}_T \mid Z \Rightarrow^* H \text{ for } Z \in \mathcal{A}\}$.

In this paper, we assume that the function Θ is computable, so that the generated languages are recursively enumerable. It should be noted that we assume terminal labels in contrast to the usual definition of centextual grammars in the string case. But this allows us more flexibility from the very beginning. The task of the Θ function is to provide variants of a type-0 hypergraph that can be merged with a chosen context. While in all following examples the originals and their variants are closely related, the very general definition of Θ admits also much more sophisticated constructions.

Example 1 (All Graphs). As a first example, we define a contextual hypergraph grammar that generates the set of all directed (unlabeled) graphs:

$$CHG_{all} = (\{*\}, \{*\}, \{empty\}, \{Vertex, Edge\}, \Theta_{all})$$

where *empty* denotes the empty graph and *Vertex*, *Edge* and Θ_{all} are given as follows.

1. *Vertex* is the type-0 graph with a single vertex without edges.
2. *Edge* is the graph with two vertices and a connecting edge whose attachment defines also the sequence of external vertices.
3. The only *Vertex*-variant of a hypergraph is the hypergraph itself, that is $\Theta_{all, Vertex}(H) = \{H\}$.
4. The *Edge*-variants of a type-0 hypergraph are given by all choices of a sequence of two distinct vertices, i.e. $\Theta_{all, Edge}(H) = \{(H, v_1 v_2) \mid v_1, v_2 \in V_H, v_1 \neq v_2\}$ where $(H, v_1 v_2) = (V_H, E_H, att_H, lab_H, v_1 v_2)$.

Given a hypergraph H (of type 0), *Vertex* $\circ H$ adds a single vertex disjointly to H, and *Edge* $\circ (H, v_1 v_2)$ adds a new edge to H connecting v_1 and v_2 for each choice of v_1v_2. If a graph G has n vertices, V_G can be generated by $Vertex^n \circ empty$ up to the naming of vertices. Then every edge of G can be added successively, connecting the proper vertices by means of the context $Edge$. In other words, CHG_{all} generates the set of all graphs.

Example 2 (Eulerian Graphs). Our second example is a contextual hypergraph grammar that generates the sets of all Eulerian graphs (where a graph is Eulerian if it has a cycle passing each edge exactly once):

$$CHG_{Euler} = (\{*\}, \{*\}, \{2Cycle^0\}, \{Vertex, 2Cycle, 2Path\}, \Theta_{Euler})$$

where $Vertex$ is the same graph as in the previous example, $2Cycle^0$ is the type-0 graph consisting of a cycle of length 2, $2Cycle$ is the same graph with its two vertices as external vertices, and $2Path$ is a graph with three vertices, say v_1, v_2, and v_3, two edges, one from v_1 to v_2 and the other from v_2 to v_3, and $v_1v_2v_3$ as sequence of external vertices. Θ_{Euler} is defined as follows.

1. $\Theta_{Euler,Vertex} = \Theta_{all,Vertex}$,
2. $\Theta_{Euler,2Cycle}(H) = \{(H, v_1v_2) \in \Theta_{all,Edge} \mid v_1 \text{ or } v_2 \text{ not isolated}\}$,
3. $\Theta_{Euler,2Path}(H) = \{(H - e, v_1v_2v_3) \mid e \in E_H, \ v_1, v_2, v_3 \text{ pairwise distinct}\}$
   where $H - e$ is the subhypergraph of H obtained by removing the hyperedge e, and $(H-e, v_1v_2v_3)$ is $H-e$ with $v_1v_2v_3$ instead of λ as sequence of external vertices.

It is not difficult to see that the merging of $Vertex$, $2Cycle$ or $2Path$ with an admitted variant of the axiom or a derived graph preserves connectivity up to isolated vertices as well as the property that indegree equals outdegree for each vertex. Conversely, every graph with these properties can be obtained from $2Cycle^0$ by a sequence of such mergings. Altogether, the grammar generates the language of Eulerian graphs according to the well-known characterization of Eulerian graphs by these two properties.

In a similar manner, we can generate the sets of all Hamiltonian graphs or all non-Hamiltonian graphs.

Example 3 (Square Grids Graphs).

$$CHG_{SG} = (\{a, c, N\}, \{a, c\}, \{E, F\}, \{Stop, Cont, Corner, Tile\}, \Theta_{SG})$$

In Figure 2 we see the grammar's hypergraphs. Θ_{SG} is defined as follows.

1. $\Theta_{SG,Stop}(H) = \{(V_H, E_H \setminus \{e_1\}, att_H|_{E_H\setminus\{e_1\}}, lab_H|_{E_H\setminus\{e_1\}}, v_1v_2v_4) \mid (v_1,$
   $v_2, v_3, v_4 \in V_H, \text{ pairwise distinct nodes}), e_1 \in E_H, condStop(H, v_1, v_2, v_3,$
   $v_4, e_1)\}; condStop(H, v_1, v_2, v_3, v_4, e_1)$ is a boolean function that is *true* when
   the following conditions hold:
   - $att_H(e_1) = v_2v_1, \ lab_H(e_1) = N$,
   - $\exists$ edges $e_2, e_3 \in E_H, att_H(e_2) = v_2v_3, \ lab_H(e_2) = a, \ att_H(e_3) = v_3v_4,$
     $lab_H(e_3) = a$,

E	F	$Stop$	$Cont$	$Corner$	$Tile$

Fig. 2. CHG_{SG} hypergraphs

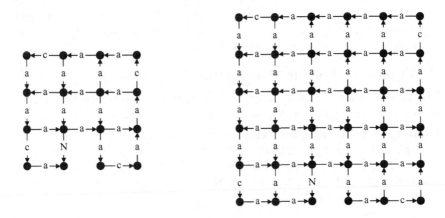

Fig. 3. $\Theta_{SG,Stop}$ returns nonempty sets for these hypergraphs

$$- \; |\{e \in H \mid att_H(e) \in V_H^* v_3 V_H^*, lab_H(e) = a\}| = 4,$$
$$- \; |\{e \in H \mid att_H(e) \in V_H^* v_4 V_H^*\}| = 2.$$

In order to understand the conditions checked by $\Theta_{SG,Stop}$ we present in Figure 3 two hypergraphs containing vertices and edges such that $condStop$ is true.

2. $\Theta_{SG,Cont} = \Theta_{SG,Stop}$.

3. $\Theta_{SG,Corner}(H) = \{(V_H, E_H \setminus \{e_1, e_2\}, att_H|_{E_H \setminus \{e_1,e_2\}}, lab_H|_{E_H \setminus \{e_1,e_2\}},$
   $v_1 v_2 v_3) \mid (v_1, v_2, v_3 \in V_H$ pairwise distinct vertices), $(e_1, e_2 \in E_H,$ distinct edges), $att_H(e_1) = v_2 v_1$, $lab_H(e_1) = N$, $att_H(e_2) = v_2 v_3$, $lab_H(e_2) = c\}$.

4. $\Theta_{SG,Tile}(H) = \{(V_H, E_H \setminus \{e_1\}, att_H|_{E_H \setminus \{e_1\}}, lab_H|_{E_H \setminus \{e_1\}}, v_1 v_2 v_4) \mid (v_1, v_2,$
   $v_3, v_4 \in V_H$ pairwise distinct nodes), $e_1 \in E_H$,
   NOT $condStop(H, v_1, v_2, v_3, v_4, e_1)$, $att_H(e_1) = v_2 v_1$, $lab_H(e_1) = N$, $\exists$ an edge $e_2 \in E_H$, $att_H(e_2) = v_2 v_3$, $lab_H(e_2) = a\}$.

It is not difficult to see the existence of only one edge labelled by N to all derived hypergraphs except the ones from the language as an invariant for all the derivations. Also all nonempty Θ sets contain hypergraphs with the external nodes in the neighborhood of the edge labelled by N.

Our example actually generates only odd sizes square grids; to get all even square grids we should add the corresponding axioms.

The graph languages generated by Eulerian, Hamiltonian, Square Grid Grammars are particularly interesting because they cannot be generated by any context-free (hyper)graph grammar (c.f., e.g., [5,6] where the generative power of hyperedge replacement grammars is discussed).

4   Simulation Results

In this section, we show that contextual hypergraph grammars generalize total contextual grammars on strings. Moreover, contextual hypergraph grammars can simulate in a natural way two well-known types of graph grammars: hyperedge replacement grammars [5,6], which provide a context-free generation mechanism for (hyper)graph sets, and hypergraph grammars in the double-pushout approach [7], which is one of the most frequently used graph transformation frameworks.

4.1   Contextual String-Hypergraph Grammars

The notion of contextual grammars evolved starting from the initial paper proposed by Solomon Marcus up to the current definition [1].

Definition 2 (Total Contextual Grammars). *A (string) total contextual grammar is a construct $TCG = (\Sigma, A, C, \varphi)$, where Σ is an alphabet, A is a finite subset of Σ^*, C is a finite subset of $\Sigma^* \times \Sigma^*$ and $\varphi : \Sigma^* \times \Sigma^* \times \Sigma^* \to \mathcal{P}(C)$. The elements of A are called* axioms, *the elements of C are called* contexts, *and φ is a* choice function.

A derivation relation $\underset{TC}{\Rightarrow}$ is defined as $x \underset{TC}{\Rightarrow} y$ if and only if $x = x_1 x_2 x_3$, $y = x_1 u x_2 v x_3$, for $x_1, x_2, x_3 \in \Sigma^*$, and $(u, v) \in C$, s.t. $(u, v) \in \varphi(x_1, x_2, x_3)$. The transitive closure of the relation $\underset{TC}{\Rightarrow}$ is denoted by $\underset{TC}{\overset{*}{\Rightarrow}}$.

The generated language is $L(TCG) = \{w \in \Sigma^* \mid a \underset{TC}{\overset{*}{\Rightarrow}} w, \text{ for } a \in A\}$.

In order to relate string contextual grammars with contextual hypergraph grammars, we need the notion of a *string hypergraph* $\overrightarrow{s}$ for a string s so that a string language L can be considered as a hypergraph language $\overrightarrow{L}$.

We will use the "·" operator to denote the concatenation of numeric symbols.

Definition 3 (string-hypergraph). *Let us consider an alphabet Σ and a finite nonempty string $s = a_1 \dots a_n$ over Σ.*

A string-hypergraph context denoted by $\underline{s} \in \mathcal{H}_{\Sigma,n}$ is the hypergraph $([n], \{ec_i \mid i \in [n]\}, (att(ec_i) = i, \text{ for all } i \in [n]), (lab(ec_i) = a_i, \text{ for all } i \in [n]), 1 \cdot \dots \cdot n)$ with all its vertices as external nodes; and let $\underline{\lambda} = empty$.

A string-hypergraph denoted by $\overrightarrow{s} \in \mathcal{H}_\Sigma$ is the hypergraph $([n+1], \{ec_i \mid i \in [n]\} \cup \{es_i \mid i \in [n]\}, (att(ec_i) = i, att(es_i) = i \cdot (i+1), \text{ for all } i \in [n]), (lab(ec_i) = a_i, lab(es_i) = *, \text{ for all } i \in [n]))$ having no external nodes but some sequential edges in order to reconstruct a string form a string-hypergraph. $\overrightarrow{\lambda} = Vertex$.

For a given string language L we denote by $\overrightarrow{L} = \{\overrightarrow{s} \mid s \in L\}$.

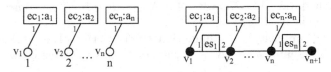

Fig. 4. A string-hypergraph context and a string-hypergraph

We can see a representation of a string-hypergraph context and a string-hypergraph in Figure 4, the rectangles are the hyperedges with their names and labels, while the numbers next to them represent the attachments to nodes.

A total contextual (string) grammar $TCG = (\Sigma, Ax, Ctx, \varphi)$ can be transformed into a contextual hypergraph grammar $CHG(TCG) = (\Sigma, \Sigma, \mathcal{A}, \mathcal{C}, \Theta_{TCG})$ in the following way. $\mathcal{A} = \{\overrightarrow{a} \mid a \in Ax\}$ that is the axioms are transformed into string-hypergraphs, $\mathcal{C} = \{\underline{u}+\underline{w} \mid (u, w) \in Ctx\}$, that is each context is transformed into the disjoint union of the left and right string-hypergraph contexts. The function $\Theta_{TCG,\underline{u}+\underline{w}}(\overrightarrow{xyz})$ takes a string-hypergraph and prepares it for the merging operation with a string-hypergraph context. Figure 5 shows the Θ-preparation, where the ec hyperedges took labels from the original string.

$$\Theta_{TCG,\underline{u}+\underline{w}}(\overrightarrow{xyz}) = \{([|xuywz| + 1], EC \cup ES, att, lab, (|x| + 1) \cdot \ldots \cdot |xu| \cdot (|xuy| + 1) \cdot \ldots \cdot |xuyw|) \mid \text{ for } (u, w) \in Ctx, x, y, z \in \Sigma^*, s.t.(u, w) \in \varphi(x, y, z)\},$$

where

- $ES = \{es_i \mid i \in [|xuywz|]\}$,
- $EC = \{ec_i \mid i \in [|xyz|]\}$,
- $att(es_i) = i \cdot (i + 1)$, $lab(es_i) = *$, for i in $[|xuywz|]$,
- $att(ec_i) = i$ for $1 \leq i \leq |x|$, $att(ec_i) = i + |u|$ for $|x| + 1 \leq i \leq |xy|$, $att(ec_i) = i + |uw|$ for $|xy| + 1 \leq i \leq |xyz|$, $lab(ec_i) = xyz(i)$ for i in $[|xyz|]$.

Using this construction, it is easy to prove the following theorem.

Theorem 1. *Let TCG be a total contextual (string) grammar and $CHG(TCG)$ the corresponding contextual hypergraph grammar. Then they generate the same string-hypergraph language, i.e. $\overrightarrow{L(TCG)} = L(CHG(TCG))$.*

4.2 Hyperedge Replacement Grammars

A *hyperedge replacement grammar* is a system $HRG = (N, T, P, S)$ where $N \subseteq \Sigma$ is a set of *nonterminal labels*, $T \subseteq \Sigma$ is a set of *terminal labels*, P is a set of

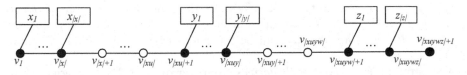

Fig. 5. An object returned by the function Θ_{TCG}

rules having the form $(A ::= R)$ with $A \in N$ and $R \in \mathcal{H}_{\Sigma,n}$ for some $n \in \mathbb{N}$, and $S \in N$ is a start symbol.

For technical simplicity, we assume that the nonterminals are *typed*, meaning that there is a mapping $type : N \to \mathbb{N}$ subject to the following conditions:

i) $type(A) = type(R)$ for each $(A ::= R) \in P$,
ii) $type(e) = type(lab_R(e))$ for each $e \in E_R$ with R being the right-hand side of some rule $(A ::= R) \in P$.

Let $H \in \mathcal{H}_\Sigma$ and $e \in E_H$ with $type(e) = type(lab_H(e))$. Then H *derives directly* $\overline{H}$ through $(lab_H(e) ::= R) \in P$ if $\overline{H} = (H - e) \circ R$. Note that the merging of $(H - e)$ and R is always defined because the types of e and R are equal and e transfers its type to $(H - e)$. Here $H - e$ denotes the *removal* of e from H yielding the hypergraph $H - e = (V_H, E_H \setminus \{e\}, att, lab, att_H(e))$ where att and lab are the restrictions of att_H and lab_H, respectively, to the set $E_H \setminus \{e\}$.

A direct derivation of $\overline{H}$ from H is denoted by $H \to \overline{H}$ and the reflexive and transitive closure of this relation by $\to^*$.

The language generated by a hyperedge replacement grammar $HGR = (N, T, P, S)$ consists of all terminal hypergraphs derivable from the start handle, i.e. $L(HRG) = \{H \in \mathcal{H}_T \mid S^\circ \to^* H\}$. Here, a *handle* of a nonterminal $A \in N$ denotes the hypergraph $A^\circ = ([type(A)], \{e_0\}, att, lab, 1 \dots type(A))$ with $att(e_0) = 1 \dots type(A)$ and $lab(e_0) = A$.

A hyperedge replacement grammar $HRG = (N, T, P, S)$ can be translated into a contextual hypergraph grammar $CHG(HRG) = (N \cup T, T, \{S^\circ\}, \{R \mid (A ::= R) \in P\}, \Theta_{HRG})$ with $\Theta_{HRG,R}(H) = \{H - e \mid e \in E_H, (lab_H(e) ::= R) \in P\}$ such that $CHG(HRG)$ generates the same language as HRG.

Theorem 2. *Let HRG be a hyperedge replacement grammar and $CHG(HRG)$ the corresponding contextual hypergraph grammar. Then HRG and $CHG(HRG)$ generate the same language, i.e. $L(HRG) = L(CHG(HRG))$.*

4.3   Hypergraph Grammars

In this subsection, we introduce hypergraph grammars in the so-called double-pushout approach. They generalize hyperedge replacement grammars in that a direct derivation does not replace a hyperedge only, but a subgraph (up to external nodes) which is a matching of a left-hand side of a rule.

A *hypergraph grammar* is a system $HG = (T, P, Z)$ where $T \subseteq \Sigma$ is a set of *terminal labels*, P is a finite set of *rules* of the form $L \supseteq K \subseteq R$ with $L, K, R \in \mathcal{H}_\Sigma$, and $Z \in \mathcal{H}_\Sigma$ is an *axiom*.

Without loss of generality, we may assume that the *gluing hypergraph* K of each rule $L \supseteq K \subseteq R$ is totally disconnected, i.e. $E_K = \emptyset$, and $V_K = [n]$ for some $n \in \mathbb{N}$. Therefore, the rule can be represented by the pair $(L, 1 \dots n) ::= (R, 1 \dots n)$.

A hypergraph $H \in \mathcal{H}_\Sigma$ *directly derives* a hypergraph $\overline{H} \in \mathcal{H}_\Sigma$ through the application of the rule $(L, 1 \dots n) ::= (R, 1 \dots n)$ if there is a hypergraph morphism $g : L \to H$ such that

(1) $att_H(e) \in (V_H \setminus (g_V(V_L) \setminus g_V([n])))^*$ for all $e \in E_H \setminus g_E(E_L)$,
(2) g_E is injective, and $g_V(v) = g_V(v')$ for $v \neq v'$ implies $v, v' \in [n]$,
(3) $X = (V_H \setminus (g_V(V_L) \setminus g_V([n]))), E_H \setminus g_E(E_L), att_X, lab_X, g_V(1) \ldots g_V(n))$ with $att_X(e) = att_H(e)$ and $lab_X(e) = lab_H(e)$ for all $e \in E_H \setminus g_E(E_L)$, and
(4) $\overline{H} = X \circ (R, 1 \ldots n)$.

A direct derivation from H to $\overline{H}$ through the rule r is denoted by $H \underset{P}{\rightarrow} \overline{H}$ if $r \in P$. The reflexive and transitive closure of the relation $\underset{P}{\rightarrow}$ is denoted by $\underset{P}{\overset{*}{\rightarrow}}$.

Let $HG = (T, P, Z)$ be a hypergraph grammar. Then the *generated language* $L(HG)$ contains all terminal hypergraphs derivable from the axiom through given rules *i.e.* $L(HG) = \{H \in \mathcal{H}_T \mid Z \underset{P}{\overset{*}{\rightarrow}} H\}$.

It should be noted that the hypergraphs H and $\overline{H}$ are *pushouts* of the intermediate hypergraph X and the left-hand side L respectively the right-hand side R using the gluing hypergraph K. H and $\overline{H}$ remain invariant whether K is totally disconnect or has got hyperedges. In the *double − pushout* approach, the hypergraphs X and $\overline{H}$ are usually constructed as *pushout complement* and *pushout* respectively. We prefer the given version because it is easier related to contextual hypergraph grammars.

A hypergraph grammar $HG = (T, P, Z)$ can be transformed into a contextual hypergraph grammar $CHG(HG) = (\Sigma, T, \{Z\}, \mathcal{C}_{HG}, \Theta_{HG})$ where $\mathcal{C}_{HG} = \{(R, 1 \ldots n) \mid ((L, 1 \ldots n) ::= (R, 1 \ldots n)) \in P\}$ and $\Theta_{HG,(R,1\ldots n)}(H)$ contains all hypergraphs X that are constructed as in Point 3 above for all rules $((L, 1 \ldots n) ::= (R, 1 \ldots n)) \in P$ and hypergraph morphisms $g : L \rightarrow H$ that fulfil Points 1 and 2.

We get $H \Rightarrow \overline{H}$ with $\overline{H} = (R, 1 \ldots n) \circ X$ for some $(R, 1 \ldots n) \in \mathcal{C}_{HG}$ and $X \in \Theta_{HG,(R,1\ldots n)}(H)$ if and only if $H \underset{P}{\rightarrow} \overline{H}$. This proves the following theorem.

Theorem 3. *Let HG be a hypergraph grammar and $CHG(HG)$ the corresponding contextual hypergraph grammar. Then HG and $CHG(HG)$ generate the same language, i.e. $L(HG) = L(CHG(HG))$.*

It is known that hypergraph grammars in the double-pushout approach generate all recursively enumerable hypergraph languages. Using the theorem, this holds for our new concept of contextual hypergraph grammars, too.

5   Conclusion

We have introduced a new type of hypergraph grammars, namely contextual hypergraph grammars, that generalize in a natural way the contextual string grammars. Using the power of our formalism, we are able to simulate already existing devices for the generation of hypergraph languages, like hyperedge replacement grammars and hypergraph grammars in the double-pushout approach. Our approach is useful to model operations with annotated documents even with multiple structures including syntactic-semantic structures, dependencies, multilingual information, etc. Further investigations are needed to study possible hierarchies within the generated languages. Also, the type and complexity of the Θ function deserve special attention. Possible classes to be studied are arbitrary functions, NP, or P. Furthermore, types of specifying a matching condition may be distinguished, such as global checking, local checking, without labels, without edges limitation, with a maximum number of connections, etc.

Acknowledgements

We would like to thank Victor Mitrana, Claudio Moraga, and the anonymous referees for their helpful comments and observations.

References

1. Păun, G.: Marcus Contextual Grammars. Kluwer Academic Publishers, Norwell, MA, USA (1997)
2. Kudlek, M., Martín-Vide, C., Mateescu, A., Mitrana, V.: Contexts and the concept of mild context-sensitivity. Linguistics and Philosophy (26) (2002) 703–725
3. Marcus, S., Păun, G., Martín-Vide, C.: Contextual grammars as generative models of natural languages. Comput. Linguist. **24**(2) (1998) 245–274
4. Kappes, M.: Combining contextual grammars and tree adjoining grammars. Grammars, A Journal of Mathematical Research on Formal and Natural Languages **3**(2-3) (2000) 175–187
5. Habel, A.: Hyperedge replacement: Grammars and languages. LNCS **643** (1992)
6. Drewes, F., Kreowski, H.J., Habel, A.: Hyperedge replacement graph grammars. In: Handbook of graph grammars and computing by graph transformation: volume I. foundations. World Scientific Publishing (1997) 95–162
7. Corradini, A., Montanari, U., Rossi, F., Ehrig, H., Heckel, R., Löwe, M. In: Algebraic approaches to graph transformation. Part I: basic concepts and double pushout approach. World Scientific, River Edge, NJ, USA (1997) 163–245
8. Courcelle, B.: The expression of graph properties and graph transformations in monadic second-order logic. In Rozenberg, G., ed.: Handbook of graph grammars and computing by graph transformations, vol. 1: Foundations. World Scientific, New-Jersey, London (1997) 313–400

End-Marked Maximal Depth-First Contextual Grammars*

K. Lakshmanan

School of Technology & Computer Science
Tata Institute of Fundamental Research
Colaba, Mumbai - 400 005, India
laksh@tifr.res.in

Abstract. In this paper, we present a few results which are of interest for the potential application of contextual grammars to natural languages. We introduce two new classes of internal contextual grammars, called *end-marked maximal depth-first* and *inner end-marked maximal depth-first* contextual grammars. We analyze the new variants with respect to the basic properties of the *mildly context sensitive* languages. With this aim, we show that (i) the three basic *non-context-free constructions* in natural languages can be realized upon using these variants, (ii) the *membership problem* for these family of languages is decidable in polynomial time algorithm, (iii) the family of languages generated by end-marked maximal depth-first grammars contains non-semilinear languages. We also solve the following open problem addressed in [3] and [1]: whether the families of languages generated by *maximal depth-first* and *maximal local* contextual grammars are *semilinear* or not?

Keywords: Internal contextual grammars, non-context-free languages, membership problem, semilinearity, mildly context sensitiveness.

1 Introduction

Contextual grammars produce languages starting from a finite set of *axioms* and adjoining *contexts*, iteratively, according to the *selector* present in the current sentential form. As introduced in [4], adjoining the contexts are done at the ends of the strings and is called *external* contextual grammars. *Internal* contextual grammars were introduced by Păun and Nguyen in 1980 [8], where the contexts are adjoined to the selector strings appeared as substrings of the derived string. Later on, many variants of contextual grammars were introduced consequently by imposing restriction in choosing the selectors, viz., *maximal* [5], *depth-first* [6] contextual grammars. The basic idea in restricting the selectors and the derivation was to obtain classes of contextual languages more appropriate from natural languages point of view. In fact, the class of languages searched for should have the following properties, which define the so-called *mildly context sensitive* (MCS) languages and mildly context sensitive formalisms:

---

* A part of the work was carried out during the author's visit to University of Turku, Finland.

O.H. Ibarra and Z. Dang (Eds.): DLT 2006, LNCS 4036, pp. 339–350, 2006.

1. The class should contain languages corresponding to the following three basic non-context-free constructions: (i) *multiple agreements:* $L = \{a^n b^n c^n | n \geq 1\}$, (ii) *crossed dependencies:* $L = \{a^n b^m c^n d^m | n, m \geq 1\}$, and (iii) *marked duplication:* $L = \{wcw | w \in \{a, b\}^*\}$.
2. All the languages in the class are *parsable in polynomial time*. In other words, the *membership problem* can be solved in polynomial time algorithm.
3. The class contains *semilinear* languages only.

Sometimes the third property is considered to be too strong and this is replaced by the following property.

3′. All the languages in the class have the *bounded growth property*. That is, for any infinite languages L, there is a constant k_L depending only on L such that, for any $n \geq 1$, if L contains words of length n, then it contains also words of some length between $n + 1$ and $n + k_L$. In other words, there should not be arbitrarily large gap between two consecutive words present in the language.

As MCS formalisms are considered to be an appropriate model for description of natural languages, it is important in formal language theory and natural language processing to obtain certain classes of languages which satisfy the above said MCS property.

The main variants of internal contextual languages (for instance, depth-first contextual grammars, where at each derivation, the selector for the next derivation must contain one of the contexts u or v which are adjoined in the previous derivation) failed to contain the non-context-free constructions itself. So, new classes of grammars were introduced to tackle this problem, for instance, maximal contextual grammars [5], where at each derivation, the selector of maximal length is chosen for the next derivation. Though they generate the basic non-context-free languages, they contain non-semilinear languages and also their membership problem either remains open or is solvable with exponential time complexity.

But, during this last decade, some attempts have been made to introduce some variants of contextual grammars by restricting the selectors to obtain certain specific classes of contextual languages which satisfy the MCS property [1, 3]. In [1], Ilie considered a new variant called *local* contextual grammars, in which, at each derivation, the contexts for the further derivations are adjoined inside or nearer to the contexts adjoined in the previous derivation. When the maximal restriction is included with this variant, the grammar is said to be *maximal local*. Ilie showed that the languages generated by maximal local grammars with regular selectors contain the basic non-context-free languages and the membership problem for the family of these languages was solvable in polynomial time, but the question of semilinearity was left open to these languages.

In [3], a variant motivated by local contextual grammars, called *absorbing right context grammars* (denoted by *arc*) was considered in order to resolve the semilinear problem for maximal local languages. The class of languages generated by *arc* grammars with regular selectors was shown to satisfy all the properties needed for MCS formalism. Also, in [3], another variant of internal contextual grammars, namely *maximal depth-first* grammars (denoted by *mdf*) were introduced, by combining the maximal and depth-first conditions. Like maximal grammars, the family of languages generated by this variant contains non-context-free languages, but the membership and semilinear problems were left open for these languages.

In this paper, we introduce two new variants of maximal depth-first contextual grammars, called *end-marked maximal depth-first* and *inner end-marked maximal depth-first* contextual grammars and analyze their relevances with natural languages in view of mildly context sensitive formalisms. We impose the following *end-marked* restriction in maximal depth-first grammars. Whenever, some contexts u and v are introduced in a derivation, the selector for the next derivation should either begin or end with one of the contexts u or v (which were introduced in the previous derivation), and the context (u or v) which should begin or end with the selector for the next derivation is specified while defining the grammar itself. The main motivation for imposing this restriction is to tackle the membership problem for these languages since handling the membership problem for maximal depth-first languages is difficult and remains open. We also show that maximal local languages contain semilinear languages only which solves an important open problem addressed by Ilie in [1].

2 Basic Definitions

We assume the readers are familiar with the basic formal language theory notions. However, we recall the following notions and definitions which are used in the proceeding sections. For more details on formal language theory, we refer to [9].

For a word $x \in V^*$, we denote by $Sub(x)$, $Pref(x)$, $Suf(x)$, the set of *subwords, prefixes*, and *suffixes* of x, respectively. Also, $|x|$ denote the *length of* x. $|x|_a$ is the number of occurrences of the symbol a in the word x. Assume that $V = \{a_1, \ldots, a_k\}$. The *Parikh mapping* of V denoted by Ψ is $\Psi : V^* \longrightarrow N^k$, $\Psi(w) = (|w|_{a_1}, \ldots, |w|_{a_k})$, $w \in V^*$. If L is a language, then its *Parikh set* is defined by $\Psi(L) = \{\Psi(w) \mid w \in L\}$. Two languages L_1 and L_2 are said to be *letter equivalent* iff $\Psi(L_1) = \Psi(L_2)$. A *linear* set is a set $M \subseteq N^k$ such that

$$M = \{v_0 + \sum_{i=1}^{m} v_i x_i \mid x_i \in N\},$$

for some $v_0, v_1, \ldots, v_m$ in N^k. A *semilinear set* is a finite union of linear sets and a *semilinear language* is a language L such that $\Psi(L)$ is a semilinear set. The families of *finite* and *regular* languages are denoted by FIN, REG respectively.

We now present some basic classes of contextual grammars. An *internal contextual grammar* is a construct

$$G = (V, A, (S_1, C_1), \ldots (S_m, C_m)), \ m \geq 1, \text{ where}$$

- V is an alphabet,
- $A \subseteq V^*$ is a finite set, called the set of *axioms*,
- $S_j \subseteq V^*$, $1 \le j \le m$, are the sets of *selectors* or *selection* or *choice*,
- $C_j \subseteq V^* \times V^*$, C_j finite, $1 \le j \le m$, are the sets of *contexts*.

The usual *derivation* in the *internal mode* is defined as $x \Longrightarrow_{in} y$ iff $x = x_1 x_2 x_3$, $y = x_1 u x_2 v x_3$, for $x_1, x_2, x_3 \in V^*$, $x_2 \in S_j$, $(u, v) \in C_j$, $1 \le j \le m$.

For, maximal and depth-first contextual grammars, we refer to the informal definition presented in the earlier section. Now, we define maximal depth-first grammars. Given a contextual grammar G as above, a *maximal depth-first derivation* (denoted by *mdf*) in G is a derivation

$$w_1 \Longrightarrow_{mdf} w_2 \Longrightarrow_{mdf} \cdots \Longrightarrow_{mdf} w_n, \ n \ge 1, \text{ where}$$

(i) $w_1 \in A$, $w_1 \Longrightarrow w_2$ in the maximal way,
(ii) For each $i = 2, 3, \ldots, n - 1$, if $w_{i-1} = z_1 z_2 z_3$, $w_i = z_1 u z_2 v z_3$ ((u, v) is the context adjoined to w_{i-1} in order to get w_i), then $w_i = x_1 x_2 x_3$, $w_{i+1} = x_1 s x_2 t x_3$, such that $x_2 \in S_j$, $(s, t) \in C_j$, for some j, $1 \le j \le m$, and x_2 contains one of the contexts u or v as a substring (satisfying depth-first condition). That is, at every derivation, the selector $x_2 \in S_j$ contains one of the contexts u or v which was adjoined in the previous derivation
(iii) For each $i = 2, 3, \ldots, n - 1$, if $w_i \Longrightarrow_{df} w_{i+1}$ then there will be no other derivation in G with $w_i \Longrightarrow_{df} w'_{i+1}$ such that $w_i = x'_1 x'_2 x'_3$, $x'_2 \in S_j$ and $|x'_2| > |x_2|$ where $x_2 \in S_j$ (satisfying maximal condition). That is, at every derivation, w_i should have no other subword x'_2 with $x'_2 \in S_j$ of greater length than $x_2 \in S_j$.

Lengthwise depth-first grammars, are the restricted versions of depth-first grammars, introduced in [2], where at each derivation, the selector for the next derivation must contain the context u or v (introduced in the previous derivation) whichever is of maximal length. When the lengths of u and v are equal, the selector is chosen non-deterministically (which contains u or v). In a similar way like maximal depth-first grammars, we can define *maximal lengthwise depth-first* grammars (denoted by *mldf*), where, at each derivation, the selector contains one of the contexts u or v whichever is of maximum length and the selector is of maximal.

The next variant we define is local contextual grammars. Given a contextual grammar G, we define the *local* mode in the following way. For $z \in A$, $z \Longrightarrow_{in} x$ such that $z = z_1 z_2 z_3$, $x = z_1 u z_2 v z_3$, $z_2 \in S_k$, $(u, v) \in C_k$, for $z_1, z_2, z_3 \in V^*$, $1 \le k \le m$, then $x \Longrightarrow_{loc} y$ is called local with respect to $z \Longrightarrow x$, iff we have $u = u'u''$, $v = v'v''$, $u', u'', v', v'' \in V^*$, $y = z_1 u' s u'' z_2 v' t v'' z_3$ for $u'' z_2 v' \in S_j$, $(s, t) \in C_j$, $1 \le j \le m$. When the maximal restriction is included with this local variant, the grammar is said to be *maximal local* (denoted by *mloc*). For more technical details and for formal definition of basic classes of contextual grammars, we refer to the monograph on contextual grammars by Păun [7].

The following assumption is made throughout this paper. As we do not discuss external contextual grammars or any other grammars in this paper, we simply refer internal contextual grammars as grammars in many occurrences. We also call maximal length as maximal in many places for the sake of brevity.

3 End-Marked Maximal Depth-First Contextual Grammars

In this section, we introduce two new classes of contextual grammars and investigate whether they are good formalisms for natural languages by analyzing the properties that define mildly context sensitive languages.

We shall now introduce the first variant. An *end-marked maximal depth-first contextual grammars* is a construct $G = (V, A, (S_1, C_1), \ldots, (S_m, C_m)), m \geq 1$, where $V, A, S_1, \ldots S_m$, are as mentioned in the definition of internal contextual grammar and $C_j \subseteq (V_{\{L,R\}}^+ \times V^*) \cup (V^* \times V_{\{L,R\}}^+)$, C_j finite, $1 \leq j \leq m$, are the set of contexts. The elements of C_j's are of the form $(u_L, v), (u_R, v), (u, v_L)$, and (u, v_R). The suffix L and R represent end marker (left and right) for the selector of the next derivation. u_L (or v_L) indicates the selector for the next derivation should start with u (or v). In other words, u (or v) is the left end of the selector for the next derivation. Similarly, u_R (or v_R) indicates the selector for the next derivation should end with the context u (or v). In other words, u (or v) is the right end of the selector for the next derivation. Given such a grammar G, an end-marked maximal depth-first derivation (denoted by *emdf*) in G is a derivation $w_1 \Longrightarrow_{emdf} w_2 \Longrightarrow_{emdf} \cdots \Longrightarrow_{emdf} w_n$, $n \geq 1$, where

- $w_1 \in A$, $w_1 \Longrightarrow w_2$ in the maximal way,
- For each $i = 2, 3, \ldots, n-1$, if $w_{i-1} = z_1 z_2 z_3$, $w_i = z_1 u z_2 v z_3$, such that $z_2 \in S_k$, $1 \leq k \leq m$, then $w_i = x_1 x_2 x_3$, $w_{i+1} = x_1 s x_2 t x_3$, such that $x_2 \in S_j$, $1 \leq j \leq m$, and x_2 will be of one of the following four cases:
  (i) $x_2 = u z_2'$, if $(u_L, v) \in C_k$, with $z_2' \in V^*$ is of maximal (i.e., there exists no $z_2'' \in V^*$, such that $u z_2'' \in S_j$, with $|z_2''| > |z_2'|$).
  (ii) $x_2 = z_1' u$, if $(u_R, v) \in C_k$, with $z_1' \in V^*$ is of maximal (i.e., there exists no $z_1'' \in V^*$, such that $z_1'' u \in S_j$, with $|z_1''| > |z_1'|$).
  (iii) $x_2 = z_2' v$, if $(u, v_R) \in C_k$, with $z_2' \in V^*$ is of maximal (i.e., there exists no $z_2'' \in V^*$, such that $z_2'' v \in S_j$, with $|z_2''| > |z_2'|$).
  (iv) $x_2 = v z_3'$, if $(u, v_L) \in C_k$, with $z_3' \in V^*$ is of maximal (i.e., there exists no $z_3'' \in V^*$, such that $v z_3'' \in S_j$, with $|z_2''| > |z_2'|$).

Now, we introduce the second variant which is a restricted version of the above introduced grammar. Given a *emdf* grammar G, we can define the *inner end-marked maximal depth-first grammar* (denoted by *iemdf*) by imposing the following changes in the grammar and derivations:

- $C_j \subseteq (V_L^* \times C^*) \cup (V^* \times V_R^*)$.
- As the elements of C_j's are of the form (u_L, v) and (u, v_R), the cases (ii) and (iv) discussed above will become void and only the cases (i) and (iii) will be valid.

– The selector for the next derivation should be inside the contexts u and v which were adjoined in the previous derivation. More precisely, if u and v are the contexts adjoined to the selector, say z_2, then the next selector, say x_2, will be a strict subword of uz_2v and x_2 should either begin with u or end with v (in order to satisfy the end-marked condition) but not both hold true.

The language generated by a grammar G in the mode β, $\beta \in \{in, mdf, mldf, mloc, emdf, iemdf\}$ is given by

$$L_\beta(G) = \{w \in V^* \mid x \Longrightarrow_\beta^* w,\ x \in A\},$$

where $\Longrightarrow_\beta^*$ is the reflexive transitive closure of the relation $\Longrightarrow_\beta$.

If all the sets of selectors $S_1, \ldots, S_m$ are in a family F of languages, then we say that the grammar G is with F selection. As usual, the family of languages $L_\beta(G)$, for G working in $\beta \in \{in, mdf, mldf, mloc, emdf, iemdf\}$ mode with F selection is given as $ICC(F), ICC_{mdf}(F), ICC_{mldf}(F), ICC_{mloc}(F), ICC_{emdf}(F)$, and $ICC_{iemdf}(F)$, respectively.

3.1   Non Context-Free Constructions in Natural Languages

This section deals all the three basic non-context-free constructions in natural languages, that is, *multiple agreements, crossed dependencies* and *duplication* can be realized using *emdf, iemdf* grammars with regular selection.

Lemma 1. *(multiple agreements)*
$L_1 = \{a^n b^n c^n \mid n \geq 1\} \in ICC_\alpha(REG),\ \alpha \in \{emdf, iemdf\}$.

Proof. Consider the following grammar

$$G_1 = (\{a, b, c\}, abc, \{(ab^+, (a, bc_R)), (b^+c, (ab_L, c))\}).$$

Any derivation in $\alpha \in \{emdf, iemdf\}$ mode of G_1 is as follows:

$$^\downarrow ab^\downarrow c \Longrightarrow_\alpha \underline{a}a^\downarrow \underline{bb}c^\downarrow c \Longrightarrow_\alpha aa^\downarrow \underline{abbb}^\downarrow \underline{c}cc \Longrightarrow_\alpha aa\underline{a}a^\downarrow bbb\underline{bc}^\downarrow ccc \Longrightarrow_\alpha^* a^n b^n c^n, n \geq 1.$$

The underlined symbols are the newly inserted contexts in the derivation. The word between the two down arrows indicates the selector for the next derivation. Starting from the axiom abc, ab is considered as the selector for the first derivation and (a, bc) is adjoined. As bc is the right end-marker, the selector for the next derivation should end with bc and of maximal. Therefore, b^2c is the selector. Now (ab, c) is adjoined and ab is the left end marker. Therefore, the selector for the next derivation is ab^3 and (a, bc) is adjoined to $aabbcc$. Continuing the derivation in this fashion, we can see that $L_\alpha(G_1) = L_1$, $\alpha \in \{emdf, iemdf\}$.   □

Lemma 2. *(crossed agreements)*
$L_2 = \{a^n b^m c^n d^m \mid n, m \geq 1\} \in ICC_\alpha(REG),\ \alpha \in \{emdf, iemdf\}$.

Proof. Consider the grammar

$$G_2 = (\{a, b, c, d\}, abcd, (b^+c^+, \{(b_L, d), (a, c_R)\})).$$

Any derivation in the mode $\alpha \in \{emdf, iemdf\}$ in G_2 is as follows:

$$a^\downarrow bc^\downarrow d \Longrightarrow_\alpha a\underline{a}^\downarrow bc\underline{c}^\downarrow d \Longrightarrow_\alpha aa^\downarrow \underline{b}bcc^\downarrow \underline{d}d \Longrightarrow_\alpha aa^\downarrow \underline{b}bbcc^\downarrow \underline{d}dd \Longrightarrow_\alpha^* a^n b^m c^n d^m.$$

Starting from the axiom $abcd$, bc is considered as the selector for the first derivation and (a, c) is adjoined (one can also adjoin (b, d) instead of (a, c) and continue the derivation accordingly). As c is the right end marker, the selector for the next derivation should end with c and of maximal. Therefore, bc^2 is the selector. Now, (b, d) is adjoined and b is the left end marker, the selector for the next derivation should begin with b. Therefore, b^2c^2 is the selector and (b, d) is adjoined to the word $aabbccdd$. Continuing the derivation, we can see that $L_\alpha(G_2) = L_2$, $\alpha \in \{emdf, iemdf\}$. □

Lemma 3. *(marked duplication)*
$L_3 = \{wcw \mid w \in \{a, b\}^*\} \in ICC_\alpha(REG)$, $\alpha \in \{emdf, iemdf\}$.

Proof. Consider the following grammar

$$G_3 = (\{a, b, c\}, c, (\{cx \mid x \in \{a, b\}^*\}, \{(a, a_R), (b, b_R)\})).$$

Any derivation in $\alpha \in \{emdf, iemdf\}$ mode in G_3 is as follows:

$$^\downarrow c^\downarrow \Longrightarrow_\alpha \underline{w_1}^\downarrow c\underline{w_1}^\downarrow \Longrightarrow_\alpha w_1 \underline{w_2}^\downarrow cw_1 \underline{w_2}^\downarrow \Longrightarrow_\alpha w_1 w_2 \underline{w_3}^\downarrow cw_1 w_2 \underline{w_3}^\downarrow \Longrightarrow_\alpha^* wcw,$$

where $w_i \in \{a, b\}$, $1 \leq i \leq |w|$, $w \in \{a, b\}^*$. The reader can easily verify that, $L_\alpha(G_3) = L_3, \alpha \in \{emdf, iemdf\}$. □

Remark. Take $c = \lambda$ in the above grammar G_3, then the derivation in the *iemdf* mode will be given as $\lambda \Longrightarrow_{iemdf} \underline{w_1}^\downarrow \underline{w_1}^\downarrow \Longrightarrow_{iemdf} w_1 \underline{w_2}^\downarrow w_1 \underline{w_2}^\downarrow \Longrightarrow_{iemdf}^* ww$. Note that in *iemdf* mode, every time, the selector for the next derivation is chosen inside the contexts adjoined in the previous derivation and of maximal. It is easy to see that $L_{iemdf}(G_3) = \{ww \mid w \in \{a, b\}^*\}$. This language is known as *non-marked duplication*. This result is important from contextual grammars point of view since no other existing contextual grammar is capable of generating this language except absorbing right context (*arc*) grammars [3].

3.2   Computational Complexity

In this section, we show that the family of all languages $ICC_\alpha(F)$, for $\alpha \in \{emdf, iemdf\}$, $F \in \{FIN, REG\}$ are parsable in polynomial time. That is, given a *emdf* or *iemdf* grammar G and a string $w \in V^*$ of length n, there is a Turing Machine M which accepts or rejects w in polynomial time.

Theorem 1. *For any end-marked maximal depth-first and inner end-marked maximal depth-first contextual grammar G with F selection, $F \in \{FIN, REG\}$, there is a polynomial time algorithm which accepts $L_\alpha(G)$, $\alpha \in \{emdf, iemdf\}$ on any input w of length n. In other words, $ICC_\alpha(F) \subseteq \mathcal{P}$ for $\alpha \in \{emdf, iemdf\}$, $F \in \{FIN, REG\}$*

Proof. We will prove the result for the case $\alpha = emdf$ and $F = REG$. The other case $\alpha = iemdf$ follows automatically.

Consider a *emdf* grammar $G = (V, A, (S_1, C_1), \ldots, (S_m, C_m))$, where $S_j \in REG$, $1 \leq j \leq m$. Given a input w of length $n, n \geq 1$, we denote the ith letter of w by w_i, $1 \leq i \leq n$. When $w = \lambda$ (i.e., $|w| = 0$), it can be easily verified by checking the axiom of G. For $x \in A$, any derivation in *emdf* mode can be given as $x = y_0 \Longrightarrow y_1 \Longrightarrow \ldots y_{k-1} \Longrightarrow y_k = w$, $k \geq 0$.

Construct a two tape non-deterministic Turing machine M which accepts the language $L_{emdf}(F)$ as follows:

- On the first tape, w is written as input and is never changed.
- On the second tape, M copies the input w from tape 1. During a computation on w, M works on tape 2 as follows:
  1. Check whether $w \in A$. If $w \in A$, then M accepts w; else, go to next step.
  2. M non-deterministically identifies four indices $p, q, r, s \in \{1, \ldots, n\}$, $p \leq q \leq r \leq s$ with the following property: $u_k = w_p \ldots w_{q-1}$, $x_k = w_q \ldots w_r$, $v_k = w_{r+1} \ldots w_s$, where $x_k \in S_{j_k}$, $(u_k, v_k) \in C_{j_k}$, $1 \leq j_k \leq m$.
     Also, whenever M tries new combination of values to identify the indices p, q, r, s (with the above said property), M rewrites the input w on tape 2 and continues the computation.
  3. If there is no such four indices with the above property, then M rejects w. Otherwise, M goes to next step.
  4. At each time, M tries to back track the derivation by deleting the contexts u_h and v_h from y_h, $k \geq h \geq 1$ (initially, $h = k$). Let the modified resultant string be y_{h-1}. By definition of the grammar, the selector at any derivation (except the first derivation) should begin or end with the contexts which were adjoined in the previous derivation. Therefore, once the selector (say x_h which is $w_q \ldots w_r$) is identified, according to the definition of the grammar, one of the following cases is possible.

     case i: $x_h = u_{h-1} z_2$, $z_2 \in V^*$, $\exists$ three indices $q_1, r_1, s_1 \geq 0$, where $q \leq q_1 \leq r_1 \leq s_1$, such that $u_{h-1} = w_p \ldots w_{q_1-1}$, $x_{h-1} = w_{q_1} \ldots w_{r_1}$, $v_{h-1} = w_{r_1+1} \ldots w_{s_1}$, $x_{h-1} \in S_{j_{h-1}}$, $(u_{h-1}, v_{h-1}) \in C_{j_{h-1}}$, $1 \leq j_{h-1} \leq m$.

     case ii: $x_h = z_1 u_{h-1}$, $z_1 \in V^*$, $\exists$ three indices $q_2, r_2, s_2 \geq 0$, where $q_2 \leq r \leq r_2 \leq s_2$, such that $u_{h-1} = w_{q_2} \ldots w_r$, $x_{h-1} = w_{r+1} \ldots w_{r_2}$, $v_{h-1} = w_{r_2+1} \ldots w_{s_2}$, $x_{h-1} \in S_{j_{h-1}}$, $(u_{h-1}, v_{h-1}) \in C_{j_{h-1}}$, $1 \leq j_{h-1} \leq m$.

     case iii: $x_h = v_{h-1} z_3$, $z_3 \in V^*$, $\exists$ three indices $q_3', q_3, r_3 \geq 0$, where $q_3' \leq q_3 \leq q \leq r_3 \leq r$, s.t $u_{h-1} = w_{q_3'} \ldots w_{q_3-1}$, $x_{h-1} = w_{q_3} \ldots w_{q-1}$, $v_{h-1} = w_q \ldots w_{r_3}$, $x_{h-1} \in S_{j_{h-1}}$, $(u_{h-1}, v_{h-1}) \in C_{j_{h-1}}$, $1 \leq j_{h-1} \leq m$.

     case iv: $x_h = z_2' v_{h-1}$, $z_2' \in V^*$, $\exists$ three indices $q_4', q_4, r_4 \geq 0$, where $q_4' \leq q_4 \leq r_4 \leq r$, s.t $u_{h-1} = w_{q_4'} \ldots w_{q_4-1}$, $x_{h-1} = w_{q_4} \ldots w_{r_4-1}$, $v_{h-1} = w_{r_4} \ldots w_r$, $x_{h-1} \in S_{j_{h-1}}$, $(u_{h-1}, v_{h-1}) \in C_{j_{h-1}}$, $1 \leq j_{h-1} \leq m$.
  5. M tries to identify three indices q_i, r_i, s_i, $i = 1, 2$, or q_l, q_l', r_l, $l = 3, 4$, with the above mentioned property. If there exists three such indices, then M goes to next step. Else, M rejects w.
  6. Check $y_{h-1} \in A$. If yes, then M accepts w. Else, M replaces the variables simultaneously as given below and goes to Step 4 for further

back tracking. $(u_h, y_h, v_h \leftarrow u_{h-1}, y_{h-1}, v_{h-1})$ and one of the following. (i) $(p, q, r, s \leftarrow p, q_1, r_1, s_1)$, (ii) $(p, q, r, s \leftarrow q_2, r+1, r_2, s_2)$, (iii) $(p, q, r, s \leftarrow q_3', q_3, q-1, r_3)$, (iv) $(p, q, r, s \leftarrow q_4', q_4, r_4, r)$.

Since M tries correct derivations only (with respect to G), it cannot accept a word not in the language generated by G, and so $L(M) \subseteq L_{emdf}(G)$. Conversely, if $w \in L_{emdf}(G)$, then there is a derivation in G yielding w from the axiom, and M guesses the right choices for the added contexts (in the reverse order), thus accepting w. Consequently, $L(M) = L_{emdf}(G)$.

The space required by M is the space needed to remember the four indices p, q, r, s, and three more indices during the computation and therefore M needed to store 7 indices all the time which is $\mathcal{O}(7log(|w|))$. Since the selectors are regular type and the contexts are finite, verifying the membership and the four cases require only finite amount of space. This follows, M operates in space $\mathcal{O}(logn)$. The remaining proof of the result follows from the well known result $NSPACE\ (log(n) \subseteq \mathcal{P})$.

3.3   Semilinear Property

In this section, we analyze the semilinear property for $ICC_\alpha(F)$, $\alpha \in \{mdf, mldf, emdf, mloc\}$, $F \in \{FIN, REG\}$.

Theorem 2. *The family of languages $ICC_{emdf}(REG)$ contains non-semilinear languages.*

Proof. Consider the *emdf* grammar $G = (\{a, b, c, d\}, cababc, P)$ with P consisting the following production rules:

1. $(ca, (\lambda, b_L))$, 2. $(bba, (\lambda, b_L))$, 3. $(bbc, (\lambda, c_L))$, 4. $(bcc, (a_R, \lambda))$,
5. $((ba)^+, (a_R, \lambda))$, 6. $(caba, (c_L, \lambda))$, 7. $(cca, (\lambda, b_L))$, 8. $(caba, (cd_L, \lambda))$,
9. $(aa, (cd_L, \lambda))$, 10. $(cdaa, (\lambda, d_R))$, 11. $(cdc(ab)^+c, (\lambda, d_R))$.

The intuition behind the construction of the above rules is the following. Between every a and b, one b is introduced from left to right of the word by rules 1 and 2. Therefore, the number of b's are doubled. Now, between every two b's, one a is introduced from right to left of the word by rules 4 and 5. Therefore, the number of a's are doubled. c is the terminal identifier of the word. Whenever, the scanning of the word is over (from left to right or right to left), c is incremented to one time (by rules 3 and 6), and the selector for further derivations move right or left in order to double the occurrences of b's or a's respectively. Rules 6 and 7 are applied to continue the derivations further and rules 8,11 or 9,10 are applied to terminate the derivation. Once, the context (λ, d_R) is introduced, no further derivation can be performed, as there is no selector end with d. We present a sample derivation in *emdf* mode as below (the number in suffix of each derivation indicates the rule which is applied). Initially, ca is assumed to be the selector for the axiom.

$$cababc \Longrightarrow_1 ca^\downarrow \underline{b}ba^\downarrow bc \Longrightarrow_2 cabba^\downarrow \underline{b}bc^\downarrow \Longrightarrow_3 cabbab^\downarrow bc\underline{c}^\downarrow \Longrightarrow_4 cab^\downarrow bab\underline{a}^\downarrow bc^2 \Longrightarrow_5$$
$$ca^\downarrow b\underline{a}^\downarrow (ba)^2 bc^2 \Longrightarrow_5 c^\downarrow a\underline{a}^\downarrow (ba)^3 bc^2 \Longrightarrow_9 c^\downarrow \underline{cd}aa^\downarrow (ba)^3 bc^2 \Longrightarrow_{10} ccdaa\underline{d}(ba)^3 bc^2.$$

Now, consider the words containing two d's. One can also terminate the derivation at any time without doubling the occurrences of a's and b's and not introducing any d's. But, we do not bother such words as our claim is that the generated language contains non-semilinear also. Now, intersect $\Phi_{\{a,b,c,d\}}(L_{emdf}(G))$ with the semilinear set $R = \{(n_1, n_2, n_3, 2) \mid n_1, n_2, n_3 \geq 1\}$, we obtain

$$R' = \{|a| = 2^i + 1, |b| = 2^i, |c| = 2i, |d| = 2 \mid i \geq 1\},$$

which is not semilinear as the family of semilinear sets is closed under intersection. This proves that $L_{emdf}(G)$ is not semilinear. $\qquad \square$

The proof does not hold for $ICC_{iemdf}(REG)$ since the contexts in the rules (1-5 and 7) are not of the form mentioned in the definition. So, the question of semilinearity is left open for $ICC_{imedf}(F)$, $F \in \{FIN, REG\}$.

Theorem 3. *The family of languages $ICC_{mldf}(REG)$ (and so $ICC_{mdf}(REG)$) contains non-semilinear languages.*

Proof. Consider the above grammar G where the contexts do not have any end-markers. With the same proof, one can easily verify that $L_{mldf}(G)$ is not semilinear. This result also hold true for $ICC_{mdf}(REG)$ since $ICC_{mldf}(F) \subseteq ICC_{mdf}(F)$, $F \in \{FIN, REG\}$ [3]. $\qquad \square$

This solves the open problem posed in [3]. In the next theorem, we solve an important open problem in affirmative, addressed by Ilie in [1].

Theorem 4. *The family of languages generated by maximal local grammars with F selection, $F \in \{FIN, REG\}$, is semilinear.*

Proof. We prove the result for local grammars and the proof can be easily extended to maximal local grammars. Let L be the language generated by the local grammar $G = (V, A, P)$ with finite or regular choice, where P is the set of production rules are of the form $(S_1, C_1), \ldots, (S_m, C_m))$, $m \geq 1$.

For all $1 \leq j, f, r, s, t \leq m$, $u_f = u'_f u''_f$, $v_f = v'_f v''_f$ such that $u'_f \in Pref(u_f)$, $u''_f \in Suf(u_f)$, and $v'_f \in Pref(v_f)$, $v''_f \in Suf(v_f)$. As the contexts are finite languages, any u_f will be of finite length, say n. Then there will be only $(n+1)$ possible u'_f (since $u'_f = Pref(u_f)$) and u''_f (since $u''_f = Suf(u_f)$) exist. Similar statement holds true for v_f also. Also, as $x \in A$ is a finite language, $x_2 \in Sub(x)$ will also be finite. Therefore, once $x_2 \in S_f$ is known, checking $u''_f x_2 v'_f \in S_r$ will only be of finite combination (in fact, $(n+1)(k+1)$ possible combinations, if we assume $|v_f| = k$).

Now, construct a *linear* grammar G' as follows:

$$G' = (\{S, S_{r,s} \mid 1 \leq r, s \leq m\}, V, \{S\}, P')$$

where P' consists of the following production rules:

(1) $\{S \to x_1 x_2 x_3 \mid x = x_1 x_2 x_3 \in A\}$
(2) $\{S \to x_1 x_2 x_3 u_j v_j \mid x_2 \in S_j, (u_j, v_j) \in C_j\}$

(3) $\{S \rightarrow x_1 x_2 x_3 u_j'' S_{j,r} v_j' \mid u_j'' x_2 v_j' \in S_r, \ (u_r, v_r) \in C_r\}$

(4) $\{S_{r,s} \rightarrow u_r' v_r'' u_s'' S_{s,t} v_s' \mid u_r' \ldots u_j' x_2 v_j' \ldots v_s' \in S_t, \ (u_t, v_t) \in C_t\}$

(5) $\{S_{r,s} \rightarrow u_r' v_r'' u_s v_s \mid u_r' \ldots u_j' x_2 v_j' \ldots v_s' \in S_t, \ (u_t, v_t) \in C_t\}$.

Rules 1,2,5 are terminal rules which can be used to stop the derivation whenever necessary. The non-terminals in rule 3,4 will be used to simulate the derivations for further steps. It is easy to verify from the structure of P' that $L_{loc}(G)$ and $L(G')$ are letter equivalent. Since $L(G')$ is linear, $L(G)$ is also semi-linear [9]. □

Note that a similar proof cannot be given to *iemdf* grammars with regular selectors and the reason is the following: If z is a regular selector in a derivation, then by definition of *iemdf* grammars, uz' (or $z'v$), $z' \in Sub(z)$ may be a possible selector for the next derivation and this cannot be represented by a linear grammar with finite rules. On the other hand, in local grammars, if z is a regular selector for a derivation, then the selector for the next derivation is $u''zv'$. As z fully presents in the selector for the next derivation (in fact, at each derivation, the selector is expanded by the contexts which are adjoined) and u'', v' are of finite length, it can be represented by a linear grammar with finite rules.

4 Conclusion

In this paper, we have introduced two new classes of contextual grammars and investigated their feasibility with respect to MCS formalism. Both the variants are capable of generating the non-context-free languages. The membership problem has been solved for these two new classes of languages. Though, the family of languages generated by *emdf* grammars does not satisfy the semilinear property, it satisfies the bounded growth property since every derivation of a contextual grammar is obtained by adjoining the contexts of finite length, any two consecutive derivations can be easily bounded by a constant. We have also solved two open problems on semilinearity for the class of languages generated by *mdf* and *mloc* grammars. Thus, we have defined two new classes of contextual grammars which are shown to be MCS if we take the condition for MCS as properties 1,2 and 3' mentioned in the introduction. The semilinear property for the family $ICC_{iemdf}(F)$ is left as open and the techniques used in this paper to generate non-semilinear languages or semilinear languages did not work for this family of languages. As the variants discussed in this paper were analyzed towards MCS point of view, analyzing the generative power and hierarchical relations of these variants remain as future work.

Acknowledgement

I sincerely thank Prof. Juhani Karhumäki for his support and encouragement during my stay at University of Turku, Finland. I would also like to thank to Dr. Krishna Shankaranarayanan for her helpful conversations.

References

1. L.Ilie, On Computational complexity of contextual languages, *Theoretical Computer Science*, **183**(1), 1997, 33–44.
2. S. Krishna, K. Lakshmanan and R. Rama, On some classes of contextual grammars, *International Journal of Computer Mathematics*, **80**(2), 2003, 151–164.
3. K. Lakshmanan, S. Krishna, R. Rama and C. Martin-Vide, Internal contextual grammars for mildly context sensitive languages, **to appear** in *Research on Language and Computation*, 2006.
4. S.Marcus, Contextual Grammars, *Rev. Roum. Pures. Appl.*, 14, 1969, 1525–1534.
5. S.Marcus, C.Martin-Vide, Gh.Păun, On Internal contextual grammars with maximal use of selectors, *Proc. 8th Conf. Automata & Formal Lang.*, Salgotarjan, 1996, Publ. Math. Debrecen, **54**, 1999, 933–947.
6. C. Martin-Vide, J. Miquel-Verges and Gh. Păun, Contextual Grammars with depth-first derivation, *Tenth Twente Workshop on Language Tech.; Algebraic Methods in Language Processing*, Twente, 1995, 225–233.
7. Gh.Păun, *Marcus Contextual Grammars*, Kluwer Academic Publishers, 1997.
8. Gh.Păun, X.M.Nguyen, On the Inner Contextual Grammars, *Rev. Roum. Pures. Appl.*, **25**, 1980, 641–651.
9. A. Salomaa, *Formal Languages*, Academic Press, 1973.

Some Examples of Semi-rational DAG Languages

Jan Robert Menzel, Lutz Priese, and Monika Schuth

Fachbereich Informatik
Universität Koblenz-Landau, Germany
{juhees, priese, moschuth}@uni-koblenz.de

Abstract. The class of semi-rational *dag* languages can be characterized by labeled Petri nets with ε-transitions, by rather simple leaf substituting tree grammars with additional non-local merge rules, or as a synchronization closure of Courcelles class of recognizable sets of unranked, unordered trees. However, no direct recognition by some magma is known. For a better understanding, we present here some examples of languages within and without the class of semi-rational *dag* languages.

1 Introduction

For languages of finite words, infinite words, finite and infinite ranked and ordered trees various concepts like recognizability, rationality, regular expressions, definability by monadic second-order logics over certain signatures, generation by right-linear grammars, etc. proved to be equivalent. However, a generalization of those equivalence results to *dags* (directed acyclic graphs) has failed. Even for finite grids the classic concept of recognizability is stronger than that of monadic second-order definability, which is stronger than that of acceptability with finite-state graph automata [1], which is again stronger than acceptability by automata over planar *dags* [2], see [1].

In contrast to trees, *dags* are usually unranked and unordered. In a ranked graph the label of a node determines its number of sons. A ranked and ordered tree is simply a correct term over a ranked alphabet. The theory of ranked, ordered trees goes back to the early works of Church, Büchi, Rabin, Doner, Thatcher, Trakhtenbrot. A good overview is given in the TATA book project [3]. Most automata theoretical results transfer also to ordered, unranked trees, see the well-known report [4]. Unranked and unordered trees have been researched by Courcelle in [5], where he presented a characterization of recognizable unranked, unordered tree languages by magmas.

In the last DLT conference 2005 a different approach to *dag* languages was presented by Priese [6]. Semi-rational and semi-regular *dag* languages have been introduced via Petri nets and merge grammars. The term "semi" indicates that a concept of infinity is involved. In the case of Petri nets this is the state space (set of reachable markings). Merge grammars are non-local, as two leaves of an arbitrary, unbounded distance may be merged.

Connections between true-concurrency Petri net behavior and graph grammars are known for over 20 years. Early papers are from Kreowsky [7], Reisig [8] - from 1981 - , Castellani and Montanari [9], Genrich, Janssen, Rozenberg and Thiagarajan [10], and Starke [11], all from 1983. Starke presents detailed proofs for a similarity

O.H. Ibarra and Z. Dang (Eds.): DLT 2006, LNCS 4036, pp. 351–362, 2006.

between his so called graph structure grammars and semi-word languages of free Petri nets without auto-concurrency. Grabowski introduced in [12] a concept ("Y-section") of synchronization of so-called partial words that became later better known as pomsets. The connection between pomsets and *dags* is very close: pomsets may be regarded as transitively closed *dags*. However, the essential difference between those well-known approaches to pomset semantics and the latter one to *dag* semantics is the use of ε-nodes in *dags* in [6].

It is shown in [6] that semi-rational *dag* languages coincide with semi-regular ones and with the synchronization closure of regular or recognizable sets of trees. Here, synchronization is a generalization of the Y-section of Grabowski.

However, no type of finite or infinite magma is known recognizing exactly the class of semi-rational *dag* languages. In the case of their existence, such magmas should also possess some feature of infinity, leading to a concept of "semi-recognizable" *dag* languages.

As a step towards such an algebraic characterization we present some examples and counterexamples for those languages to deepen their understanding.

2   DAGs with ε-Nodes

Σ denotes a finite, non-empty alphabet of *labels*, ε the empty word, empty graph, and also the 'empty label'. We assume that ε is not a symbol in the alphabet Σ and define $\Sigma_\varepsilon := \Sigma \cup \{\varepsilon\}$. An unranked and unordered *graph* γ over Σ is a triple $\gamma = (N, E, \lambda)$ of two finite sets N of *nodes* and $E \subseteq N \times N$ of *edges* and a labeling mapping $\lambda : N \to \Sigma$. We use the notation ${}^{\bullet}v := \{v' \in N | (v', v) \in E\}$, $v^{\bullet} := \{v' \in N | (v, v') \in E\}$. Two graphs $\gamma_i = (N_i, E_i. \lambda_i)$ are *isomorphic* if there exists a bijective mapping $h : N_1 \to N_2$ with $(v, v') \in E_1 \Leftrightarrow (h(v), h(v')) \in E_2$ and $\lambda_1(v) = \lambda_2(h(v))$ holds for all v, v' in N_1.

A graph over Σ_ε may possess nodes labeled with ε (ε-*nodes*). We treat ε-nodes like the empty word, where $uv = u\varepsilon v$ holds: Let α' result from a graph α over Σ_ε by removing one ε-node v and declaring all sons of v in α to be in α' the sons of each father of v in α. In this case α' is called a *reduct* of α. The ε-*equivalence* of graphs over Σ_ε is the reflexive, symmetric and transitive closure of the reduct-relation. Isomorphic and ε-equivalent graphs are identified. Thus, Figure 1 shows two representations of the same *dag*.

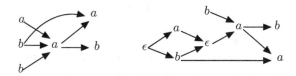

Fig. 1. Two representations of the same *dag*

Directed acyclic graphs (*dags*) and trees are special types of graphs. A pomset in concurrency theory is nothing else than a transitive closed *dag*. Reducing a tree over

Σ_ε may result in a forest over Σ. As we allow ε-nodes we have to be a little bit careful in defining concepts as an *in-degree, out-degree, level*, etc., in a *dag*. The corresponding nodes labeled with a or b in the two *dags* in Figure 1 may possess different levels, in-degrees, or out-degrees. Let α be a *dag* over Σ_ε then $_\varepsilon\alpha$ denotes some ε-free representative of α. The level of a node v in α is the length of the maximal path from a root to v in $_\varepsilon\alpha$. The *level i* of α is the set of all nodes of level i. Also, we define the *In-degree* or *Out-degree* of a node v in α as the in- or out-degree of v in a representative $_\varepsilon\alpha$. These definitions are independent of the chosen representative. We write $\underline{In}$- or $\underline{Out}$-degree whenever we refer to the degrees of an ε-free representative. The allowance of ε-nodes is essential in this paper. Without such a treatment of ε-nodes the set of all (unranked, unordered !) trees over $\{a\}$ could not be generated by a finite regular grammar.

A *Petri net* $\mathcal{N}$ over Σ is a tuple $\mathcal{N} = (P, T, \mathcal{F}, \phi, s_0, F)$ of two finite sets P of *places* and T of *transitions*, $P \cap T = \emptyset$, a multi set $\mathcal{F}$ over $T \times P \cup P \times T$ of *arcs*, a *labeling mapping* $\phi : T \to \Sigma_\varepsilon$, an *initial* state $s_0 \in \mathbb{N}^P$, and a finite set $F \subseteq \mathbb{N}^P$ of *terminal* states. A *state* s of $\mathcal{N}$ is a multi set over P. A transition t with $\phi(t) = a \in \Sigma_\varepsilon$ is also called an a-transition. This general concept of a Petri net includes the possibility of ε-transitions, unbounded places, multiple arcs, and auto-concurrency.

The standard interleaving semantics of a Petri net is given by the labelings of all firing sequences from the initial to some terminal state. In a true concurrency semantics firing sequences are replaced by processes. A process π is a possible concurrent execution of a firing sequence. It is a *dag* over $P \cup T$, with an in- and out-degree ≤ 1 for the P-nodes. The roots (leaves) of π are the multi sets s (s', respectively). A path from a root to a leaf is the life span of a single token in $s \Rightarrow^x s'$. The (terminal) *process semantics* $P_t(\mathcal{N})$ of a Petri net $\mathcal{N}$ is given by all processes from the initial to some terminal state. The (terminal) *dag semantics* $D_t(\mathcal{N})$ of $\mathcal{N}$ is $h_\lambda(D_t(\mathcal{N}))$. $h_\lambda(\pi)$ is an abstraction of the process π where h_λ maps all labels from P into ε and any label $t \in T$ into $\phi(t)$. Thus, all places are dropped and all transitions are replaced by their labels.

A *dag* language D is *semi-rational* if $D = D_t(\mathcal{N})$ holds for some Petri net $\mathcal{N}$. In this case we also say that $\mathcal{N}$ *accepts* D.

3   A Connection Between Rational Word and Semi-rational DAG Languages

$\Sigma^\dagger$ denotes the set of all *dags* over Σ. For $\alpha \in \Sigma^\dagger$ we denote by $path(\alpha)$ the set of all (labeling of) paths from some root to some leaf in α. For $D \subseteq \Sigma^\dagger$ $path(D) := \bigcup_{\alpha \in D} path(\alpha)$ is the *projection* of D from $\Sigma^\dagger$ into Σ^*. Vice versa, for $L \subseteq \Sigma^*$ $D_L := \{\alpha \in \Sigma^\dagger | path(\alpha) \subseteq L\}$ is the *embedding* of L from Σ^* into $\Sigma^\dagger$. We shall study here a property of embedding.

Let $A = (S, \Sigma, \Delta, s_1, F)$ be a finite ε-automaton over an alphabet Σ with a finite set S of states, an initial state $s_1 \in S$, a set $F \subseteq S$ of final states, and a next-state relation $\Delta \subseteq S \times \Sigma_\varepsilon \times S$, accepting the language $L_A := \{w \in \Sigma^* \mid \Delta^*(s_1, w) \cap F \neq \emptyset\}$. To A we define the *corresponding Petri net* $\mathcal{N}_A$ to consist of S as places, a transition $t_{s,a,s'}$ labeled with a and one arc from place s and one to place s' for any s, s', a with $s' \in \Delta(s, a)$, plus a further transition $t'_{s,\varepsilon}$ labeled with ε with an arc from s, but without any outgoing arc, for any terminal state $s \in F$. The initial state of $\mathcal{N}_A$ is $1 \cdot s_1$ and the empty marking 0 is the only terminal state. Obviously, the interleaving semantics of $\mathcal{N}_A$ is $L(A)$.

An *i-j-a-loop* at a place p in a Petri net $\mathcal{N}$ is a transition labeled with a with exactly i arcs from p and exactly j arcs to p. The *ε-completion* $\mathcal{N}^\varepsilon$ of a Petri net $\mathcal{N}$ is defined by adding an additional 1-2-ε- and 2-1-ε-loop to each place in $\mathcal{N}$. For readability, we present a place with an attached 1-2-ε- and 2-1-ε-loop by a circle with a black bottom and drop in the graphical presentation both loops. Figure 2 visualizes this.

It is rather obvious that $D_t(\mathcal{N}_A^\varepsilon) \subseteq L(A)$ holds for any finite automaton A. However, $\supseteq$ doesn't hold in general. Figure 3 also presents the Petri net $\mathcal{N}_{A_1}^\varepsilon$ with $1 \cdot p_1$ as initial and 0 as terminal state.

Fig. 2. Graphical representation of a place with a 1-2-ε and 2-1-ε-loop

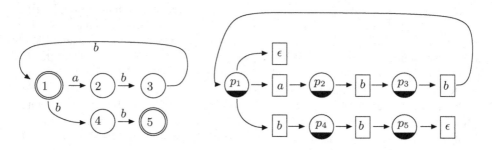

Fig. 3. An automaton A_1 and the $\mathcal{N}_{A_1}^\varepsilon$

Nevertheless, the following theorem holds:

Theorem 1. $L \subseteq \Sigma^*$ *rational* $\implies$ D_L *semi-rational*.

Sketch of Proof. Let L be a regular language over some alphabet Σ. Let A_L be the minimal deterministic and complete ε-free finite automaton accepting L. The Petri net $\mathcal{N}_{A_L}^\varepsilon$ will in general not accept D_L. We have to modify A_L.

Without restriction we assume that for any ε-automaton $E = (S, \Sigma, \Delta, s_1, F)$ $|\Delta(s, a)| = 1$ holds for $a \in \Sigma, s \in S$. Thus, nondeterminism is restricted to ε-transitions.

Two different states s_1, s_2 of some finite ε-automaton E are called *w-equivalent* for some word $w \in \Sigma^*$ if a path from s_1 labeled with w can reach a terminal state and a possibly different path from s_2 also labeled with w can reach some terminal state. The *weak completion* E^c of E is a finite ε-automaton where from any set $M \subseteq S$ of mutually *w*-equivalent states there exists an ε-transition from each $s' \in M$ to a common path $p_{M,w}$ with the labeling w to a terminal state. This weak completion may be easily constructed as follows:

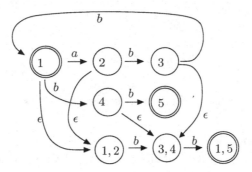

Fig. 4. The weak completion A_1^c of A_1

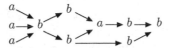

Fig. 5. A *dag* with its paths in L_1, recognized by $\mathcal{N}^\varepsilon_{A_1^c}$ but not by $\mathcal{N}^\varepsilon_{A_1}$

Set $\hat{E} := (2^S, \Sigma, \hat{\Delta}, \{s_1\}, 2^F)$, where for all $M, M' \subseteq S, a \in \Sigma$ there holds:
$\hat{\Delta}(M, a) := \bigcup_{s \in M} \Delta(s, a)$,
$M' \in \hat{\Delta}(M, \varepsilon)$ iff $M \subseteq M'$.
Set $E^c := \hat{E}^t$, the trim sub-automaton of $\hat{E}$, consisting only of the reachable and co-reachable states. Note, $|\Delta(M, a)| = 1$ holds for all $M \subseteq S, a \in \Sigma$.

Now, $D_L = D_t(\mathcal{N}^\varepsilon_{A_L^c})$ holds in general. ∎

Figure 4 shows the weak completion of A_1 in Figure 3. For readability we dropped the (unnecessary but harmless) ε-arcs from 1 and 5 to state $\{1, 5\}$.

Obviously, the *dag* in Figure 5 is in not in $D_t(\mathcal{N}^\varepsilon_{A_1})$ but in $D_t(\mathcal{N}^\varepsilon_{A_1^c})$.

We suppose that the opposite is also true, namely:

D semi-rational $\implies L$ rational.

However, we have no proof. A construction similar to the previous one will not work as the set of reachable states in a Petri net is in general infinite. On the other hand, Petri nets accepting any *dag* whose paths are in L seem to be so restrictive that one might be able to construct a finite automaton accepting *path*(D). Some kind of normal form for such "path-closed" Petri nets would be welcome.

4 Further Examples of Semi-rational DAG Languages

4.1 Topological Constraints

We firstly examine some sets of *dags* with constraints on the neighborhood of nodes. Thus, we are not interested in the names $\lambda(v)$ of a node v and will operate with the trivial alphabet $\{a\}$.

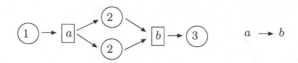

Fig. 6. a, b with different in- or out-degrees in a process and its *dag*

Fixed In- and Out-Degree

We regard the *dag* language $D^{i,o}$ of all *dags* over $\{a\}$ whose nodes possess only In-degree i or 0 and Out-degree o or 0.

There is a difficulty in finding a Petri net accepting $D^{i,o}$ as the topological structures of a process, even without ε-transitions, and its ε-free *dag* may be rather different. Figure 6 presents an example.

Thus, to get a node with Out-degree 2 in some *dag* the 2 token generated from a corresponding transition in the Petri net must not be both consumed by a single transition. To get In-degree 2 in some *dag* the two token consumed by a corresponding transition in the Petri net must not be generated by a single transition. This situation leads to a combinatorial explosion and we only know of a Petri net of size $O(2^{i+o})$ that accepts $D^{i,o}$.

Theorem 2. $D^{i,o}$ *is semi-rational for all values of i and o.*

Sketch of Proof. This is trivial for $i = 1$ or $o = 1$. Thus, $i \geq 2$ and $o \geq 2$. We define the Petri net $\mathcal{N}^{i,o} := (P, T, \mathcal{F}, \phi, s_0, F)$ over $\{a\}_\varepsilon$ by setting

$P := \{s, m, r_1, \ldots, r_x\}$, for $x := i \cdot (o - 1) + 1$,

$T := \{S_\epsilon, S_d, SM, M_\epsilon, E_1, \ldots, E_x, A_1, \ldots, A_y\}$, for $y = (x - i)! \cdot x! \, / \, i!$,

$\mathcal{F} := \mathcal{F}_1 + \mathcal{F}_2$, with $\mathcal{F}_1 := 1 \cdot (s, S_\epsilon) + 1 \cdot (s, S_d) + 2 \cdot (S_d, s) + 1 \cdot (s, SM) + 1 \cdot (SM, m) + 1 \cdot (m, M_\epsilon) + \sum_{1 \leq j \leq x}(m, E_j) + \sum_{1 \leq j \leq y}(A_j, m) + o \cdot \sum_{1 \leq j \leq x}(E_j, r_j)$
and $\mathcal{F}_2$ is defined below.

$\phi(X) := a$, for $= X \in \{A_i | 1 \leq i \leq y\} \cup \{SM\}$ and

$\phi(X) := \varepsilon$ for all other transitions X,

$s_0 := 1 \cdot s$, and

$F := \{0\}$.

By 2_i^M we denote the set of all subsets of M that contain exactly i elements. As $|2_i^{\{r_1, \ldots, r_x\}}|$
$= y$ holds we can choose a bijection f from $\{r_1, \ldots, r_x\}$ into $\{A_1, \ldots, A_y\}$. Now, $\mathcal{F}_2$ defines an arc from any element of the j-th set in $2_i^{\{r_1, \ldots, r_x\}}$ to the transition $A_{f(j)}$.

Figure 7 shows the Petri net for $i = o = 2$.

The 1-2-ε-loop at s can create any number of tokens on s. Each of them leads to a root in the *dag* by a firing of SM and is transferred on place m in the Petri net. Any transition A_j puts exactly o token on the j-th place r_j in the *ring* r_1 to r_x. Thus, an a-transition in the Petri net that has created a token t on place m, and t is fired by an E_j-transition, leads to a node in the *dag* with an Out-degree o. A firing of one of the transitions A_1 to A_y removes exactly i token from i different ring places. Each of those i token on different ring places was generated by a different transition in the Petri net. Thus, a node with an In-degree i is created in the *dag*. There must be one A_j transition for every combination of i different places in the ring. Thus, $D_t(\mathcal{N}^{i,o}) \subseteq D^{i,o}$ is rather obvious.

$D^{i,o} \subseteq D_t(\mathcal{N}^{i,o})$: We firstly present an algorithm that adds to any node of a given ε-free representative $_\varepsilon\alpha$ of a *dag* $\alpha \in D^{i,o}$ a label from $\{\varepsilon, 1, ..., x\}$ in such a way that no node v in $_\varepsilon\alpha$ possesses two father nodes with the same label. v is labeled with ε iff it is a leaf. Let $_\varepsilon\alpha$ possess the levels 0 up to n. The algorithm is simple:

Step 1: Add label ε to all nodes with out-degree 0.
Step 2: Create a list *blacklist(v)* with empty content for any node v in α. (*blacklist(v)* tells which labels are forbidden for v.)
Step 3: For all i down from $n - 1$ to 0 do
        for all nodes v of level i do
            add a label $l(v)$ not in *blacklist(v)* to v
                for all sons v' of v do
                    for all fathers v'' of v' do
                        add $l(v)$ to *blacklist(v'')*.

Any node v has 0 or i sons, and any of its sons has exactly $o - 1$ fathers different from v. Thus $x = i \cdot (o - 1) + 1$ labels are sufficient.

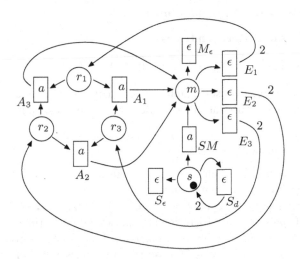

Fig. 7. Petri net $\mathcal{N}^{2,2}$ for $D^{2,2}$

We now can generate α by a process π of $\mathcal{N}^{i,o}$: A node v labeled with ε in α results from a firing of M_ε in π. A node v labeled with j results from a firing of A_j. Add in α the label r_j to all arcs leaving a node labeled with j. This indicates that a firing of A_j puts all o token on the place r_j. As any node $v \in \alpha$ possesses no or differently labeled father nodes all its incoming arcs possess different labels. Thus, v is generated in π by a firing of $E_{l(v)}$ and all token required to fire $E_{l(v)}$ are generated by different transitions

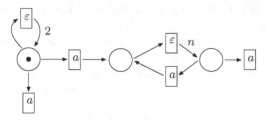

Fig. 8. Petri net for $D^{1,mod\,n}$

in $\mathcal{N}^{i,o}$. Thus, the situation as described in Figure 6 doesn't apply and the abstraction of the process π is $_\varepsilon\alpha$ with the correct number of in- and outgoing arcs. ∎

Out-Degree Modulo a Constant

$D^{c,mod\,n} := \{\alpha \in \{a\}^\dagger | \alpha = (N, E, \lambda) \wedge \forall v \in N(|v^\bullet| \bmod n = 0 \wedge (|^\bullet v| = c \vee |^\bullet v| = 0))\}$ defines the language of all *dag*s over a single letter with an In-degree 0 or c and an Out-degree 0 modulo n. Figure 8 shows a Petri net accepting $D^{1,mod\,n}$. Thus

Theorem 3. $D^{1,mod\,n}$ *is semi-rational.*

However, it seems that $D^{c,mod\,n}$ is no longer semi-rational for $c > 1$.

4.2   Constraints on Topology and Alphabet

Number of Labels for Fathers and Sons

$$D_s^f := \{\alpha \in \Sigma_\varepsilon^\dagger |_\varepsilon \alpha = (N, E, \lambda) \implies \forall v \in N(|\lambda(^\bullet v)| \leq 1 \wedge |\lambda(v^\bullet)| \leq 1)\}$$

defines the *dag* language over some alphabet Σ_ε where in any ε-free representant of a *dag* any node possesses only fathers with a common label and only sons also with a common label.

Theorem 4. D_s^f *is semi-rational.*

We present an example. The Petri net $\mathcal{N}_s^f$ of Figure 9 accepts D_s^f over the alphabet $\{a, b, c\}$ with 0 as initial and terminal state.

It is easily seen that $\mathcal{N}_s^f$ accepts D_s^f over $\{a, b, c\}$. A token on a place $p_{x,y}$ for $x, y \in \{a, b, c\}$ results from firing an x-transition and allows only the firing of an y-transition. The 2-1-ε-loop at $p_{x,y}$ can melt thus only token resulting from x-transitions. Thus, any ε-free representant of a *dag* in $D_t(\mathcal{N}_s^f)$ possesses only nodes with all fathers of a unique label. The 1-2-ε-loop at $p_{x,y}$ can multiply a token only into further token that can only fire a y-transition, therefor any ε-free representant of a *dag* in $D_t(\mathcal{N}_s^f)$ possesses only nodes with fathers of a unique label and sons of a unique label. Obviously, $\mathcal{N}_s^f$ can generate any *dag* in D_s^f. ∎

Number of Labels for Fathers

$$D_f := \{\alpha \in \{a, b, c\}^\dagger |_\varepsilon \alpha = (N, E, \lambda,) \implies \forall v \in N((\lambda(v) \in \{a, b\} \Leftrightarrow |^\bullet v| = |v^\bullet| \leq 1)$$

$$\wedge (\lambda(v) = c \Rightarrow |\{v' \in {}^\bullet v \mid \lambda(v') = a\}| = |\{v' \in {}^\bullet v \mid \lambda(v') = b\}|))\}.$$

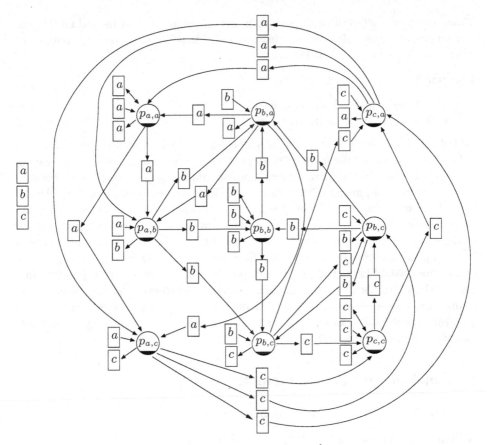

Fig. 9. A Petri net accepting D_s^f

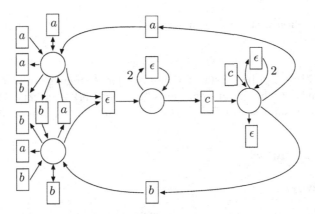

Fig. 10. Petri net for D_f

is the language of all of those *dags* that possess only nodes a, b of In- and Out-degree ≤ 1 and that possess only ε-free representatives whose nodes have the same number of fathers labeled with a as labeled with b.

Theorem 5. D_f *is semi-rational.*

Sketch of Proof. The Petri net of Figure 10
   with 0 as initial and terminal state accepts D_f. ■

With the same idea it should be possible to construct a Petri net for versions of D_f with a weaker restriction on the In- and Out-degrees of a- and b-nodes. However, if we put no restriction on those degrees the languages should not stay semi-rational. We have no proof, only an argument: The In-degree of c-nodes is not bounded, nevertheless, it has to be ensured that the numbers of ingoing arcs from a- and from b-nodes are equal. Thus, the Petri net must not send to any c-transition two token that result from a firing of the same a-transition and reach c only via some further ε-transitions (Condition $2a \nrightarrow c$). In a process those two paths from one a-transition to the c-transition via ε-transitions would be melted into a single arc from a to c in the corresponding *dag*, compare Figure 6. Thus, even if an equal number of paths from a- and from b-transitions to any c-transition is ensured in a process, it might no longer be true in the corresponding *dag*. But how should a Petri net ensure condition $2a \nrightarrow c$ if the Out-degree of the a-transitions is also unbounded?

5   Candidates for Counterexamples

5.1   Number of Fathers Equals Number of Sons

With the same argument as in 4.2 we are convinced that

$$D_{f=s} := \{\alpha \in \{a\}^{\dagger}|_{\varepsilon} \alpha = (N, E, \lambda) \wedge v \in N \implies |{}^{\bullet}v| = |v^{\bullet}|\}$$

cannot be semi-rational, but we have no proof.

5.2 $a^n b^n$

Let $L := \{a^n b^n | n \in \mathbb{N}\}$ be the standard non-regular cf language of words. It is well-known that L and even the non-cf language $\{a^n b^n c^n | n \in \mathbb{N}\}$ are accepted by Petri-nets. However, we again suppose that $D_L = \{\alpha \in \{a, b\}|path(\alpha \subseteq L)\}$ is not semi-rational.

6   Counterexamples

6.1   Connectivity

An undirected path in a *dag* is a sequence of nodes $v_1, ..., v_n$ with $(v_i, v_{i+1}) \in E$ or $(v_{i+1}, v_i) \in E$ holds for $1 \leq i < n$. A *dag* is *connected* if any two nodes can be connected by an undirected path.

Theorem 6. $D_{conn} := \{\alpha \in \{a\}^{\dagger}_{\varepsilon}|\alpha$ *is connected*$\}$ *is not semi-rational.*

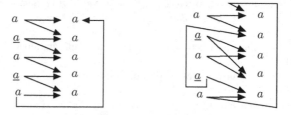

Fig. 11. Two *dags* resulting from $\mathcal{N}_c$

Sketch of Proof. We define for $n \in \mathbb{N}$ a cyclic *dag* $\alpha_{cyc,n}$ of length $2n$ in D_{conn} over $\{a\}$ as

$\alpha_{cyc,n} = (N_{cyc,n}, E_{cyc,n}, \lambda_{cyc,n})$ with
$N_{cyc,n} = \{v_{i,j} | i \in \{1, ..., n\} \wedge j \in \{1, 2\}\}$
$E_{cyc,n} = \{(v_{i_1,1}, v_{i_2,2}) | i_1 = i_2 \vee i_2 = (i_1 + 1) mod\ n\}$.

Let us assume a Petri net $\mathcal{N}_c$ with m transitions labeled with a accepts D_{conn}. In $\alpha_{cyc,n}$ for n large enough there must be a transition t such that a process π for $\mathcal{N}_c$ generating $\alpha_{cyc,n}$ must possess four nodes v_i, v_i', v_j, v_j' where v_i' is a son of v_i, v_i' is a son of v_j, and v_i, v_j possess the same label t, and also v_i', v_j' possess the same label, say p', a place of $\mathcal{N}_c$. There results a second process π' from π where v_i gets v_j' as a son and v_j gets v_i'. π' also is a correct process for $\mathcal{N}_c$. However, π' now generates an unconnected *dag*. Figure 11 illustrates the situation. We assume the underlined labels to result from the same transition. ∎

7   Conclusion

The paper presents some example *dag* languages with rather different local and global properties. We have seen that some of these properties can easily be recognized by a petri net, while others can not. This should give us an idea for the characteristics of semi-rational languages and help us to proceed in examining further concepts.

References

1. W. Thomas. Finite-state recognizability of graph properties. In D. Krob, editor, *Theorie des Automates et Applications*, volume 172, pages 147–159. l'Université de Rouen, France, 1992.
2. T. Kamimura, G. Slutzki. Parallel and two-way automata on directed ordered acyclic graphs. *Inf. Control*, 49:10–51, 1981.
3. H. Comon, M. Daucher, R. Gilleron, S. Tison, M. Tommasi. Tree automata techniques and application. Available on the Web from 13ux02.univ-lille.fr in directoty tata, 1998.
4. A. Brüggemann-Klein, M. Murata, D. Wood. Regular tree and hedge languages of unranked alphabets. Theor. Comp. Science Center Report HKUST-TCSC 2001-5,pp29, 2001.
5. B. Courcelle. On recognizable sets and tree automata. In H. Aït-Kaci, M. Nivat, editor, *Resolution of Equations in Algebraic Structures*, volume 1, pages 93–126. Academic Press, 1989.

6. Lutz Priese. Semi-rational sets of dags. In *Developments in Language Theory: 9th International Conference, DLT 2005, Palermo, Italy*, LNCS 3572, page 385. Springer Verlag, 6 2005.
7. H.J. Kreowski. A comparison between Petri-nets and graph grammars. *LNCS 100*, pages 306–317, 1981.
8. W. Reisig. A graph grammar representation of nonsequential processes. *LNCS 100*, pages 318–325, 1981.
9. I. Castellani, H. Montanara. Graph grammars for distributed systems. *LNCS 153*, pages 20–38, 1983.
10. H.J. Genrich, D. Janssen, G. Rozenberg, P.S. Thiagarajan. Petri nets and their relation to graph grammars. *LNCS 153*, pages 15–129, 1983.
11. P. Starke. Graph grammars for Petri net processes. *EIK*, 19:199–233, 1983.
12. J. Grabowski. On partial languages. *Annales Societatis Mathematicas Polonae, Fundamenta Informaticae IV.2*, pages 428–498, 1981.

Finding Lower Bounds for Nondeterministic State Complexity Is Hard

(Extended Abstract)

Hermann Gruber and Markus Holzer

Institut für Informatik, Technische Universität München,
Boltzmannstraße 3, D-85748 Garching bei München, Germany
{gruberh, holzer}@informatik.tu-muenchen.de

Abstract. We investigate the following lower bound methods for regular languages: The fooling set technique, the extended fooling set technique, and the biclique edge cover technique. It is shown that the maximal attainable lower bound for each of the above mentioned techniques can be algorithmically deduced from a canonical finite graph, the so called *dependency graph* of a regular language. This graph is very helpful when comparing the techniques with each other and with nondeterministic state complexity. In most cases it is shown that for any two techniques the gap between the best bounds can be arbitrarily large. Moreover, we show that deciding whether a certain lower bound w.r.t. one of the investigated techniques can be achieved is in most cases computationally hard, i.e., PSPACE-complete and hence is as hard as minimizing nondeterministic finite automata.

1 Introduction

Finite automata are one of the oldest and most intensely investigated computational models. It is well known that deterministic and nondeterministic finite automata are computationally equivalent, and that nondeterministic finite automata can offer exponential state savings compared to deterministic ones [16]. Nevertheless, some challenging problems of finite automata are still open. For instance, to estimate the size, in terms of the number of states, of a minimal nondeterministic finite automaton for a regular language is stated as an open problem in [1] and [9]. This is contrary to the deterministic case, where for a given n-state deterministic automaton the minimal automaton can be efficiently computed in $O(n \log n)$ time. Observe that computing a state minimal nondeterministic finite automaton is known to be PSPACE-complete [12].

Several authors have introduced communication complexity methods for proving such lower bounds; see, e.g., [3, 6, 8]. Although the bounds provided by these techniques are not always tight and in fact can be arbitrarily worse compared to the nondeterministic state complexity, they give good results in many cases. In this paper we investigate the fooling set technique [6], the extended fooling set technique [3, 8], and the biclique edge cover technique. Note that the latter method is an alternative representation of the nondeterministic message

O.H. Ibarra and Z. Dang (Eds.): DLT 2006, LNCS 4036, pp. 363–374, 2006.
© Springer-Verlag Berlin Heidelberg 2006

complexity [8]. One drawback of all these methods is that getting such a good estimate seems to require conscious thought and "clever guessing." However, we show for the considered techniques that this is in fact *not* the case. In order to achieve this goal, we present a unified view of these techniques in terms of bipartite graphs. This setup allows us to show that there is a canonical bipartite graph for each regular language, which is independent of the considered method, such that the best attainable lower bound can be determined algorithmically for each method. This canonical bipartite graph is called the *dependency graph* of the language.

The dependency graph is a tool that allows us to compare the relative strength of the methods, and to determine whether they provide a guaranteed relative error w.r.t. the nondeterministic state complexity. Following [1], no lower bound technique is known to have such a bounded error, but a lower bound can be obtained by noticing that the numbers of states in minimal deterministic automata and in minimal nondeterministic automata are at most exponentially apart from each other. We are able to prove that the biclique edge cover technique always gives an estimate at least as good as this trivial lower bound, whereas the other methods cannot provide any guaranteed relative error. On the other hand, we give evidence that the guarantee for the biclique edge cover technique is essentially optimal. In turn we improve a result of [10, 13] on the gap between nondeterministic message complexity and nondeterministic state complexity.

Finally, we also address computational complexity issues and show that deciding whether a certain lower bound w.r.t. one of the investigated techniques can be achieved is in most cases computationally hard, i.e, PSPACE-complete and hence these problems are as hard as minimizing nondeterministic finite automata. Here it is worth mentioning that the presented algorithms for the upper bounds also rely on the dependency graph, whose vertices are the equivalence classes of the Myhill-Nerode relation for the language L and its reversal L^R. Hence, doing the computation on this object in a straightforward manner would result in an exponential space algorithm. This is due to the fact that the index of the Myhill-Nerode equivalence relation for L^R can be exponential in terms of the index of the Myhill-Nerode relation for L, or equivalently to the size of the minimal deterministic finite automaton accepting L. Nevertheless, by cleverly encoding the equivalence classes we succeed in implicitly representing the dependency graph, which finally results in PSPACE-algorithms for the problems under consideration.

2   Definitions

We assume the reader to be familiar with the basic notations in formal language and automata theory as contained in [7]. In particular, let Σ be an alphabet and Σ^* the set of all words over the alphabet Σ, including the empty word λ. The length of a word w is denoted by $|w|$, where $|\lambda| = 0$. The reversal of a word w is denoted by w^R and the reversal of a language $L \subseteq \Sigma^*$ by L^R, which equals the set $\{ w^R \mid w \in L \}$.

A *nondeterministic finite automaton* is a 5-tuple $A = (Q, \Sigma, \delta, q_0, F)$, where Q is a finite set of states, Σ is a finite set of input symbols, $\delta : Q \times \Sigma \to 2^Q$ is the transition function, $q_0 \in Q$ is the initial state, and $F \subseteq Q$ is the set of accepting states. The transition function δ is extended to a function from $\delta : Q \times \Sigma^* \to 2^Q$ in the natural way, i.e., $\delta(q, \lambda) = \{q\}$ and $\delta(q, aw) = \bigcup_{q' \in \delta(q,a)} \delta(q', w)$, for $q \in Q$, $a \in \Sigma$, and $w \in \Sigma^*$. The *language accepted* by the finite automaton A is $L(A) = \{ w \in \Sigma^* \mid \delta(q_0, w) \cap F \neq \emptyset \}$. Two automata are equivalent if they accept the same language. A nondeterministic finite automaton $A = (Q, \Sigma, \delta, q_0, F)$ is *deterministic*, if $|\delta(q, a)| = 1$ for every $q \in Q$ and $a \in \Sigma$. In this case we simply write $\delta(q, a) = p$ instead of $\delta(q, a) = \{p\}$. By the powerset construction one can show that every nondeterministic finite automaton can be converted into an equivalent deterministic finite automaton by increasing the number of states from n to 2^n; this bound is known to be sharp [15]. Thus, deterministic and nondeterministic finite automata are equally powerful.

For a regular language L, the deterministic (nondeterministic, respectively) state complexity of L, denoted by $sc(L)$ ($nsc(L)$, respectively) is the minimal number of states needed by a deterministic (nondeterministic, respectively) finite automaton accepting L. Observe that the minimal deterministic finite automaton is isomorphic to the deterministic finite automaton induced by the Myhill-Nerode equivalence relation $\equiv_L$, which is defined as follows: For $u, v \in \Sigma^*$ let $u \equiv_L v$ if and only if $uw \in L \iff vw \in L$, for all $w \in \Sigma^*$. Hence, the number of states of the minimal deterministic finite automaton accepting the language $L \subseteq \Sigma^*$ equals the index, i.e., the cardinality of the set of equivalence classes, of the Myhill-Nerode equivalence relation $\equiv_L$. The set of all equivalence classes w.r.t. $\equiv_L$ is referred to $\Sigma^*/\equiv_L$ and we denote the equivalence class of a word u w.r.t. the relation $\equiv_L$ by $[u]_L$. Moreover, we define the relation $_L\equiv$ as follows: For $u, v \in \Sigma^*$ let $u _L\equiv v$ if and only if $wu \in L \iff wv \in L$, for all $w \in \Sigma^*$. The set of all equivalence classes w.r.t. $_L\equiv$ is referred to $\Sigma^*/_L\equiv$ and we denote the equivalence class of a word u w.r.t. the relation $_L\equiv$ by $_L[u]$.

Finally, we recall two remarkably simple lower bound techniques for the nondeterministic state complexity of regular languages. Both methods are commonly called *fooling set* techniques and were introduced in [3] and [6].

Theorem 1 (Fooling Set and Extended Fooling Set Technique). *Let $L \subseteq \Sigma^*$ be a regular language and suppose there exists a set of pairs $S = \{ (x_i, y_i) \mid 1 \leq i \leq n \}$ with the following properties:*

1. *If (i) $x_i y_i \in L$ for $1 \leq i \leq n$, (ii) $x_i y_j \notin L$, for $1 \leq i, j \leq n$, and $i \neq j$, then any nondeterministic finite automaton accepting L has at least n states, i.e., $nsc(L) \geq n$. Here S is called a* fooling set *for L.*
2. *If (i) $x_i y_i \in L$ for $1 \leq i \leq n$, (ii) and $i \neq j$ implies $x_i y_j \notin L$ or $x_j y_i \notin L$, for $1 \leq i, j \leq n$, then any nondeterministic finite automaton accepting L has at least n states, i.e., $nsc(L) \geq n$. Here S is called an* extended fooling set *for L.*

Note that the lower bounds provided by these techniques are not always tight and in fact can be arbitrarily bad compared to the nondeterministic state complexity. Nevertheless, they give good results in many cases—for the fooling set technique see the examples provided in [6].

3   Lower Bound Techniques and Bipartite Graphs

In this section we develop a unified view of fooling sets and extended fooling sets in terms of bipartite graphs and introduce a technique that leverages the shortcomings of the fooling set techniques. We need some notations from graph theory.

A *bipartite graph* is a 3-tuple $G = (X, Y, E)$, where X and Y are the (not necessarily finite, or disjoint) sets of vertices, and $E \subseteq X \times Y$ is the set of edges. A bipartite graph $H = (X', Y', E')$ is a *subgraph* of G, if $X' \subseteq X$, $Y' \subseteq Y$, and $E' \subseteq E$. The subgraph H' is *induced* if $E' = (X' \times Y') \cap E$. Given a set of edges E', the *subgraph induced* by E' w.r.t. E is the smallest induced subgraph containing all edges in E'.

The relation between fooling sets and graphs is quite natural, because a (extended) fooling set S can be interpreted as the edge set of a bipartite graph $G = (X, Y, S)$ with $X = \{ x \mid$ there is a y such that $(x, y) \in S \}$ and $Y = \{ y \mid$ there is a x such that $(x, y) \in S \}$. In case S is a fooling set, the induced bipartite graph is nothing other than a ladder, i.e., a collection of pairwise vertex-disjoint edges. More generally, the notation of (extended) fooling sets carries over to bipartite graphs as follows: Let $G = (X, Y, E)$ be a bipartite graph.

1. Then a set $S \subseteq E$ is a fooling set for G, if for every two different edges e_1 and e_2 in S, the subgraph *induced* by the edges e_1 and e_2 w.r.t. E is the rightmost graph of Figure 1,
2. and a set $S \subseteq E$ is an extended fooling set for G, if for every two different edges e_1 and e_2 in S, the subgraph *induced* by the edges e_1 and e_2 w.r.t. E is one of the graphs depicted in Figure 1.

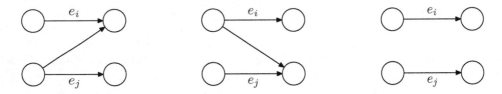

Fig. 1. Three important bipartite (sub)graphs

Now let us associate to any language $L \subseteq \Sigma^*$ and sets $X, Y \subseteq \Sigma^*$ a bipartite graph $G = (X, Y, E_L)$, where $(x, y) \in E_L$ if and only if $xy \in L$, for every $x \in X$ and $y \in Y$. Then it is easy to see that the following statement holds—we omit the straight forward proof.

Theorem 2. *Let $L \subseteq \Sigma^*$ be a regular language. Then the set S is a fooling set for L if and only if the edge set $S \subseteq E_L$ is a fooling set for the bipartite graph $G = (\Sigma^*, \Sigma^*, E_L)$. An analogous statement holds for extended fooling sets.* □

For the lower bound technique to come we need the notion of a biclique edge cover for bipartite graphs. Let $G = (X, Y, E)$ be a bipartite graph. A set $C =$

$\{H_1, H_2, \ldots\}$ of non-empty bipartite subgraphs of G is an *edge cover* of G, if every edge in G is present in at least one subgraph. An edge cover C of the bipartite graph G is a *biclique edge cover* if every subgraph in C is a biclique, where a *biclique* is a bipartite graph $H = (X, Y, E)$ satisfying $E = X \times Y$. The *bipartite dimension* of G is denoted $d(G)$ and is defined to be the size of the smallest biclique edge cover of G if it exists and is infinite otherwise. Then the biclique edge cover technique reads as follows—this technique is a reformulation of the nondeterministic message complexity method [8] in terms of graphs:

Theorem 3 (Biclique Edge Cover Technique). *Let $L \subseteq \Sigma^*$ be a regular language and suppose there exists a bipartite graph $G = (X, Y, E_L)$ with $X, Y \subseteq \Sigma^*$ (not necessarily finite) for the language L. Then any nondeterministic finite automaton accepting L has at least as many states as the bipartite dimension of G, i.e., $nsc(L) \geq d(G)$.*

Proof. Let $A = (Q, \Sigma, \delta, q_0, F)$ be any nondeterministic finite automaton accepting L. We show that every finite automaton induces a finite size biclique edge cover of the bipartite graph G. For each state $q \in Q$ let $H_q = (X_q, Y_q, E_q)$ with $X_q = X \cap \{ w \in \Sigma^* \mid \delta(q_0, w) \ni q \}$, $Y_q = Y \cap \{ w \in \Sigma^* \mid \delta(q, w) \cap F \neq \emptyset \}$, and $E_q = X_q \times Y_q$. We claim that $C = \{ H_q \mid q \in Q \}$ is a biclique edge cover for G. By definition each H_q, for $q \in Q$, is a biclique. Moreover, each bipartite graph H_q is a subgraph of G. Since by construction $X_q \subseteq X$ and $Y_q \subseteq Y$ it remains to show that $E_q \subseteq E_L$. To this end assume that $x \in X_q$ and $y \in Y_q$. Then the word xy belongs to the language L because $q \in \delta(q_0, x)$ and $\delta(q, y) \cap F \neq \emptyset$. But then (x, y) is an edge of G. Finally, we must prove that C is an edge cover. Let (x, y) be an edge in G, for $x \in X$ and $y \in Y$. Then the word xy is in L and since the nondeterministic finite automaton A accepts the language L, there is a state q in Q such that $q \in \delta(q_0, x)$ and $\delta(q, y) \cap F \neq 0$. Therefore $x \in X_q$ and $y \in Y_q$ and moreover (x, y) is an edge in E_q, because H_q is a biclique. This proves that C is a biclique edge cover of G.

Now assume that there is a nondeterministic finite automaton accepting L whose number of states is strictly less than the bipartite dimension of G. Then this automaton induces a biclique edge cover C of G, whose size is bounded by the number of states and thus is also strictly less than the bipartite dimension of G. This is a contradiction because the bipartite dimension is defined to be the size of the smallest biclique edge cover. Therefore any nondeterministic finite automaton accepting L has at least as many states as the bipartite dimension of G. □

By the previous theorem we obtain the following corollary.

Corollary 4. *Let $L \subseteq \Sigma^*$ be a regular language. Then the bipartite graph $G = (\Sigma^*, \Sigma^*, E_L)$ has finite bipartite dimension.* □

4   The Dependency Graph of a Language

In applying the lower bound theorems from the previous section to any particular language it is necessary to choose pairs (x_i, y_i) or sets X and Y appropriately. For fooling sets a heuristic,[1] which of course also applies to the other techniques, was proposed in [6] and seems to work well in most cases. In fact, we show that such a heuristic is *not* needed. To this end we define the following bipartite graph:

Definition 5. *Let* $L \subseteq \Sigma^*$. *Then the* dependency graph *for the language* L *is defined to be the bipartite graph* $G_L = (X, Y, E_L)$, *where* $X = \Sigma^*/\equiv_L$ *and* $Y = \Sigma^*/_L\equiv$ *and* $([x]_L, {}_L[y]) \in E_L$ *if and only if* $xy \in L$.

It is easy to see that the dependency graph G_L for a language L is independent of the chosen representation of the equivalence classes. Hence all these graphs are isomorphic to each other. Moreover, it is worth mentioning that the dependency graph of a language was implicitly defined in [14]. Now we are ready to state the main lemma of this section.

Lemma 6. *Let* $L \subseteq \Sigma^*$ *be a regular language and* $G = (\Sigma^*, \Sigma^*, E_L)$ *its associated bipartite graph.*

1. *The maximum size of a (extended, respectively) fooling set for* G *is* n *if and only if the maximum size of a (extended, respectively) fooling set for the dependency graph* G_L *equals* n.
2. *The bipartite dimension of* G *is* n *if and only if the bipartite dimension of the dependency graph* G_L *equals* n.

Proof (Sketch). The idea is to replace the "left" vertices in the bipartite graph $G = (\Sigma^*, \Sigma^*, E_L)$ with the Myhill-Nerode equivalence classes $\Sigma^*/\equiv_L$, (see also, e.g., [8]), and *simultaneously* the "right" vertices with the classes $\Sigma^*/_L\equiv$. Then one can show that $S = \{(x_i, y_i) \mid 1 \leq i \leq n\}$ is a fooling set for G if and only if $\{([x_i]_L, {}_L[y_i]) \mid 1 \leq i \leq n\}$ is a fooling set for G_L, and that similar facts hold for the other techniques.   □

An immediate consequence of the previous lemma is that finding the best possible lower bound for the technique under consideration is indeed solvable in an algorithmic manner. For instance, a fooling set corresponds to an *induced matching* [5] in G_L, and an extended fooling set to a *cross-free matching* in G_L, and *vice versa*. Their associated decision problems are easily seen to be solvable in nondeterministic time polynomial in terms of the size of G_L. The drawback, however, is that this size can be exponential in terms of the state complexity of the deterministic finite automaton for the language [17].

5   How Good Are the Lower Bounds Induced by These Techniques?

We compare the introduced techniques with each other w.r.t. the lower bounds that can be obtained in the best case and to the nondeterministic state complexity.

---

[1] In [6] the following heuristic is proposed: "Construct a nondeterministic finite automaton $A = (Q, \Sigma, \delta, q_0, F)$ accepting L, and for each state q in Q let x_q be the shortest string such that $\delta(q_0, x_q) = q$, and let y_q be the shortest string such that $\delta(q, y_q) \cap F \neq \emptyset$. Then choose the set S to be some *appropriate subset* of the pairs $\{(x_q, y_q) \mid q \in Q\}$."

It is noted in [10] that the bipartite dimension generalizes the fooling set techniques and can sometimes yield exponentially better bounds than these. In fact we show that the gaps can be arbitrary, also between the fooling set technique and its simple extension. Due to the lack of space we omit the proof of the following theorem—the first statement is from [6].

Theorem 7. *There is a sequence of languages $(L_n)_{n \geq 1}$ such that the nondeterministic state complexity of L_n is at least n, i.e., $nsc(L_n) \geq n$, but any fooling set for L has size at most c, for some constant c. An analogous statement holds for extended fooling sets versus fooling sets, nondeterministic state complexity versus extended fooling sets, and the bipartite dimension versus extended fooling sets.*   □

As the reader may have noticed, the comparison between bipartite dimension and nondeterministic state complexity is missing in the previous theorem. The following theorem shows that the bipartite dimension of a regular language is a measure of descriptional complexity.

Theorem 8. *Let $L \subseteq \Sigma^*$ be a regular language and G_L the dependency graph for L. Then $2^{d(G_L)}$ is greater or equal to the deterministic state complexity of L, i.e., $2^{d(G_L)} \geq sc(L)$.*

Proof (Sketch). Let $G_L = (X, Y, E_L)$ and assume that the bipartite dimension of G_L equals k. Then the edge set of G_L can be covered by a set of bicliques $C = \{H_1, H_2, \ldots, H_k\}$. For $x \in \Sigma^*$, let $B(x) \subseteq C$ be the set of bicliques where x occurs as a "left vertex." Then define $x \sim x'$ with $x, x' \in \Sigma^*$ if and only if $B(x) = B(x')$. This equivalence relation induces $2^{|C|}$ equivalence classes, and is a refinement of the Myhill-Nerode relation. Thus we have shown that $2^{|C|}$ is greater or equal to the deterministic state complexity of L.   □

Hence, $d(G_L) \geq \log sc(L)$ and $d(G_L) \geq \log nsc(L)$. By Corollary 4 and Theorem 8 we obtain a characterization of regular languages in terms of bipartite dimension: A language $L \subseteq \Sigma^*$ is regular if and only if $d(G)$ is finite, where $G = (\Sigma^*, \Sigma^*, E_L)$. The above result is essentially optimal. In [10, 13] it was shown that the nondeterministic state complexity can be $\Omega(2^{\sqrt{d}})$, where d is the bipartite dimension of the dependency graph. We significantly improve on this result using graph-theoretic methods.

Theorem 9. *There is a sequence of languages $(L_n)_{n \geq 1}$ over a one letter alphabet such that $nsc(L_n) = \Omega\left(d_n^{-1/2} \cdot 2^{d_n}\right)$, where d_n is the bipartite dimension of G_{L_n}.*

Proof (Sketch). Let $L_m = \{w \in 0^* \mid |w| \neq 0 \mod m\}$ and define $\overline{G}_{L_m} := (X, Y, (X \times Y) \setminus E_L)$ to be the bi-complement of G_{L_m}. Observe that the bipartite graph $\overline{G}_{L_m}$ is an induced matching with n edges. By a recent result in graph theory [2], the bipartite dimension d_m of $\overline{G}_{L_m}$ equals k, where k is the smallest integer such that $m \leq \binom{k}{\lfloor \frac{k}{2} \rfloor}$, and thus $m = \Omega\left(d_m^{-1/2} \cdot 2^{d_m}\right)$. It remains to be shown that there are infinitely many m such that $nsc(L_m) \geq m$.

We show that this is the case, whenever m is a prime number and thus taking the sequence $(L_{p_n})_{n \geq 1}$, where p_n is the nth prime number, will prove the stated result. This follows by closer inspection of the languages together with a result on unary cyclic languages [11, Corollary 2.1]. $\qquad \square$

6 Computational Complexity of Lower Bound Techniques

To determine the nondeterministic state complexity of a regular language is known to be a computationally hard task, namely PSPACE-complete [12]. In this section we consider three decision problems based on the lower bound techniques presented so far. The *fooling set problem* is defined as follows:

- Given a deterministic finite automaton A and a natural number k in binary, i.e., an encoding $\langle A, k \rangle$.
- Is there a fooling set S for the language $L(A)$ of size at least k?

The *extended fooling set* and the *biclique edge cover problem* are analogously defined. We start our investigations with the fooling set problem.

Theorem 10. *The fooling set problem is NP-hard and contained in PSPACE.*

Proof. For the NP-hardness we reduce the NP-complete induced matching problem on bipartite graphs [5] to the problem under consideration. We omit the proof due to lack of space.

The containment in PSPACE is seen as follows: Let $\langle A, k \rangle$ be an instance of the fooling set problem, where $A = (Q, \Sigma, \delta, q_0, F)$ is a deterministic finite automaton and k an integer. If S is a fooling set, then one can assume w.l.o.g. that for every $(x, y) \in S$ we have $|x| \leq |Q|$ and $|y| \leq 2^{|Q|}$. Moreover we note that the size of S cannot exceed $|Q|$. This gives the idea to the following algorithm: A polynomially space bounded nondeterministic Turing machine can guess k different words x_i with $|x_i| \leq |Q|$ and store k copies of the states $q_i = \delta(q_0, x_i)$ in a k^2-vector. Then the Turing machine simultaneously guesses pairwise different words y_i of length at most $2^{|Q|}$ letter by letter, for $1 \leq i \leq k$, thereby simulating the automaton A on all inputs $x_i y_j$ using the k^2-vector of states. It verifies that the y_i are indeed pairwise different by remembering that they are of different length or eventually differ in at least one position. Finally it checks that $\delta(q_0, x_i y_j) \in F$ if and only if $i = j$. $\qquad \square$

The following theorem exactly classifies the computational complexity of the extended fooling set problem.

Theorem 11. *The extended fooling set problem is PSPACE-complete.*

Proof. The upper bound for the fooling set problem shown in Theorem 10 easily transfers to extended fooling sets. The details are left to the reader.

The hardness is shown by a reduction from the PSPACE-complete deterministic finite automaton union universality problem: Given a list of deterministic

finite automata $A_1, A_2, \ldots, A_n$ over a common alphabet Σ, is $\bigcup_{i=1}^{n} L(A_i) = \Sigma^*$?
For technical reasons we assume w.l.o.g. that all words of length at most one
are in $\bigcup_{i=1}^{n} L(A_i)$. The construction to come relies on a definition of a special
language L commonly specified by the multiple deterministic finite automata—
recall the proof given in [12, Theorem 3.2]. Let $\langle A_1, A_2, \ldots, A_n \rangle$ be the instance of the union universality problem for deterministic finite automata, where
$A_i = (Q_i, \Sigma, \delta_i, q_{i1}, F_i)$, for $1 \leq i \leq n$ is a deterministic finite automaton with
state set $Q_i = \{q_{i1}, q_{i,2}, \ldots q_{i,t_i}\}$. We assume that $Q_i \cap Q_j = \emptyset$ for $i \neq j$. Define the language $P(i, j) = \{w \in \Sigma^* \mid \delta(q_{i1}, w) = q_{ij}\}$. We introduce a new
symbol a_i for each automaton A_i, and a new symbol b_{ij} for each state q_{ij} in
$\bigcup_{i=1}^{n} Q_i$. In addition, we have new symbols c, d and f. The let $P(i)$ be the
marked version of the language accepted by A_i, i.e., $P(i) = \bigcup_{j=1}^{t_i} [a_i \cdot P(i, j) \cdot b_{ij}]$.
The language $Q(i) = \{wb_{ij} \mid w \in (\Sigma \cup \lambda) \text{ and } \delta(q_{i1}, w) = q_{ij}\}$ consists of short
prefixes of words in $L(A_i)$, which are marked at the end. Let B be the set
of symbols b_{ij} introduced above. Then the auxiliary language R is given by
$R = (\{c\} \cup \Sigma)(d \cup \Sigma)\Sigma^*(\{f\} \cup B)$. Lastly, let

$$L = \bigcup_{i=1}^{n} [P(i) \cup a_i L(A_i) \cup Q(i)] \cup R \cup \Sigma^*. \tag{1}$$

As noted in [12], a deterministic finite automaton accepting L can be obtained
in polynomial time from the finite state machines $A_1, A_2, \ldots, A_n$.

Let $k = 4 + \sum_{i=1}^{n} |Q_i|$. Then one can show the following claim: There is an
extended fooling set of cardinality at least $k + 1$ if and only if $\bigcup_{i=1}^{n} L(A_i) \neq \Sigma^*$.
To this end we argue as follows: (1) The set $S = S' \cup S''$ with

$$S' = \{(a_i w_{ij}, b_{ij}) \mid 1 \leq i \leq n \text{ and } 1 \leq j \leq t_i\}$$

and

$$S'' = \{(\lambda, a_1 b_{11}), (a_1 b_{11}, \lambda), (c, df), (cd, f)\},$$

where w_{ij} is any word in $P(i, j)$, for each $1 \leq i \leq n$ and $1 \leq j \leq t_i$, is an
extended fooling set of size k. (2) If $\bigcup_{i=1}^{n} L(A_i) = \Sigma^*$, then S is optimal [12,
Claim 3.2], i.e., $|S| = \mathrm{nsc}(L)$. Otherwise one can show that S can be extended
by any pair (x, y) with $|x| \geq 1$ and $|y| \geq 1$ satisfying $xy \in \Sigma^* \setminus \bigcup_{i=1}^{n} L(A_i)$. The
details are omitted.   □

Finally, we show that that deciding the biclique edge cover problem is also
PSPACE-complete, although the dependency graph of the given language can
be of exponential size in terms of the input.

Theorem 12. *The biclique edge cover problem is PSPACE-complete.*

Proof. The PSPACE-hardness follows along the lines of the proof of Theorem 11.
In the case where L as defined in Equation (1) is not universal, there is an extended fooling set of size $k + 1$, and since the bipartite dimension cannot be
lower we have $d(G_L) \geq k + 1$. In the other case, the nondeterministic state complexity equals k, and matches the size of a the extended fooling set S for L. But
the bipartite dimension of the graph G_L is sandwiched between both measures.
Thus, $d(G_L) \leq k$.

The containment in PSPACE is seen as follows: We present a PSPACE algorithm deciding on input $\langle A, k \rangle$ whether there is a biclique edge cover of size at most k for $G_{L(A)}$. Since PSPACE is closed under complement, this routine can also be used to decide whether there is no biclique edge cover of size at most $k - 1$, and moreover that the bipartite dimension of the graph is at least k.

Due to the space constraints, keeping the dependency graph $G_{L(A)}$ in memory is ruled out, since the index of $\equiv_{L(A)}$ can be exponential in the size of the given deterministic finite automaton. Recall that the vertex sets of $G_{L(A)}$ can be chosen to correspond to the equivalence classes of $\equiv_{L(A)}$ and $_{L(A)}\equiv$. So the first vertex set is in one-to-one correspondence with the state set Q of the automaton A, while by Brzozowski's theorem [4], the second vertex set corresponds one-to-one to a certain subset of 2^Q. Namely, for $A = (Q, \Sigma, \delta, q_0, F)$ let $A^R = (Q, \Sigma, \delta^R, F, \{q_0\})$, where $p \in \delta^R(q, a)$ if and only if $\delta(p, a) = q$, be a finite automaton with multiple initial states, the so called reversed automaton of A. Moreover, let $D(A^R)$ be the automaton obtained by applying the "lazy" subset construction to the automaton A^R, that is we generate only the subsets reachable from the set of start states of the finite state automaton A^R. Then these subsets of Q correspond to the equivalence classes of $_{L(A)}\equiv$. Since this automaton can be of size exponential in $|Q|$, however, it cannot be kept in the working memory, too. Nevertheless, assuming $Q = \{q_0, q_1, \ldots q_{n-1}\}$, we can represent the subsets of Q as binary string of length n in a natural fashion. By these mappings, we may assume now that $G_{L(A)} = (X, Y, E_{L(A)})$ with $X = Q$, $Y = \{0,1\}^n$, and the suitably induced edge relation $E_{L(A)}$. Thus, we have established a compact representation of the vertices in the dependency graph. Next, we need a routine to decide membership in the edge set of $G_{L(A)}$.

Given the implicit representation of $G_{L(A)}$ in terms of a n-state deterministic finite automaton $A = (Q, \Sigma, \delta, q_0, F)$, there is a PSPACE algorithm deciding, given a state q of A and a subset address $s = a_0 a_1 \ldots a_{n-1}$, if $(q, s) \in E_{L(A)}$. Assume x to be a word satisfying $\delta(q_0, x) = q$. As $|x| \leq n$, it can be determined and stored to the work tape without affecting the space bounds. If s corresponds to a reachable subset M in $D(A^R)$, then we can guess on the fly a word y of length at most 2^n, and verify that M is reached in $D(A^R)$ by reading y. Now, (q, s) is an edge in $G_{L(A)}$ if and only if xy^R is in $L(A)$. This is the case if and only if $(xy^R)^R = yx^R$ is accepted by $D(A^R)$. Recall that the word y may be of exponential length and cannot be directly stored on the work tape. But $D(A^R)$ is in the state set M after reading y, and we only have to verify that we reach an accepting state if we continue by reading x^R. This is the desired subroutine for deciding whether $(q, s) \in E_{L(A)}$, which runs in (nondeterministic) polynomial space.

The next obstacle is that, although there surely exists a biclique edge cover of cardinality at most n for $G_{L(A)}$, a single biclique in this cover can be of exponential size. Thus we have to reformulate the biclique edge cover problem in a suitable manner. Let $G = (X, Y, E)$ be a bipartite graph, and for $y \in Y$ define $\Gamma(y) = \{ x \in X \mid (x, y) \in E \}$. Then the formula

$$\exists C \subseteq 2^X : |C| \leq k \wedge \big(\forall(x, y) \in E : \exists c \in C : x \in c \wedge c \subseteq \Gamma(y)\big) \qquad (2)$$

is a statement equivalent to the biclique edge cover problem. This is seen as follows: Assume C is a set of at most k subsets of X satisfying the above conditions. We construct a set of $|C|$ bicliques covering all edges in G. For $c \in C$, let c' be the set of vertices in Y such that $\Gamma(y) \supseteq c$. Then (c, c') induces a biclique in G, since every vertex in c is adjacent to all vertices in c'. Furthermore, the condition on C ensures that every edge is a member of least one such biclique, and we have obtained a biclique edge cover of size at most k. Conversely, assume that $\{H_1, H_2, \ldots, H_k\}$ is a biclique edge cover of size k for G, where $H_i = (c_i, c'_i, c_i \times c'_i)$ for $1 \leq i \leq k$. We set $C = \{c_1, c_2, \ldots c_k\}$. Then for every edge (x, y) in G, there is a $c \in C$ such that $x \in c$ and $c \subseteq \Gamma(y)$. If H_i is a biclique covering of the edge (x, y), then obviously $x \in c_i$ and y is adjacent to all vertices in c_i. This proves the stated claim on Equation (2).

Now let us come back to the input $\langle A, k \rangle$, where $A = (Q, \Sigma, \delta, q_0, F)$. The reformulated statement can be checked in PSPACE by guessing a set C of at most k subsets of Q, and then the Turing machine checks the following for each pair $(x, y) \in X \times Y$, where X and Y is chosen as described above: If $(x, y) \notin E_{L(A)}$ it goes to the next pair. Otherwise, it guesses a subset $c \in C$ and verifies that both $x \in C$ and that for every $x' \in c$, $(x', y) \in E_{L(A)}$. By our previous investigations it is easy to see that this algorithm can be implemented on a nondeterministic polynomial space bounded Turing machine. This proves that the biclique edge cover problem belongs to PSPACE.   □

Finally, let us mention that the complexity of the fooling set and the extended fooling set problem does not increase, if the regular language is specified as a nondeterministic finite automaton. The proofs for the upper bounds on the complexity carry over to this setup with minor modifications. Currently, we do not know whether this also true for the biclique edge cover technique, if the regular language is given as a nondeterministic finite automaton. The best upper bound we are aware of is co-NEXPTIME, obtained by explicit construction of G_L and verifying that there is no biclique edge cover of size at most k.

Acknowledgments. Thanks to Martin Kutrib for some discussion on the subject during the early stages of the paper, and to the referees for hints and suggestions.

References

1. H. N. Adorna. Some descriptional complexity problems in finite automata theory. In R. P. Saldaña and C. Chua, editors, *Proceedings of the 5th Philippine Computing Science Congress*, pages 27–32, Cebu City, Philippines, March 2005. Computing Society of the Philippines.
2. S. Bezrukov, D. Fronček, S. J. Rosenberg, and P. Kovář. On biclique coverings. Preprint, 2005.
3. J.-C. Birget. Intersection and union of regular languages and state complexity. *Information Processing Letters*, 43:185–190, 1992.
4. J. A. Brzozowski. Mathematical theory of automata. In *Canonical Regular Expressions and Minimal State Graphs for Definite Events*, volume 12 of *MRI Symposia Series*, pages 529–561. Polytechnic Press, NY, 1962.

5. K. Cameron. Induced matchings. *Discrete Applied Mathematics*, 24:97–102, 1989.
6. I. Glaister and J. Shallit. A lower bound technique for the size of nondeterministic finite automata. *Information Processing Letters*, 59:75–77, 1996.
7. J. E. Hopcroft and J. D. Ullman. *Formal Languages and Their Relation to Automata*. Addison-Wesley, 1968.
8. J. Hromkovič. *Communication Complexity and Parallel Computing*. Springer, 1997.
9. J. Hromkovič. Descriptional complexity of finite automata: Concepts and open problems. *Journal of Automata, Languages and Combinatorics*, 7(4):519–531, 2002.
10. J. Hromkovič, J. Karhumäki, H. Klauck, G. Schnitger, and S. Seibert. Measures of nondeterminism in finite automata. Report TR00-076, Electronic Colloquium on Computational Complexity (ECCC), 2000.
11. T. Jiang, E. McDowell, and B. Ravikumar. The structure and complexity of minimal NFAs over a unary alphabet. *International Journal of Foundations of Computer Science*, 2(2):163–182, June 1991.
12. T. Jiang and B. Ravikumar. Minimal NFA problems are hard. *SIAM Journal on Computing*, 22(6):1117–1141, December 1993.
13. G. Jirásková. Note on minimal automata and uniform communication protocols. In C. Martín-Vide and V. Mitrana, editors, *Grammars and Automata for String Processing*, volume 9 of *Topics in Computer Mathematics*, pages 163–170. Taylor and Francis, 2003.
14. T. Kameda and P. Weiner. On the state minimization of nondeterministic finite automata. *IEEE Transactions on Computers*, C-19(7):617–627, 1970.
15. F. R. Moore. On the bounds for state-set size in the proofs of equivalence between deterministic, nondeterministic, and two-way finite automata. *IEEE Transaction on Computing*, C-20:1211–1219, 1971.
16. M. O. Rabin and D. Scott. Finite automata and their decision problems. *IBM Journal of Research and Development*, 3:114–125, 1959.
17. A. Salomaa, D. Wood, and S. Yu. On the state complexity of reversals of regular languages. *Theoretical Computer Science*, 320(2–3):315–329, 2004.

Lowering Undecidability Bounds for Decision Questions in Matrices

Paul Bell and Igor Potapov

Department of Computer Science,
University of Liverpool, Chadwick Building,
Peach St, Liverpool L69 7ZF, U.K.
{pbell, igor}@csc.liv.ac.uk

Abstract. In this paper we consider several reachability problems such as vector reachability, membership in matrix semigroups and reachability problems in piecewise linear maps. Since all of these questions are undecidable in general, we work on lowering the bounds for undecidability. In particular, we show an elementary proof of undecidability of the reachability problem for a set of 7 two-dimensional affine transformations. Then, using a modified version of a standard technique, we also prove the vector reachability problem is undecidable for two (rational) matrices in dimension 16. The above result can be used to show that the system of piecewise linear functions of dimension 17 with only two intervals has an undecidable set-to-point reachability problem. We also show that the "zero in the upper right corner" problem is undecidable for two integral matrices of dimension 18 lowering the bound from 23.

Keywords: Theory of computing, membership, vector reachability, matrix semigroups, piecewise linear maps.

1 Introduction

The significant property of iterative maps as well as dynamical systems in general is that the slightest uncertainty about the initial state leads to very large uncertainty after some time. With such initial uncertainties, the system's behaviour can only be predicted accurately for a short amount of time into the future. Many fundamental questions about iterative maps are closely related to reachability problems in different abstract structures. The question of whether these problems have an algorithmic solution or not is important since it gives an answer to many practical questions related to the analysis of model systems, whose components are governed by simple laws, but whose overall behaviour is highly complex.

Let us start with an example of Iterative Function Systems (IFS). IFS can be defined as a set of affine transformations that are iteratively applied (equiprobably or with assigned probability) starting from some initial state. A fascinating property of IFS is that they have the ability to create incredibly complex images by very small sets of functions. A spleenwort fern (Figure 1, left) can be generated by 4 planar affine transformations and the well-known Sierpinski gasket

O.H. Ibarra and Z. Dang (Eds.): DLT 2006, LNCS 4036, pp. 375–385, 2006.

(Figure 1, centre) is an attractor for only 3 affine transformations. The "Dragon Fractal" is another example that is generated by affine transformations and is shown in figure 1 on the right.

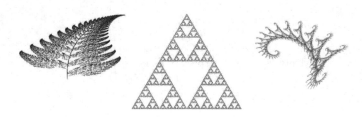

Fig. 1. Fractals generated by two-dimensional affine maps

The fact that we can encode Post's Correspondence Problem (PCP) shows that IFS with 7 transformations are in fact so complex that they can be used as a computational device. This is not a first attempt to encode PCP into an IFS framework. However, in contrast to the previous research [6], which shows only the undecidability of parameterized reachability, we show the undecidability of point-to-point reachability for a set of non-deterministic affine maps.

The question as to whether the reachability problem is decidable for a set of one dimensional affine transformations applied in a non-deterministic way is currently an open problem. It is also not clear whether there exists any class of more complex functions for which the reachability problem in non-deterministic maps becomes undecidable in one dimension. Similar open problems are stated in [1, 10] for *piecewise* affine or polynomial maps. It is known that the reachability question for piecewise maps in the case of semi-algebraic [12] or analytic maps [11] is undecidable even in one dimension. In the first part of this paper we show an elementary proof of the undecidability of the reachability problem for two-dimensional affine transformations. This leads us to the undecidability result for the vector reachability problem in 3×3 rational matrix semigroups.

We then extend the result to show the undecidability of the vector reachability problem in the case of a matrix semigroup generated by two rational matrices of dimension 16. Using a similar encoding, we improve the bound for the "zero in the upper right corner problem". It was shown that for a semigroup generated by two integral matrices of dimension 24, the problem of determining if some matrix in the semigroup has zero in the upper right corner is undecidable [5]. This bound was later improved to 23 [7]. We show that the problem is still undecidable for two integral matrices of dimension 18.

We also show how to lower the dimensions in related problems concerning control systems that are defined by a system of piecewise transformations. We apply our result for the undecidability of the vector reachability problem in dimension 16 to show that set-to-point reachability is undecidable for piecewise linear functions of dimension 17 with only two intervals. The natural problem we can thus state for iterative maps is whether we can find any undecidable problems for lower dimensions concerning maps defined by only two transformations.

2 Lowering Undecidability Bounds

2.1 Reachability in Affine and Linear Transformations

In this subsection we show an elementary proof of the undecidability of the reachability problem for two-dimensional affine transformations and the vector reachability problem for 3×3 rational matrix semigroups.

Let us define $\Sigma = \{a, b\}$ to be a binary alphabet and Σ^* to be the free monoid generated by Σ. For any word, $w = w_1 w_2 \cdots w_k$ we denote the reverse of the word by $w^R = w_k \cdots w_2 w_1$. Define a mapping $\psi' : \Sigma \cup \{\epsilon\} \to \mathbb{Q}^{2 \times 2}$ by

$$\psi'(\epsilon) = \begin{pmatrix} 1 & 0 \\ 0 & 1 \end{pmatrix} \quad \psi'(a) = \begin{pmatrix} 1 & 1 \\ 0 & 2 \end{pmatrix} \quad \psi'(b) = \begin{pmatrix} 1 & 2 \\ 0 & 2 \end{pmatrix}$$

where $\psi'(\epsilon)$ is the identity matrix I_2. We can now define $\psi : \Sigma^* \to \mathbb{Q}^{2 \times 2}$ by:

$$\psi(w) = \psi'(w_1) \times \psi'(w_2) \times \cdots \times \psi'(w_n) \quad | \, w = w_1 w_2 \cdots w_n \in \Sigma^*$$

It is easy to check that the matrices $\psi'(a)$ and $\psi'(b)$ generate a free semigroup [4] and the mapping ψ is an isomorphism between Σ^* and the monoid generated by $\{\psi'(\epsilon), \psi'(a), \psi'(b)\}$. We also define the mapping ϕ' using the inverse matrices:

$$\phi'(\epsilon) = \begin{pmatrix} 1 & 0 \\ 0 & 1 \end{pmatrix} \quad \phi'(a) = \begin{pmatrix} 1 & -\frac{1}{2} \\ 0 & \frac{1}{2} \end{pmatrix} \quad \phi'(b) = \begin{pmatrix} 1 & -1 \\ 0 & \frac{1}{2} \end{pmatrix}$$

and similarly we have the related morphism $\phi : \Sigma^* \to \mathbb{Q}^{2 \times 2}$ given by:

$$\phi(w) = \phi'(w_1) \times \phi'(w_2) \times \cdots \times \phi'(w_n) \quad | \, w = w_1 w_2 \cdots w_n \in \Sigma^*$$

Notice that for any word $w \in \Sigma^*$, $\psi(w)$ and $\phi(w)$ will have a matrix representation of the following form:

$$\begin{pmatrix} 1 & x \\ 0 & y \end{pmatrix} \quad | \, x, y \in \mathbb{Q}$$

Post's correspondence problem (in short, PCP) is formulated as follows: Given a finite alphabet Γ and a finite (ordered) set of pairs of words in Γ^*: $\{(u_1, v_1), \ldots, (u_k, v_k)\}$. Does there exist a finite sequence of indices $(i_1, i_2, \ldots, i_m)$ with $1 \le i_j \le k$ for $1 \le j \le m$, such that $u_{i_1} \cdot \ldots \cdot u_{i_m} = v_{i_1} \cdot \ldots \cdot v_{i_m}$. It is easily shown that this problem is undecidable even with a binary alphabet Γ. PCP(n) denotes the problem with a set of n pairs. We write n_p for the minimum size PCP is known to be undecidable (currently 7, see [13]).

Problem 1. Decide whether a point $(x, y) \in \mathbb{Q}^2$ can be mapped to itself by non-deterministically applying a sequence of two-dimensional affine transformations from a given finite set.

We shall now show that this problem is undecidable.

Theorem 1. *PCP(n) can be reduced to Problem 1 with a set of n affine transformations.*

Proof. Given a set of pairs of words over a binary alphabet Σ:

$$\{(u_1, v_1), (u_2, v_2), \ldots, (u_n, v_n)\} \qquad | \, u_i, v_i \in \Sigma^+$$

Let us construct a set of pairs of 2×2 matrices using the two mappings ϕ and ψ, i.e. : $\{(\phi(u_1), \psi(v_1)), \ldots, (\phi(u_n), \psi(v_n))\}$.

Instead of the equation $u = v$, we consider a concatenation of two words $u^R \cdot v$, which is a palindrome in the case where $u = v$. In fact by using inverse elements for the u word, then u is equal to v if and only if $u^R \cdot v = \epsilon$. We call this an *inverse palindrome* since it equals the identity iff u^R is the reverse and inverse of v. Initially we take an empty word and for every pair (u_i, v_i) that we use, we concatenate the reverse of word u_i (using inverse elements) to the left and word v_i from the right.

Let us consider now a matrix interpretation of the PCP problem. We associate a 2×2 matrix

$$C = \begin{pmatrix} 1 & x \\ 0 & y \end{pmatrix}$$

with a word w of the form $u^R \cdot v$. Initially C is an identity matrix corresponding to an empty word. The extension of a word w by a new pair of words (u_i, v_i) (i.e. that gives us $w' = u_i^R \cdot w \cdot v_i$) corresponds to the following matrix multiplication

$$C_{w'} = C_{u_i^R \cdot w \cdot v_i} = \phi(u_i^R) \times C_w \times \psi(v_i) \tag{1}$$

Equation (1) is therefore written:

$$\begin{pmatrix} 1 & x' \\ 0 & y' \end{pmatrix} = \begin{pmatrix} 1 & p_1 \\ 0 & p_2 \end{pmatrix} \times \begin{pmatrix} 1 & x \\ 0 & y \end{pmatrix} \times \begin{pmatrix} 1 & q_1 \\ 0 & q_2 \end{pmatrix} \tag{2}$$

If we multiply the matrices in (2) we have a very simple transformation from a word $v_i^R \cdot w \cdot u_i$ to a matrix:

$$\begin{pmatrix} 1 & x' \\ 0 & y' \end{pmatrix} = \begin{pmatrix} 1 & q_2 x + q_2 p_1 y + q_1 \\ 0 & q_2 p_2 y \end{pmatrix} \tag{3}$$

Thus we can state that PCP has a solution iff we can get the identity matrix by using by a set of transformations defined for each pair of words from PCP.

In fact we can rewrite (3) as a two-dimensional affine transformation:

$$\begin{cases} x' = q_2 x + q_2 p_1 y + q_1 \\ y' = \qquad p_2 q_2 y \end{cases}$$

where we define one such affine transformation for each pair in the PCP. It can now be seen that the problem is reduced to the question about reaching the point $x = 0, y = 1$ starting from the point $x = 0, y = 1$. This follows since $x = 0, y = 1$ corresponds to the identity matrix (see [15]) in the above calculations.

Corollary 1. *Problem 1 is undecidable for a set of seven affine transformations of dimension two.*

Proof. PCP(7) was shown to be undecidable in [13].

We now show that the vector reachability problem is undecidable for rational matrices of dimension 3. The stronger result about integer matrices was included in [7] but unfortunately the proof does not cover all cases and cannot be used for the vector reachability problem. We therefore use a different idea for rational matrices of dimension 3 and the problem remains open over integral matrices.

Theorem 2. *The vector reachability problem in rational matrix semigroups of dimension 3 is undecidable.*

Proof. Let us convert each two-dimensional affine transformation into a three-dimensional linear transformation as follows:

$$\begin{cases} x' = q_2 x + q_2 p_1 y + q_1 \\ y' = \qquad\qquad p_2 q_2 y \end{cases} \Rightarrow \begin{pmatrix} x' \\ y' \\ 1 \end{pmatrix} = \begin{pmatrix} q_2 & q_2 p_1 & q_1 \\ 0 & p_2 q_2 & 0 \\ 0 & 0 & 1 \end{pmatrix} \begin{pmatrix} x \\ y \\ 1 \end{pmatrix}$$

Thus for a set of n affine functions, this conversion gives us a set of matrices $\{M_1, M_2, \ldots, M_n\}$, where $M_i \in \mathbb{Q}^{3 \times 3}$ for $1 \leq i \leq n$.

From the proof of Theorem 1 follows that the problem to decide whether there exists a product $M = M_{i_1} M_{i_2} \cdots M_{i_k}$ where $1 \leq i_j \leq n$ for $1 \leq j \leq k$ such that $Mv = v$ where $v = (0, 1, 1)^T$ is undecidable. It was stated that PCP(7) is undecidable in Corollary 1 thus seven matrices are needed.

It is also possible to get a symmetric result by converting the additive form of linear transformations into multiplicative form. In this case we will obtain the undecidability of the reachability problem for two dimensional transformations of the following form:

$$\begin{cases} x = 2^{q_1} x^{q_2} y^{q_2 p_1} \\ y = \qquad y^{p_2 q_2} \end{cases}$$

Theorem 3. *The vector reachability problem is undecidable for two matrices of dimension $2n_p + 2$ (where n_p is the instance size for which PCP is undecidable).*

Proof. We use a modification of a standard technique for converting membership problems from a set of matrices into one defined by just two matrices (see, for example, [5] or [3]). We first obtain the undecidability for two matrices of dimension 21 then show how this can be reduced to just 16.

Given a set of matrices $\{M_1, M_2, \ldots, M_n\}$ where $M_i \in \mathbb{Q}^{m \times m}$. Let us define two block diagonal matrices A' and T' by:

$$A' = \begin{pmatrix} M_1 & 0 & 0 & 0 \\ 0 & M_2 & 0 & 0 \\ 0 & 0 & \ddots & 0 \\ 0 & 0 & 0 & M_n \end{pmatrix}, \qquad T' = \begin{pmatrix} 0 & I_m \\ I_{n(m-1)} & 0 \end{pmatrix}$$

where 0 denotes a submatrix with zero elements. Clearly the dimension of both of A' and T' is nm. Further, it can be seen that for any $1 \leq j \leq n$ then $T'^{n-j+1}A'T'^{j-1}$ cyclically permutes the blocks of A' so that the direct sum of $T'^{n-j+1}A'T'^{j-1}$ is $M_j \oplus M_{j+1} \oplus \cdots \oplus M_n \oplus M_1 \oplus \cdots \oplus M_{j-1}$. We can also note that $A' \sim T'^{n-j+1}A'T'^{j-1}$ (i.e. this is a similarity transform) since $T'^{n-j+1} \cdot T'^{j-1} = T'^n = I_n$. It is therefore apparent that any product of the matrices can thus occur and in fact can appear in the first block of the nm matrix product.

Let us define a vector $x = (v^T, 0, 0, \cdots, 0)^T \in \mathbb{Q}^{nm \times 1}$ where $v = (0, 1, 1)^T$ as before. It is easily observed that there exists a matrix product $M = M_{i_1} M_{i_2} \cdots M_{i_t}$ satisfying $Mv = v$ iff there exists a product $R' = \{A', T'\}^+$ satisfying $R'x = x$. From theorem (2), this establishes the undecidability of vector reachability over 2 matrices of dimension $3 \cdot 7 = 21$.

We can observe however that $(M_i)_{[3,3]} = 1$ and M_i is upper triangular for all $1 \leq i \leq n$. Let us now construct two new matrices of dimension $2n + 2$:

$$
A = \begin{pmatrix}
(q_2)_1 & (q_2 p_1)_1 & 0 & 0 & \cdots & (q_1)_1 & 0 \\
0 & (p_2 q_2)_1 & 0 & 0 & \cdots & 0 & 0 \\
0 & 0 & (q_2)_2 & (q_2 p_1)_2 & \cdots & (q_1)_2 & 0 \\
0 & 0 & 0 & (p_2 q_2)_2 & \cdots & 0 & 0 \\
\vdots & \vdots & \vdots & \vdots & \ddots & \vdots & 0 \\
0 & 0 & 0 & 0 & 0 & 1 & 0 \\
0 & 0 & 0 & 0 & 0 & 0 & 0
\end{pmatrix}, T = \begin{pmatrix}
0 & I_2 & 0 & 0 \\
I_{2n-2} & 0 & 0 & 0 \\
0 & 0 & 1 & 0 \\
0 & 0 & 0 & 1
\end{pmatrix}
$$

where 0 denotes either the number zero or a submatrix with zero elements, I_k is the k dimensional identity matrix and $(x)_i$ denotes the element x from matrix M_i. Straight forward calculation shows that $T^{n-j+1}AT^{j-1}$ permutes the pairs of *rows* in A and using a similar argument as before, we thus can form any product of matrices in the first two rows of this matrix. We define a $2n + 2$ dimensional vector $w = (0, 1, 0, \cdots, 0, 1, 1)^T$ to act in the same way as v did previously. Now define a vector $s = (0, 1, 0, \cdots, 0, 1, 0)^T$ needed to avoid the pathological case where only T matrices are used. Finally we see that there exists a solution $Mv = v$ to PCP iff there exists a product $R = \{A, T\}^+$ satisfying $Rw = s$. Note that in this construction we have two matrices of dimension $2n + 2$.

Since PCP(7) is undecidable, the dimension of the two matrices for which the vector reachability is undecidable is therefore $(2 \cdot 7) + 2 = 16$ as required:

Corollary 2. *The vector reachability problem is undecidable for two matrices of dimension* 16.

Next we show a related problem but instead of considering vector reachability in a semigroup, we instead choose the next matrix to apply in a piecewise manner depending upon the current position of an element in the vector. This realates non-deterministic and piecewise reachability questions in a natural way.

Theorem 4. *Given a pair of matrices $M_1, M_2 \in \mathbb{Q}^{k \times k}$ and a vector $y' \in \mathbb{Q}^{k \times 1}$, the problem of deciding if there exists a vector $w' \in \mathbb{Q}^{k \times 1}$ of a special form which* can be mapped to y' by (left) multiplication by M_1 or M_2 chosen in a iteratively piecewise way is undecidable for $k = 2n_p + 3$.

Proof. We extend the matrices $A, T \in \mathbb{Q}^{m \times m}$ defined in the previous proof by 1 dimension therefore $k = m + 1$. Let:

$$\Upsilon = \begin{pmatrix} A & z \\ d_1 & 10 \end{pmatrix}, \qquad W = \begin{pmatrix} T & z \\ d_2 & 10 \end{pmatrix}$$

where $z = (0, 0, \ldots, 0)^T \in \mathbb{Q}^{(k-1) \times 1}$ is the zero vector, $d_1 = (0, 0, \ldots, -1)$, $d_2 = (0, 0, \ldots, -2)$ and $d_1, d_2 \in \mathbb{Q}^{1 \times (k-1)}$. We also extend the vector w used in the previous proof to $w' = (w^T, x)^T = (0, 1, 0, 0, \ldots, 1, x)^T \in \mathbb{Q}^{k \times 1}$ where $x \in \mathbb{Q}$ and $x \in (0, 1]$. We will choose which matrix (Υ or W) to apply at each step depending upon the current x value. Define $y' = (0, 1, 0, 0, \ldots, 1, 0)^T \in \mathbb{Q}^{n \times 1}$.

Let vector w' be given where $0 < x \leq 1$. We choose which matrix to multiply by on the next step by a simple rule. Let $t = (0, 0, \ldots, 1)^T \in \mathbb{Q}^{n \times 1}$. Then at each step of the computation we update w' according to:

$$w' = \begin{cases} \Upsilon w' & ; \text{ if } \frac{1}{10} \leq w'^T \cdot t < \frac{2}{10} \\ W w' & ; \text{ if } \frac{2}{10} \leq w'^T \cdot t < \frac{3}{10} \end{cases}$$

and the next step if undefined outside of these regions. Note that $\Upsilon w'$ is the same as in the previous proof except for the last element of the vector. This is equal to $10x - 1$, which is equivalent to shifting the decimal representation of x to the left and removing the integer part (since the first decimal digit is a 1). Similarly, $(Ww')_{[n,1]} = 10x - 2$ which is equivalent to shifting the decimal one place left and subtracting the integer part.

This is applied in an iterative manner until either $0 < x < \frac{1}{10}$ or $x \geq \frac{3}{10}$ (which is undefined so we halt) or else $x = 0$ in which case, if $w' = y'$ then there exists a correct solution to the PCP as in the previous proof. If $w' \neq y'$ then x did not correspond to a correct solution and we halt because the next step is undefined (when $x = 0$).

The previous theorem is a set-to-point piecewise reachability question. If there exists some x with the desired property (e.g. of the form $x = 0.1221121 \ldots$) then we choose the next matrix depending on whether we have a 1 or a 2 in the next decimal position (e.g. $\Upsilon WW\Upsilon\Upsilon W\Upsilon$). Note that in this case, the decimal part of x at the start of the computation corresponded to a sequence of indices giving a solution to PCP.

2.2 Zero in the Upper Right Corner Problem

We now move to a different problem which has been studied in the literature and is related to Skolem's problem.

Problem 2. Given a set of matrices $M_1, M_2, \ldots, M_n$ of dimension m generating a semigroup $\mathcal{S}$, is it decidable if there exists a matrix $M \in \mathcal{S}$ such that $M_{[1,m]} = 0$. I.e. does there exist a matrix $M \in \mathcal{S}$ with a 0 in the upper right corner.

This problem has been studied for two main reasons. Firstly it is related to the mortality problem (Does the zero matrix belong to a semigroup?). Actually the upper *left* corner is used in the proof but the upper right corner problem is used in related problems. See for example [9].

Secondly the problem is related to Skolem's problem of finding zeros in linear recurrent sequences. It can be easily shown that Skolem's problem can be reformulated in terms of a single matrix R and the question, "Does there exist a $k \in \mathbb{Z}^+$ such that $(R^k)_{[1,m]} = 0$ where m is the degree of the linear recurrence?". For an overview of this problem and a proof of the decidability for degree 5 linear recurrence sequences, see the recent paper [8].

In terms of undecidability, it was shown that for two integral matrices of dimension 24 the zero upper right corner problem is undecidable [5]. This was improved to dimension 23 in [7]. Using a similar idea to that shown above and the technique used in [5], we show how to improve the bound to two integral matrices of dimension 18.

Theorem 5. *Given two matrices $A, B \in \mathbb{Z}^{n \times n}$ forming a semigroup $\mathcal{S}$, it is undecidable if there exists a matrix $X \in \mathcal{S}$ with a zero in the upper right corner for $n = 2n_p + 4$ (currently 18).*

Proof. It can be seen that there exists an injective morphism between words over a binary alphabet $\Gamma = \{L_1, L_2\}$ and three dimensional matrices. In fact, one such morphism, which was originally used by Paterson to prove undecidability of the mortality problem [14], is $\lambda' : \Gamma^* \times \Gamma^* \mapsto \mathbb{Z}^{3 \times 3}$ defined by:

$$\lambda'(u, v) = \begin{pmatrix} 3^{|u|} & 0 & \sigma(u) \\ 0 & 3^{|v|} & \sigma(v) \\ 0 & 0 & 1 \end{pmatrix}$$

for two words $u = L_{i_1} L_{i_2} \cdots L_{i_r}$ and $v = L_{j_1} L_{j_2} \cdots L_{j_s}$, with $i_1, \cdots, i_r, j_1, \cdots, j_s \in \{1, 2\}$, where we define

$$\sigma(u) = \sum_{k=1}^{|u|} i_k 3^{|u|-k}$$

and similarly for v. This matrix is the unique 3-adic representation of the binary words $u, v \in \Gamma^*$. Now consider the following matrix:

$$H = \begin{pmatrix} 1 & -1 & 0 \\ 0 & -1 & 0 \\ 0 & 0 & -1 \end{pmatrix}$$

which is self-inverse since $HH = I_3$. We can thus define a similarity transform $H\lambda'(u, v)H$ which gives us the alternate (but still injective) morphism:

$$\lambda(u, v) = H\lambda'(u, v)H = \begin{pmatrix} 3^{|u|} & 3^{|v|} - 3^{|u|} & \sigma(v) - \sigma(u) \\ 0 & 3^{|v|} & \sigma(v) \\ 0 & 0 & 1 \end{pmatrix}$$

Notice that $u = v$ iff $\lambda(u,v)_{[1,3]} = 0$ since the top right element of the matrix is the subtraction of the 3-adic representations of u, v.

Using the same idea as before, we can now encode n such matrices into a single matrix, B, of size $2n + 1$ and use a second matrix T which is identical to the permutation matrix defined previously. Therefore, given n pairs of words $\{(u_1, v_1), (u_2, v_2), \ldots, (u_n, v_n)\}$ we define:

$$B = \begin{pmatrix} 3^{|u_1|}\, 3^{|v_1|} - 3^{|u_1|} & 0 & 0 & 0 & \cdots & \sigma(v_1) - \sigma(u_1) \\ 0 & 3^{|v_1|} & 0 & 0 & \cdots & \sigma(v_1) \\ 0 & 0 & 3^{|u_2|}\, 3^{|v_2|} - 3^{|u_2|} & \cdots & \sigma(v_2) - \sigma(u_2) \\ 0 & 0 & 0 & 3^{|v_2|} & \cdots & \sigma(v_2) \\ \vdots & \vdots & \vdots & \vdots & \ddots & \vdots \\ 0 & 0 & 0 & 0 & \cdots & 1 \end{pmatrix}$$

We can see that a product containing *both* B and T has a zero in the upper right corner iff there exists a solution to the PCP however T has a zero upper right corner on its own. We can apply the encoding technique used in [5] so that the case with a power of only T matrices can be avoided. Define:

$$B' = \begin{pmatrix} 0 & 1 & x & 1 \\ 0 & 0 & 0 & 1 \\ 0 & 0 & B & z \\ 0 & 0 & \cdots & 0 \end{pmatrix}, \qquad T' = \begin{pmatrix} 0 & 1 & x & 1 \\ 0 & 1 & 0 & 1 \\ 0 & 0 & T & z \\ 0 & 0 & \cdots & 0 \end{pmatrix}$$

where $x = (1, 0, \cdots, 0)$, $z = (0, 0, \cdots, 1)^T$, with $x \in \mathbb{Z}^{1 \times k}$, $z \in \mathbb{Z}^{k \times 1}$ and k is the dimension of matrix B (and T). It is clear that the sub-matrices B, T are multiplied in the same way as before and unaffected by this extension. Notice the $[2, 2]$ element is 0 in B' and 1 in T'. This is used to avoid the pathological case of a matrix product with only T' matrices.

Consider a product of these matrices $Q = Q_1 Q_2 \cdots Q_m$ where $Q_i \in \{B', T'\}$ for $1 \leq i \leq m$. It is easily seen that if $m \leq 2$ then the top right element of Q equals 1 for any Q_1, Q_2. Let us thus assume $m \geq 3$ and write this multiplication as $Q = Q_1 C Q_m$ where $C = Q_2 Q_3 \cdots Q_{m-1}$,

$$C = \begin{pmatrix} 0 & * & * & * \\ 0 & \lambda & 0 & * \\ 0 & 0 & C' & * \\ 0 & 0 & \cdots & 0 \end{pmatrix}$$

where $*$ denotes unimportant values, $\lambda = \{0, 1\}$ and C' is a submatrix equal to some product of B, T matrices.

Now we will compute the top right element of Q. Let r denote the dimension of matrix C (or Q). The first row of $Q_1 C$ equals $(0, \lambda, C'_{1,1}, C'_{1,2}, \cdots, C'_{1,k}, *)$ where again $*$ is unimportant. Note that this vector contains the top row of the C' submatrix. We can now easily see that $Q_{[1,r]} = (Q_1 C Q_m)_{[1,r]}$ equals $(0, \lambda, C'_{1,1}, C'_{1,2}, \cdots, C'_{1,k}, *) \cdot (1, 1, z^T, 0)^T = \lambda + C'_{1,k}$. It is clear that $\lambda = 1$ iff

$C = (T')^{m-2}$ i.e. C is a power of only T' matrices. In this case, note that $(C'^{m-2})_{[1,k]} = 0$ since this is a power of matrix T. Thus $Q_{[1,r]} = 1 + 0 = 1$.

In the second case, $\lambda = 0$ whenever C' contains a factor B'. Therefore $Q_{[1,r]} = 0 + C'_{[1,k]} = C'_{[1,k]}$ which is exactly the top right element of C' as required. This equals 0 iff there exists a solution to the PCP.

We require 3 extra rows and columns for this encoding therefore the problem is undecidable for dimension $2n_p + 1 + 3 = 2n_p + 4$ (currently 18).

3   Conclusion

We showed that point to point reachability is undecidable for n_p two-dimensional affine transformations. A simple extension allowed us to prove that the vector reachability problem for n_p rational matrices of dimension 3 is undecidable and also for 2 matrices of dimension $2n_p+2$. We then showed an undecidable result for a piecewise vector reachability question where the next matrix applied depends upon the current value of an element in the vector. Finally we improved the bounds on the "zero in the upper right corner problem" from $3n_p + 2$ (currently 23) to $2n_p + 4$ (currently 18) using two integral matrices.

It is interesting to consider the lowest dimensions possible for these types of reachability and membership questions, especially when we have just two transformations. As far as the authors know, 16 is currently the smallest dimension for an undecidablity result over two matrices. It would be interesting to consider whether there exists other problems related to matrices with undecidbility over some smaller dimension. We showed a connection between semigroup membership and piecewise reachability, however we believe this connection can be strengthened to show other undecidability results in piecewise functions.

Acknowledgement

We would like to thank the referees for their helpful suggestions and especially the referee who informed us of a missing case in one of our proofs.

References

1. E.Asarin, G.Schneider. Widening the Boundary between Decidable and Undecidable Hybrid Systems. CONCUR 2002: 193-208, 2002.
2. J.Bestel and J. Karhumäki. Combinatorics on Words - A Tutorial, Bulletin of the EATCS, 178 - 228, February 2003.
3. V.Blondel, J.Tsitsiklis, When is a pair of matrices mortal? IPL 63, 283-286, 1997.
4. J. Cassaigne, T. Harju and J. Karhumäki, On the decidability of the freeness of matrix semigroups. In the special issue of International Journal of Algebra and Computation, 9, 295-305, 1999.
5. J.Cassaigne, J.Karhumäki, Examples of undecidable problems for 2-generator matrix semigroups. Theoretical Computer Science, 204, 29-34, 1998.

6. S.Dube. Undecidable problems in fractal geometry. Complex Systems 7, no. 6, 423–444, 1993.
7. S.Gaubert, R.Katz, Reachability problems for products of matrices in semirings, Manuscript, 2003.
8. V.Halava, T.Harju, M.Hervensalo, J.Karhumäki, Skolem's Problem - On the Border Between Decidability and Undecidability, TUCS Technical Report, 2005.
9. V.Halava, T.Harju, Mortality in matrix semigroups, Amer. Math. Monthly, 2001.
10. P. Koiran, My favourite problems, http://www.ens-lyon.fr/~koiran/problems.html.
11. P.Koiran, C.Moore. Closed-form Analytic Maps in One and Two Dimensions can Simulate Universal Turing Machines. Theor. Comput. Sci. 210(1), 217-223, 1999.
12. O.Kurganskyy, I.Potapov. Computation in One-Dimensional Piecewise Maps and Planar Pseudo-Billiard Systems. Unconventional Computation, LNCS 3699, 169-175, 2005.
13. Y.Matiyasevich, G.Senizergues. Decision problems for semi-Thue systems with a few rules. Theoretical Computer Science, 330(1), 145-169, 2005.
14. M.S. Paterson. Unsolvability in 3×3 matrices, Studies in Appl. Math., v 49, 105-107, 1970.
15. I.Potapov. From Post Systems to the Reachability Problems for Matrix Semigroups and Multicounter Automata. Developments in Language Theory, LNCS 3340, 345-356, 2004.

Complexity of Degenerated Three Dimensional Billiard Words

J.-P. Borel[*]

XLim, UMR CNRS 6172, - Université de Limoges, 123 avenue Albert Thomas
F-87060 Limoges Cedex - France
borel@unilim.fr

Abstract. We consider Billiard Words on alphabets with $k = 3$ letters: such words are associated to some 3-dimensional positive vector $\overrightarrow{\alpha} = (\alpha_1, \alpha_2, \alpha_3)$. The language of these words is already known in the usual case, i.e., when the α_j are linearly independent over $\mathbb{Q}$, and so for the α_j^{-1}'s. Here we study the language of these words when there exist some linear relations. We give the complexity of Billiard Words in any case, which has asymptotically a "polynomial-like" form, with degree less or equal to 2. These results are obtained by geometrical methods.

Keywords: Languages, Billiard Words, Complexity.
AMS Classification. 68R15.

1 k-Dimensional Billiard Words

1.1 Standard Billiard Words

Let $\mathcal{D}$ be the half-line of origin O, in the k-dimensional space $\mathbb{R}^k$, and parallel to the positive vector $\overrightarrow{\alpha} := (\alpha_1, \alpha_2, \ldots, \alpha_k)$. Then we define the associated Standard Billiard word, or *cutting sequence*, (starting from O) denoted by $c_{\overrightarrow{\alpha}} = c_{\alpha_1, \alpha_2, \ldots, \alpha_k}$ on the alphabet $\mathcal{A} = \{a_1, a_2, \ldots, a_k\}$ as shown in Fig.1.

In dimension 2, this can be made using the three following methods:

1. encoding by a_1 the black horizontal unitary segment and by a_2 the black vertical unitary segment (see Fig.1.a). Then $a_1 c_{\alpha_1, \alpha_2}$ encodes the discrete path immediately below the half-line, hence $c_{\alpha_1, \alpha_2} = a_2 a_1 a_2 a_1 a_2 a_2 a_1 a_2 a_1 \ldots$ in Fig.1.a. The infinite word $a_1 c_{\alpha_1, \alpha_2}$ is the well-known *Christoffel word*. However this method cannot be generalized in higher dimensions;
2. moving from the origin to infinity, encode the sequence of intercepts between $\mathcal{D}$ and the grid, using a_1 for a vertical line and a_2 for an horizontal one (black points on Fig.1.a);
3. by looking at the sequence of the centers (white points) of the unit squares crossed by $\mathcal{D}$. Two consecutive centers correspond to joining squares, so that the vector joining these two points is one of the two vectors of the canonical basis $(\overrightarrow{e_1}, \overrightarrow{e_2})$. Then encode by a_j the vector $\overrightarrow{e_j}$, $j = 1, 2$.

---

[*] Partially supported by Region Limousin.

O.H. Ibarra and Z. Dang (Eds.): DLT 2006, LNCS 4036, pp. 386–396, 2006.

In higher dimension $k \geq 3$, both methods 2 and 3 can be generalized. We consider now the *facets* of the unit k-cubes crossed by $\mathcal{D}$, instead of the sides of the unit squares (see Fig.1.b, with $k = 3$). In this figure, the crossing points are the black points and the centers of unit k-cubes are the white ones, as in Fig.1.a. In both cases, we encode the vectors $\vec{e_j}$ ($1 \leq j \leq k$) of the canonical basis by the letters a_j, and a crossed facet by its orthogonal direction, and we get in Fig.1.b the billiard word $c_{\alpha_1, \alpha_2, \alpha_3} = a_1 a_2 a_3 a_1 a_2 a_3 \ldots$.

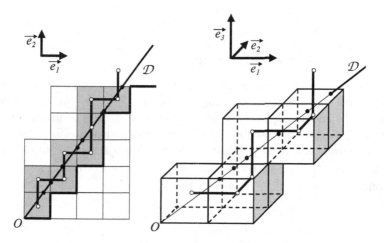

Fig. 1. Standard Billiars Words in dimension 2 (left part **1.a**) and in dimension 3 (right part **1.b**)

This works as long as the half-line $\mathcal{D}$ crosses each facet in its interior (so we can define *consecutive* crossed unit k-cubes), i.e., $\mathcal{D}$ does not contain any 2-integer point, except for O: a *2-integer point* in $\mathbb{R}^k$ is a point with at least two integer coordinates, see [9]. Geometrically, a 2-integer point is a point which belongs to more than two unit k-cubes of the grid. This property for $\mathcal{D}$ corresponds to the following condition which is (almost) always assumed in the sequel:

Hypothesis 1. *All ratios $\dfrac{\alpha_i}{\alpha_j}$ are irrational numbers, $1 \leq i < j \leq k$.*

This condition already holds in the usual case:

Hypothesis 2. *The real numbers $\alpha_1, \alpha_2, \ldots, \alpha_k$, are $\mathbb{Q}$-linearly independent.*

This strong hypothesis has always been made in the former works in this topic, and was not sufficient: it has been pointed out in [5] when $k = 3$. In this paper, we also consider the following hypothesis:

Hypothesis 3. *The real numbers $\dfrac{1}{\alpha_1}, \dfrac{1}{\alpha_2}, \ldots, \dfrac{1}{\alpha_k}$, are $\mathbb{Q}$-linearly independent.*

Note that Hyp.1, Hyp.2 and Hyp.3 are identical when $k = 2$. When Hyp.2 or Hyp.3 are not satisfied, there exist some relations over $\mathbb{Z}$:

$$\sum_{j=1}^{k} m_j \alpha_j = 0 \tag{1}$$

or

$$\sum_{j=1}^{k} \frac{m_j}{\alpha_j} = 0 \tag{2}$$

with coprime integers m_j, $1 \le j \le k$.

1.2   Billiard Words with Intercept

The same construction can be made with the half-line $\mathcal{D}$ starting from any point S and parallel to the positive vector $\vec{\alpha}$. To simplify we can choose S in the subspace $\vec{\alpha}^{\perp}$, and by periodicity we can assume that S is in the orthogonal projection $\mathcal{P}$ of the unit k-cube centered at the origin onto $\vec{\alpha}^{\perp}$. Thus $\mathcal{P}$ is a $(k-1)$-dimensional convex polyhedron, and the projection O' of the center of this unit k-cube is a symmetry center for $\mathcal{P}$. We denote by $\vec{b_j}$, $1 \le j \le k$, the orthogonal projections of $\vec{e_j}$ onto $\vec{\alpha}^{\perp}$. For $k = 3$, $\mathcal{P}$ is an hexagon whose edges

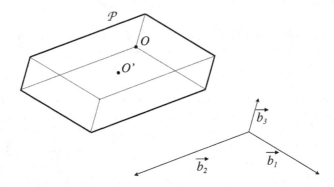

Fig. 2. The hexagon $\mathcal{P}$ and the three vectors b_1, b_2 and b_3 when $k = 3$

correspond to the three vectors $\vec{b_1}$, $\vec{b_2}$ and $\vec{b_3}$, see Fig.2. O is the starting point of the Standard Billiard word, and $\overrightarrow{O'O} = -\dfrac{1}{2}\sum_{j=1}^{k} \vec{b_j}$.

The starting point $S \in \mathcal{P}$ is called a *nonambiguous point* in $\mathcal{P}$ when the line $\mathcal{L}$ parallel to $\vec{\alpha}$ and passing by S does not contain any 2–integer point. Then the Billiard Word starting from S is well defined, as we have seen before (we only need that the positive part $\mathcal{D}$ of $\mathcal{L}$ does not contain any 2-integer point). This word is denoted by $c_{\vec{\alpha},S} = c_{\alpha_1,\alpha_2,\ldots,\alpha_k,S}$.

We must also consider those starting points S such that $\mathcal{L}$ contains 2-integer points: the existence and position of these points is the keypoint of the main theorem. We call these points a *d-ambiguous point* when $\mathcal{L}$ contains exactly d 2-integer points, or a ∞-*ambiguous point* when $\mathcal{L}$ contains infinitely many 2-integer points.

The set of ambiguous points is neglectible in $\mathcal{P}$, in the sense of the Lebesgue measure of $\mathbb{R}^{k-1}$. However, except for some vectors $\overrightarrow{\alpha}$, this set is dense in $\mathcal{P}$.

2   Already Known Results on Factors of Billiard Words

Billiard Words and Sturmian Words have been intensively studied, see [1], [7], [8], [14] or [15], for general surveys, and many results are known, concerning the *language* of these words, i.e., the set of all finite factors.

In the usual case, i.e., with Hyp.2 and Hyp.3, the complexity function is known, and so for the palindromic complexity function: the *complexity function* $p_u(n)$ (resp. $p_{\mathcal{U}}(n)$) of an infinite word u (resp. of a set $\mathcal{U}$ of words u) is the number of distinct factors v of length n of u (resp. of some word $u \in \mathcal{U}$), see [17], [3] and [13] for a general exposure concerning complexity..

In the generic case, the language of the Billiard word $c_{\overrightarrow{\alpha},S}$ depends only on the vector $\overrightarrow{\alpha}$, and the complexity function $p_k(n) = p_{k,\overrightarrow{\alpha},S}(n)$ depends only on the dimension k:

$$p_2(n) = n + 1$$

$$p_3(n) = n^2 + n + 1$$

$$p_k(n) = \sum_{i=0}^{min(k-1,n)} i! \binom{k-1}{i} \binom{n}{i}$$

These results come from the original works on Sturmian Words in dimension $k = 2$, from [2] (dimension $k = 3$, with the additional hypothesis that Hyp.3 is also satisfied as it was noticed in [5]; this result was first conjectured in [16]), and [4] (for any k). In this last case, the property that the complexity function does not depend on $\overrightarrow{\alpha}$ is the main key of the proof which is both technical and complicated. When Hyp.2 is true, but not for Hyp.3, it can be proved that $An^2 \leq p_3(n) \leq Bn^2$ for positive constants A and B, see [6].

We consider in the following the case $k = 3$, and the two complexity functions $p(n) := p_3(n)$ of the Billiard Word $c_{\overrightarrow{\alpha},S}$ and $p^+(n) := p_3^+(n)$ of the set of all Billiard Words $c_{\overrightarrow{\alpha},S}$ corresponding to a given vector $\overrightarrow{\alpha}$, S being any nonambiguous starting point.

3   Main Result

3.1   Some Notations

In this section we have $k = 3$. We consider the six distinct possibilities:

Case 1. Hyp.2 and Hyp.3 are true.

Case 2. Hyp.2 is true and Hyp.3 is false.

Case 3. Hyp.2 is false and Hyp.3 is true.

Case 4. Both Hyp.2 and Hyp.3 are false, and Hyp.1 is true.

Case 5. Hyp.1 is false (and so for Hyp.2 and Hyp.3), but only one of the ratios $\dfrac{\alpha_1}{\alpha_2}, \dfrac{\alpha_2}{\alpha_3}, \dfrac{\alpha_3}{\alpha_1}$ is a rational number.

Case 6. The three ratios $\dfrac{\alpha_1}{\alpha_2}, \dfrac{\alpha_2}{\alpha_3}, \dfrac{\alpha_3}{\alpha_1}$ are rational numbers.

(when two of these ratios are rational numbers, so is the third one).

Denote by d_+ and d_- the dimension of the $\mathbb{Q}$-vector space generated by $(\alpha_1, \alpha_2, \alpha_3)$ and $(\dfrac{1}{\alpha_1}, \dfrac{1}{\alpha_2}, \dfrac{1}{\alpha_3})$ respectively. We have $(d_+, d_-) = (3,3), (3,2),$ $(2,3), (2,2), (2,2), (1,1)$ respectively, in the six cases above.

When $d_+ = 2$, we denote $p_1\alpha_1 + p_2\alpha_2 + p_3\alpha_3 = 0$ the only relation (1) with integer coprime numbers p_j (this relation is defined up to a factor ± 1, we set $P := |p_1| + |p_2| + |p_3|$), and when $d_- = 2$, $\dfrac{q_1}{\alpha_1} = \dfrac{q_2}{\alpha_2} + \dfrac{q_3}{\alpha_3}$ the only relation (2) with coprime positive integers q_j (this gives the general case, up to a circular permutation of the coordinates).

In Case 5 we define coprime positive integers by $\dfrac{\alpha_1}{\alpha_2} = \dfrac{r_1}{r_2}$ when it is a rational number, which can be supposed up to a circular permutation of coordinates. In Case 6 we define positive coprime integers r_j by $(\alpha_1, \alpha_2, \alpha_3) = \lambda(r_1, r_2, r_3)$.

We consider a given vector $\overrightarrow{\alpha}$. Then the compexity function $p_k(n)$ of the Billiard Word depends only on the starting point S, which needs to be a non-ambiguous point.

3.2   The Main Theorem

Theorem 1. *The complexity functions $p(n)$ and $p^+(n)$ have the following forms:*

– *(Case 1, see [2]) for all n and all S :*

$$p^+(n) = p(n) = n^2 + n + 1$$

– *(Case 2) for all S :*

$$p^+(n) = p(n) = (1 - \frac{1}{2q_1} \frac{\alpha_1}{\alpha_1 + \alpha_2 + \alpha_3})n^2 + O(n)$$

– *(Case 3) for all n :*

$$p^+(n) = n^2 + n + 1$$

and, only for large n:

$$p(n) = Pn + P \text{ in the general case for } S$$

$$p(n) = (P-1)n + P'' \text{ for special } S$$

with $P'' \geq P$ except $P'' = P - 1$ in a very special case;

– *(Case 4) for all n :*

$$p^+(n) = (1 - \frac{1}{2q_1}\frac{\alpha_1}{\alpha_1 + \alpha_2 + \alpha_3})n^2 + O(n)$$

and, only for large n :

$$p(n) = Pn + P \text{ in the general case for } S$$

$$p(n) = P'n + P'' \text{ for special } S$$

with $P' = P - 1$ or $P' = P - 2$, and $P'' \geq P$, and $P'' \geq P$;
– *(Case 5)*

$$p^+(n) = \frac{\alpha_3}{\alpha_1 + \alpha_2 + \alpha_3}n^2 + O(n)$$

and for large n :

$$p(n) = Rn + R \text{ in the general case for } S$$

$$p(n) = (R - 1)n + R'' \text{ for special } S$$

with $R'' \geq R := r_1 + r_2$;
– *(Case 6) for large n:*

$$p^+(n) = (R - 1)R$$

$$p(n) = R$$

with $R := r_1 + r_2 + r_3$, *the first formula is valid for pairwise coprime integers* r_j.

The special starting points S can be explicitely given as we see further. In Case 2 the result improves Theorem 8 in [6]. In Case 6, a formula for $p^+(n)$ can be given in the general case, using the $r_{ij} := \text{pgcd}(r_i, r_j)$ for $i \neq j$. The formula for $p(n)$ in Case 6 may be viewed as a classical unwritten result. Remark that the ratio $\dfrac{\alpha_j}{\alpha_1 + \alpha_2 + \alpha_3}$ is the asymptotic frequency of occurence of the letter a_j in the Billiard Word.

4   Some Details on the Proofs

4.1   Some New Results: On 2-Integer Points on a Line

We consider a line $\mathcal{L}$ parallel to $\overrightarrow{\alpha}$ and passing by the point S in $\mathcal{P}$.

Lemma 1. *([11]) With Hyp.1:*

– *for any $1 \leq j \neq j' \leq k$, there is at most one point M in $\mathcal{L}$ with rational coordinates m_j and $m_{j'}$;*
– *there is at most $\dfrac{n(n-1)}{2}$ 2-integer points in $\mathcal{L}$;*

- *when $\mathcal{L}$ contains an integer point, there exist no other 2-integer point in $\mathcal{L}$;*
- *when $\mathcal{L}$ contains a point with rational non integer coordinates, there exist no 2-integer point in $\mathcal{L}$, i.e., S is a nonambiguous point.*

The proof is very easy, as we have $\dfrac{m_{2j} - m_{1j}}{m'_{2j'} - m'_{1j'}} = \dfrac{\lambda\alpha_j}{\lambda\alpha_{j'}} = \dfrac{\alpha_j}{\alpha_{j'}} \notin \mathbb{Q}$ when

$\overrightarrow{M'M} = \lambda\overrightarrow{\alpha}$. The second item is an easy consequence of the first one.

We said that the finite sequence $j_1, j_2, \ldots, j_{d+1}$ of indices is a *chain* of length d if there exist a sequence $M_1, M_2, \ldots, M_d$ of distinct points in $\mathcal{L}$ such that M_i has rational coordinates of indices $j_i \neq j_{i+1}$. This chain is called a *circular chain* when $j_{d+1} = j_1$. We denote by d_- the dimension of the $\mathbb{Q}$-vector space generated by $(\dfrac{1}{\alpha_1}, \dfrac{1}{\alpha_2}, \ldots, \dfrac{1}{\alpha_k})$.

Lemma 2

Circular chain exactly corresponds to some linear relation $\displaystyle\sum_{i=1}^{d} \dfrac{n_i}{\alpha_{j_i}} = 0$ with nonzero integer coefficients.

The proof is based on $\overrightarrow{M_iM_{i+1}} = \dfrac{m_{i+1,j} - m_{i,j}}{\alpha_{i+1}}\overrightarrow{\alpha}$ with $j := j_{i+1}$. Using lemma 2 with $k = 3$, we may have at most two 2-integer points on a line $\mathcal{L}$ with Hyp.3 and at most three with Hyp.1.

4.2   Previous Results

The following results can be proved for any $k \geq 3$ and most of them are proved in [11]. As we used them for computing the complexity functions when $k = 3$, we give their expressions in this context, which is rather simple.

Grids and Tilings of the Plane. The hexagon $\mathcal{P}$ is a fundamental domain associated to the lattice $\mathcal{L}_0$ generated in $\overrightarrow{\alpha}^\perp$ by the vectors $\overrightarrow{b_3} - \overrightarrow{b_1}$ and $\overrightarrow{b_2} - \overrightarrow{b_1}$. For $i \geq 0$ we denote by $\mathcal{M}_i$ the sets of the hexagons $T_{\overrightarrow{b}}\mathcal{P}$ translated from $\mathcal{P}$ by the vectors $\overrightarrow{b} = n_1\overrightarrow{b_1} + n_2\overrightarrow{b_2} + n_3\overrightarrow{b_3}$, with relative integers such that $n_1 + n_2 + n_3 = -i$ (see Fig.3 for the tilings $\mathcal{M}_0, \mathcal{M}_1, \mathcal{M}_2$). We get regular tilings of $\overrightarrow{\alpha}^\perp$, and we have $\mathcal{M}_{i+1} = T_{-\overrightarrow{b_j}}\mathcal{M}_i$ for any $1 \leq j \leq 3$. We consider in the following the pieces of $\mathcal{P}$ obtained by cutting $\mathcal{P}$ by all the grids corresponding to the tilings $\mathcal{M}_i$, $1 \leq i \leq n$. In Fig.3 we give these thirteen pieces corresponding to $n = 3$, i.e., the possible factors of length 3 for the Billiard Words. We denote by $\mathcal{P}_{(n)}$ a generic piece.

On the Projection of the Integer Points. We denote by $\mathcal{G}$ the closure of the set of the projection of the integer points in $\mathbb{R}^k$ onto $\overrightarrow{\alpha}^\perp$. Using a classical result on closed subgroups in vector spaces $\mathbb{R}^d$, see for example [12], TG VII.5, Theorem 2, we describe in [11] the set $\mathcal{G}$, which is in the case $k = 3$ the set of those points (x_1, x_2, x_3) in $\overrightarrow{\alpha}^\perp$ such that $m_1x_1 + m_2x_2 + m_3x_3 \in \mathbb{Z}$ for any integers (m_1, m_2, m_3) satisfying (1). It implies that $\mathcal{G}$ is:

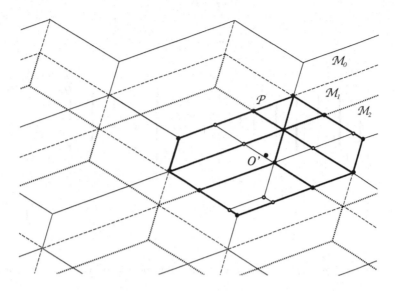

Fig. 3. Grids and pieces in the hexagon

- when $d_+ = 3$, the whole plane $\overrightarrow{\alpha}^\perp$;
- when $d_+ = 2$, a set of equidistant parallel lines, perpendicular to the vector $\overrightarrow{p} := (p_1, p_2, p_3)$;
- when $d_+ = 1$, a lattice of the plane $\overrightarrow{\alpha}^\perp$.

On the Set of Edges of "b-Walks". We introduce in [10] the notion of b-walk, which consists to start from some point $S = S_0$ in $\mathcal{P}$, and make some translation to get a new point $S_1 := S + \overrightarrow{b_j}$ in $\mathcal{P}$. When S is a nonambiguous point, there exist only one index $1 \leq j_1 \leq 3$ such that $S + \overrightarrow{b_j}$ still remains in $\mathcal{P}$, and we can iterate to get an infinite b-walk, coded by the sequence of indices $j_1, \ldots$. We prove in [11] that the infinite word coding this walk is the Billiard Word $c_{\overrightarrow{\alpha}, S}$, and that the closure $\mathcal{H}_S$ of the set of the edges $S_0, S_1, \ldots$ of the infinite b-walk starting from S is the intersection between $\mathcal{P}$ and $S + \mathcal{G}$. So we get ($\equiv_1$ means that the difference is an integer):

Proposition 1. *When $k = 3$ the set $\mathcal{H}_S$ is:*

- *the whole hexagon $\mathcal{P}$ in Cases 1 and 2;*
- *the set of those points (x_1, x_2, x_3) such that $p_1 x_1 + p_2 x_2 + p_3 x_3 \equiv_1 p_1 s_1 + p_2 s_2 + p_3 s_3$ in Cases 3 and 4, which is a set of P parallel segments, except when $p_1 s_1 + p_2 s_2 + p_3 s_3 = 0$, when it contains only $P - 1$ parallel segments (see Fig.4.a and 4.b resp.);*
- *a set of R equidistant segments parallel to $\overrightarrow{b_\ell}$ in Case 5;*
- *a set of R points in Case 6.*

In Fig.4, $(p_1, p_2, p_3) = (3, 1, -2)$ and O' belongs to $\mathcal{G}$ as the sum $P = p_1 + p_2 + p_3$ is even. We have $\overrightarrow{p} = 3\overrightarrow{b_1} + \overrightarrow{b_2} - 2\overrightarrow{b_3} \in \overrightarrow{\alpha}^\perp$, and the set $\mathcal{H}_S$ is the union of six (resp. five) vertical thick segments.

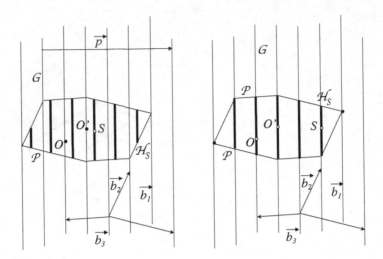

Fig. 4. The set $\mathcal{H}_S$, generic case (**4.a** on the left) and special case (**4.b** on the right)

On the Complexity Functions. We prove in [11] that each polygonal piece $\mathcal{P}_{(n)}$ of $\mathcal{P}$ is the set $\mathcal{P}_v$ of the starting points of the b walks of length n encoded by some word v of length n. Then the complexity function $p^+(n)$ is the number $\Pi(n)$ of pieces of $\mathcal{P}$ under the tilings $\mathcal{M}_1, \mathcal{M}_2, \ldots, \mathcal{M}_n$.

In the same way, we prove that $\mathcal{H}_S$ is the closure set of the edges of the infinite b-walk starting from S, hence the complexity function $p(n)$ is the number $\Pi_S(n)$ of pieces of $\mathcal{P}$ under the tilings $\mathcal{M}_1, \mathcal{M}_2, \ldots, \mathcal{M}_n$, whose the interior intersects the set $\mathcal{H}_S$.

Moreover, except in Case 6, the diameter of the pieces $\mathcal{P}_v$ tends to zero as the length $|v|$ of v tends to infinity. So we get:

- when $d_+ = 3$, $p(n) = \Pi_S(n) = \Pi(n)$;
- when $d_+ = 2$, $p(n) = \Pi_S(n)$ for large n.

4.3   Computation of $\Pi(n)$ and $\Pi_S(n)$ When $k = 3$

The Number of Pieces of $\mathcal{P}$. The keypoint is to observe that two grids $\mathcal{M}_i$ and $\mathcal{M}_j$ have exactly two intersecting points in $\mathcal{P}$ for $i < j$, except for $j = i+1$ (only one, which is the new edge due to $\mathcal{M}_i$) and $i = 0$ (four intersecting points) (see the white points in Fig.3, which correspond to the intersecting points between $\mathcal{M}_3$ and $\mathcal{M}_j$, $j = 0, 1, 2$, respectively $4, 0, 2$ points, and the one corresponding to the new edge), and that these points are always distinct, except if it corresponds to 3-ambiguous points.

These exceptional points appear only in Cases 2 and 4.

When they do not appear, we get $n^2 + 2n + 4$ edges and $2n^2 + 3n + 4$ vertices for positive n, and so $n^2 + n + 1$ pieces, this is the main result of [2].

As one 3-ambiguous point appears the number of edges decreases of 2 and the number of vertices decreases of 1, and we loose one piece. Hence the complexity is equal to $n^2 + n + 1 - m$, where $m = m(n)$ is the number of exceptional points, i.e., of common points of three grids $\mathcal{M}_{i_1}, \mathcal{M}_{i_2} \mathcal{M}_{i_3}$, $0 \le i_1 < i_2 < i_3 \le n$. The following lemma, whose proof is based on Lemma 2, gives the anounced result in this case.

Lemma 3. *When* $d_- = 2$, *the number of common points of three grids* $\mathcal{M}_{i_1}, \mathcal{M}_{i_2} \mathcal{M}_{i_3}$, $0 = i_1 < i_2 < i_3 \le n$, *is the maximal integer* κ *such that*

$$\kappa q_1 + \left\lfloor \kappa q_1 \frac{\alpha_2}{\alpha_1} \right\rfloor + \left\lfloor \kappa q_1 \frac{\alpha_3}{\alpha_1} \right\rfloor \le n.$$

In particular, we have $\kappa = \dfrac{1}{q_1} \dfrac{\alpha_1}{\alpha_1 + \alpha_2 + \alpha_3} n + O(1)$.

The Number of Pieces of $\mathcal{H}_S$. When $d_+ = 2$ the number of intersecting points between the $\mathcal{M}_i$'s and $\mathcal{H}_S$ modulo $\mathcal{P}$ is equal to the number of segments composing $\mathcal{H}_S$. This number is equal to P or $P-1$. If we consider only the number e_n of new intersecting points, i.e., those in $\mathcal{M}_n$ but not in the preceeding ones, then we have:

- $e_0 = P$ or $P - 1$;
- $e_{n-1} - e_n$ is the number of common points of $\mathcal{H}_S$, $\mathcal{M}_0$ and $\mathcal{M}_n$.

Lemma 4. *(For $d_+ = 2$) A common point (x_1, x_2, x_3) of $\mathcal{M}_0$ and $\mathcal{M}_m$ satisfies*

$$p_1 x_1 + p_2 x_2 + p_3 x_3 \equiv_1 \left\{ -N \frac{p_i \alpha_i}{\alpha_\ell} \right\} = \left\{ N \frac{p_j \alpha_j}{\alpha_\ell} \right\}.$$

Any $\{i, j, \ell\} = \{1, 2, 3\}$ and $N \in \mathbb{Z}^*$ appears exactly one time, for some value $m \in \mathbb{N}^*$.

In Case 3, these scalar products are all distincts. This proves that either $e_n = e_0$ for all n, or $e_n = e_0$ for $n \le n_0$, and $e_n = e_0 - 1$ for $n > n_0$. This gives the annouced form of the complexity function $p(n)$ in this case, the exceptional starting points being those such that the scalar product $p_1 s_1 + p_2 s_2 + p_3 s_3$ is modulo one a nonzero multiple of some ratio $\dfrac{p_i \alpha_i}{\alpha_j}$.

In Case 4, the scalar products are not always distincts, and a given value can be obtained one or two times. So there also exist points S such that $e_n = e_0$ for small n, and then on some interval $e_n = e_0 - 1$, and then $e_n = e_0 - 2$ for large n.

In Case 5, the same value may appears infinitely many in these scalar products modulo one. The lines of $\mathcal{G}$ are parallel to $\vec{b_3}$, the set $\mathcal{H}_S$ is the union of $r_1 + r_2$ segments parallel to $\vec{b_3}$, and we get easily $\Pi_S(n) = (r_1 + r_2)(n + 1)$, as each new tiling $\mathcal{M}_m$, $m \ge 1$, crosses $(r_1 + r_2)$ time exactly the set $\mathcal{H}_S$, except when there exist 2-ambiguous points on $\mathcal{H}_S$. Then the result is the same as in Case 3. However, the computation of $\Pi(n)$ is rather different.

Case 6 is a very special case and cannot be treated in the same way. As the integers r_j are coprime, the Billiard Word $c_{\overrightarrow{\alpha},S}$ is periodic with a smallest period equal to R, so that $p(n) = R$ for $n \geq R$. The computation of $\Pi(n)$ is more technical. When the r_j are paiwise coprime integers and for large n, the hexagon is cutted by $r_1 + r_2 - 1$ segments (resp. $r_1 + r_3 - 1$, $r_2 + r_3 - 1$) in the three directions parallel to its sides. Then we cut the hexagon, to build a new fundamental domain of the lattice $\mathcal{L}_0$, which is a parallelogram. The number of pieces can be computed more easily in this parallelogram, and we get the formula $p^+(n) = (R-1)R$.

References

1. J.-P. Allouche, J. Shallit, *Automatic sequences: Theory and Applications*, Cambridge University Press, Cambridge, 2003.
2. P. Arnoux, C. Mauduit, I. Shiokawa, J. I. Tamura, Complexity of sequences defined by billiard in the cube, Bull. Soc. Math. France **122** (1994) 1–12.
3. P. Arnoux, G. Rauzy, Représentation géomtrique de suites de complexité $2n + 1$, Bull. Soc. Math. France **119** (1991) 199–215.
4. Y. Baryshnikov, Complexity of trajectories in rectangular billiards, Comm. Math. Phys. **174** (1995) 43–56.
5. N. Bédaride, Etude du billard dans un polyhédre, PhD Thesis, Univ. Aix-Marseille, 2005.
6. N. Bédaride, Billiard complexity in rational polyhedra, Regul Chaotic Dyn, **8**, 2003, 97–104.
7. J. Berstel, A. de Luca, Sturmian words, Lyndon words and trees, Theoret. Comput. Sci. **178** (1997) 171–203.
8. J. Berstel, P. Séébold, Sturmian words, in M. Lothaire, *Algebraic combinatorics on words*, Cambridge University Press, 2002.
9. J.-P. Borel, C. Reutenauer, Palindromic factors of billiard words, Theoret. Comput. Sci. **340-2** (2005) 334–348.
10. J.-P. Borel, C. Reutenauer, Some new results on palindromic factors of billiard words, in DLT 2005, De Felice - Restivo ed., Lecture Notes Comput. Sci. **3572** (2005) 180–188.
11. J.-P. Borel, Complexity and palindromic complexity of billiard words, in WORDS'05 5th Intern. Conf. on Words, Brlek - Reutenauer ed., Publ. LACIM Montral **36** (2005) 175–184.
12. N. Bourbaki, *Topologie générale, Chap. 5-10*, Herman, Paris, 1974.
13. J. Cassaigne, Complexité et facteurs spéciaux, Bull. Belg. Math. Soc. Simon Stevin **4** (1997) 67–88.
14. A. de Luca, Sturmian words: structure, combinatorics, and their arithmetics, Theoret. Comput. Sci. **183**, (1997) 45–82.
15. A. de Luca, Combinatorics of standard Sturmian words, in Structures in Logic and Computer Science, Lecture Notes Comput. Sci. **1261** (1997) 249–267.
16. G.Rauzy, Des mots en arithmetique, in Avignon Conf. on Language Theory and Algorithmic Complexity 1983, Univ. Claude-Bernard, Lyon (1984), 103–113.
17. G. Rauzy, Sequences with terms in a finite alphabet, *Seminar on Number Theory Univ. Bordeaux I* (1983) exp. n25, 16pp.

Factorial Languages of Low Combinatorial Complexity

Arseny M. Shur*

Ural State University
Ekaterinburg, Russia
Arseny.Shur@usu.ru

Abstract. The complexity (or *growth*) functions of some languages are studied over arbitrary nontrivial alphabets. The attention is focused on the languages determined by a finite set of forbidden factors, called an *antidictionary*. Let m be a nonnegative integer. Examples of languages of complexity $\Theta(n^m)$ with finite (and even symmetric finite) antidictionaries are given. For any integer s such that $1 \leq s \leq m$, a sequence of languages with finite antidictionaries and $\Theta(n^m)$ complexities, converging to the language of $\Theta(n^s)$ complexity, is exhibited. Some languages of intermediate complexity are shown to be the limits of sequences of languages with finite antidictionaries and polynomial complexities.

The combinatorial complexity of a formal language (just *complexity* throughout the paper) is a function that shows the diversity of the language. The most well-known and intensively studied particular case of complexity is the subword complexity of an infinite word (see Sect. 9 of [3], for example). Also, some attention is drawn to complexity of languages of power-free words. The first result in this direction was obtained in [2], where some bounds were given for complexity of two important languages, namely, the language of binary cube-free words, and the one of ternary square-free words.

At the same time, complexity is an important characteristics of any language. So, there are good reasons to study complexity of languages in a more general framework, moving from the question "what complexity has the given language?" to the question "what complexities can the languages from a given class have?". So far, a satisfactory classification of complexity classes is known only for infinite words, generated by iterations of morphisms (cf. [3]), and this particular result is already highly nontrivial. In this paper we continue the study of complexity classes for rational languages, initiated in [7], and show that polynomial complexities of any degree are permitted for an interesting and important subclass of rational languages.

The mentioned subclass consists of *languages with finite antidictionaries* (see the definition below), and plays a very important role in the study of complexity

---

* Supported by Federal Science and Innovation Agency of Russia under the grant 02.442.11.7521, and by Russian Foundation for Basic Research under the grant 05-01-00540.

O.H. Ibarra and Z. Dang (Eds.): DLT 2006, LNCS 4036, pp. 397–407, 2006.
© Springer-Verlag Berlin Heidelberg 2006

of languages. The complexity of a language with finite antidictionary can be effectively evaluated. The finite automaton, recognizing such a language, can be consructed using the algorithm of [4]. The structure of such an automaton allows us to decide whether the language is exponential or polynomial, and to determine the precise degree of the polynomial in the latter case ([7], see Theorem 1 below). In the former case the growth rate of the language (the base of the exponential function) is equal to the Frobenius root of the adjacency matrix of the automaton (cf. [5]).

Thus, to evaluate the complexity of a language, we can approximate this function by complexities of languages with finite antidictionaries. To do this, we construct a sequence of languages with finite antidictionaries, converging to the target language. In this paper we consider both natural ways of approximation: from above and from below. For the case of the above approximation we consider the relation between the polynomial complexity of the target language and complexities of its approximations. The approximation from below is used to prove that some languages defined over non-trivial finite alphabets under some natural conditions have intermediate complexities.

The paper is organized as follows. The definitions and some necessary preliminaries are given in Sect. 1. In Sect. 2 two countable families of finite automata are introduced. Sections 3–5 contain main results, the proofs of which are based on the properties of those automata.

1   Preliminaries

We begin with some notation and definitions on words, finite automata and complexity functions. For more background see, e.g., [6].

An *alphabet* Σ is a non-empty set the elements of which are called *letters*. A *word* is a finite sequence of letters, say $W = a_1 \ldots a_n$. A word U is a *factor* of the word W if W can be written as PUQ for some (possibly empty) words P and Q. The *power factorization* of a word W is its factorization $W = a_{i_1}^{n_1} \ldots a_{i_t}^{n_t}$ to the minimal number of factors with all factors being powers of a single letter. As usual, we write Σ^n for the set of all n-letter words, and Σ^* for the set of all words over Σ. The subsets of Σ^* are called *languages*. A language is *factorial* if it is closed under taking factors of its words, and *antifactorial* if no one of its words is a factor of another one. If X is a word over an auxiliary alphabet Δ, W is a word over Σ, and for any substitution $f : \Delta \to \Sigma^* \backslash \{\lambda\}$ the word $f(X)$ is not a factor of W, then W is said to *avoid* the *pattern* X.

A *deterministic finite automaton* (DFA) is a 5-tuple $(\Sigma, Q, \delta, s, T)$ consisting of a finite input alphabet Σ, a finite set of states (vertices) Q, a partial transition function $\delta : Q \times \Sigma \to Q$, one initial state s, and a set of terminal states T. The underlying digraph of the automaton contains states as vertices and transitions as directed labeled edges. Then every path in this digraph is labeled by a word. We make no difference between a DFA and its underlying digraph. An *accepting path* is any path from the initial to a terminal vertex. A DFA *recognizes* the language which is the set of all labels of the accepting paths, and is *consistent* if each its vertex is contained in some accepting path. A *trie* is a DFA whose underlying digraph is a tree such that the initial vertex is its root and the set of terminal vertices is the set of all its leaves.

For an arbitrary language L over a finite alphabet Σ the *complexity function* (or simply *complexity*) is defined by $C_L(n) = |L \cap \Sigma^n|$. We are interested in the growth rate rather than the precise form of the complexity function. As usual, we call a complexity function *polynomial* if it is $O(n^p)$ for some $p \geq 0$ (bounded from above by a polynomial of degree p), and *exponential* if it is $\Omega(\alpha^n)$ for some $\alpha > 1$ (bounded from below by an exponential function at base α). We also write $\Theta(n^p)$ for the function which is bounded from above and from below by polynomials of degree p. A complexity function is said to be *intermediate* if it is bounded neither by a polynomial from above nor by an exponential function from below. Alternatively, it can be said that such a function is *superpolynomial* and *subexponential*.

The definition of Ω (and, hence, of Θ) suits well only for increasing functions. So, if the complexity function is not increasing, we estimate its fastest increasing subsequence. In this paper we deal only with factorial languages (but Theorem 1 works for any rational language). For a factorial language the complexity is known to be either bounded by a constant or strictly increasing (cf. [3], and also [1] for the proof in the general case). We also note that the complexity of the language of all finite factors of an infinite word W is well-known in combinatorics of words under the name of *subword complexity* of the word W.

The following useful theorem characterizes complexities of rational languages.

Theorem 1 ([7]). *Let a language L be recognized by a consistent DFA $\mathcal{A}$. Then*

   1) If $\mathcal{A}$ is acyclic, then L is finite;

   2) If $\mathcal{A}$ contains two cycles sharing one vertex, then L is exponential;

   3) If $\mathcal{A}$ contains a cycle, and all cycles in $\mathcal{A}$ are disjoint, then L is polynomial, and its complexity function is $\Theta(n^{m-1})$, where m is the maximum number of cycles encountered by an accepting path.

A word W is *forbidden* for the language L if it is a factor of no element of L. A forbidden word is minimal if all its proper factors are not forbidden. The sets of all minimal forbidden words for given languages were intensively studied by different authors. One of the most significant early works is [5]. Following [4], we refer to such sets as *antidictionaries* of languages. The antidictionary is always antifactorial. If a factorial language L over the alphabet Σ has the antidictionary AD, then the following equalities holds:

$$L = \Sigma^* \setminus AD \cdot \Sigma^* \cdot AD, \qquad AD = \Sigma \cdot L \cap L \cdot \Sigma \cap \Sigma^* \setminus L.$$

We see that any antidictionary determines a unique factorial language, which is rational if the antidictionary is also rational. In particular, the factorial languages with finite antidictionaries form a proper subclass of the class of rational languages. The role of this subclass in the study of complexity was already mentioned in the introduction. In order to build an automaton recognizing a language with a known finite antidictionary, we use the algorithm of [4] which is shortly described here in a suitable notation.

Algorithm.
Input: an antidictionary AD.
Output: a DFA $\mathcal{A}$ recognizing the factorial language L with the antidictionary AD.
Step 1. Construct a trie $\mathcal{T}$, recognizing AD. ($\mathcal{T}$ is actually the digraph of the prefix order on the set of all prefixes of AD.)
Step 2. Associate each vertex of $\mathcal{T}$ with the word labeling the accepting path ending in this vertex. (Now the set of vertices is the set of all prefixes of AD.)
Step 3. Add all possible edges to $\mathcal{T}$, following the rule:
the edge (U, V) labeled by a should be added if
   U is not terminal, and
   U has no outgoing edge labeled by a, and
   V is the longest suffix of Ua which is a vertex of $\mathcal{T}$.
(These edges are called *backward* while the edges of the trie are called *forward*.)
Step 4. Remove terminal vertices and mark all remaining vertices as terminal to get $\mathcal{A}$.

It is easy to check that $\mathcal{A}$ is consistent, and hence can be studied using Theorem 1.

2   Web-Like Automata

The results of this paper are derived from the properties of deterministic finite automata of a special form, recognizing factorial languages with finite antidictionaries.

   Let Σ be a k-letter alphabet with $k \geq 2$. To define the first family of DFA, suppose that Σ is endowed with a cyclic order $\prec$. Let $\bar{a}$ denote the successor of a in this order. The family of finite antidictionaries $\{AD_{m,\prec}\}_{m \geq 1}$ over Σ is defined by

$$AD_{m,\prec} = \{ab \mid a, b \in \Sigma, \, b \neq a, \, b \neq \bar{a}\} \cup$$
$$\{a^2 \bar{a} \bar{\bar{a}}, a^3 \bar{a}^2 \bar{\bar{a}}, \ldots, a^m \bar{a}^{m-1} \bar{\bar{a}} \mid a \in \Sigma\} \cup \{a^{m+1} \bar{a} \mid a \in \Sigma\}. \tag{1}$$

For any m, we use the Algorithm to construct the DFA recognizing the factorial language $L_{m,\prec}$ with the antidictionary $AD_{m,\prec}$ (see Fig. 1). We denote this DFA by $\mathcal{W}_{k,m}$ and call it a *web-like automaton*.

   This automaton contains k loops, and exactly one cycle of length sk for any $s = 1, \ldots, m$, which is referred to as the *level s cycle*. It is clear that all cycles of $\mathcal{W}_{k,m}$ are disjoint. An accepting path encounters at most $m+1$ cycles, which are level 1 to level m cycles, and a loop. Thus, $L_{m,\prec}$ has $\Theta(n^m)$ complexity by Theorem 1. The internal structure of the words of $L_{m,\prec}$ is clarified by the following lemma.

Lemma 1. *The language $L_{m,\prec}$ consists of all words with the power factorization $U = a_{i_1}^{t_1} \ldots a_{i_n}^{t_n}$ such that*
   1) $a_{i_j} \prec a_{i_{j+1}}$ for all $j = 1, \ldots n-1$,
   2) $t_j \leq t_{j+1}$ for all $j = 1, \ldots n-2$, and
   3) $t_j \leq m$ for all $j = 1, \ldots n-1$.

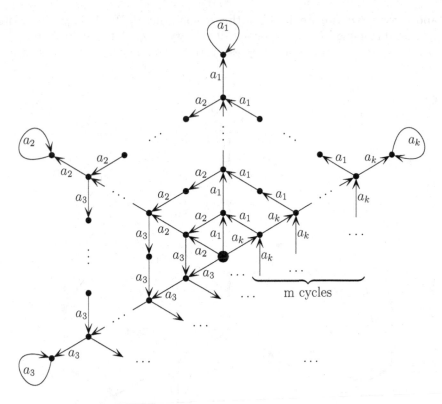

Fig. 1. The web-like automaton $\mathcal{W}_{k,m}$. The cyclic order is $a_1 \prec a_2 \prec \ldots \prec a_k \prec a_1$. The larger circle represents the initial vertex.

Proof. The first, second and third sets of words in the antidictionary $AD_{m,\prec}$ (see (1)) provide exactly the conditions 1), 2), and 3) of the lemma, respectively.   □

Now define the second family of DFA. It consists of more complicated automata, which we call *generalized web-like automata*. Such an automaton can be considered as a special construction of a large finite number of isomorphic web-like automata.

Consider a family of symmetric finite antidictionaries $\{AD_m\}_{m \geq 1}$ over Σ, where AD_m is the minimal symmetric (that is, stable under all permutations of Σ) language containing the following set of words:

$$\{ \; a_1^{m+1} a_2,$$
$$a_1^m a_2^m a_1, \qquad a_1^m a_2^m a_3^m a_1, \qquad \ldots, a_1^m a_2^m \ldots a_{k-1}^m a_1,$$
$$a_1^m a_2^{m-1} a_3,$$
$$a_1^{m-1} a_2^{m-1} a_1, \; a_1^{m-1} a_2^{m-1} a_3^{m-1} a_1, \ldots, a_1^{m-1} a_2^{m-1} \ldots a_{k-1}^{m-1} a_1,$$
$$a_1^{m-1} a_2^{m-2} a_3,$$
$$\ldots$$
$$a_1^2 a_2 a_3,$$
$$a_1 a_2 a_1, \qquad a_1 a_2 a_3 a_1, \qquad \ldots, a_1 a_2 \ldots a_{k-1} a_1 \qquad \}.$$

$$(2)$$

As above, for any m, we use the Algorithm to construct the DFA $\mathcal{G}_{k,m}$ recognizing the factorial language L_m with the antidictionary AD_m. The automaton $\mathcal{G}_{3,2}$ is shown in Fig. 2 (this is the most representative example which can be drawn in one page).

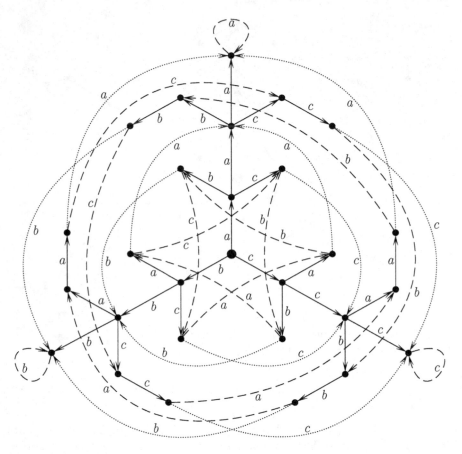

Fig. 2. The generalized web-like automaton $\mathcal{G}_{3,2}$. The forward edges are drawn by usual lines, while the backward ones are represented by dash or dotted lines depending on whether they belong to a cycle or not. The larger circle represents the initial vertex.

The cycles of this automaton are exhausted by k loops, and $(k-1)!$ level s cycles of length sk for any $s = 1, \ldots, m$. In Fig. 2 one can see level 1 cycles labeled by abc (vertices bc, ca, ab), and acb (vertices cb, ba, ac); level 2 cycles are labeled by $aabbcc$ (vertices $bbcc$, cca, $ccaa$, aab, $aabb$, bbc), and $aaccbb$ (vertices $ccbb$, bba, $bbaa$, aac, $aacc$, ccb). All cycles of $\mathcal{G}_{k,m}$ are disjoint, and an accepting path encounters at most $m+1$ cycles. The corresponding proofs are omitted for the sake of brevity. Thus, L_m has $\Theta(n^m)$ complexity by Theorem 1. The internal structure of the words of L_m is a bit more complicated, than the one of the words of $L_{m,\prec}$, as the following lemma shows.

Lemma 2. *The language L_m consists of all words with the power factorization* $U = a_{i_1}^{t_1} \ldots a_{i_n}^{t_n}$ *such that*

1) *there exist cyclic orders* $\prec_1, \ldots, \prec_m$ *on* Σ *such that* $t_j = s$ *implies* $a_{i_j} \prec_s a_{i_{j+1}}$ *for all* $j = 1, \ldots n-1$,

  2) $t_j \leq t_{j+1}$ *for all* $j = 1, \ldots n-2$, *and*

  3) $t_j \leq m$ *for all* $j = 1, \ldots n-1$.

Proof. The words of the antidictionary that are situated in the odd rows of (2) clearly provide the conditions 2) and 3). To prove 1) we construct the required cyclic order $\prec_s$ for any $s = 1, \ldots, m$. Fix the number s and consider the segment $U_s = a_{i_l}^{t_l} \ldots a_{i_r}^{t_r} a_{i_{r+1}}^{t_{r+1}}$ of the power factorization of U such that

$$t_l = t_r = s,$$
$$t_{l-1} < s \text{ or } l = 1,$$
$$t_{r+1} > s \text{ or } r+1 = n.$$

Since the antidictionary contains the words

$$a_1^s a_2^s a_1, \; a_1^s a_2^s a_3^s a_1, \; \ldots, \; a_1^s a_2^s \ldots a_{k-1}^s a_1,$$

together with all their cyclic permutations, we obtain the following. If U_s consists of at most k powers of letters, then all these letters are different. Hence we can order the existing letters as $a_{i_l} \prec \ldots \prec a_{i_r} \prec a_{i_{r+1}}$, and complete this partial order to a cyclic order on Σ in an arbitrary way. Suppose that U_s consists of more than k powers of letters. We see that any k successive powers are that of different letters. Then $a_{i_{l+k}} = a_{i_l}$, $a_{i_{l+k+1}} = a_{i_{l+1}}$, and so on. Therefore we can define the required cyclic order as

$$a_{i_l} \prec_s a_{i_{l+1}} \prec_s \ldots \prec_s a_{i_{l+k-1}} \prec_s a_{i_l}.$$

It can be directly verified that any word with the power factorization satisfying the conditions 1)–3) has no factors from AD_m and hence belongs to L_m.   □

3   Polynomial Languages with Finite Antidictionaries

Since the languages with finite antidictionaries are very important in the study of complexity functions, it is very natural to ask the following question: *what kind of complexity functions can languages with finite antidictionaries have?*

Some well-known properties of words are stable under all permutations of the alphabet. The language of all words possessing such a property is closed under all permutations of the alphabet, i.e., *symmetric*. For example, the language of all words avoiding a given pattern is symmetric. A symmetric language surely has the symmetric antidictionary. So, it is natural to consider a resctricted version of the above question as well: *what kind of complexity functions can languages with symmetric finite antidictionaries have?*

Here we give a partial answer to both of the above questions.

Theorem 2. *For arbitrary alphabet Σ such that $|\Sigma| > 1$, and arbitrary nonnegative integer m there exists a factorial language over Σ with a non-symmetric finite antidictionary and the complexity $\Theta(n^m)$.*

Proof. For the binary alphabet this statement was proved in [7], Theorem 4.1. If $|\Sigma| \geq 3$, and $\prec$ is an arbitrary cyclic order on Σ, then the language $L_{m,\prec}$ has the required properties. $\square$

Theorem 3. *For arbitrary alphabet Σ such that $|\Sigma| > 1$, and arbitrary nonnegative integer m there exists a factorial language over Σ with a symmetric finite antidictionary and the complexity $\Theta(n^m)$.*

Proof. The language L_m has the required properties. $\square$

4   Factorial Languages: Approximations from Above

The complexity of any factorial language can be estimated using the following approximation scheme. Let $L \in \Sigma^*$ be a factorial language with the (infinite) antidictionary M. Consider a family $\{M_i\}$ of finite subsets of M such that

$$M_1 \subseteq M_2 \subseteq \ldots \subseteq M_n \subseteq \ldots \subseteq M, \qquad \bigcup_{i=1}^{\infty} M_i = M.$$

(It is often convinient to take for M_i the set of all words of M with lengths at most i.) Denote by L_i the factorial language over Σ having the antidictionary M_i. One has

$$L \subseteq \ldots \subseteq L_i \subseteq \ldots \subseteq L_1,$$

and for any n, there is i such that $L \cap \Sigma^n = L_i \cap \Sigma^n$. Then for any n

$$C_L(n) = \ldots = C_{L_i}(n) \leq \ldots \leq C_{L_1}(n). \tag{3}$$

Hence the sequence $C_{L_i}(n)$ converges to $C_L(n)$ from above. It is clear that with larger i we obtain better approximations, but the computational complexity in most cases depends on i exponentially (actually it depends polynomially on the total length of the words in the antidictionary). The sequence $\{L_i\}$ will be called an *above approximating sequence* of the language L.

Assuming that L is polynomial we study the possibilities for complexities of the languages L_i. If all but finitely many of them belong to the same complexity class, we say that the sequence $\{L_i\}$ has the *final complexity* of this class. In view of Theorem 1 it is clear that only two cases may occur. Either all L_i are exponential, or all but finitely many of them have polynomial complexity of the same degree m. It is clear that all approximation sequences of the same language have the same final complexity. Indeed, the elements of any approximation sequence can be placed in between the languages of the "standard" approximation sequence $\{L_i\}$ with the antidictionaries defined by $M_i = M \cap (\Sigma \cup \ldots \cup \Sigma^i)$.

It was proved in [7] that the above approximating sequences of the Thue-Morse language have exponential final complexity, while the complexity of the language itself is known to be linear. S.V. Avgustinovich and A.E. Frid gave an example of $\Theta(n^2)$ ternary language, the above approximating sequences of which belong to the same complexity class (personal communication). Therefore, the second case mentioned above is possible as well. Moreover, it takes place in very different forms, as the theorem below shows.

Theorem 4. *For any non-trivial alphabet Σ and any positive integers s and m, $s \leq m$, there exists a factorial language over Σ having an infinite antidictionary and $\Theta(n^s)$ complexity, while its above approximating sequences have the final complexity $\Theta(n^m)$.*

The remaining special case $s = 0$ is covered by the following proposition.

Proposition 1. *Let L be a factorial language over a non-trivial alphabet Σ with an infinite antidictionary and $\Theta(1)$ complexity. Then the above approximating sequences of L have the final complexity $\Theta(1)$.*

Proof. The complexity of an arbitrary factorial language is either bounded by a constant or exceeds n for all n (see [1], and also [3], Theorem 9.1). Suppose that the complexity function $C_L(n)$ is bounded by a constant N. Fix an $n \geq N$. By (3) we have $C_L(n) = C_{L_i}(n)$ for some approximation L_i of L. Consequently, we have $C_{L_i}(n) \leq N \leq n$, and thus $C_{L_i}(n)$ is bounded by a constant. Since the sequence $C_{L_j}(n)$ is decreasing and converges to $C_L(n)$ from above, the statement follows. □

To proof Theorem 4, we take the language $L_{m,\prec}$ of complexity $\Theta(n^m)$, and exclude words from it, expanding the antidictionary $AD_{m,\prec}$. We add words to $AD_{m,\prec}$ in an infinite sequence of steps, finite set of words per one step. On each step $O(n^{m-1})$ words are excluded from the current language, but as the result of the entire procedure, $L_{m,\prec}$ loses $\Theta(n^m)$ words. The expansion of $AD_{m,\prec}$ can be organized in such a way, that the resulting language will have the complexity $\Theta(n^s)$ for any $1 \leq s \leq m$. The description of the corresponding expansion procedures is omitted here.

5   Factorial Languages: Approximations from Below

The approximations from below of factorial languages by languages with finite antidictionaries are much more restrictive. For many languages all such approximations are trivial, due to the following lemma.

Lemma 3. *Let L be the set of all words avoiding a given set of patterns. Then L has no infinite subset with finite antidictionary.*

Proof. Assume that L' is an infinite subset of L with finite antidictionary. Then its recognizing automaton has a cycle. Hence, any power of the label V of this cycle is a factor of some word of L'. Take an avoided pattern X and substitute V for each letter of X to get a contradiction. □

Nevertheless, approximations from below provide another useful tool for studying complexity, as our next result shows.

Consider the language L of all words $U \in \Sigma^*$ with the power factorization $U = a_{i_1}^{t_1} \ldots a_{i_n}^{t_n}$ satisfying $t_j \leq t_{j+1}$ for all $j = 1, \ldots, n-2$. Thus, the powers of letters in U are non-decreasing, with the last letter being the only possible exception. This exception is necessary to make L factorial. Now try to determine the complexity of L.

The language L is obviously exponential if $k \geq 3$. For example, it contains the language of all square-free words over Σ, which is known to be exponential (cf. [2]). However, there is no evidence about the complexity of L in the case of binary alphabet. It appears to be intermediate, as a partial case of a more general result on subsets of L for arbitrary k.

We introduce two subsets of L. For the first one we fix a cyclic order $\prec$ on Σ, and consider all words $U \in L$ with the power factorizations satisfying the additional condition $a_{i_j} \prec a_{i_{j+1}}$ for all $j = 1, \ldots, n-1$. We denote the obtained language by $L_{\prec}$.

As to the second mentioned subset of L, we can informally say that all its words *locally* satisfy the same additional condition. More precisely, this subset, denoted by $\bar{L}$, consists of all words $U \in L$ with the power factorizations satisfying the following condition. There exists an infinite sequence $\{\prec_m\}$ of cyclic orders on Σ such that for any $m \in \mathbb{N}$ the statement (*) below holds true. A *segment* of the power factorization is a product of several consecutive factors.

(*) Suppose that $a_{i_l}^m \ldots a_{i_r}^m$ is a segment of the power factorization such that $t_{l-1} < m$ or $l = 1$, and $t_{r+1} > m$ or $r+1 = n$. Then
$$a_{i_l} \prec_m \ldots \prec_m a_{i_r} \prec_m a_{i_{r+1}}.$$

Note that $L_{\prec} \subseteq \bar{L}$ for any cyclic order $\prec$, and $L_{\prec} = \bar{L} = L$ for the binary alphabet.

Theorem 5. *The languages $L_{\prec}$ and $\bar{L}$ have intermediate complexity.*

Proof (of superpolynomiality). By Lemmas 1 and 2 we have

$$L_{1,\prec} \subset \ldots \subset L_{m,\prec} \subset \ldots \subset L_{\prec}; \qquad L_1 \subset \ldots \subset L_m \subset \ldots \subset \bar{L};$$
$$\bigcup_{m=1}^{\infty} L_{m,\prec} = L_{\prec}; \qquad\qquad \bigcup_{m=1}^{\infty} L_m = \bar{L}.$$

Since the complexity of the languages $L_{m,\prec}$ and L_m is $\Theta(n^m)$, we immediately conclude that the complexity of $L_{\prec}$ and $\bar{L}$ is not bounded from above by a polynomial. $\square$

The proof of subexponentiality is long and based on detailed study of the automata $\mathcal{W}_{k,m}$ and $\mathcal{G}_{k,m}$, so we omit it here. The complexity functions of both languages are shown to be $O(n^{\sqrt{n}})$.

Acknowledgement

The author is greatful to Yu. Karhumaki for fruitful discussions.

References

1. J. Balogh, B. Bollobas, *Hereditary properties of words*, RAIRO – Inf. Theor. Appl., **39** (2005), 49-65.
2. F.-J. Brandenburg, *Uniformly growing k-th power free homomorphisms*, Theor. Comput. Sci., **23** (1983), 69-82.
3. C. Choffrut, J. Karhumäki, *Combinatorics of words*, in: G. Rosenberg, A. Salomaa (Eds.), Handbook of formal languages, v.1, Ch.6, Springer, Berlin, 1997, 329-438.
4. M. Crochemore, F. Mignosi, A. Restivo, *Automata and forbidden words*, Inform. Processing Letters, **67**(3) (1998), 111-117.
5. Y. Kobayashi, *Repetition-free words*, Theor. Comp. Sci. **44** (1986), 175-197.
6. M. Lothaire. Combinatorics on words. Addison-Wesley, 1983.
7. A.M. Shur, *Combinatorial complexity of rational languages*, Discr. Anal. and Oper. Research, Ser. 1, **12**(2) (2005), 78-99 (Russian).

Perfect Correspondences Between Dot-Depth and Polynomial-Time Hierarchy

Christian Glaßer, Stephen Travers, and Klaus W. Wagner

Theoretische Informatik
Julius-Maximilians Universität Würzburg
Am Hubland
97074 Würzburg, Germany
{glasser, travers, wagner}@informatik.uni-wuerzburg.de

Abstract. We introduce the polynomial-time tree reducibility (ptt-reducibility). Our main result establishes a one-one correspondence between this reducibility and inclusions between complexity classes. More precisely, for languages B and C it holds that B ptt-reduces to C if and only if the *unbalanced* leaf-language class of B is robustly contained in the unbalanced leaf-language class of C. Formerly, such correspondence was only known for *balanced* leaf-language classes [Bovet, Crescenzi, Silvestri, Vereshchagin].

We show that restricted to regular languages, the levels 0, 1/2, 1, and 3/2 of the dot-depth hierarchy (DDH) are closed under ptt-reducibility. This gives evidence that the correspondence between the dot-depth hierarchy and the polynomial-time hierarchy is closer than formerly known.

Our results also have applications in complexity theory: We obtain the first gap theorem of leaf-language definability above the Boolean closure of NP. Previously, such gap theorems were only known for P, NP, and Σ_2^P [Borchert, Kuske, Stephan, and Schmitz].

1 Introduction

In their pioneering work for the leaf-language approach, Bovet, Crescenzi, and Silvestri [4] and Vereshchagin [18] independently introduced the notion of polylog-time reducibility (plt-reducibility for short). This reducibility allows an amazing translation between two seemingly independent questions.

1. Are given complexity classes separable by oracles?
2. Are given languages plt-reducible?

Leaf Languages. The translation mentioned above uses the concept of *leaf languages*. Let M be a nondeterministic polynomial-time bounded Turing machine such that every computation path outputs one letter from a fixed alphabet. Let $M(x)$ denote the computation tree of M on input x. Let $\beta_M(x)$ be the concatenation of all leaf-symbols of $M(x)$. For a language B, let $\mathrm{Leaf}_\mathrm{u}^\mathrm{P}(B)$ be the class of

O.H. Ibarra and Z. Dang (Eds.): DLT 2006, LNCS 4036, pp. 408–419, 2006.

languages L such that there exists a nondeterministic polynomial-time-bounded Turing machine M as above such that for all x,

$$x \in L \iff \beta_M(x) \in B.$$

We refer to $\mathrm{Leaf}_u^p(B)$ as the *unbalanced leaf-language class* of B. Call a non-deterministic polynomial-time-bounded Turing machine M *balanced* if on all inputs, it produces a balanced computation tree in the following sense: On input x, there exists a path p with length $l(x)$ such that all paths on the left of p have length $l(x)$, and all paths on the right have length $l(x) - 1$. This is equivalent to demanding that there exists a polynomial-time computable function that on input (x, n) computes the n-th path of $M(x)$. If we assume M to be balanced in the definition above, then this defines the class $\mathrm{Leaf}_b^p(B)$ which we call the *balanced leaf-language class* of B. For any class of languages $\mathcal{C}$ let $\mathrm{Leaf}_u^p(\mathcal{C}) = \bigcup_{B \in \mathcal{C}} \mathrm{Leaf}_u^p(B)$ and $\mathrm{Leaf}_b^p(\mathcal{C}) = \bigcup_{B \in \mathcal{C}} \mathrm{Leaf}_b^p(B)$. Call a complexity class $\mathcal{D}$ *unbalanced leaf-language definable* if there exists $\mathcal{C}$ such that $\mathcal{D} = \mathrm{Leaf}_u^p(\mathcal{C})$. Analogously define *balanced leaf-language definability*. We will also consider relativized leaf-language classes which are denoted by $\mathrm{Leaf}_b^p(\mathcal{C})^O$, where the superscript O indicates that the nondeterministic machine is allowed to query oracle O. For a survey on leaf-languages we refer to [19].

BCSV-Theorem. Suppose for given complexity classes $\mathcal{D}_1$ and $\mathcal{D}_2$, there exist languages L_1 and L_2 such that $\mathcal{D}_1 = \mathrm{Leaf}_b^p(L_1)$ and $\mathcal{D}_2 = \mathrm{Leaf}_b^p(L_2)$. Bovet, Crescenzi, Silvestri, and Vereshchagin showed:

$$L_1 \leq_m^{\mathrm{plt}} L_2 \iff \forall O \big(\mathrm{Leaf}_b^{pO}(L_1) \subseteq \mathrm{Leaf}_b^{pO}(L_2) \big) \tag{1}$$

Here $\leq_m^{\mathrm{plt}}$ denotes polylog-time reducibility (Definition 1). For this equivalence it is crucial that balanced leaf-language classes are used. The theorem does not hold for the unbalanced model: Observe that languages $L, L' \subseteq \{0,1\}^*$ with $L =_{\mathrm{def}} \{w \mid |w| \text{ is odd}\}$, $L' =_{\mathrm{def}} 0\{0,1\}^*$ form a counterexample, since $\mathrm{Leaf}_u^p(L) = \oplus P$ is not robustly contained (i.e., relative to all oracles) in $\mathrm{Leaf}_u^p(L') = P$ though L plt-reduces to L'. In this paper we introduce a new reducibility (ptt-reducibility) which allows us to prove:

$$L_1 \leq_m^{\mathrm{ptt}} L_2 \iff \forall O \big(\mathrm{Leaf}_u^{pO}(L_1) \subseteq \mathrm{Leaf}_u^{pO}(L_2) \big) \tag{2}$$

Besides the scientific interest of a Bovet-Crescenzi-Silvestri-Vereshchagin-like theorem (BCSV-theorem for short) for the unbalanced case, further motivation comes from a connection between complexity theory and the theory of finite automata: We show that on the lower levels, the dot-depth hierarchy *perfectly corresponds* to the polynomial-time hierarchy when we consider *unbalanced* leaf-languages. Below, after the introduction of both hierarchies, we will explain the term *perfect correspondence*.

Dot-Depth Hierarchy. *Starfree regular languages* (starfree languages for short) are regular languages that can be build up from single letters by using Boolean operations and concatenation (so iteration is not allowed). SF denotes the class of starfree languages. Brzozowski and Cohen [8, 5] introduced the *dot-depth hierarchy* (DDH for short) which is a parameterization of the class of starfree languages. The dot-depth counts the minimal number of nested alternations between Boolean operations and concatenation that is needed to define a language. The classes of the dot-depth hierarchy consist of languages that have the same dot-depth. For a class of languages $\mathcal{C}$, let $\mathrm{Pol}(\mathcal{C})$ denote $\mathcal{C}$'s closure under finite union and finite concatenation. Let $\mathrm{BC}(\mathcal{C})$ denote the Boolean closure of $\mathcal{C}$. The classes (or levels) of the dot-depth hierarchy are defined as:

$$\mathcal{B}_0 =_{\mathrm{def}} \{L \subseteq A^* \mid A \text{ is a finite alphabet with at least two letters and } L$$
$$\text{is a finite union of terms } vA^*w \text{ where } v, w \in A^*\}$$
$$\mathcal{B}_{n+\frac{1}{2}} =_{\mathrm{def}} \mathrm{Pol}(\mathcal{B}_n)$$
$$\mathcal{B}_{n+1} =_{\mathrm{def}} \mathrm{BC}(\mathcal{B}_{n+\frac{1}{2}})$$

The dot-depth of a language L is defined as the minimal m such that $L \in \mathcal{B}_m$ where $m = n/2$ for some integer n. All levels of the dot-depth hierarchy are closed under union, under intersection, under taking inverse morphisms, and under taking residuals [13, 1, 14]. The dot-depth hierarchy is strict [6, 17] and exhausts the class of starfree languages [9].

Polynomial-Time Hierarchy. For a complexity class $\mathcal{D}$ let $\mathrm{co}\mathcal{D} = \{\overline{L} \mid L \in \mathcal{D}\}$. Let $\exists \cdot \mathcal{D}$ denote the class of languages L such that there exists a polynomial p and $B \in \mathcal{D}$ such that $x \in L \Leftrightarrow \exists y, |y| \leq p(|x|), (x, y) \in B$. Let $\forall \cdot \mathcal{D} = \mathrm{co}\exists \cdot \mathrm{co}\mathcal{D}$. Define $\exists! \cdot \mathcal{D}$ and $\forall! \cdot \mathcal{D}$ similarly by using $\exists!$ and $\forall!$ instead of $\exists$ and $\forall$. Stockmeyer [16] introduced the polynomial-time hierarchy (PH for short). We use a definition which is due to Wrathall [20].

$$\Sigma_0^{\mathrm{P}} = \Pi_0^{\mathrm{P}} =_{\mathrm{def}} \mathrm{P}, \quad \Sigma_{n+1}^{\mathrm{P}} =_{\mathrm{def}} \exists \cdot \Pi_n^{\mathrm{P}}, \quad \Pi_{n+1}^{\mathrm{P}} =_{\mathrm{def}} \forall \cdot \Sigma_n^{\mathrm{P}}$$

Connection between DDH and PH. Hertrampf et al. [11], and Burtschick and Vollmer [7] proved that the levels of the polynomial-time hierarchy are connected with the levels of the dot-depth hierarchy. For $n \geq 1$,

$$L \in \mathcal{B}_{n-1/2} \Rightarrow \forall O(\mathrm{Leaf}_{\mathrm{b}}^{\mathrm{p}O}(L) \subseteq \Sigma_n^{\mathrm{P}O}), \tag{3}$$

$$L \in \mathcal{B}_{n-1/2} \Rightarrow \forall O(\mathrm{Leaf}_{\mathrm{u}}^{\mathrm{p}O}(L) \subseteq \Sigma_n^{\mathrm{P}O}). \tag{4}$$

In particular, the attraction of this connection comes from the fact that both hierarchies are prominent and well-studied objects. Even more, with the P-NP problem and the dot-depth problem, they represent two of the most fundamental problems in theoretical computer science.

Gap Theorems. For certain lower levels of the dot-depth hierarchy, much closer connections than those in (3) and (4) are known. The following theorem is due to Borchert, Kuske, Stephan, and Schmitz:

Theorem 1 ([2, 3, 15]). *Let L be a regular language.*

1. *[2] If $L \in \mathcal{B}_0$, then $\mathrm{Leaf}_u^p(L) \subseteq \mathrm{P}$. If $L \notin \mathcal{B}_0$, then $\mathrm{Leaf}_u^p(L) \supseteq \mathrm{NP}$ or $\mathrm{Leaf}_u^p(L) \supseteq \mathrm{coNP}$ or $\mathrm{Leaf}_u^p(L) \supseteq \mathrm{MOD}_p\mathrm{P}$ for a prime p.*
2. *[3] If $L \in \mathcal{B}_{1/2}$, then $\mathrm{Leaf}_u^p(L) \subseteq \mathrm{NP}$. If $L \notin \mathcal{B}_{1/2}$, then $\mathrm{Leaf}_u^p(L) \supseteq \mathrm{coNP}$ or $\mathrm{Leaf}_u^p(L) \supseteq \mathrm{co1NP}$ or $\mathrm{Leaf}_u^p(L) \supseteq \mathrm{MOD}_p\mathrm{P}$ for a prime p.*
3. *[15] If $L \in \mathcal{B}_{3/2}$, then $\mathrm{Leaf}_u^p(L) \subseteq \Sigma_2^P$. If $L \notin \mathcal{B}_{3/2}$, then $\mathrm{Leaf}_u^p(L) \supseteq \forall \cdot \mathrm{UP}$ or $\mathrm{Leaf}_u^p(L) \supseteq \mathrm{co}\exists! \cdot \mathrm{UP}$ or $\mathrm{Leaf}_u^p(L) \supseteq \mathrm{MOD}_p\mathrm{P}$ for a prime p.*

For instance, by (4), for all $L \in \mathcal{B}_{1/2}$ it holds that $\mathrm{Leaf}_u^p(L)$ is robustly contained in NP. Theorem 1 states that the languages in $\mathcal{B}_{1/2}$ are in fact the *only* regular languages having this property. This means that for $\mathcal{B}_{1/2}$ and regular L, even the converse of (4) holds. So $\mathcal{B}_{1/2}$ and NP perfectly correspond:

$$L \in \mathcal{B}_{1/2} \Leftrightarrow \forall O(\mathrm{Leaf}_u^{pO}(L) \subseteq \mathrm{NP}^O) \tag{5}$$

By our main result (2) this is equivalent to the following:

Restricted to regular languages, $\mathcal{B}_{1/2}$ is closed under ptt-reducibility.   (6)

Here and in the following, this formulation means that $\mathcal{R}_m^{ptt}(\mathcal{B}_{1/2}) \cap \mathrm{REG} = \mathcal{B}_{1/2}$ where $\mathcal{R}_m^{ptt}(\mathcal{B}_{1/2})$ denotes $\mathcal{B}_{1/2}$'s closure under ptt-reducibility.

Perfect Correspondence. The example above shows that we can utilize (1) and (2) to make the notion of perfect correspondence precise:

1. A class of regular languages $\mathcal{C}$ and a complexity class $\mathcal{D}$ *perfectly correspond with respect to balanced leaf-languages* if (restricted to regular languages) $\mathcal{C}$ is closed under plt-reducibility and $\mathrm{Leaf}_b^p(\mathcal{C}) = \mathcal{D}$.
2. A class of regular languages $\mathcal{C}$ and a complexity class $\mathcal{D}$ *perfectly correspond with respect to unbalanced leaf-languages* if (restricted to regular languages) $\mathcal{C}$ is closed under ptt-reducibility and $\mathrm{Leaf}_u^p(\mathcal{C}) = \mathcal{D}$.

Due to (2), we obtain the following perfect correspondences from known results [2, 3, 15].

- $\mathcal{B}_0$ perfectly corresponds to P with respect to unbalanced leaf-languages.
- $\mathcal{B}_{1/2}$ perfectly corresponds to NP with respect to unbalanced leaf-languages.
- $\mathcal{B}_{3/2}$ perfectly corresponds to Σ_2^P with respect to unbalanced leaf-languages.

In other words, we show that restricted to regular languages, the classes $\mathcal{B}_0$, $\mathcal{B}_{1/2}$, and $\mathcal{B}_{3/2}$ are closed under ptt-reducibility.

Furthermore, we show that restricted to regular languages, $\mathcal{B}_1$ is closed under ptt-reducibility. From this we obtain a new perfect correspondence:

– $\mathcal{B}_1$ perfectly corresponds to the Boolean closure of NP with respect to unbalanced leaf-languages.

Consequently, the following holds for every regular language L :

$$L \in \mathcal{B}_0 \Leftrightarrow \forall O(\mathrm{Leaf}_u^{pO}(L) \subseteq P^O) \tag{7}$$

$$L \in \mathcal{B}_{1/2} \Leftrightarrow \forall O(\mathrm{Leaf}_u^{pO}(L) \subseteq \mathrm{NP}^O) \tag{8}$$

$$L \in \mathcal{B}_1 \Leftrightarrow \forall O(\mathrm{Leaf}_u^{pO}(L) \subseteq \mathrm{BC(NP)}^O) \tag{9}$$

$$L \in \mathcal{B}_{3/2} \Leftrightarrow \forall O(\mathrm{Leaf}_u^{pO}(L) \subseteq \Sigma_2^{pO}) \tag{10}$$

As the dot-depth hierarchy perfectly corresponds to the polynomial-time hierarchy on the lower levels, we consider this as evidence that restricted to regular languages, *all* levels of the dot-depth hierarchy might be closed under ptt-reducibility. This would turn (4) into an equivalence (for regular L).

Remarkably, this correspondence does not hold for *balanced* leaf-language classes: It is known that the reverse of (3) does not hold, even if we demand L to be starfree: For every $n \geq 1$, there exists a starfree regular language $L_n \notin \mathcal{B}_{n-1/2}$ such that L_n plt-reduces to a language in $\mathcal{B}_{1/2}$ [10]. So by (1), $\forall O(\mathrm{Leaf}_b^{pO}(L_n) \subseteq \mathrm{NP}^O)$, but $L_n \notin \mathcal{B}_{n-1/2}$. This shows that the levels of the dot-depth hierarchy are not closed under plt-reducibility even if we restrict ourselves to starfree regular languages.

A New Gap Theorem. From our studies of the ptt-closure of $\mathcal{B}_1$, we obtain a gap in unbalanced leaf-language definability above the Boolean hierarchy over NP: If $\mathcal{D} = \mathrm{Leaf}_u^p(\mathcal{C})$ for some class $\mathcal{C}$ of regular languages, then $\mathcal{D} \subseteq \mathrm{BC(NP)}$ or there exists an oracle O such that $\mathcal{D}^O \not\subseteq \mathrm{P}^{\mathrm{NP}[\epsilon \cdot \log n]^O}$ for all $\epsilon < 1$.

Remarks. Our investigations of the ptt-reducibility further show the following phenomenon: While we can (unconditionally) prove that level 0 of the dot-depth hierarchy is closed under ptt-reducibility, we can show the similar property for higher levels only if we restrict ourselves to regular languages. We can construct a language $B \in \mathrm{NP} \setminus \mathrm{REG}$ that is ptt-reducible to a language in $\mathcal{B}_{1/2}$. The exception of level 0 allows to improve the correspondence between $\mathcal{B}_0$ and P even further: Not only that $\mathcal{B}_0$ and P perfectly correspond, but in fact it even holds that for any language $L \notin \mathcal{B}_0$ (this includes all nonregular languages) there exists an oracle O such that $\mathrm{Leaf}_u^{pO}(L) \supsetneq P^O$.

Besides the new perfect correspondence stated in (9), dot-depth one also is closely related to a certain class of arithmetic circuits. Maciel, Peladeau, and Thérien [12] use a concept of *non-uniform* leaf languages to establish a one-one connection between these classes.

Organization of the Paper. Section 3 defines ptt-reducibility. In section 4 we formulate the central result of this paper. Section 5 studies the ptt-closure of classes of the dot-depth hierarchy, and it shows that on some lower levels, the dot-depth hierarchy perfectly corresponds to the polynomial-time hierarchy.

2   Preliminaries

For a machine or automaton M, let $L(M)$ denote the accepted language. For a finite alphabet Σ, the *initial word relation* $\sqsubseteq$ on Σ^* is defined by $u \sqsubseteq v \overset{df}{\Longleftrightarrow} \exists w (w \in \Sigma^* \wedge uw = v)$.

We write $u \sqsubset v$ if and only if $u \sqsubseteq v$ and $u \neq v$. The lexicographical order on $\{0,1\}^*$ is defined by $x \preceq y \overset{df}{\Longleftrightarrow} x \sqsubseteq y \vee \exists u (u0 \sqsubseteq x \wedge u1 \sqsubseteq y)$.

The quasi-lexicographical order on $\{0,1\}^*$ is defined by $x \leq y \overset{df}{\Longleftrightarrow} |x| < |y| \vee (|x| = |y| \wedge \exists u (u0 \sqsubseteq x \wedge u1 \sqsubseteq y)) \vee x = y$.

In what follows we identify the set $\{0,1\}^*$ with the set $\mathbb{N}$ of natural numbers according to the quasi-lexicographical order. So $\{0,1\}^*$ inherits operations like $+$ from the natural numbers. Furthermore, we identify a set $O \subseteq \mathbb{N}$ with the characteristic sequence $c_O(0)c_O(1)c_O(2) \cdots \in \{0,1\}^\omega$ where c_O is the characteristic function of O. A language $L \subseteq \Sigma^*$ is called *nontrivial* if $L \neq \emptyset$ and $L \neq \Sigma^*$.

The following theorem shows the aforementioned close relation between the dot-depth hierarchy and the polynomial-time hierarchy. Here $NP(n)$ denotes level n of the Boolean hierarchy over NP.

Theorem 2 ([11, 7, 3]). *The following holds for $n \geq 1$ and relative to all oracles.*

1. $P = \mathrm{Leaf}_b^P(\mathcal{B}_0) = \mathrm{Leaf}_u^P(\mathcal{B}_0)$
2. $\Sigma_n^P = \mathrm{Leaf}_b^P(\mathcal{B}_{n-1/2}) = \mathrm{Leaf}_u^P(\mathcal{B}_{n-1/2})$
3. $\Pi_n^P = \mathrm{Leaf}_b^P(co\mathcal{B}_{n-1/2}) = \mathrm{Leaf}_u^P(co\mathcal{B}_{n-1/2})$
4. $BC(\Sigma_n^P) = \mathrm{Leaf}_b^P(\mathcal{B}_n) = \mathrm{Leaf}_u^P(\mathcal{B}_n)$
5. $NP(n) = \mathrm{Leaf}_b^P(\mathcal{B}_{1/2}(n)) = \mathrm{Leaf}_u^P(\mathcal{B}_{1/2}(n))$

Bovet, Crescenzi, and Silvestri [4] and Vereshchagin [18] introduced the notion of polylog-time reducibility and showed that it is related to balanced leaf-language definable classes.

Definition 1. *A function $f : \Sigma^* \to \Sigma^*$ is polylog-time computable if there exist two polynomial-time-bounded oracle transducers $R : \Sigma^* \times \mathbb{N} \to \Sigma$ and $l : \Sigma^* \to \mathbb{N}$ such that for all x, $f(x) = R^x(|x|,1)R^x(|x|,2) \cdots R^x(|x|,l^x(|x|))$ where R and l access the input x as an oracle.*

A language B is polylog-time reducible (plt-reducible) to a language C, $B \leq_m^{plt} C$ for short, if there exists a polylog-time computable f such that for all x, $x \in B \Leftrightarrow f(x) \in C$.

Theorem 3 ([4, 18]). *For all languages B and C,*

$$B \leq_m^{plt} C \iff \text{for all oracles } O, \ \mathrm{Leaf}_b^{pO}(B) \subseteq \mathrm{Leaf}_b^{pO}(C).$$

3   Polynomial-Time Tree Reducibility

The idea. With polynomial-time tree reducibility (ptt-reducibility for short) we introduce the unbalanced analog of polylog-time reducibility (plt-reducibility).

For the representation of a balanced computation tree it suffices to think of a leaf-string such that each symbol is accessible in polylog-time in the length of the leaf-string. Representations of unbalanced computation trees are more complicated. Here the particular structure of the tree must be taken into account. This makes it necessary to define suitable representations of trees. Intuitively, a language B ptt-reduces to a language C, if there exists a polynomial-time (in the height of the tree) computable function f that transforms trees such that for every tree t, the leaf-string of t belongs to B if and only if the leaf-string of $f(t)$ is in C.

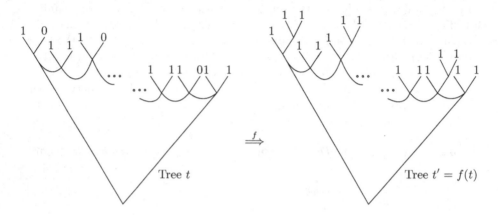

Fig. 1. An example of a tree function f

Example 1. Let $\Sigma_1 = \{0, 1\}$ and $\Sigma_2 = \{1\}$ be alphabets, and let $L_1 \subseteq \Sigma_1^*$ be defined as $L_1 = (0^*10^*1)^*0^*$, the language of all words w over $\{0, 1\}^*$ such that w contains an even number of 1's. Let $L_2 \subseteq \Sigma_2^*$ be defined as $L_2 = (11)^*$, the language of all words w over $\{1\}^*$ such that w has even length. Then L_1 ptt-reduces to L_2.[1] This can easily be seen: Imagine M to be a nondeterministic, polynomial-time Turing machine that outputs words from Σ_1^* as leaf-strings. Roughly speaking, in order to prove that L_1 ptt-reduces to L_2, we have to transform the computation tree t of M into a tree t' whose leaf-string $\beta(t')$ is a word from Σ_2^* and $\beta(t') \in L_2$ if and only if $\beta(t) \in L_1$. Since the desired tree-function f transforms trees and not machines, it needs to be independent of the program of M, i.e., it also has to work with any other machine that outputs words from Σ_1^* as leaf-strings. In our case, the transformation can be described as follows: For all paths in t that output 1, do not change anything. For all paths that output 0, do not output 0 but branch nondeterministically into two paths and output 1 on both paths. This is shown in Figure 1.

Formalization. This idea of ptt-reducibility can be formalized as follows: We start with representations of trees. Let Σ be a finite alphabet. A triple $t = (T, h, m)$ is called a Σ-*tree* if $T \subseteq \{0, 1\}^*$ is finite, $h : T \to \Sigma$, and $m \in \mathbb{N}$ such

---

[1] Note that L_1 does not plt-reduce to L_2.

that $\forall z \forall u((u \sqsubseteq z \wedge z \in T) \rightarrow u \in T)$ and $\forall z(z \in T \rightarrow |z| \leq m)$. Let T_Σ be the set of all Σ-trees. A *leaf* of t is a $z \in T$ such that there is no $u \in T$ with $z \sqsubset u$. For a Σ-tree $t = (T, h, m)$, we define the *leaf word* of t as $\beta(t) =_{\mathrm{def}} h(z_1)h(z_2) \cdots h(z_s)$ where $\{z_1, z_2, \ldots, z_s\}$ is the set of all leaves of t and $z_1 \prec z_2 \prec \cdots \prec z_s$. Hence, the labels of the inner nodes have no effect on the leaf word.

We describe how a Σ-tree can be encoded as a language: Choose $r \geq 1$ such that $|\Sigma| \leq 2^r$, and let $e : \Sigma \rightarrow \{0,1\}^r$ be an injective mapping. A Σ-tree $t = (T, h, m)$ is encoded by the set $O_t =_{\mathrm{def}} \{ze(h(z)) \,|\, z \in T\}$ and the number $m_t =_{\mathrm{def}} m$.

Now let us define functions that transform unbalanced computation trees.

Definition 2. *Let Σ_1 and Σ_2 be finite alphabets. A function $f : T_{\Sigma_1} \rightarrow T_{\Sigma_2}$ is called a* polynomial-time tree function *(ptt-function for short) if there exist $k > 0$ and functions $g_1 : T_{\Sigma_1} \times \{0,1\}^* \times \mathbb{N} \rightarrow \{0,1\}$ and $g_2 : T_{\Sigma_1} \times \{0,1\}^* \times \mathbb{N} \rightarrow \Sigma_2$ such that:*

- *There exists a polynomial $p(\cdot, \cdot)$ such that $g_1(t, z, m)$ and $g_2(t, z, m)$ are computable in time $p(|z|, m)$ where the tree t is accessed as the oracle O_t.*
- *It holds that $f(t) = (T', h', m_t^k + k)$ where $T' =_{\mathrm{def}} \{z \,|\, g_1(t, z, m_t) = 1\}$ and $h'(z) =_{\mathrm{def}} g_2(t, z, m_t)$.*

We will also write $g_1^{O_t}(z, m)$ and $g_2^{O_t}(z, m)$ instead of $g_1(t, z, m)$ and $g_2(t, z, m)$, respectively. Finally, we define polynomial-time tree reducibility.

Definition 3. *For $L_1 \subseteq \Sigma_1^*$ and $L_2 \subseteq \Sigma_2^*$, we define $L_1 \leq_m^{\mathrm{ptt}} L_2$ (L_1 is ptt-reducible to L_2) if there exists a ptt-function $f : T_{\Sigma_1} \rightarrow T_{\Sigma_2}$ such that for all $t \in T_{\Sigma_1}$,*

$$\beta(t) \in L_1 \leftrightarrow \beta(f(t)) \in L_2.$$

Proposition 1. *1. $\leq_m^{\mathrm{ptt}}$ is reflexive and transitive.*
2. $\leq_m^{\mathrm{ptt}}$ and $\leq_m^{\mathrm{plt}}$ are incomparable.

Proof. We sketch a proof for the second statement. From Example 1 we know that $\leq_m^{\mathrm{ptt}}$ does not imply $\leq_m^{\mathrm{plt}}$. For the other direction, let $\Sigma = \{1\}$. It is easy to see that $(11)^* \leq_m^{\mathrm{plt}} (1)$, but $(11)^* \not\leq_m^{\mathrm{ptt}} (1)$. Hence, neither of the two reducibilities implies the other. $\square$

Remark. Although plt- and ptt-reducibility are incomparable, the following straightforward modification of Definition 3 yields plt-reducibility: On the one hand, we let the tree function in Definition 3 assume that the input tree is balanced, while on the other hand we require it to output a balanced tree. So this modification is neither a restriction nor a generalization.

4   The BCSV-Theorem for Unbalanced Leaf Languages

We are now ready to prove the main theorem. Let B and C be languages. Bovet, Crescenzi, and Silvestri [4] and Vereshchagin [18] proved that B polylog-time reduces to C if and only if for all oracles O, $\mathrm{Leaf}_b^{pO}(B) \subseteq \mathrm{Leaf}_b^{pO}(C)$. So plt-reducibility corresponds to robust inclusions of balanced leaf-language classes. Our central theorem states that ptt-reducibility and unbalanced leaf-language classes share the same connection:

Theorem 4. *For nontrivial $L_1 \subseteq \Sigma_1^*$ and $L_2 \subseteq \Sigma_2^*$ the following are equivalent:*

(1) $L_1 \leq_m^{ptt} L_2$
(2) $\forall O(\mathrm{Leaf}_u^{pO}(L_1) \subseteq \mathrm{Leaf}_u^{pO}(L_2))$

5   ptt-Reducibility and the Dot-Depth Hierarchy

By Theorem 2, the levels of the dot-depth hierarchy and the levels of the polynomial-time hierarchy are closely related. Note that this connection exists for both models, balanced and unbalanced leaf-languages. In this section we discuss evidence showing that for the unbalanced model this connection is much closer than that stated in Theorem 2.

Definition 4. *A class of regular languages C and a complexity class D perfectly correspond with respect to balanced leaf-languages if (restricted to regular languages) C is closed under plt-reducibility and $\mathrm{Leaf}_b^p(C) = D$. A class of regular languages C and a complexity class D perfectly correspond with respect to unbalanced leaf-languages if (restricted to regular languages) C is closed under ptt-reducibility and $\mathrm{Leaf}_u^p(C) = D$.*

Perfect correspondences are connections closer than those stated in Theorem 2: For a class of regular languages C and a complexity class D that perfectly correspond with respect to unbalanced leaf-languages, we know that the languages in C are *precisely* those whose unbalanced leaf-language classes are robustly contained in D. Therefore, there can be no regular language L' outside C such that $\mathrm{Leaf}_u^p(L')$ is robustly contained in D.

Proposition 2. *If C perfectly corresponds to D with respect to balanced leaf-languages, then for every regular $L \notin C$ there exists an oracle relative to which $\mathrm{Leaf}_b^p(C) \not\subseteq D$. The similar statement holds for unbalanced leaf-languages.*

The levels of the dot-depth hierarchy and the levels of the polynomial-time hierarchy do not perfectly correspond with respect to balanced leaf-languages. In particular, for $n \geq 1$, $\mathcal{B}_{n/2}$ is not closed under plt-reducibility even if we restrict ourselves to starfree regular languages.

Theorem 5. *For every $n \geq 1$, $\mathcal{B}_{n-1/2}$ does not perfectly correspond to Σ_n^P with respect to balanced leaf-languages.*

In contrast the classes $\mathcal{B}_0$, $\mathcal{B}_{1/2}$, $\mathcal{B}_1$, and $\mathcal{B}_{3/2}$ are closed under ptt-reducibility, restricted to regular languages. In particular, these classes perfectly correspond to the classes of the polynomial-time hierarchy. While for $\mathcal{B}_0$, $\mathcal{B}_{1/2}$, and $\mathcal{B}_{3/2}$ the latter can be derived from known results [2, 3, 15], this is a new result for $\mathcal{B}_1$. The class $\mathcal{B}_0$ is closed under ptt-reducibility even without the restriction to regular languages, this does not hold for the higher levels (see Corollary 2).

Theorem 6. *1. $\mathcal{R}_m^{ptt}(\mathcal{B}_0) = \mathcal{B}_0$, so $\mathcal{B}_0$ and P correspond perfectly.*
2. $\mathcal{R}_m^{ptt}(\mathcal{B}_{1/2}) \cap \mathrm{REG} = \mathcal{B}_{1/2}$, so $\mathcal{B}_{1/2}$ and NP correspond perfectly.
3. $\mathcal{R}_m^{ptt}(\mathcal{B}_{3/2}) \cap \mathrm{REG} = \mathcal{B}_{3/2}$, so $\mathcal{B}_{3/2}$ and Σ_2^P correspond perfectly.

The special case of $\mathcal{B}_0$ yields an even tighter connection of $\mathcal{B}_0$ and P: Not only that $\mathcal{B}_0$ and P perfectly correspond, but in fact it even holds that for any language $L \notin \mathcal{B}_0$ (this includes all nonregular languages) there exists an oracle O such that $\mathrm{Leaf}_u^{pO}(L) \supsetneq \mathrm{P}^O$.

Lemma 1. *Let $L \in \mathrm{REG} \smallsetminus \mathcal{B}_1$. Then there exists an oracle B such that $\mathrm{Leaf}_u^{pB}(L) \not\subseteq \mathrm{P}^{\mathrm{NP}[\epsilon \cdot \log n]^B}$ for all $\epsilon < 1$.*

Utilizing Theorem 4, we can translate this oracle separation into a statement about the ptt-closure of $\mathcal{B}_1$.

Theorem 7. *$\mathcal{R}_m^{ptt}(\mathcal{B}_1) \cap \mathrm{REG} = \mathcal{B}_1$.*

We consider the results of Theorems 6 and 7 as evidence that restricted to regular languages, all levels of the dot-depth hierarchy are closed under ptt-reducibility and therefore, perfectly correspond to the levels of the polynomial-time hierarchy. In addition, Theorem 7 enables us to prove the first gap theorem of leaf-language definability above the Boolean closure of NP.

Corollary 1. *Let $\mathcal{D} = \mathrm{Leaf}_u^p(\mathcal{C})$ for some $\mathcal{C} \subseteq \mathrm{REG}$. Then $\mathcal{D} \subseteq \mathrm{BC}(\mathrm{NP})$ or there exists an oracle O such that $\mathcal{D}^O \not\subseteq \mathrm{P}^{\mathrm{NP}[\epsilon \cdot \log n]^O}$ for all $\epsilon < 1$.*

Recall that by Theorem 6.1, $\mathcal{B}_0$ is closed under ptt-reducibility, and by Theorems 6.2, 6.3, and 7, classes $\mathcal{B}_{1/2}$, $\mathcal{B}_{3/2}$, and $\mathcal{B}_1$ are closed under ptt-reducibility if we restrict ourselves to regular languages. We now prove that the restriction to regular languages is in fact crucial.

Theorem 8. *There exists $B \in \mathrm{NP} \smallsetminus \mathrm{REG}$ such that $\mathrm{Leaf}_u^p(B) \subseteq \mathrm{NP}$.*

In particular, for $k \geq 1$, the ptt-closure of $\mathcal{B}_{k/2}$ contains non-regular languages. By Theorem 6.1, this does not hold for the ptt-closure of $\mathcal{B}_0$.

Corollary 2. *1. There exists $B \in \mathrm{NP} \smallsetminus \mathrm{REG}$ such that $B \in \mathcal{R}_m^{ptt}(\mathcal{B}_{1/2})$.*
2. For every $k \geq 1$, $\mathcal{B}_{k/2}$ is not closed under $\leq_m^{ptt}$-reducibility.

We state an upper bound for the $\leq_m^{ptt}$-closure of regular languages.

Theorem 9. *$\mathcal{R}_m^{ptt}(\mathrm{REG}) \subseteq \bigcup_{k \geq 1} \mathrm{DSPACE}(\log^k n)$.*

Due to this theorem, we can now specify the complexity of nonregular sets C such that $\mathrm{Leaf}_u^p(C) \subseteq \mathrm{NP}$. (Recall that for regular sets, we already know by Theorem 6.2 that only languages in $\mathcal{B}_{1/2}$ come into question.)

Corollary 3. *Let C be a set. If for all oracles O, $\mathrm{Leaf}_u^{pO}(C) \subseteq \mathrm{NP}^O$, then $C \in \bigcup_{k \geq 1} \mathrm{DSPACE}(\log^k n)$.*

Since $\mathrm{PSPACE} = \mathrm{Leaf}_u^p(\mathrm{REG})$ [11], the last corollary remains valid if we replace NP by PSPACE.

6   Conclusions and Open Questions

We have shown that the new ptt-reducibility is very useful in terms of unbalanced leaf-language classes. It is in a sense the counterpart of plt-reducibility and allows us to prove an analogue of the well known result by Bovet, Crescenzi, and Silvestri [4], and Vereshchagin [18] for unbalanced leaf-languages.

Interestingly, the ptt-reducibility furthermore indicates that the connection between the levels of the dot-depth hierarchy and the levels of the polynomial-time hierarchy via *unbalanced leaf-languages* is much closer than the formerly known connection via balanced leaf-languages. We have shown that the lower levels of the DDH and the PH *perfectly correspond*. Whether the DDH and the PH perfectly correspond on *all* levels (which we believe to hold true) remains a challenging open question.

Acknowledgments. We thank Bernd Borchert, Heinz Schmitz, Victor Selivanov, and Pascal Tesson for helpful discussions about leaf languages.

References

1. M. Arfi. Opérations polynomiales et hiérarchies de concaténation. *Theoretical Computer Science*, 91:71–84, 1991.
2. B. Borchert. On the acceptance power of regular languages. *Theoretical Computer Science*, 148:207–225, 1995.
3. B. Borchert, D. Kuske, and F. Stephan. On existentially first-order definable languages and their relation to NP. *Theoretical Informatics and Applications*, 33:259–269, 1999.
4. D. P. Bovet, P. Crescenzi, and R. Silvestri. A uniform approach to define complexity classes. *Theoretical Computer Science*, 104:263–283, 1992.
5. J. A. Brzozowski. Hierarchies of aperiodic languages. *RAIRO Inform. Theor.*, 10:33–49, 1976.
6. J. A. Brzozowski and R. Knast. The dot-depth hierarchy of star-free languages is infinite. *Journal of Computer and System Sciences*, 16:37–55, 1978.
7. H.-J. Burtschick and H. Vollmer. Lindström quantifiers and leaf language definability. *International Journal of Foundations of Computer Science*, 9:277–294, 1998.
8. R. S. Cohen and J. A. Brzozowski. Dot-depth of star-free events. *Journal of Computer and System Sciences*, 5:1–16, 1971.
9. S. Eilenberg. *Automata, languages and machines*, volume B. Academic Press, New York, 1976.
10. C. Glaßer. Polylog-time reductions decrease dot-depth. In *Proceedings 22nd Symposium on Theoretical Aspects of Computer Science*, volume 3404 of *Lecture Notes in Computer Science*. Springer Verlag, 2005.
11. U. Hertrampf, C. Lautemann, T. Schwentick, H. Vollmer, and K. W. Wagner. On the power of polynomial time bit-reductions. In *Proceedings 8th Structure in Complexity Theory*, pages 200–207, 1993.
12. A. Maciel, P. Péladeau, and D. Thérien. Programs over semigroups of dot–depth one. *Theoretical Computer Science*, 245:135–148, 2000.
13. D. Perrin and J. E. Pin. First-order logic and star-free sets. *Journal of Computer and System Sciences*, 32:393–406, 1986.

14. J. E. Pin and P. Weil. Polynomial closure and unambiguous product. *Theory of computing systems*, 30:383–422, 1997.
15. H. Schmitz. *The Forbidden-Pattern Approach to Concatenation Hierarchies*. PhD thesis, Fakultät für Mathematik und Informatik, Universität Würzburg, 2001.
16. L. Stockmeyer. The polynomial-time hierarchy. *Theoretical Computer Science*, 3:1–22, 1977.
17. W. Thomas. An application of the Ehrenfeucht–Fraïssé game in formal language theory. *Société Mathématique de France, mémoire 16*, 2:11–21, 1984.
18. N. K. Vereshchagin. Relativizable and non-relativizable theorems in the polynomial theory of algorithms. *Izvestija Rossijskoj Akademii Nauk*, 57:51–90, 1993. In Russian.
19. K. W. Wagner. Leaf language classes. In *Proceedings International Conference on Machines, Computations, and Universality*, volume 3354 of *Lecture Notes in Computer Science*. Springer Verlag, 2004.
20. C. Wrathall. Complete sets and the polynomial-time hierarchy. *Theoretical Computer Science*, 3:23–33, 1977.

Language Equations with Complementation*

Alexander Okhotin[1] and Oksana Yakimova[2]

[1] Department of Mathematics, University of Turku, Finland
alexander.okhotin@utu.fi
[2] Max-Planck-Institut für Mathematik, Bonn, Germany yakimova@mpim-bonn.mpg.de

Abstract. Systems of language equations of the form $X_i = \varphi_i(X_1, \ldots, X_n)$ $(1 \leqslant i \leqslant n)$ are studied, in which every φ_i may contain the operations of concatenation and complementation. The properties of having solutions and of having a unique solution are given mathematical characterizations. As decision problems, the former is *NP*-complete, while the latter is in *co-RE* and its decidability remains, in general, open. Uniqueness becomes decidable for a unary alphabet, where it is *US*-complete, and in the case of linear concatenation, where it is *L*-complete. The position of the languages defined by these equations in the hierarchy of language families is established.

1 Introduction

Systems of equations with formal languages as unknowns have been studied since the early 1960s, when Ginsburg and Rice [3] found an equivalent representation of context-free grammars as resolved systems of the form $X_i = \alpha_{i1} \cup \ldots \cup \alpha_{i\ell_i}$ $(1 \leqslant i \leqslant n)$, where $\ell_i > 0$ and every α_{ij} is a concatenation of variables and symbols of the alphabet. For example, the equation $X = aXb \cup \{\varepsilon\}$ with the unique solution $\{a^n b^n \mid n \geqslant 0\}$ corresponds to a context-free grammar with two rules, $S \to aSb$ and $S \to \varepsilon$. This equational semantics became an important tool in the study of the context-free languages [5].

The study of more general types of language equations began only in the 1990s. In particular, equations of Ginsburg and Rice equipped with all Boolean operations have been considered: Charatonik [2] showed the undecidability of solution existence for such equations, later Okhotin [10] carried out a detailed study of the hardness of different decision problems. The languages representable by unique solutions of these equations are exactly the recursive languages [10].

Besides the cases of union only and of all Boolean operations, other interesting variants of equations of Ginsburg and Rice have been considered. Equations with union and intersection were found to share many theoretical properties of those with union only [9]. This study led to natural generalizations of context-free grammars known as conjunctive grammars and Boolean grammars [11, 12].

Systems of equations of the same form $X_i = \varphi_i(X_1, \ldots, X_n)$ $(1 \leqslant i \leqslant n)$, in which φ_i use *concatenation and complementation* only, form another natural

---

* Supported by Academy of Finland grant 206039, CRDF Grant RM1–2543–MO–03 and RFBI Grant 05–01–00988.

O.H. Ibarra and Z. Dang (Eds.): DLT 2006, LNCS 4036, pp. 420–432, 2006.

case. These equations have been considered by Leiss [8], in the special case
of a unary alphabet and right-hand sides of a restricted form that guarantees
existence and uniqueness of a solution. He proved that this unique solution is
always context-sensitive, but it need not be context-free, as demonstrated by the
following equation

$$X = \{a\} \cdot \overline{\overline{\overline{X}^2}^2}^2 \,, \tag{1}$$

which has a unique solution $\{a^n \mid \exists k \geqslant 0, \text{ such that } 2^{3k} \leqslant n < 2^{3k+2}\}$ [8].

The goal of this paper is to study these equations in the general case, without the simplifying assumptions made by Leiss [8]. In Section 3, we consider the
property of having a solution and establish a key technical result, that a solution
modulo any finite nonempty language can be extended to a solution. This leads
us to a simple characterization of solution existence and allows us to establish its
NP-completeness. We give a mathematical characterization of solution uniqueness in Section 4 and show that this problem is co-r.e. (cf. Π_2-completeness in
the case of all Boolean operations [10]), but it remains open whether it is decidable. For a unary alphabet, the problem is complete for the class US (*unique
satisfiability*) studied by Blass and Gurevich [1], which consists of all languages
representable as $\{w \mid \exists! x : R(w, x)\}$ for a polynomial-time predicate R.

The rest of the paper is devoted to the study of the family of languages representable as unique solutions of these equations. Recent results on computational
universality in language equations of an extremely simple form, see Kunc [6, 7],
make one suspect universality even in this restricted case. However, we show
that these equations are not universal, though their expressive power is not too
weak either. We show nonrepresentability of some concrete languages by our
equations in Section 5, and use these examples in Section 6, where we determine
that our equations are strictly weaker than Boolean grammars. We study the
case of equations with linear concatenation in Section 7, and find interesting
relations to linear context-free and linear conjunctive languages.

2   Definitions and Notation

Let $n \geqslant 1$ and let $(X_1, \ldots, X_n)$ be a vector of variables, which assume values of
languages over Σ. Consider a system of equations resolved with respect to its
unknowns as follows:

$$\begin{cases} X_1 = \varphi_1(X_1, \ldots, X_n) \\ \quad \vdots \\ X_n = \varphi_n(X_1, \ldots, X_n) \end{cases} \tag{2}$$

In general, each φ_i is an expression that contains variables and constant languages from some predefined language family, connected with arbitrarily nested
concatenation and Boolean set-theoretic operations. The equations studied in
this paper may use only one Boolean operation, the complementation, and we

shall either consider arbitrary regular constants, or restrict them to be singletons $\{w\}$ ($w \in \Sigma^*$), which will be denoted by just w. Some restrictions on concatenation will be considered: concatenation in (2) is said to be *linear*, if for every subexpression $\xi \cdot \eta$ in every φ_i either ξ or η is a constant language; further, if it is always ξ (or always η) that is a constant, the concatenation is *one-sided linear*.

A vector of languages $(L_1, \ldots, L_n)$ is called a solution of (2), if a substitution of L_j for X_j for all j turns each equation into an equality. We shall also consider partial solutions modulo a given language. Two languages $K, L \subseteq \Sigma^*$ are said to be *equal modulo* a third language $M \subseteq \Sigma^*$ if $K \cap M = L \cap M$; this is denoted $K = L \pmod{M}$. A language $L \subseteq \Sigma^*$ is called *subword-closed*, if for every word $w \in L$ all its subwords are also in L (u is a subword of w if $\exists x, y \in \Sigma^*: w = xuy$). For any subword-closed language $M \subseteq \Sigma^*$, a vector $(L_1, \ldots, L_n)$, where $L_j \subseteq M$ for all j, is a *solution modulo* M of (2), if the same substitution turns each equation into an equality modulo M. Two vectors of languages, $(L_1, \ldots, L_n)$ and $(L'_1, \ldots, L'_n)$, are said to be equal modulo M, if $L_i = L'_i \pmod{M}$. Uniqueness of a solution of (2) modulo M is defined with respect to this equivalence.

A system (2) is said to have a *strongly unique solution*, if its solution is unique modulo every finite subword-closed language; this implies the uniqueness of solution in the general sense [10]. Let us say that a solution $(L_1, \ldots, L_n)$ modulo M can be extended to a solution modulo $M' \supset M$, if there exists a solution $(L'_1, \ldots, L'_n)$ modulo M', such that $(L_1, \ldots, L_n) = (L'_1, \ldots, L'_n) \pmod{M}$.

For convenience we shall often assume, without loss of generality, that every equation in (2) is $X_i = \overline{X_j X_k}$ or $X = const$. It is easy to see that every system can be equivalently transformed to this form.

We shall use the logical dual of concatenation, defined as $K \odot L = \overline{\overline{K} \cdot \overline{L}}$ or, equivalently, as $K \odot L = \{w \mid \forall u, v: w = uv \Rightarrow u \in K \text{ or } v \in L\}$ [12].

3   Existence of a Solution

A language equation may have or not have solutions. For equations with complementation the property of having solutions is nontrivial even in the most restricted case: consider an equation $X = \overline{aX}$ with the unique solution $(a^2)^*$, another equation $X = X$ with every language as a solution, and one more equation $X = \overline{X}$ that has no solutions.

The following criterion of solution existence holds in a more general case:

Proposition 1 (Okhotin [10]). *A system $X_i = \psi_i(X_1, \ldots, X_n)$ ($1 \leqslant i \leqslant n$) with concatenation and all Boolean operations has a solution if and only if it has a solution modulo every finite subword-closed language.*

Existence of a solution modulo any given finite language can be checked by a straightforward search, so in order to test the existence of a solution it is sufficient to repeat this procedure for countably many languages. It has also been shown that this infinite search is in some sense necessary, because the problem is undecidable [2, 10] — to be exact, Π_1-complete [10].

The following statement is the key element in the proof of Proposition 1:

Proposition 2 (Okhotin [10]). *Let $X_i = \varphi_i(X_1, \ldots, X_n)$ $(1 \leqslant i \leqslant n)$ be a system with concatenation, all Boolean operations, and with any constant languages. Let $(L_1, \ldots, L_n)$ be a solution modulo a finite subword-closed language $M \subset \Sigma^*$, such that for every subword-closed language $M' \supset M$ there exists a solution $(L'_1, \ldots, L'_n)$ modulo M', which coincides with $(L_1, \ldots, L_n)$ modulo M. Then the system has a solution $(\widehat{L}_1, \ldots, \widehat{L}_n)$ that equals $(L_1, \ldots, L_n)$ modulo M.*

The existence of an extension to every M' is essential here. For instance, the system $\{X = X, Y = \overline{Y} \cap aX\}$ [10] has a solution $(\{a\}, \varnothing)$ modulo $\{\varepsilon, a\}$, but none of its solutions modulo $\{\varepsilon, a, a^2\}$ contain a in the X component. If we consider language equations with monotone operations only, then such wrong partial solutions cannot exist, and a solution modulo M can be extended to any M'. Thus Proposition 2 degenerates to the following unconditional statement:

Proposition 3 (Okhotin [12]). *Let $X_i = \varphi_i(X_1, \ldots, X_n)$ $(1 \leqslant i \leqslant n)$ be a system with concatenation, dual concatenation, union and intersection, and with any constant languages. Let M be a finite possibly empty subword-closed language, let $(L_1, \ldots, L_n)$ be a solution modulo M. Then the system has a solution $(\widehat{L}_1, \ldots, \widehat{L}_n)$ that coincides with $(L_1, \ldots, L_n)$ modulo M.*

Our equations with complementation appear to have nothing in common with this monotone case, and one could naturally expect the same difficulties as in Proposition 2. On the contrary, we obtain a statement almost like Proposition 3:

Lemma 1. *If a system $X_i = \varphi_i(X_1, \ldots, X_n)$ $(1 \leqslant i \leqslant n)$ with concatenation and complementation and with any constant languages has a solution $(L_1, \ldots, L_n)$ modulo some finite nonempty subword-closed language M, then it has a solution $(\widehat{L}_1, \ldots, \widehat{L}_n)$ that is equal to $(L_1, \ldots, L_n)$ modulo M.*

The only difference between Proposition 3 and Lemma 1 is the requirement of nonemptiness of M. While Proposition 3 allows one to obtain a solution "out of the air" by taking $M = \varnothing$, Lemma 1 requires a solution modulo $\{\varepsilon\}$, which serves as a basis of induction on $|M|$. The inductive construction proceeds as follows. For every M and for $w \notin M$, such that all proper subwords of w are in M, let $(L_1, \ldots, L_n)$ be a solution modulo M and substitute it into the system. Its extension to $M \cup \{w\}$ can be represented by a system of Boolean equations in variables $(x_1, \ldots, x_n)$, where each $x_i \in \{0, 1\}$ determines the membership of w in X_i. A solution of this system can be constructed, which gives a required solution modulo $M \cup \{w\}$ of the system of language equations. Finally, by Proposition 2, these finite extensions imply the existence of an infinite extension to a solution.

Now we can prove a necessary and sufficient condition of having a solution for our class of language equations. As compared to the case of equations of the general form, see Proposition 1, here it is sufficient to check existence of a solution modulo just one finite language:

Theorem 1. *A system* $X_i = \varphi_i(X_1, \ldots, X_n)$ *(*$1 \leqslant i \leqslant n$*) with concatenation and complementation and with arbitrary constant languages has a solution if and only if it has a solution modulo* $\{\varepsilon\}$*.*

Proof. The $\Rrightarrow$ implication is trivial, while $\Lleftarrow$ is given by Lemma 1 for $M = \{\varepsilon\}$.

This leads us to the computational complexity of this problem:

Theorem 2. *Fix any finite nonempty alphabet* Σ *and any set of constant languages, such that the membership of* ε *in constants is polynomially decidable. Consider systems of language equations* $X_i = \varphi_i(X_1, \ldots, X_n)$ *(*$1 \leqslant i \leqslant n$*) with concatenation and complement and with the above constants. Then the problem of testing whether a given system of this form has a solution is* NP-*complete.*

The membership in NP is by guessing a solution modulo $\{\varepsilon\}$ and verifying that it is indeed a solution modulo $\{\varepsilon\}$; this, by Theorem 1, shows that a solution exists. The proof of NP-hardness is based upon the following construction:

Lemma 2. *Let* $f : \{0,1\}^n \to \{0,1\}$ *be a Boolean function and for every* $L \subseteq \Sigma^*$ *denote* $e(L) = 1$ *if* $\varepsilon \in L$*,* $e(L) = 0$ *otherwise. Then there exists an expression* $\varphi(X_1, \ldots, X_n)$ *with concatenation and complementation, such that* $\varepsilon \in \varphi(L_1, \ldots, L_n)$ *if and only if* $f(e(L_1), \ldots, e(L_n)) = 1$*.*

Assuming that the formula $f(x_1, \ldots, x_n)$ uses two propositional connectives, conjunction and negation, $\varphi(X_1, \ldots, X_n)$ can have same structure as f: conjunction, negation and variables x_i are represented with concatenation, complementation and variables X_i, respectively.

For the proof of Theorem 1 we reduce the satisfiability problem for f. Take the variables $X_1, \ldots, X_n, T$ and use equations $X_i = X_i$ $(1 \leqslant i \leqslant n)$ and $T = T \cdot \overline{\varphi(X_1, \ldots, X_n)}$. The latter equation, unless $\varepsilon \notin \varphi(e(X_1), \ldots, e(X_n))$, expresses a contradiction of the form "$\varepsilon \in T$ if and only if $\varepsilon \notin T$".

4   Uniqueness of Solution

Let us now devise a necessary and sufficient condition of having a unique solution. For language equations of a more general form the following criterion of solution uniqueness is known:

Theorem 3 (Okhotin [10]). *A system with concatenation and all Boolean operations has a unique solution if and only if for every finite subword-closed language* M *there exists a finite subword-closed language* $M' \supset M$*, such that all solutions modulo* M' *coincide modulo* M*.*

For our systems, Lemma 1 simplifies this condition to the following:

Corollary 1. *A system with concatenation and complementation has a unique solution if and only if it has a unique solution modulo every finite subword-closed language* M*.*

Indeed, if there are multiple solutions modulo some M, then each of them can be extended to a full solution, and these solutions are distinct.

Let us define a more precise uniqueness criterion. Consider a system of equations $X_i = \varphi_i(X_1, \ldots, X_n)$ $(1 \leqslant i \leqslant n)$. Assume that it has a unique solution $(L_1^\varepsilon, \ldots, L_n^\varepsilon)$ modulo $\{\varepsilon\}$ and that this solution is defined by a Boolean vector $(y_1, \ldots, y_n)$ in the sense that $L_i^\varepsilon = \{\varepsilon\}$ if $y_i = 1$ and $L_i^\varepsilon = \varnothing$ otherwise. By Theorem 1, the system has a solution. Let $(L_1', \ldots L_n')$ be a solution modulo $\{\varepsilon, a\}$, for any $a \in \Sigma$. Since each $L_i' \cap \{\varepsilon\}$ is determined by y_i, $(L_1', \ldots, L_n')$ is uniquely defined by a Boolean vector $(x_1, \ldots, x_n)$, which satisfies the following equations:

$$x_i = \overline{(y_j \wedge x_k) \vee (y_k \wedge x_j)}, \quad \text{for all } \varphi_i(X_1, \ldots X_n) = \overline{X_j \cdot X_k};$$
$$x_j = c_j, \qquad\qquad\qquad\quad \text{for all } \varphi_i(X_1, \ldots X_n) = const.$$

Here we have used the facts that $L_j \cdot L_k \subset \{\varepsilon\}$ and $a \notin L_j \cdot L_k$.

This system of Boolean equations can be represented by an oriented graph $\Gamma = \Gamma(\{\varepsilon\}, a)$ with vertices $\{1, \ldots, n\}$, where each vertex i corresponds to a variable x_i and an arc (i, j) belongs to Γ if and only if there is k such that $y_k = 1$ and $\varphi_i(X_1, \ldots X_n)$ equals either $\overline{X_j X_k}$ or $\overline{X_k X_j}$. Let us say that the initial system of language equations is *rigid* if is has a unique solution modulo $\{\varepsilon\}$ and Γ contains no cycles. This gives the following purely syntactical sufficient condition of solution uniqueness:

Lemma 3. *Every rigid system has a unique solution.*

The proof is by a reduction to a system of Boolean equations. As in Lemma 1, it is proved that the solution modulo every finite subword-closed M is unique, because the Boolean system describing the extension from M to $M \cup \{w\}$ always has a unique solution. However, this sufficient condition is not necessary, and a non-rigid system with a cyclic graph Γ can still have a unique solution.

Example 1. Take any language L_0, such that $\{\varepsilon\} \subseteq L_0 \subseteq \Sigma^*$. Then the system

$$\begin{aligned} X_1 &= \overline{YX_2} & Y &= \{\varepsilon, a\} \\ X_2 &= \overline{UX_3} & Z &= \varepsilon \cup ba^* \\ X_3 &= \overline{X_4 Z} & U &= L_0 \\ X_4 &= \overline{X_1 U} & T &= \overline{TX_1} \end{aligned}$$

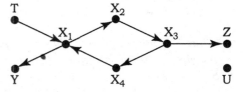

has a unique solution modulo $\{\varepsilon\}$, which is $(\varnothing, \{\varepsilon\}, \varnothing, \{\varepsilon\}, \{\varepsilon\}, \{\varepsilon\}, \{\varepsilon\}, \{\varepsilon\})$. The corresponding graph Γ shown here contains a cycle $X_1 \to X_2 \to X_3 \to X_4 \to X_1$, and hence the system is not rigid. However, it still has a unique solution $(\varnothing, \Sigma^*, \varnothing, \Sigma^*, \{\varepsilon, a\}, \varepsilon \cup ba^*, L_0, \Sigma^*)$.

It is enough to prove that for every finite subword-closed M there is a unique solution $(L_1, \ldots, L_8)$ modulo M, such that $L_1 = L_3 = \varnothing$ and $L_2 = L_4 = M$. This is proved inductively on $|M|$. Given $M' = M \cup \{w\}$, where $w \notin M'$ and all subwords of w are in M, one can show that $w \notin L_1, L_3$ and $w \in L_2, L_4$ by the following argument: (i) if $w = au$ for some $u \in M$, then $u \in L_2$, $w \in aL_2$, and, thereby, $w \notin L_1$; (ii) if $w = uba^i$ for $u \in M$ and $i \geqslant 0$, then $u \in L_4$, $w \in L_4 \cdot ba^*$, and $w \notin L_3$. Every nonempty word over $\{a, b\}$ meets one of these conditions.

Note that the above argument is based upon the fact that every word $w \in \Sigma^+$ has a nonempty prefix from Y or a nonempty suffix from Z; in other words, $Y^*Z^* = \Sigma^*$. We shall now see that the exact condition of having a unique solution is that an equality of this kind holds for all such cycles.

A variable X_i is said to be *perishable* if Γ contains a path from i to any cycle; in the above example, X_1, X_2, X_3, X_4 and T are perishable. A solution $(L_1, \ldots, L_n)$ is *distinguished* if $L_i \in \{\Sigma^*, \varnothing\}$ for each perishable variable X_i.

Lemma 4. *If a system has a unique solution modulo ε, then it has a unique distinguished solution.*

To show that such a solution exists, substitute $\varnothing$ or Σ^* for all perishable variables according to the solution modulo $\{\varepsilon\}$. The resulting system in variables $(X_1, \ldots, X_m)$ has the same unique solution modulo $\{\varepsilon\}$, and this system is rigid.

Corollary 2. *If $(L_1, \ldots, L_n)$ is the unique solution of a system, then $L_i \in \{\Sigma^*, \varnothing\}$ for each perishable variable X_i.*

Corollary 3. *For every system that has a unique solution there exists and can be effectively constructed a rigid system with the same unique solution, which additionally uses constants $\varnothing$ and Σ^*.*

Let us say that a minimal cycle S in Γ is *appropriate* if there is no arc (i, j), such that $j \notin S$ and j is perishable. For every such cycle, define

$$J_\ell(S) := \{j \notin S \mid \exists i \in S : \varphi_i(X_1, \ldots, X_n) = \overline{X_j X_k} \text{ and } y_k = 1\},$$
$$J_r(S) := \{j \notin S \mid \exists i \in S : \varphi_i(X_1, \ldots, X_n) = \overline{X_k X_j} \text{ and } y_k = 1\}.$$

The only cycle in Example 1 is appropriate and $J_\ell(S) = \{Y\}$, $J_r(S) = \{Z\}$. Note that, for every S, $J_\ell(S) \cup J_r(S) = \{j \mid j \notin S, \exists i \in S : \Gamma \text{ contains } (i, j)\}$.

Lemma 5. *Suppose a system has a unique solution modulo $\{\varepsilon\}$ and $(L_1, \ldots, L_n)$ is its distinguished solution. Let $S \subset \Gamma$ be an appropriate cycle with $J_\ell(S) = \{\ell_1, \ldots, \ell_t\}$ and $J_r(S) = \{r_1, \ldots, r_s\}$. Suppose that $K := (L_{\ell_1} L_{\ell_2} \ldots L_{\ell_t})^* (L_{r_1} L_{r_2} \ldots L_{r_s})^* \neq \Sigma^*$, $w \notin K$, and M is the set of the proper subwords of w. Then there are at least two distinct solutions modulo $M \cup \{w\}$.*

The proof is again by reduction to a Boolean system, which is shown to have two distinct solutions.

Theorem 4. *A system $X_i = \varphi_i(X_1, \ldots, X_n)$ $(1 \leqslant i \leqslant n)$ with a unique solution modulo $\{\varepsilon\} \cup \Sigma$ has a unique solution if and only if for the distinguished solution $(L_1, \ldots, L_n)$ and for each appropriate cycle $S \subset \Gamma$ with $J_\ell(S) = \{\ell_1, \ldots, \ell_t\}$, $J_r(S) = \{r_1, \ldots, r_s\}$ we have $(L_{\ell_1} L_{\ell_2} \ldots L_{\ell_t})^* (L_{r_1} L_{r_2} \ldots L_{r_s})^* = \Sigma^*$.*

The forward implication is by Lemma 5: if the condition is violated, then the system is bound to have multiple solutions. The converse is proved by a reduction to a Boolean system.

An important question is whether the condition of Theorem 4 can be tested algorithmically. Testing it requires deciding whether $(K_1 \ldots K_m)^*(L_1 \ldots L_n)^* = \Sigma^*$ for any given languages K_i, L_j, and the languages are given by unique solutions of rigid systems. Our present knowledge on the form of these unique solutions is insufficient to resolve this question.

Let us consider a particular case, in which the decidability can be established. This is the case of a unary alphabet: here our criterion of uniqueness is simplified to the following clear condition that reminds of Theorem 1:

Theorem 5. *Let $\Sigma = \{a\}$. A system $X_i = \varphi_i(X_1, \ldots, X_n)$ $(1 \leqslant i \leqslant n)$ over Σ, with concatenation and complementation and with arbitrary constants, has a unique solution if and only if it has a unique solution modulo $\{\varepsilon, a\}$.*

The proof is based upon the observation that whenever $L_S = (L_{\ell_1} L_{\ell_2} \ldots L_{\ell_t})^* (L_{r_1} L_{r_2} \ldots L_{r_s})^* \neq a^*$, it implies $a \notin L_S$, and then by Lemma 5, the solution modulo $\{\varepsilon, a\}$ is not unique.

Let us determine the computational complexity of the problem in this case.

Theorem 6. *Let $\Sigma = \{a\}$. Fix any set of constants, for which the membership of ε and a can be decided in polynomial time. Then the problem whether a system $X_i = \varphi_i(X_1, \ldots, X_n)$ $(1 \leqslant i \leqslant n)$ with concatenation and complement and with these constants has a unique solution is US-complete.*

Membership in US is proved by guessing the unique solution modulo $\{\varepsilon, a\}$. Hardness is established by a slight elaboration of the method used for Theorem 2.

5 Nonrepresentable Languages

Consider the family of languages that occur in unique solutions of language equations with concatenation and complementation. Actually, there are two families corresponding to the cases of regular and singleton constants (we denote them by N^{Reg} and N), and their distinctness is one of our results.

In order to prove that some languages are not representable by these equations, consider their factorizations. Let us say that a decomposition $L = L_1 L_2$ is *trivial* if either L_1 or L_2 equals $\{\varepsilon\}$. A language L is called *prime* [13], if $L \neq \{\varepsilon\}$ and every decomposition $L = L_1 L_2$ is trivial.

Lemma 6. *Suppose $(L_1, \ldots, L_n)$ is the unique solution of a system $X_i = \varphi_i(X_1, \ldots, X_n)$, where every equation is of the form $X = \overline{YZ}$ or $X = const$. Suppose that for some $L = L_i$ such that $L \notin \{\Sigma^*, \varnothing\}$ there are no non-trivial decompositions $L = L_j L_k$ or $\overline{L} = L_p L_q$. Then one of the languages $L, \overline{L}$ must be among the constant languages used in the system.*

Assume that $\varepsilon \in L$ and let $(L, \ldots, L, \overline{L}, \ldots, \overline{L}, L_m, \ldots, L_n)$ be the solution of the system. If neither L nor $\overline{L}$ are among the constants, then it can be proved that $(\Sigma^*, \ldots, \Sigma^*, \varnothing, \ldots, \varnothing, L_m, \ldots, L_n)$ is a solution as well, which contradicts the assumption of solution uniqueness.

Corollary 4. *If $L \subseteq \Sigma^*$ and its complement $\overline{L}$ are primes, and either of them is among the components of a unique solution, then one of them must be among the constants.*

Example 2. Let $\Sigma = \{a, b\}$. The regular language $L = a\Sigma^*b \cup b\Sigma^*a \cup \varepsilon$ and its complement $\overline{L} = a\Sigma^*a \cup b\Sigma^*b \cup a \cup b$ are primes. Therefore, there exists no system of language equations with concatenation and complementation such that it has a unique solution, L or $\overline{L}$ is among the components of that solution, and neither L nor $\overline{L}$ is among constant languages used in the system.

Example 3. Let $\Sigma = \{a, b\}$. The language $L = (a\Sigma^*b \cup b\Sigma^*a \cup \varepsilon) \setminus \{a^n b^n \mid n > 1\}$ and its complement $\overline{L}$ are primes. Therefore, in particular, L is not representable by equations with concatenation, complementation and regular constants.

Let us turn to the case of a unary alphabet. Unfortunately, there is no $L \subset a^*$ such that both L and $\overline{L}$ are primes: one of them is bound to be divisible by $\{a\}$. Thereby our construction of a nonrepresentable language is rather complicated.

Example 4. Let $L_1 = \{a^n \mid \exists i \geqslant 0 : 2^{3i} \leqslant n < 2^{3i+2}\}$, $L_2 = a(a^2)^*$ and $L_3 = \{a^n, a^{n+1} \mid n = 2^{3i+1}, i \geqslant 0\}$. The language $L = L_1 \triangle L_2 \triangle L_3$ is not representable using regular constants.

To prove its nonrepresentability, we define *shifts* M_k, N_k of L and $\overline{L}$ by the formulas $M_k = L \cdot \{a^k\}^{-1}$, $N_k = \overline{L} \cdot \{a^k\}^{-1}$. Note that $N_k = \overline{M_k}$. We then obtain

Lemma 7. *All decompositions of M_k and N_k are of the form $M_k = \{a^r\} \cdot M_{k+r}$, $N_k = \{a^r\} \cdot N_{k+r}$.*

Now suppose there exists a system with a unique solution $(K_1, \ldots, K_n)$, such that L or $\overline{L}$ are among $\{K_i\}$'s. Let k be the greatest number such that M_k or N_k is among $\{K_i\}$'s. According to Lemma 7, there are no non-trivial decompositions $M_k = K_j K_t$ or $N_k = K_q K_s$. Hence, by Lemma 6, either M_k or N_k should be among the constants. Since both are nonregular, we get a contradiction. This shows the correctness of Example 4, i.e., $L_1 \triangle L_2 \triangle L_3 \notin N^{Reg}$.

6   Expressive Power of Unique Solutions

Let us continue the study of the family of languages representable by unique solutions of our equations. Following is a general upper bound for this family:

Lemma 8. *Consider a system with concatenation and complementation and with constants generated by Boolean grammars. If it has any solutions, then for one of them, $(L_1, \ldots, L_n)$, every L_i is generated by a Boolean grammar. If the concatenation is linear, while the constants are regular, the components are linear conjunctive. The corresponding grammars can be effectively constructed.*

Using Corollary 3 we can construct a rigid system that has one of the solutions of the original system. Then, as stated in Corollary 1, it has a strongly unique solution and therefore can be reformulated as a Boolean grammar [11]. The exact transformations are due to the normal form theorems for conjunctive and Boolean grammars [11]. On the other hand, the next example shows that even with regular constants our equations are weaker than Boolean grammars.

Example 5. Consider $L_1, L_2, L_3 \subseteq a^*$ given in Example 4. Then the system

$$X = a \cdot \overline{\overline{X}^2}^2 \qquad Y = a^2 Y \cup a \qquad Z = \overline{X}^2 \cap a^2 \cdot \overline{\overline{X}^2}^2 \qquad U = X \,\Delta\, Y \,\Delta\, Z$$

has the unique solution $(L_1, L_2, L_3, L_1 \,\Delta\, L_2 \,\Delta\, L_3)$. This solution is strongly unique, and therefore the system can be transcribed as a Boolean grammar that generates $L_1 \,\Delta\, L_2 \,\Delta\, L_3$.

The equation for X is from Leiss [8], who proved that L_1 is its unique solution. The rest can be verified by substituting L_1 into the equations for Y, Z and U.

Since, as shown in Example 4, $L_1 \,\Delta\, L_2 \,\Delta\, L_3$ is not representable by our equations, it separates these language families, establishing the following hierarchy:

Theorem 7. $N \subset N^{Reg} \subset Bool$, *where* Bool *is the family of languages generated by Boolean grammars.*

The first inclusion is obvious, and its strictness is shown by Example 2. Lemma 8 establishes the second inclusion, and it is proper due to Example 5.

Consider the case of a unary alphabet. Here every regular language can be represented using complementation, one-sided concatenation and constant $\{a\}$:

Lemma 9. *For every DFA A over $\Sigma = \{a\}$ one can effectively construct a system $\{X = \varphi(Y), Y = \psi(Y)\}$ with one-sided concatenation and complementation and with the constant language $\{a\}$, such that the first component of its unique solution is $L(A)$.*

Recall the form of a unary DFA, which starts with a *prefix* and then enters a *loop*. In the constructed system, φ represents the prefix and ψ represents the loop, while the accepting states are encoded in the alternation of concatenation of a and complementation.

Theorem 8. *Consider equations over the alphabet $\{a\}$ with concatenation and complementation. The classes of languages representable using singleton or regular constants coincide, and this class is properly contained between regular languages and the languages generated by Boolean grammars.*

The proof follows from Lemma 9, Examples 4 and 5, Theorem 7, and (1).

7   The Case of Linear Concatenation

Let us consider a restricted class of equations, in which, for every occurrence of concatenation in the right-hand sides, one of the operands must be a constant. The general form of such equations is

$$X_i = \overline{K_{i1} \cdot \overline{K_{i2} \cdot \ldots \cdot \overline{K_{im_i} \cdot X_{j_i} \cdot L_{im_i}} \cdot \ldots \cdot L_{i2}} \cdot L_{i1}} \quad \text{or} \quad X_i = const \quad (3)$$

Note that each variable directly depends on at most one variable. The dependencies of variables in this system can be represented by a graph with variables as vertices and with arcs labelled by $\{+, -\}$. This graph contains an arc $(X_i, X_{j_i}, \mathfrak{S})$ for a non-constant equation (3) if $\varepsilon \in K_j, L_j$ for all j and $\mathfrak{S} = +$ if m_i is even, $\mathfrak{S} = -$ otherwise. Using this graph, the solution existence and uniqueness problems for the system of language equations can be characterized as follows:

Lemma 10. *A system (3) has a solution if and only if the constructed graph has no cycles with an odd number of negative arcs. It has a unique solution if and only if the graph has no cycles at all.*

Since the constructed graph is of out-degree one, its cycles can be analyzed in deterministic logarithmic space, and thus the properties of the system of language equations can be decided efficiently, as long as it is computationally easy to determine the membership of ε in constant languages.

Theorem 9. *Consider any set of constants, for which the membership of ε is decidable in L (deterministic logarithmic space). Then, given a system $X_i = \varphi_i(X_1, \ldots, X_n)$ $(1 \leqslant i \leqslant n)$ with linear concatenation and complementation and with the above constants, the problems of whether it has a solution and whether it has a unique solution are L-complete wrt. one-way logspace reductions [4].*

This applies to singleton constants, to regular constants given by DFAs, to constants given by context-free grammars in Chomsky normal form, etc.

Let us turn to the expressive power of our equations in the case of linear concatenation. Denote the family of languages by *LinN* and *LinN*^{Reg} for singleton and regular constants. If the constants are singletons, there is a way to eliminate complementation in the equations as follows:

$$\overline{u_1 u_2 L v_2 v_1} = \overline{u_1 \Sigma^* v_1} \cup u_1 u_2 \overline{L} v_2 v_1 \quad (\forall\, u_1, u_2, v_1, v_2 \in \Sigma^* \text{ and } L \subseteq \Sigma^*) \quad (4)$$

The identity (4) can be used for equivalent transformation of equations:

Example 6. The equations $X = \overline{a\overline{X}b}$ and $X = \overline{b\Sigma^*} \cup aXb$, share the unique solution $L_0 = \{a^n w b^n \mid w = \varepsilon \text{ or } w \in b\Sigma^*\}$. Since $L_0 \cap a^* b^* = \{a^m b^n \mid m \leqslant n\}$, this solution is nonregular.

This method is in fact applicable to every one-variable equation.

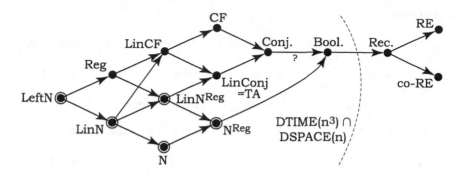

Fig. 1. Hierarchy of families of languages defined by language equations

Lemma 11. *Let $X = \varphi(X)$ be a one-variable language equation with linear concatenation and complementation and with singleton constants that has a unique solution L. Then $L = \{u^n xv^n \mid n \geqslant 0, x \in R\}$ for some words $u, v \in \Sigma^*$ and for some regular language R, such that $R \cap u\Sigma^* v = \varnothing$.*

If φ contains an even number of complementations, then the transformation (4) applied appropriately many times converts the equation to the form $X = uXv \cup R$. If the number of complementations is odd, consider the equivalent equation $X = \varphi(\varphi(X))$. These constructions can be used to obtain the following.

Theorem 10. *If a system of language equations $X_i = \varphi_i(X_1, \ldots, X_n)$ ($1 \leqslant i \leqslant n$) with linear concatenation and complementation and with singleton constants has a unique solution, then all of its components are linear context-free, and the corresponding grammars can be effectively constructed.*

In Theorem 10 it is essential that constants are singletons. If regular constants are allowed, the languages need not be context-free:

Example 7. Let L_0 be the unique solution of the equation $X = \overline{a\overline{X}b}$ (see Example 6), and let $R = b^+ \cup a^* b^+ ab^+$. Then $(L_0 \odot R) \cap a^+ b^+ ab^+ = \{a^i b^j ab^k \mid i \geqslant 1, i \leqslant j, i \leqslant k\}$ and hence $L_0 \odot R$ is not context-free. The system $\{X = \overline{a\overline{X}b}, Y = \overline{X} \cdot \overline{R}\}$ thus has a unique solution with a non-context-free component Y.

Denote the families of (linear) context-free and (linear) conjunctive languages by *LinCF*, *CF*, *LinConj* and *Conj*. The following properties have been established:

Theorem 11. **(I)** *LinN* $\subset$ *LinCF*; **(II)** *LinN* $\subset$ *LinN^{Reg}* $\subset$ *LinConj*; **(III)** *LinN* *is incomparable with* *Reg*; **(IV)** *LinN^{Reg}* *is incomparable with* *LinCF* *and with* *CF*.

Together with Theorem 7, this establishes the place of *N*, *N^{Reg}*, *LinN* and *LinN^{Reg}* among the known families defined by language equations, shown in Figure 1.

The rest of the properties of these equations are left for future research. Let us emphasize a particular open problem: is the condition of Theorem 4 decidable?

References

1. A. Blass, Yu. Gurevich, "On the unique satisfiability problem", *Information and Control*, 55 (1982), 80–88.
2. W. Charatonik, "Set constraints in some equational theories", *Information and Computation*, 142 (1998), 40–75.
3. S. Ginsburg, H. G. Rice, "Two families of languages related to ALGOL", *Journal of the ACM*, 9 (1962), 350–371.
4. J. Hartmanis, N. Immerman, S. Mahaney, "One-way log-tape reductions", *FOCS 1978*, 65–71.
5. W. Kuich, "Semirings and formal power series: their relevance to formal languages and automata", in: Rozenberg, Salomaa (Eds.), *Handbook of Formal Languages*, Vol. 1, Springer-Verlag, 1997, 609–677.
6. M. Kunc, "The power of commuting with finite sets of words", *STACS 2005*.
7. M. Kunc, "On language inequalities $XK \subseteq LX$", *DLT 2005*.
8. E. L. Leiss, "Unrestricted complementation in language equations over a one-letter alphabet", *Theoretical Computer Science*, 132 (1994), 71–93.
9. A. Okhotin, "Conjunctive grammars and systems of language equations", *Programming and Computer Software*, 28 (2002), 243–249.
10. A. Okhotin, "Decision problems for language equations with Boolean operations", *ICALP 2003*.
11. A. Okhotin, "Boolean grammars", *Inform. Comput.*, 194:1 (2004), 19–48.
12. A. Okhotin, "The dual of concatenation", *Theoretical Computer Science*, 345 (2005), 425–447.
13. A. Salomaa, S. Yu, "On the decomposition of finite languages", *DLT 1999*.

Synchronizing Automata
with a Letter of Deficiency 2[*]

D.S. Ananichev, M.V. Volkov, and Yu. I. Zaks

Department of Mathematics and Mechanics,
Ural State University, 620083 Ekaterinburg, Russia
{Dmitry.Ananichev, Mikhail.Volkov}@usu.ru,
zaksjulia@r66.ru

Abstract. We present two infinite series of synchronizing automata with a letter of deficiency 2 whose shortest reset words are longer than those for synchronizing automata obtained by a straightforward modification of Černý's construction.

1 Background and Motivation

Let $\mathscr{A} = \langle Q, \Sigma, \delta \rangle$ be a *deterministic finite automaton* (DFA), where Q is the state set, Σ stands for the input alphabet, and $\delta : Q \times \Sigma \to Q$ is the transition function defining an action of the letters in Σ on Q. The action extends in a unique way to an action $Q \times \Sigma^* \to Q$ of the free monoid Σ^* over Σ; the latter action is still denoted by δ. The DFA $\mathscr{A}$ is called *synchronizing* if there exists a word $w \in \Sigma^*$ whose action resets $\mathscr{A}$, that is leaves the automaton in one particular state no matter which state in Q it started at: $\delta(q_1, w) = \delta(q_2, w)$ for all $q_1, q_2 \in Q$. Any word w with this property is said to be a *reset word* for the DFA.

It is rather natural to ask how long a reset word for a given synchronizing automaton may be. The problem is known to be NP-complete (see, e.g., [11, Section 6]), but on the other hand, there are some upper bounds on the minimum length of reset words for synchronizing automata with a given number of states. The best such bound known so far is due to J.-E. Pin [10] (it is based on a combinatorial theorem conjectured by Pin and then proved by P. Frankl [5]): for each synchronizing automaton with n states, there exists a reset word of length at most $(n^3 - n)/6$. In 1964 J. Černý [2] produced for each $n > 1$ a synchronizing automaton $\mathscr{C}_n$ with n states whose shortest reset word has length $(n - 1)^2$ and conjectured that these automata represent the worst possible case, that is, every synchronizing automaton with n states can be reset by a word of length $(n-1)^2$. By now this simply looking conjecture is arguably the most longstanding open problem in the combinatorial theory of finite automata. The reader is referred to the survey [8] for an interesting overview of the area and its relations to multiple-valued logic and symbolic dynamics; applications of synchronizing automata to robotics are discussed in [4].

---

[*] Supported by the Russian Foundation for Basic Research, grant 05-01-00540.

O.H. Ibarra and Z. Dang (Eds.): DLT 2006, LNCS 4036, pp. 433–442, 2006.
© Springer-Verlag Berlin Heidelberg 2006

There are many papers where the Černý conjecture is proved for various restricted classes of synchronizing automata (cf. [3, 7, 1, 12], to mention a few recent advances only). On the other hand, there are only very few examples of "slowly" synchronizing automata, that is automata whose shortest reset words have lengths close to the Černý bound. In fact, it seems that the only infinite series of n-state synchronizing automata with shortest reset words of length $O(n^2)$ that appeared in the literature so far is the Černý series $\mathscr{C}_n$, $n = 2, 3, \ldots$. Of course, one can obtain more examples by some slight modifications of the Černý automata (we shall discuss this later) but in general "slowly" synchronizing automata turn out to be rather exceptional. This observation is supported not only by numerous experiments (see [13] for a description of certain noteworthy experimental results in the area) but also by probabilistic arguments. Indeed, if Q is an n-element set (with n large enough), then, on average, any product of $2n$ randomly chosen transformations of Q is known to be a constant map, cf. [6]. Being retold in automata-theoretic terms, this fact implies that a randomly chosen DFA with n states and a sufficiently large input alphabet tends to be synchronizing, and moreover, the length of its shortest reset word does not exceed $2n$.

In the present paper we construct two new infinite series of "slowly" synchronizing automata. In contrast with the Černý series, in our automata one of the letters acts as a transformation of deficiency 2. (Recall that the *deficiency* of a transformation φ of a finite set Q is the difference $|Q| - |\varphi(Q)|$.) Since, in the presence of such a letter, synchronization speeds up, one cannot expect the lengths of shortest reset words for our automata to reach the Černý bound. However, surprisingly enough, our examples turn out to synchronize slower than automata with a letter of deficiency 2 derived in a natural way from the Černý automata.

Besides enlarging our supply of examples, there are various additional motivations for studying synchronizing automata with a letter of deficiency 2. For instance, we recall that the best upper bound known so far for the minimum length $\ell(n)$ of reset words for synchronizing automata with n states is cubic. Clearly, finding a quadratic upper bound for $\ell(n)$ would constitute a major step towards a proof of the Černý conjecture. It can be easily verified that if a quadratic in n function $f(n)$ provides an upper bound for the minimum length of reset words for n-state synchronizing automata with a letter of deficiency 2, then the function $4f(n)$ can serve as an upper bound for $\ell(n)$. Thus, approaching the problem through automata with a letter of deficiency 2 might be a reasonable strategy. However we shall not touch this approach in the present paper.

2   Main Results and a Discussion

Let $\mathscr{A} = \langle Q, \Sigma, \delta \rangle$ be a DFA with $|Q| \geq 3$. If a letter $a \in \Sigma$ is such that the transformation $\delta(\_, a) : Q \to Q$ has deficiency 2, then exactly one of the two following situations happens.

1. There exist four different states $q_1, q_2, q_3, q_4 \in Q$ such that

$$\delta(q_1, a) = \delta(q_2, a) \neq \delta(q_3, a) = \delta(q_4, a).$$

In this situation we say that a is a *bactrian letter*.

2. There exist three different states $q_1, q_2, q_3 \in Q$ such that

$$\delta(q_1, a) = \delta(q_2, a) = \delta(q_3, a).$$

In this case we call a a *dromedary letter*.

Figure 1 illustrates these notions and explains the terminology.

The action of a bactrian letter   The action of a dromedary letter

Fig. 1. Two types of letters of deficiency 2

An easy way to obtain slowly synchronizing automata with a letter of deficiency 2 of either type consists in modifying the Černý automata. Namely, consider the Černý automaton $\mathscr{C}_{n-1}$ whose states are the residues modulo $n-1$ and whose input letters a and b act as follows:

$$\delta(0, a) = 1, \ \delta(m, a) = m \text{ for } 0 < m < n-1, \ \delta(m, b) = m+1 \pmod{n-1}.$$

We add to $\mathscr{C}_{n-1}$ an extra state denoted $n-1$ and then extend the transition function by letting $\delta(n-1, a) = 2$, $\delta(n-1, b) = n-1$. This will give an n-state automaton $\mathscr{C}_n'$ in which a becomes a bactrian letter. Similarly, if we extend δ by defining $\delta(n-1, a) = 1$, $\delta(n-1, b) = n-1$, we obtain another n-state automaton $\mathscr{C}_n''$ in which a is a dromedary letter. Both these modifications are shown on Fig. 2.

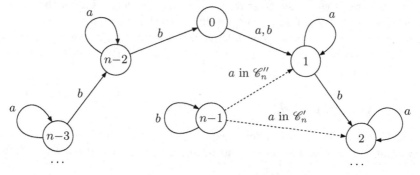

Fig. 2. The automata $\mathscr{C}_n'$ and $\mathscr{C}_n''$

It can be verified that the word $(ab^{n-2})^{n-3}a$, which resets the automaton $\mathscr{C}_{n-1}$, resets also both $\mathscr{C}'_n$ and $\mathscr{C}''_n$ and is in fact the shortest reset word for each of these automata. Hence $(n-2)^2$, i.e. the length of this word, turns out to be a lower bound for the minimum length of reset words for n-state synchronizing automata with a letter of deficiency 2 of either type. By analogy with the Černý conjecture, one may think that the bound is tight. However, as our results show, this is not the case.

Our first result significantly improves the lower bound for synchronizing automata with a bactrian letter:

Theorem 1. *For each odd $n > 3$, there exists a synchronizing automaton $\mathscr{B}_n$ with n states and two input letters one of which is bactrian such that the shortest reset word of $\mathscr{B}_n$ is of length $(n-1)(n-2)$.*

The proof of Theorem 1 is presented in Section 2. In our opinion, this proof is of independent interest as it involves a trick which, to the best of our knowledge, has not appeared in synchronization proofs so far.

It seem that the restriction on the parity of the quantity of states in Theorem 1 is essential. If n is even, then the construction used to design the automaton $\mathscr{B}_n$ still works but produces an automaton which is not synchronizing. For $n = 6$ we have found a synchronizing automaton with two input letters including one bactrian and with the shortest reset word of length $(6-1)(6-2) = 20$ but already for $n = 8$ our best bactrian example has the shortest reset word of length $39 < (8-1)(8-2) = 42$.

Now consider the dromedary case. Here it appears that the 'Černý-like' example $\mathscr{C}''_n$ is indeed optimal for the two-letter alphabet. However over three letters we are able to slightly improve the lower bound:

Theorem 2. *For each $n > 4$, there exists a synchronizing automaton $\mathscr{D}_n$ with n states and three input letters one of which is dromedary such that the shortest reset word of $\mathscr{D}_n$ is of length $(n-2)^2 + 1$.*

The proof of Theorem 2 shares some ideas with the proof of Theorem 1 but is more bulky. Due to space limitations, it will be published elsewhere. The automata $\mathscr{D}_n$ are presented in Section 3.

For $n = 5$ and $n = 6$, we have found some dromedary examples (again with three input letters) whose shortest reset words are one letter longer than those of respectively $\mathscr{D}_5$ and $\mathscr{D}_6$. These examples indicate that there may exist a series of n-state synchronizing automata with three input letters including one dromedary whose shortest reset words are of length $(n-2)^2 + 2$ but we have not managed to find such a series so far.

3   The Automata $\mathscr{B}_n$

Let $n = 2k + 1$ be an odd number greater than 3. The states of the automaton $\mathscr{B}_n$ are the residues modulo $n - 1$ and its input letters a and b act as follows:

$$\delta(m, a) = \begin{cases} m - 2 \pmod{n} & \text{for } m = 0, 1, \\ m & \text{for } 1 < m < n, \end{cases} \qquad \delta(m, b) = m - 1 \pmod{n}.$$

Observe that a is a bactrian letter in $\mathscr{B}_n$. The smallest automaton in the series is shown on Fig. 3.

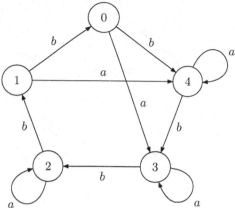

Fig. 3. The automaton $\mathscr{B}_5$

The next fact can be straightforwardly checked and we omit its proof.

Lemma 1. *Let* $n = 2k + 1$, $k > 1$. *Then the word*

$$(ab^{2k-1})^{k-1}ab^{2k-2}(ab^{2k-1})^{k-1}a \tag{1}$$

is a reset word for the automaton $\mathscr{B}_n$.

The length of the word (1) is $2k(k-1) + 2k - 1 + 2k(k-1) + 1 = 2k(2k-1) = (n-1)(n-2)$. We observe in passing that $\mathscr{B}_n$ has yet another reset word of the same length.

To complete the proof of Theorem 1, it remains to show that the length of each reset word for $\mathscr{B}_n$ is at least $(n-1)(n-2)$. For this, we use a solitaire-like game on the underlying graph of $\mathscr{B}_n$. Assume that some of the states of $\mathscr{B}_n$ are covered with pairwise distinct *coins* as shown on Fig. 4. Each *move*, that is the action of a letter $c \in \{a, b\}$, makes the coins slide along the arrows labelled c so that a state m will be covered with a coin after the move if and only if there exists a state ℓ such that $\delta(\ell, c) = m$ and ℓ was covered with a coin before the move. If two coins happen to arrive at the same state m, then from the structure of $\mathscr{B}_n$ we conclude that $c = a$, $m = n - 1$ or $m = n - 2$ and both m and $m + 2$ (mod n) held coins before the move. Then we retain the coin that had covered m before the move and delete the coin arriving from $m + 2 \pmod{n}$. Figure 5 demonstrates how the position shown on Fig. 4 changes after a single action of a letter.

Suppose that initially all the states of the automaton $\mathscr{B}_n$ are covered with coins and let a word $w \in \{a, b\}^*$ (that is the sequence of its letters) act on this initial position. It is easy to realize that after completing this action coins cover precisely the states in the image of the transformation $\delta(\_, w)$. In particular, if w is a reset word for $\mathscr{B}_n$, then after the action of w only one coin survives.

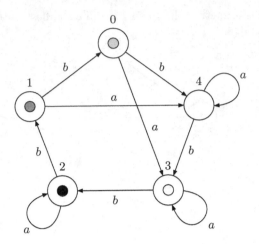

Fig. 4. A position on $\mathscr{B}_5$

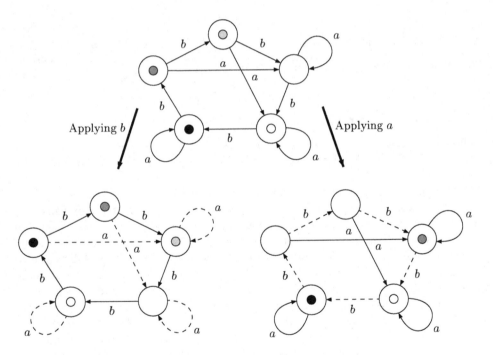

Fig. 5. Redistributing coins under the actions of b (left) and a (right)

Now we can explain the idea of our proof of Theorem 1. Given a reset word w and an initial distribution P_0 of n coins on the states of $\mathscr{B}_n$, let P_i $(0 \le i \le |w|)$ stand for the position that arises when we apply the prefix of w of length i to the position P_0. We shall assign each position P_i an integer parameter $\mathrm{wg}(P_i)$

(called the *weight* of the position) such that the following three conditions are satisfied:

(i) $\mathrm{wg}(P_0) \geq (n-1)^2$;
(ii) $\mathrm{wg}(P_{|w|}) \leq n - 1$;
(iii) for each $i = 1, \ldots, |w|$, the action of the ith letter of w decreases the weight of P_{i-1} by 1 at most, that is, $1 \geq \mathrm{wg}(P_{i-1}) - \mathrm{wg}(P_i)$.

Clearly, if such a weight function indeed exists, then summing up all the inequalities in (iii) and utilizing (i) and (ii), we obtain

$$|w| = \sum_{i=1}^{|w|} 1 \geq \sum_{i=1}^{|w|} \big(\mathrm{wg}(P_{i-1}) - \mathrm{wg}(P_i)\big) = \mathrm{wg}(P_0) - \mathrm{wg}(P_{|w|}) \geq$$
$$(n-1)^2 - (n-1) = (n-1)(n-2),$$

as required.

It remains to construct a weight function satisfying (i)–(iii). This is by no means an easy task because some moves can delete two coins at once. It is to overcome this difficulty that we let our coins be distinguishable from each other—this allows us to make weight functions depend on reset words while a 'uniform' weight function serving all reset words simultaneously may not exist.

Thus, let us fix a reset word w and an initial distribution P_0 of n coins on the states of $\mathscr{B}_n$. As mentioned, the action of w on P_0 removes $n-1$ coins. We call the only coin that remains after the action the *golden* coin and denote it by G. Now fix a position P_i ($0 \leq i \leq |w|$). For any coin C that is present in this position, let $m_i(C)$ be the state covered with C. We denote by $d_i(C)$ the least non-negative integer such that $\delta\big(m_i(C), b^{2d_i(C)}\big) = m_i(G)$. In the 'visual' terms, $d_i(C)$ is the number of double steps on the 'main circle' of $\mathscr{B}_n$ (measured clockwise) from the state covered with C to the state covered with the golden coin. We define the *weight* of C in the position P_i as

$$\mathrm{wg}(C, P_i) = (n-1) \cdot d_i(C) + m_i(C).$$

(Observe that here we multiply and add integers and not residues modulo n.) In order to illustrate this definition, assume that the black coin in the position shown on Fig. 4 is the golden coin. Then the weight of the white coin in this position is equal to $4 \cdot 3 + 3 = 15$ because the white coin covers the state 3 and from this state one needs 3 double steps in the clockwise direction in order to reach the state 2 covered with the golden coin. Similarly, the weight of the dark-grey coin on Fig. 4 is $4 \cdot 2 + 1 = 9$ and the weight of the light-grey coin is $4 \cdot 4 + 0 = 16$. As for the black (=golden) coin, its weight is $4 \cdot 0 + 2 = 2$ because, by the definition, the weight of the golden coin in any position is equal to the state it covers.

Now we define the *weight* $\mathrm{wg}(P_i)$ of the position P_i as the maximum of the weights of the coins present in this position. For instance, the weight of the position shown on Fig. 4 is 16 (if, as above, one assumes that the black coin is the golden one). It remains to verify that this weight function satisfies Conditions (i)–(iii).

Condition (i): $\mathrm{wg}(P_0) \geq (n-1)^2$. In the initial position all states are covered with coins. Consider the coin C that covers the state $m_0(G) - 2 \,(\mathrm{mod}\ n)$, that is the state in one double step clockwise after the state covered with the golden coin. Then it is easy to see that $d_0(C) = n - 1$ whence $\mathrm{wg}(C, P_0) = (n-1) \cdot (n-1) + m_0(C) \geq (n-1)^2$. Since the weight of a position is not less that the weight of any coin in this position, we conclude that $\mathrm{wg}(P_0) \geq (n-1)^2$, as required.

Condition (ii): $\mathrm{wg}(P_{|w|}) \leq n-1$. In the final position only the golden coin G remains, whence the weight of this position is the weight of G. We already have observed that $\mathrm{wg}(G, P_i) = m_i(G)$ for any position P_i and, clearly, $m_i(G) \leq n-1$.

Condition (iii): $\mathrm{wg}(P_{i-1}) - \mathrm{wg}(P_i) \leq 1$ for $i = 1, \ldots, |w|$. Let us fix a coin C of maximum weight in P_{i-1}. First consider the case when the letter that causes the transition from P_{i-1} to P_i is b. Recall that $\delta(m, b) = m - 1 \,(\mathrm{mod}\ n)$. This implies that $d_i(C) = d_{i-1}(C)$ (because the relative location of the coins does not change) and

$$
m_i(C) = \begin{cases} m_{i-1}(C) - 1 & \text{if } m_{i-1}(C) > 0, \\ n - 1 & \text{if } m_{i-1}(C) = 0. \end{cases}
$$

We see that

$$
\mathrm{wg}(P_i) \geq \mathrm{wg}(C, P_i) = (n-1) \cdot d_i(C) + m_i(C) \geq
$$
$$
(n-1) \cdot d_{i-1}(C) + m_{i-1}(C) - 1 = \mathrm{wg}(C, P_{i-1}) - 1 = \mathrm{wg}(P_{i-1}) - 1,
$$

as required.

Next suppose that the transition from P_{i-1} to P_i is caused by the action of the letter a. Recall that a moves the states 0 and 1 to the states $n-2$ and $n-1$ respectively (that is one double step clockwise) and fixes all other states. If the coin C covers neither 0 nor 1, then it does not move whence $m_i(C) = m_{i-1}(C)$ and

$$
d_i(C) = \begin{cases} d_{i-1}(C) & \text{if the golden coin } G \text{ covers neither 0 nor 1,} \\ d_{i-1}(C) + 1 & \text{if } G \text{ covers either 0 or 1.} \end{cases}
$$

We conclude that

$$
\mathrm{wg}(P_i) \geq \mathrm{wg}(C, P_i) = (n-1) \cdot d_i(C) + m_i(C) \geq
$$
$$
(n-1) \cdot d_{i-1}(C) + m_{i-1}(C) = \mathrm{wg}(C, P_{i-1}) = \mathrm{wg}(P_{i-1}).
$$

Thus, here the transition from P_{i-1} to P_i does not decrease the weight.

It remains to consider the subcase when the coin C covers either 0 or 1. As these two possibilities are analyzed with precisely the same argument, we assume that C covers 0. Then in the position P_i the state $n-2$ holds a coin C' (which may or may not coincide with C). If in the position P_{i-1} the golden coin G covers either 0 or 1, then $d_i(C') = d_{i-1}(C)$ whence

$$\mathrm{wg}(P_i) \geq \mathrm{wg}(C', P_i) = (n-1) \cdot d_i(C') + n - 2 >$$
$$(n-1) \cdot d_{i-1}(C) = \mathrm{wg}(C, P_{i-1}) = \mathrm{wg}(P_{i-1}).$$

We see that here the weight even increases. Finally, if the coin G covers neither 0 nor 1, it does not move whence $d_i(C') = d_{i-1}(C) - 1$. Therefore

$$\mathrm{wg}(P_i) \geq \mathrm{wg}(C', P_i) = (n-1) \cdot d_i(C') + n - 2 =$$
$$(n-1) \cdot (d_{i-1}(C) - 1) + n - 2 = (n-1) \cdot d_{i-1}(C) - 1 =$$
$$\mathrm{wg}(C, P_{i-1}) - 1 = \mathrm{wg}(P_{i-1}) - 1,$$

as required.

Thus, we have verified that our weight function satisfies Conditions (i)–(iii), and this completes the proof of Theorem 1.

It is very tempting to conjecture that the expression $(n-1)(n-2)$ gives the exact value for the minimum length of reset words for n-state synchronizing automata with a letter of deficiency 2 when $n \geq 5$ is odd. So far we have been able to confirm this only for $n = 5$ (thus solving a question mentioned in J.-E. Pin's early survey [9]).

4   The Automata $\mathscr{D}_n$

Take an $n > 4$ and let $\mathscr{D}_n$ be the DFA with the state set $\{1, 2, \ldots, n\}$, the input alphabet a, b, c and the transition function δ defined by the following table:

m	1	2	3	4	5	...	n
$\delta(m, a)$	1	1	1	4	5	...	n
$\delta(m, b)$	1	1	2	4	5	...	n
$\delta(m, c)$	4	1	4	5	6	...	3

Thus, both a and b fix each state m with $4 \leq n \leq m$ and c acts on the set $\{3, 4, \ldots, n\}$ as a cyclic shift. The automaton $\mathscr{D}_n$ is shown on Fig. 6.

Verifying the following fact amounts to a straightforward calculation:

Lemma 2. *Let $n > 4$. Then the word*

$$c^2(bc^{n-1})^{n-4}bc^2 \tag{2}$$

is a reset word for the automaton $\mathscr{D}_n$.

The length of the word (2) is $n(n-4) + 5 = (n-2)^2 + 1$. and this is in fact the minimum length of a reset word for $\mathscr{D}_n$. Observe that the word (2) does not involve the letter a, and therefore, it also resets the DFA obtained from $\mathscr{D}_n$ by omitting a. Thus, we see (and it seems to be somewhat surprising) that adding a letter o f deficiency 2 to a synchronizing automaton in which all letters have deficiency 1 may not decrease the minimum length of reset words.

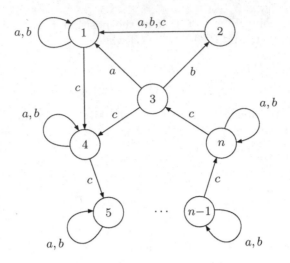

Fig. 6. The automaton $\mathcal{D}_n$

References

1. D. S. Ananichev, M. V. Volkov, Synchronizing generalized monotonic automata, Theoret. Comput. Sci. **330** (2005) 3–13.
2. J. Černý, Poznámka k homogénnym eksperimentom s konecnými automatami, Mat.-Fyz. Cas. Slovensk. Akad. Vied. **14** (1964) 208–216 [in Slovak].
3. L. Dubuc, Sur le automates circulaires et la conjecture de Černý, RAIRO Inform. Theor. Appl., **32** (1998) 21–34 [in French].
4. D. Eppstein, Reset sequences for monotonic automata, SIAM J. Comput. **19** (1990) 500–510.
5. P. Frankl, An extremal problem for two families of sets, Eur. J. Comb. **3** (1982) 125–127.
6. P. M. Higgins, The range order of a product of i transformations from a finite full transformation semigroup, Semigroup Forum **37** (1988) 31–36.
7. J. Kari, Synchronizing finite automata on Eulerian digraphs, Theoret. Comput. Sci. **295** (2003) 223–232.
8. A. Mateescu, A. Salomaa, Many-valued truth functions, Černý's conjecture and road coloring, EATCS Bull. **68** (1999) 134–150.
9. J.-E. Pin, Le probléme de la synchronisation et la conjecture de Černý, in A. De Luca (ed.), Non-commutative Structures in Algebra and Geometric Combinatorics [Quaderni de la Ricerca Scientifica **109**], CNR, Roma, 1981, 37–48 [in French].
10. J.-E. Pin, On two combinatorial problems arising from automata theory, Ann. Discrete Math. **17** (1983) 535–548.
11. A. Salomaa, Composition sequences for functions over a finite domain, Theoret. Comput. Sci. **292** (2003) 263–281.
12. A. N. Trahtman, The Černý conjecture for aperiodic automata, J. Automata, Languages and Combinatorics, accepted.
13. A. N. Trahtman, Noticeable trends and some examples concerning the Černý conjecture, unpublished manuscript.

On Some Variations of Two-Way Probabilistic Finite Automata Models

Bala Ravikumar

Department of Computer Science,
Sonoma State University, Rohnert Park, CA 94928, USA

Abstract. Rabin [21] initiated the study of probabilistic finite automata (PFA). Rabin's work showed a crucial role of the gap in the error bound (for accepting and non-accepting computations) in the power of the model. Further work resulted in the identification of qualitatively different error models (one-sided error, bounded and unbounded errors, no error etc.) Karpinski and Verbeek [16] and Nisan [20] studied a model of probabilistic automaton in which the tape containing random bits can be read by a two-way head. They presented results comparing models with one-way vs. two-way access to randomness. Dwork and Stockmeyer [5] and Condon et al. [4] studied a model of 2-PFA with nondeterministic states (2-NPFA). In this paper, we present some results about the above mentioned variations of probabilistic finite automata, as well as a model of 2-PFA augmented with a pebble studied in [22]. Our observations indicate that these models exhibit subtle variations in their computational power. We also mention many open problems about these models. Complete characterizations of their power will likely provide deeper insights about the role of randomness is space-bounded computations.

1 Introduction

Randomness has been understood to be a crucial artifact for an efficient solution of a wide range of computational problems. In a pioneering work, Rabin [21] showed that a 1-way probabilistic finite automaton (1-pfa) in which the acceptance probability is bounded away from $1/2$ is no more powerful than a determinstic finite automaton, i.e., both accept the same class of (regular) languages. In contrast, he also showed that allowing the error probability to be arbitrarily close to $1/2$ makes a probabilistic automaton accept non-regular languages. Freivalds [9] considered bounded error model, but allowed the input tape to be 2-way. He showed that this model (2-pfa) can accept non-regular languages. Specifically, Frievalds showed that a 2-pfa can accept $L_{eq} = \{0^n 1^n | n \geq 1\}$ with arbitrarily small error probability. Following Freivalds' work, several papers [13], [7], [3], [22], [4] studied this model and presented various results about the power of 2-pfa.

In this work, we study some variations of the 2-pfa model. Specifically, we are interested the following variations of the 2-pfa model:

O.H. Ibarra and Z. Dang (Eds.): DLT 2006, LNCS 4036, pp. 443–454, 2006.

1. The standard 2-pfa model has a coin-tossing mechanism in the finite control. Specifically, for some fixed k, its transition function provides k different options (just like a nondeterministic machine) from any specific configuration. At each step, one of these k options is assumed to be chosen randomly (with uniform probability) and the move is executed. However, the actual random choice made is not recorded for future reference. We can thus think of the randomization process as being implemented using a tape in which an arbitrarily long string over the alphabet $\{1, 2, ..., k\}$ is written and this tape is being read by the device using a tape-head that always moves in one direction to generate coin tosses. In contrast to this model, a 2-way random tape model is one in which the read-head on the tape can move in both directions. This model was first introduced by [2] and has been studied in [20], [16],[17] etc.

2. Interactive proof systems in which the verifier is a finite-state automaton was studied by Dwork and Stockmeyer [5], [6], [7] and Condon and Lipton [3], among others. This model is the finite-state analog of the interactive Turing machines introduced by Goldwasser, Micali and Rackoff [11] in a celebrated paper that extended the concept of a nondeterministic verification to include two new ingredients, namely: interaction and randomness. Since the model of interactive proofs we consider are such that the verifier's coin tosses are public, it is easier to model such automata as 2-pfa's with nondeterministic states (2-NPFA). We relate this model to the deterministic counter machine model in which the counters are reversal-bounded.

3. The third model we consider is the extension of the 2-pfa model by adding a pebble. This model was introduced in [22]. We consider 2-NPFA augmented with a pebble and present some results about the power of such automata.

The rest of this paper is organized as follows. In Section 2, we anwer an open question from [17] by showing that the class BP*TISP(poly, O(1)), the class of languages accepted by a 2-way PFA with bounded error that runs in expected polynomial time and with two-way access to random tape can accept non-regular languages. We show that a subclass of this class, namely ZP*TISP(poly, O(1)) (which is not allowed to have any error) can already accept non-regular languages. In Section 3, we show that the class of languages accepted by a one reveral deterministic multi-counter machine can be accepted by a 2-NPFA. In Section 4, we consider the 2-NPFA model augmented by a pebble and present examples of languages accepted by 2-NPFA(pebble) not known to be accepted by other weaker models. We present many open problems throughout the paper.

Because of space limitation, some of the proofs are not included in the extended abstract, but can be found in an appendix.

2   Two-Way PFA with Two-Way Access to the Random Tape

In the "standard" model of a probabilistic machine, the source of randomness is assumed to be a sequence of independent and unbiased coin tosses. We can view such a sequence as being stored in a read-only tape ("random-tape") which can only be accessed from left to right. A natural extension of this model is to allow the random-tape head to move bidirectionally. Such a model has been studied in [2], [16], [17], [20] and others. We will briefly describe some of the previous results about probabilistic models with two-way access to coin tosses.

Let BPTISP$(T(n), S(n))$ be the class of languages accepted by a bounded error probabilistic Turing machine operating with $O(S(n))$ space (on all paths) and expected time $O(T(n))$ on inputs of length n. As always, the expectation is over the coin tosses for every fixed input, and *not by averaging over various inputs of length n*. Following Nisan's notation in [20], the corresponding probabilistic class with two-way access to random tape is defined as BP*TISP$(T(n), S(n))$ etc. We also use the notation BSPACE$(S(n))$ to denote a bounded-error probabilistic Turing machine in which space bound $S(n)$ holds on every computational path, and in which the access to random tape is 1-way. The analoguous model with two way access to random tape will be denote by B*SPACE$(S(n))$. BSPACE$(O(1))$ will be denoted by the more familiar name 2-PFA.

The first question comparing probabilistic machine with two-way and one-way access to random tape was raised by Borodin et al. [2]. It was answered by [16] by showing that there is a language in B*SPACE$(S(n))$ (where $S(n) \geq log\ n$) that is not in DSPACE$(S(n)^k)$ for any k. This is in contrast to the result of Borodin et al. that every language in $BSPACE(S(n))$ (where $S(n) \geq log\ n$) is in DSPACE$(S(n)^2)$, a generalization of Savitch's theorem. The work [17] presents stronger results about probabilistic models with two-way access to random tape. Let Z*TISP(poly, log) denote probabilistic polynomial time and log space bounded Turing machine languages with two-way access to random tape. Nisan [20] showed that BPTISP(poly,log) $\subseteq$ ZP*TISP(poly,log). But no results were known about the classes BP*TISP(poly, $O(1)$) or ZP*TISP(poly, $O(1)$) or any other complexity classes in which the space bound is constant. (It should be noted that the finite-state analog of Nisan's theorem is trivially true since the class BPTISP(poly, $O(1)$) is known to contain only regular languages [5].)

Our first result involves an answer to an open problem in [17], namely, whether the class BP*TISP(poly,$O(1)$) contains a non-regular language. We answer these questions below.

First, we need the following lemmas.

Lemma 1. *Let s be a binary string generated by successively choosing each bit randomly (with probability 1/2 for both 0 and 1). Let E_t denote the expected number of bits that need to be generated before a string of t 0's is observed. Then, $E_t \leq t2^t$. This claim also holds if t 0's is replaced by any fixed t-bit string σ.*

Proof. The proof is simple. Consider the occurrences of a string of t 0's whose starting position is a multiple of t (where the positions are counted from 0). Thus we are looking at a sequence of t tosses as a single event and we stop as soon as the first success occurs which corresponds to the occurrence of t O's in a row. The probability of success is 2^{-t} and the expected number of trials before success (because it is a geometric distribution) is 2^t. Thus, the expected number of tosses before the first success is at most $t2^t$ since each event involves t tosses.

Lemma 2. *Suppose there are $n + 1$ points numbered 1 to $n + 1$ on a line. If a random walk starts in position 2, and moves with equal probability to either neighboring point until the walk ends at 1 or $n+1$. The probability that the walk ends in $n + 1$ is $1/n$.*

We omit the proof of the above lemma which involves the well-known one-dimensional random walk with absorbing barriers.

The next lemma is due to Alt and Mehlhorn [1] and can be shown from the prime number formula:

Lemma 3. *If n, m are two positive integers such that $n \neq m$, then there exists an integer $k \leq 4log\ (n + m)$ such that $n \not\equiv m\ (mod\ k)$.*

We next show that there is a non-regular language in BP*TISP(poly, O(1)).

Theorem 1. *The language $L = \{0^n 1^n | n \geq 1\}$ is in BP*TISP(poly, O(1)).*

Proof. We design a 2-PFA M with two-way access to randomness as follows: On input x, M first checks that x is of the form $O^n 1^m$, else the input is rejected. If the string is of this form, it proceeds as follows: M moves the head on the random tape looking for a substring of the form $10^r 1$. Whenever such a substring is found, M checks that $n \equiv m\ (mod\ r)$ as follows.

It moves the random tape head back to the leftmost 0, and its input head on the leftmost 0, and starts moving its input head and the random tape head simultenously until the random tape head reaches a 1 while the input tape is still reading a 0. (The case in which the input head reaches a 1 before random tape head reaches a 1 is exponentially rare and is, in any event, covered by our construction. This will become clear later on.) Now, M resets the random tape back to the first 0 of the block by moving its random tape back until it reaches a 1, and by moving right one step. This cycle is repeated as many times as needed, until the input head is away from 1 by t for some $0 \leq t < r$ at the start of a cycle. It is clear that $t = n\ mod\ r$. At this point, one more cycle is executed with both heads advancing to the right. M's input head will reach a 1 before the random tape's head reaches a 1. Now, the random tape is exactly reading the t-th leftmost 0 of the block $10^r 1$. Next, as the random tape head is moved back until it reaches a 1, the input head is moved to the right over the block of 1's. If the right end-marker on the input tape is reached before the random tape reaches the 1 on the block $10^r 1$, then clearly $n \not\equiv m\ (mod\ t)$ so the input is rejected and the computation halts. Assuming that this does not happen, when the random tape head reaches a 1, the input head is reading the t-th 1 (from the left-end) on the input tape. From now on, a series of cycles similar to that over the block of 0's is repeated, namely, the input head is moved over a block of r 1's over each sweep on the random tape. If at the end of a cycle when the random tape is reading the right 1 of the block $10^t 1$, the input head reaches the right end-marker, then it is clear that $n \equiv m\ (mod\ r)$. On the other hand, when the input tape reaches the right-end marker, if the random tape did not reach the right 1 of the block 10^t1, then $n \not\equiv m \pmod{r}$ so the input is rejected and the computation ends.

If the computation does not end in a previous phase of the computation as described in the last paragraph, the random tape head is moved to the next block of the form 10^r1 and another phase of computation is repeated with the new r. It is clear that, if $n \neq m$, eventually an r will be found such that $n \not\equiv m \pmod{r}$ and the input will be rejected. However, on inputs of the form 0^n1^n, such an r will never be found. So, we need a different mechanism to halt the computations. This is described below. Before the next phase is started, we execute a random process that has a probability of success $= \Theta(1/N^d)$ for a carefully chosen integer constant d (the choice of which will be described soon), where $N = n + m$ is the input length. Such a process can be simulated using lemma 2 as follows: The input head is placed on the second symbol on the input tape, and a random walk is executed until the head reaches one of the endmarkers. If in d consecutive executions of this random process, the head reaches the right end-marker each time, then we say that the random process succeeds. If the random process succeeds, then M accepts the input and stops. Else, it continues with the next phase by choosing the next block 10^r1 and performs the test "Is $n \equiv m \pmod{r}$?" for this new r as described above.

To show that the above construction is correct, we need to show the following: (a) On all inputs of length N, M halts in expected time bounded by a polynomial in N. It is clear that each cycle takes $O(N)$ time to execute (by a slightly more efficient way to implement each cycle than the one described above). The expected number of phases is given by $O(N^d)$ since the expected number of cycles executed before the random process executed in between successive cycles succeeds is $\Theta(1/N^d)$. Thus, it is clear that the M halts in average polynomial time. (b) We need to show that M accepts the language $L_{eq} = \{0^n1^n | n \geq 1\}$. This involves showing that M accepts (rejects) every string (not) in L with probability $1 - \epsilon$ for a given $\epsilon < 1/2$. It is clear that if the string is in L, it is never rejected since the only way to reject the input is to find a string of the form 10^r1 on the random tape such that $n \not\equiv m \pmod{r}$. Such an r can never be found if the input is in L, and hence M in fact, has only error on one side. Suppose the input string 0^n1^m is not L. Then, by lemma 3, there is a $k \leq 4log\,N$ such that $n \not\equiv m \pmod{k}$. By lemma 1, the expected number of moves that need to be made on the random tape before a string of k 0's is observed is at most $k2^k = O(N^c)$ for some c. Thus, by choosing an integer constant $d > c$, we can make the probability that a string of the form 10^k1 is observed in the random tape of length N^d to be smaller than $1 - \epsilon$ for any given fixed $\epsilon > 0$. Thus, the probability that M will accept the input is at most ϵ since the only way the input will be accepted in this case is if such a k is never encountered during N^d phases of execution.

Since BP*TISP(poly, O(1)) is closed under complement, we have the following corollary:

Corollary 1. $\{0^n1^m | n \neq m\}$ is in BP*TISP(poly, O(1)).

Next we address whether L_{eq} is in ZP*TISP(poly, O(1)), which is a subset of BP*TISP(poly, O(1)). ZP*TISP(poly, O(1)) denotes the class of languages which can be accepted by a 2-pfa (with two-way access to random tape) in *expected* polynomial time which halts with probability at least 1/2 on all inputs and never makes a mistake. We do not know the answer to this question, although the above construction shows that the error is *one-sided*, only on strings not in the language which means that the *yes* answer is always correct. (For the language $\{0^n 1^m | n \neq m\}$, the error is on the other side.)

Since $\{0^n 1^n | n \geq 1\}$ and $\{0^n 1^m | n \neq m\}$ are both in 2-PFA, a natural question is whether every language in 2-PFA is in BP*TISP(poly, O(1)). There is no evidence to make such a conjecture and we do not believe that this claim is true and suggest the language $\{0^n 1^m | n \leq m\}$, shown in [22] to be in 2-PFA, as a potential candidate to separate the classes 2-PFA and BP*TISP(poly, O(1)).

Next we address the question whether ZP*TISP(poly, O(1)) contains a non-regular language.

Theorem 2. *There is a non-regular language in ZP*TISP(poly, O(1)).*

Proof. Let $L = \{x_1 \# ... \# x_k | k \geq 1, x_i$ is the binary representation of i with leading 1 $\}$. This is a well-known non-regular language introduced by Hartmanis, Stearns and Lewis [14]. We describe a 2-pfa M which accepts L in polynomial average time on every input of lentgh n. The idea behind such an M is as follows: We describe the construction inductively on the block number i. Suppose the correctness of the first i blocks has been checked. i.e., M has checked that x_j is the binary representation of j for all $1 \leq j \leq i$. We will show how to check the correctness of the next block, namely x_{i+1}.

Assume that the input head is scanning the the leftmost symbol of x_i. To check the correctness of the next block x_{i+1}, M proceeds as follows: It tries to find a substring on the random tape that exactly matches x_i. It does it in the most obvious way by moving both heads to the right so long as there is a match. When the match fails, it moves both heads back until the input head reaches the # symbol, then it advances both heads by one position (to allow the string matching to start at the position immediately to the right of previously attempted matching) and the cycle is repeated. When the matching succeeds, we have the string x_i on the random tape. Now, both heads are reversed until x_i is reading the # symbol, while the random tape is reading the symbol immediately to the left of the matching position. Now the head on the random tape is moved one position to the right, and the input head moved all the way to the next # symbol. Now a matching between x_i and x_{i+1} is attempted. This is easy to do and we omit the details. If this attempt fails, the input is rejected. Otherwise, the computation proceeds to $i+2$. After all the blocks are correctly checked, the input is accepted. It is clear that the above algorithm does not make any errors. We will now show that M terminates in polynomial average time on inputs of length n where n is the input length.

Let n be the length of the input string. It can be shown that the length of the block x_i is $O(log\ n)$ for each i so the expected number of moves needed to find a copy of x_i on the random tape is $O(n^c)$ for some c as seen from lemma

1. Thus, the total time required to find the sequence of such strings in various cycles is at most $O(n^{c+1}log\ n)$. (The mutiplicative factor $log\ n$ occurs due to the backtracking after each failure.) The rest of the computation involves a sequence of string matchings and this involves $O(n)$ time. Thus the expected time of M on inputs of length n is bounded by a polynomial in n.

It should be noted that both languages decribed in the above theorems can be accepted in $log\ log\ n$ space. The latter language is in DSPACE($log\ log\ n$) [14]. The former language is actually not in NSPACE($log\ log\ n$), but its complement is in NSPACE($log\ log\ n$), and L_{eq} would also be recognizable in $log\ log\ n$ space if a work-tape of length $\lceil log\ log\ n \rceil$ is marked at the beginning of the computation [15]. Do these results suggest that perhaps every language in DSPACE($log\ log\ n$) is in BSPACE(poly, $O(1)$)? There is no evidence to make such a claim. In fact, consider the unary language $L = \{a^n |$ the smallest r that does not divide n is a power of 2 $\}$. This language is known to be in DSPACE($log\ log\ n$) [1], but it is not clear that it is in BP*TISP(poly, $O(1)$).

We propose as an interesting area to investigate the connections between the classes in 2-PFA, BP*TISP(poly,$O(1)$), and ZP*TISP(poly,$O(1)$). At this point, out knowledge of these classes is quite limited. In the next section, we discuss another interesting class of probabilistic automata.

3   Two-Way PFA with Nondeterministic States

Dwork and Stockmeyer [7] introduced the model of two-way probabilistic finite automaton with nondeterministic states (2-NPFA) as the finite-state analog of the Arthur-Merlin games. We will informally describe how a 2-NPFA works. A 2-NPFA has states partitioned into nondeterministic states and probabilistic states. The input head can move left or right and change it state based the current state and the current input scanned. In the case of probabilistic state, if there are k options, any one of them is chosen with probability k. In the case of nondeterministic state, any one of the successor moves is chosen. The actual choice is only relevant in defining the probability of acceptance of the input string: To determine the probability of acceptance on an input string x, we create a computation tree in which all the children of probabilistic states are retained with weights, while for nondeterministic state, we pick one of the next possible moves. Thus, there are many computation trees associated with an input string x. To determine the probability of acceptance of the string w.r.to a fixed tree, we associate a probability of acceptance for the tree as the weighted sum of all the paths that reach an accepting leaf. The probability of acceptance of a string is the $maximum$ probability over all possible computation trees. A 2-NPFA accepts a language L if it accepts every string $x \in L$ with probability at least 2/3 and accepts every string not in L with probability at most 1/3. It is obvious that the class of languages accepted by 2-PFA is a subset of the class of languages accepted by 2-NPFA and hence the latter includes non-regular languages (because of the results of [9], [22] etc.) With a slight abuse of the notation, 2-NPFA will be used to denote a class of machines as well as the class of languages accepted by these machines.

The main result of the section is the following theorem:

Theorem 3. *Let L be accepted by a 1-way deterministic reversal-bounded multicounter machine. Then L can be accepted by a 2-NPFA.*

Because of space limitations, we omit the proof of this theorem. It is presented in the Appendix.

We can also show that many languages accepted by a 1-way nondeterministic 1-reversal counter can be accepted by a 2-NPFA. Specifically, we can show that the following language can be so accepted: $L_1 = \{0^t \#0^{i_1}\# \ldots \#0^{i_k} \mid$ for some subset S of $\{1, ..., k\}$, $\Sigma_{j \in S} i_j = t\}$.

4   2-NPFA Augmented with a Pebble

A pebble is a marker that is initially in the finite control. The transition function of the automaton depends, in addition to the current state and the input symbol, also on whether the pebble is in the finite control. If this is true, then in the next move it is possible for the pebble to be left on the current cell on the input tape. Similarly, when a cell with a pebble is visited, a possible next move is to collect the pebble and return it to finite control etc. The study of pebble augmenting the finite control of a finite automaton goes back to Hennie who showed that a 2-DFA augmented with a pebble accepts only regular languages. Since then, much work has been done to show that a pebble can add power to computational devices, especially when the space bound is below *log n*, see e.g. [15]. In [22], the question of whether 2-PFA(pebble) is more powerful than 2-PFA was addressed. Although some evidence for such power was provided, no proof to this effect was given, and this problem is still open. Here, we will compare the powers of the machines 2-NPFA, 2-PFA and 2-NPFA(pebble). While our results do not lead to separation of 2-PFA(pebble) from 2-PFA or 2-NPFA(pebble) from 2-NPFA, they offer new candidates for proving such a separation. Specifically, we show that some languages can be accepted by 2-NPFA(pebble) that seem not likely to be accepted by 2-PFA(pebble) or 2-NPFA. We also show some conditional separation results.

Theorem 4. *The language Let $L = \{0^n 1^m \mid n \neq m\}$. L^* can be accepted by a 2-NPFA with a pebble.*

Proof. (sketch) Let ϵ be the error tolerance. Choose δ, a real number $(0 < \delta < 1)$ and a positive integer d such that $2.(1/2)^d < \epsilon$ and $(1 - \delta)^d > 1 - \epsilon$.

We view the input to be of the form $x_1...x_k$ where each x_i is in L. Let M_δ denote the 2-pfa that Freivalds constructed to accept L_{eq} with error tolerance δ. Note that M_δ has a constant acceptance probability on strings in L_{eq}. We now design a 2-pfa M for L^*. M conducts "competitions" as in Freivalds' construction of 2-pfa for L_{eq}. However, each block in our case is of the form $0^n 1^m$. We use the pebble to place it on a cell of the tape, so that on the left or the right side of the block, we have the string 0^k1^k for some k. (For example, if the string is 0^81^{10}, M_δ will place the pebble on the 9-th one.) A "macroprocessing" of the input is a sequential simulation of M_δ on each block of the form 0^j1^j once. Note that each such block is created nondeterministically by placing the pebble on the appropriate square of the block. The macroprocessing is positive for x if M_δ accepts all x_i's. Let $x' = 0^k$. Note that m is the number of x_i's in x. Let d' be the integer chosen in Freivalds' lemma to satisfy the inequality $2.(1/2)^{d'} < \delta$. A macroprocessing of x' involves tossing a sequence of biased coins with $\Pr(\text{Head})$ $= 2.(1/2)^{d'}$, and $\Pr(\text{Tail}) = 1 - 2.(1/2)^d$. A macroprocessing of x' is said to positive if all m coins turn up H. A competition is a macroprocessing of x and x' once. We say that it is a decisive competition if exactly one of the outcomes is positive, and the one with positive outcome is said to have won. M conducts a sequence of competitions until exactly d decisive competitions result. If at this time, both x and x' have won at least one match, then M accepts the input, else it rejects it. It can be shown that M accepts L^* with error at most ϵ.

Similarly, we can show the following:

Theorem 5. *Let L_{bal} be the set of balanced parentheses over a two-letter alphabet $\{[,]\}$. The complement of L_{bal} can be accepted by a 2-NPFA augmented with a pebble.*

The proof is presented in the Appendix.

Dwork and Stockmeyer [8] conjecture that L_{bal} is not in 2-PFA. The conjecture implies that its complement is also not in 2-PFA (since 2-PFA is closed under complement). A proof of this conjecture would therefore imply that 2-NPFA(pebble) is more powerful than 2-PFA. It is not clear if either of the languages described in the previous two theorems can be accepted by a 2-NPFA (without a pebble).

5   Conclusions

We have presented several results about variants of 2-PFA, an important model of computation. We also presented a solution to an open problem in [17] about the power of a 2-PFA with bounded error and with 2-way access to random tape. Our understanding of these variants of 2-PFA is quite limited at this time. We have stated many open problems throughout the paper about these models. We hope that our work will stimulate interest in these problems.

References

1. Alt, H. and K. Mehlhorn, "Lower bounds for the space complexity for context-free recognition", in *Proc. of ICALP*, 1976, 338-351.
2. A. Borodin, S. Cook and N. Pippenger, "Parallel computation for well-endowed rings and space-bounded probabilistic machines", *Information and Control*, (1983), 113-136.

452   B. Ravikumar

3. Condon, A. and R. Lipton, "On interactive proofs with space-bounded verifiers", *Proc. of 30th IEEE Annual Symp. on Found. of Comp. Science*, (1989), 462-467.

4. Condon, A. et al. "On the power of finite automata with both nondeterministic and probabilistic states", *SIAM Journal on Computing*, (1998), 739-762.

5. Dwork, C. and L. Stockmeyer, "Interactive proof systems with finite state verifiers", *IBM Report RJ 6262*, (1988).

6. Dwork, C. and Stockmeyer, L., "On the Power of 2-way Probabilistic Finite State Automata", *Proc. of 30th IEEE Annual Symp. on Found. of Comp. Science*, (1989), 480 - 485.

7. Dwork, C. and L. Stockmeyer, "A time complexity gap for two-way probabilistic finite automata", *SIAM Journal on Computing*, (1990), 1011-1023.

8. Dwork, C and L. Stockmeyer, "Finite state verifiers I: The power of interaction", *Journal of the ACM*, (1992), 800-828.

9. Freivalds, R., "Probabilistic Two-way Machines", *Proceedings of Mathematical Foundations of Computer Science*, Lecture Notes in Computer Science, **Vol 118**, Springer-Berlin (1981), 33 - 45.

10. Freivalds, R.,"Why Probabilistic Algorithms Can be More Effective in Certain Cases?" (*Invited Talk*), *Math. Foundations of Computer Science* (1986), 1-14.

11. Goldwasser, S., S. Micali and C. Rackoff, "The knowledge complexity of interactive proof systems", *Proc. of 17th Annual ACM Symp. on Theory of Computing*, (1985), pp. 291-304.

12. Goldwasser S. and Sipser M., "Private coins vs. public coins in interactive proof systems", *Proc. of 18th ACM Symp. on Theory of Computing*, (1986), 59-68.

13. Greenberg, A. and A. Weiss, "A Lower Bound for Probabilistic Algorithms for Finite State Machines", *Jl. of Comp. and Sys. Sciences*, (1986) 88 - 105.

14. Hopcroft, J. and Ullman, J., *Introduction to Automata, Languages and Computation*, Addison-Wesley. Reading, MA (1979).

15. Ibarra, O. and B. Ravikumar, "Sublogarithmic space Turing machines, nonuniform space complexity and closure properties", in *Mathematical Systems Theory*, (1988), 1-21.

16. Karpinski, M. and R. Verbeek, "There is no polynomial deterministic simulation of probabilistic space with two-way random-tape generator", *Information and Computation*, (1985), 158-162.

17. Karpinski, M. and R. Verbeek, "On the power of Two-way Random Generators and the Impossibility of Deterministic Poly-Space Simulation", *Information and Control* **71**, (1986) 131 - 142.

18. Kaneps, J., "Regularity of one-letter languages acceptable by 2-way finite probabilistic automata" *Proc. of Fundamentals of Computation Theory*, 1991, 287-296.

19. Minsky, M., "Recursive unsolvability of Post's problem of tag and other topics in the theory of Turing machines", *Annals of Mathemetics*, (1961), 570 - 581.

20. Nisan, N., "On Read-Once vs. Multiple Access to Randomness in Logspace", *Proc. of Fifth IEEE Structure in Complexity Theory, Barcelona, Spain*, (1990), 179 - 184.

21. Rabin, M., "Probabilistic finite automata", *Information and Control*, 1963, 230 - 245.

22. Ravikumar, B., "On the power of probabilistic finite automata with bounded error probability", in *Proc. of the Foundations of Sofware Technology and Theoretical Computer Science*, 1992, 392-403.

A   Appendix

To show the main result of section 3, we need the following lemmas. The first one was shown by Freivalds [9].

Lemma 4. $L_{eq} = \{0^n 1^n \mid n \geq 1\}$ *can be accepted by a 2-PFA.*

Our next lemma shows that a closely related language $L_{le} = \{0^m 1^n | m \leq n\}$ can be accepted by a 2-PFA. This lemma is due to [22].

Lemma 5. $L_{le} = \{0^n 1^m \mid m \leq n\}$ *can be accepted by a 2-PFA.*

In [22], it was shown that 2-PFA can accept 1-way blind reversal-bounded counter machines. These machines are like reversal-bounded counter machines, but are weaker in the sense that the next move of the device depends only on the current state and the input symbol scanned, but not on whether the counter value is 0.

In the next theorem, we show that every language accepted by a reversal-bounded 1-way determinstic counter machine can be accepted by a 2-NPFA.

Theorem 3. *Let L be accepted by a 1-way deterministic reversal-bounded multicounter machine. Then L can be accepted by a 2-NPFA.*

Proof. (sketch) We present the proof in terms of (P, V), a prover-verifier system in which the verifier uses public coins. It is known that such systems are equivalent to 2-NPFA machines.

Let M be a 1-way reversal-bounded deterministic k-counter machine. We describe a system (P, V) (where V is a public coin 2-pfa) to simulate M. Let the input to M be $x\#$. We assume without loss of generality that each counter exactly reverses once and on all accepted inputs, M halts with all counters set to 0. Each counter thus goes through an increment phase followed by a decrement phase. We will show the simulation for $k = 1$. Extension to arbitrary k is not difficult. The basic idea behind the construction is as follows: Let M' be the simulating machine. M''s verifier starts simulating M on an input x by making successive moves of M. Note that the initial phase of the simulation is the increment phase. During this phase, the verifier does not need interaction with the prover. As the simulation proceeds, it also needs to maintain the counter. However, its finite control is not enough the maintain a counter. The way this issue is dealt with is as follows: Suppose, the counter reaches the value I on input x during the increment phase. Since the counter value is 0 at the end of the computation, the counter is decremented exactly $D = I$ times during the decrement phase. Instead of explicitly simulating the counter, M' attempts to verify that $I = D$ using lemma 4. However, there is a problem with the decrement phase since the moves of M during this phase may depend on whether the counter is 0 or not. M' does not have this information. This is where an interaction with the prover is required.

In the decrement phase, each time M makes a move, the verifier asks the prover if the counter value if 0, and the prover P responds. In addition, as explained above, the verifier simulates two variables I and D corresponding to the values of the counter in the increment and decrement phase. When M's head reaches the end-marker, if M enters an accepting state while the counter is still positive, then V checks that $d \leq i$. But if the counter becomes zero prior to acceptance, then V checks that $d = i$. These checks are done using lemma 4 and lemma 5 respectively. The prover will tell the verifier at the beginning of the computation whether the checking to be done is equality checking or inequality checking. Note that in order to do the checking, the variables I and D need to be created many times. The details of this simulation are quite similar to Theorem 2.1 in [3] where a simulation of a two counter machine is described. The essential differences are that the counter machine simulated in [4] is not reversal bounded, and the prover transmits the successive configurations of M, and that their system uses private coins. In our case, i is created by the verifier itself, so it is always correct. The proof of correctness of the construction is similar to that of [3] and so we omit the details.

We now prove the theorem stated in section 4 without proof:

Theorem 5. *Let L_{bal} be the set of balanced parentheses over a two-letter alphabet $\{[,]\}$. The complement of L_{bal} can be accepted by a 2-NPFA augmented with a pebble.*

Proof. Let x be in the complement of L_{bal}. Then either x does not have an equal number of ['s and]'s, or there is a prefix y of x in which there are more]'s than ['s. The 2-NPFA machine M guesses one of these options and verifies it as follows: The former involves using lemma 4. The latter involves first placing a pebble to mark off y on the tape. Then, it uses lemma 5 to check the desired property.

Author Index

Lecture Notes in Computer Science

For information about Vols. 1–3956

please contact your bookseller or Springer

Vol. 3998: T. Calamoneri, I. Finocchi, G.F. Italiano (Eds.), Algorithms and Complexity. XII, 394 pages. 2006.

Vol. 3997: W. Grieskamp, C. Weise (Eds.), Formal Approaches to Software Testing. XII, 219 pages. 2006.

Vol. 3996: A. Keller, J.-P. Martin-Flatin (Eds.), Self-Managed Networks, Systems, and Services. X, 185 pages. 2006.

Vol. 3995: G. Müller (Ed.), Emerging Trends in Information and Communication Security. XX, 524 pages. 2006.

Vol. 3994: V.N. Alexandrov, G.D. van Albada, P.M.A. Sloot, J. Dongarra (Eds.), Computational Science – ICCS 2006, Part IV. XXXV, 1096 pages. 2006.

Vol. 3993: V.N. Alexandrov, G.D. van Albada, P.M.A. Sloot, J. Dongarra (Eds.), Computational Science – ICCS 2006, Part III. XXXVI, 1136 pages. 2006.

Vol. 3992: V.N. Alexandrov, G.D. van Albada, P.M.A. Sloot, J. Dongarra (Eds.), Computational Science – ICCS 2006, Part II. XXXV, 1122 pages. 2006.

Vol. 3991: V.N. Alexandrov, G.D. van Albada, P.M.A. Sloot, J. Dongarra (Eds.), Computational Science – ICCS 2006, Part I. LXXXI, 1096 pages. 2006.

Vol. 3990: J. C. Beck, B.M. Smith (Eds.), Integration of AI and OR Techniques in Constraint Programming for Combinatorial Optimization Problems. X, 301 pages. 2006.

Vol. 3989: J. Zhou, M. Yung, F. Bao, Applied Cryptography and Network Security. XIV, 488 pages. 2006.

Vol. 3987: M. Hazas, J. Krumm, T. Strang (Eds.), Location- and Context-Awareness. X, 289 pages. 2006.

Vol. 3986: K. Stølen, W.H. Winsborough, F. Martinelli, F. Massacci (Eds.), Trust Management. XIV, 474 pages. 2006.

Vol. 3984: M. Gavrilova, O. Gervasi, V. Kumar, C.J. K. Tan, D. Taniar, A. Laganà, Y. Mun, H. Choo (Eds.), Computational Science and Its Applications - ICCSA 2006, Part V. XXV, 1045 pages. 2006.

Vol. 3983: M. Gavrilova, O. Gervasi, V. Kumar, C.J. K. Tan, D. Taniar, A. Laganà, Y. Mun, H. Choo (Eds.), Computational Science and Its Applications - ICCSA 2006, Part IV. XXVI, 1191 pages. 2006.

Vol. 3982: M. Gavrilova, O. Gervasi, V. Kumar, C.J. K. Tan, D. Taniar, A. Laganà, Y. Mun, H. Choo (Eds.), Computational Science and Its Applications - ICCSA 2006, Part III. XXV, 1243 pages. 2006.

Vol. 3981: M. Gavrilova, O. Gervasi, V. Kumar, C.J. K. Tan, D. Taniar, A. Laganà, Y. Mun, H. Choo (Eds.), Computational Science and Its Applications - ICCSA 2006, Part II. XXVI, 1255 pages. 2006.

Vol. 3980: M. Gavrilova, O. Gervasi, V. Kumar, C.J. K. Tan, D. Taniar, A. Laganà, Y. Mun, H. Choo (Eds.), Computational Science and Its Applications - ICCSA 2006, Part I. LXXV, 1199 pages. 2006.

Vol. 3979: T.S. Huang, N. Sebe, M.S. Lew, V. Pavlović, M. Kölsch, A. Galata, B. Kisačanin (Eds.), Computer Vision in Human-Computer Interaction. XII, 121 pages. 2006.

Vol. 3978: B. Hnich, M. Carlsson, F. Fages, F. Rossi (Eds.), Recent Advances in Constraints. VIII, 179 pages. 2006. (Sublibrary LNAI).

Vol. 3977: N. Fuhr, M. Lalmas, S. Malik, G. Kazai (Eds.), Advances in XML Information Retrieval and Evaluation. XII, 556 pages. 2006.

Vol. 3976: F. Boavida, T. Plagemann, B. Stiller, C. Westphal, E. Monteiro (Eds.), Networking 2006. Networking Technologies, Services, and Protocols; Performance of Computer and Communication Networks; Mobile and Wireless Communications Systems. XXVI, 1276 pages. 2006.

Vol. 3975: S. Mehrotra, D.D. Zeng, H. Chen, B.M. Thuraisingham, F.-Y. Wang (Eds.), Intelligence and Security Informatics. XXII, 772 pages. 2006.

Vol. 3973: J. Wang, Z. Yi, J.M. Zurada, B.-L. Lu, H. Yin (Eds.), Advances in Neural Networks - ISNN 2006, Part III. XXIX, 1402 pages. 2006.

Vol. 3972: J. Wang, Z. Yi, J.M. Zurada, B.-L. Lu, H. Yin (Eds.), Advances in Neural Networks - ISNN 2006, Part II. XXVII, 1444 pages. 2006.

Vol. 3971: J. Wang, Z. Yi, J.M. Zurada, B.-L. Lu, H. Yin (Eds.), Advances in Neural Networks - ISNN 2006, Part I. LXVII, 1442 pages. 2006.

Vol. 3970: T. Braun, G. Carle, S. Fahmy, Y. Koucheryavy (Eds.), Wired/Wireless Internet Communications. XIV, 350 pages. 2006.

Vol. 3969: Ø. Ytrehus (Ed.), Coding and Cryptography. XI, 443 pages. 2006.

Vol. 3968: K.P. Fishkin, B. Schiele, P. Nixon, A. Quigley (Eds.), Pervasive Computing. XV, 402 pages. 2006.

Vol. 3967: D. Grigoriev, J. Harrison, E.A. Hirsch (Eds.), Computer Science – Theory and Applications. XVI, 684 pages. 2006.

Vol. 3966: Q. Wang, D. Pfahl, D.M. Raffo, P. Wernick (Eds.), Software Process Change. XIV, 356 pages. 2006.

Vol. 3965: M. Bernardo, A. Cimatti (Eds.), Formal Methods for Hardware Verification. VII, 243 pages. 2006.

Vol. 3964: M. Ü. Uyar, A.Y. Duale, M.A. Fecko (Eds.), Testing of Communicating Systems. XI, 373 pages. 2006.

Vol. 3963: O. Dikenelli, M.-P. Gleizes, A. Ricci (Eds.), Engineering Societies in the Agents World VI. XII, 303 pages. 2006. (Sublibrary LNAI).

Vol. 3962: W. IJsselsteijn, Y. de Kort, C. Midden, B. Eggen, E. van den Hoven (Eds.), Persuasive Technology. XII, 216 pages. 2006.

Vol. 3960: R. Vieira, P. Quaresma, M.d.G.V. Nunes, N.J. Mamede, C. Oliveira, M.C. Dias (Eds.), Computational Processing of the Portuguese Language. XII, 274 pages. 2006. (Sublibrary LNAI).

Vol. 3959: J.-Y. Cai, S. B. Cooper, A. Li (Eds.), Theory and Applications of Models of Computation. XV, 794 pages. 2006.

Vol. 3958: M. Yung, Y. Dodis, A. Kiayias, T. Malkin (Eds.), Public Key Cryptography - PKC 2006. XIV, 543 pages. 2006.